Biological Approaches and Evolutionary Trends in Plants

Biological Approaches and Evolutionary Trends in Plants

edited by

SHOICHI KAWANO
Department of Botany,
Kyoto University, Kyoto;
Institute of Genetic Ecology,
Tohoku University, Sendai, Japan

ACADEMIC PRESS
Harcourt Brace Jovanovich, Publishers
London San Diego New York
Boston Sydney Tokyo Toronto

ACADEMIC PRESS LTD.
24/28 Oval Road,
London NW1 7DX

United States Edition published by
ACADEMIC PRESS INC.
San Diego, California 92101-4311

This book is printed on acid-free paper

British Library Cataloguing in Publication Data

Is available

ISBN 0-12-402960-4

Printed in Great Britain by Galliard (Printers) Ltd, Great Yarmouth, Norfolk

This book is dedicated to the founders of
the International Organization of Plant Biosystematists (IOPB),
Áskell and Doris Löve.

Contributors

Michael L. Arnold: *Department of Biochemistry, Louisiana State University and Louisiana Agricultural Experiment Station, Louisiana State University Agricultural Center, Baton Rouge, LA 70803, USA*

Mary T. Kalin Arroyo: *Departamento de Biología, Facultad de Ciencias, Universidad de Chile, Casilla 653, Santiago, Chile*

Spencer C. H. Barrett: *Department of Botany, University of Toronto, Toronto, Ontario, Canada M5S 3B2*

Paulette Bierzychudek: *Department of Biology, Pomona College, Claremont, CA 91711, USA*

Cornelis W.P.M. Blom: *Department of Experimental Botany, Catholic University, Toernooiveld 6525 ED Nijmegen, The Netherlands*

Steven B. Broyles: *Department of Botany, University of Georgia, Athens, GA 30602, USA*

Russell L. Chapman: *Department of Botany, Louisiana State University and Louisiana Agricultural Experiment Station, Louisiana State University Agricultural Center, Baton Rouge, LA 70803, USA*

Julie M. Dowd: *Department of Botany, University of Adelaide, Box 498, G.P.O. Adelaide, South Australia, 5001*

Stephen R. Downie: *Department of Biology, Indiana University, Bloomington, IN 47405, USA*

Christopher G. Eckert: *Department of Botany, University of Toronto, Toronto, Ontario, Canada M5S 3B2*

R. Keith Hamby: *Department of Biochemistry, Louisiana State University and Louisiana Agricultural Experiment Station, Louisiana State University Agricultural Center, Baton Rouge, LA 70803, USA*

Herbert Hurka: *Spezielle Botanik, University of Osnabrück, Osnabrück, Federal Republic of Germany*

Terunobu Ichimura: *Institute of Applied Microbiology, University of Tokyo, Tokyo 113, Japan*

Kazuyuki Itoh: *Tropical Agriculture Research Center, MAFF, Tsukuba, Ibaraki 305, Japan*

Yoh Iwasa: *Department of Biology, Faculty of Science, Kyushu University, Fukuoka 812, Japan*

Robert K. Jansen: *Department of Ecology and Evolutionary Biology, University of Connecticut, Storrs, CT 06268, USA*

Naoki Kachi: *Division of Environmental Biology, National Institute for Environmental Studies, Tsukuba, Ibaraki 305, Japan*

Fumie Kasai: *Microbial Culture Collection, National Institute for Environmental Studies, Tsukuba, Ibaraki 305, Japan*

Paul A. Keddy: *Department of Biology, University of Ottawa, Ottawa, Canada*
Christian Knaak: *Department of Biochemistry, Louisiana State University and Louisiana Agricultural Experiment Station, Louisiana State University Agricultural Center, Baton Rouge, LA 70803, USA*
Hisao Kobayashi: *Department of Agronomy, Faculty of Agriculture, Yamaguchi University, Yamaguchi 753, Japan*
Monique D. LeBlanc: *Department of Biochemistry, Louisiana State University and Louisiana Agricultural Experiment Station, Louisiana State University Agricultural Center, Baton Rouge, LA 70803, USA*
Peter G. Martin: *Department of Botany, University of Adelaide, Box 498, G.P.O. Adelaide, South Australia, 5001*
Shooichi Matsunaka: *Laboratory of Pesticide Science, Faculty of Agriculture, Kobe University, Kobe 657, Japan*
Helen J. Michaels: *Department of Biology, University of Michigan, Ann Arbor, MI 48109, USA*
Richard G. Olmstead: *Department of Biology, Indiana University, Bloomington, IN 47405, USA*
Jeffrey D. Palmer: *Department of Biology, Indiana University, Bloomington, IN 47405, USA*
S. N. Raina: *Department of Botany, University of Delhi, Delhi, India*
Sadao Sakamoto: *Plant Germ-plasm Institute, Faculty of Agriculture, Kyoto University, Kyoto 617, Japan*
Douglas E. Soltis: *Department of Botany, Washington State University, Pullman, WA 99164, USA*
Pamela S. Soltis: *Department of Botany, Washington State University, Pullman, WA 99164, USA*
Francisco Squeo: *Departamento de Biología, Universidad de La Serena, La Serena, Chile*
Krystyna M. Urbanska: *Geobotanical Department, Swiss Federal Institute of Technology Zürich, Switzerland*
Leonard W.D. Van Raamsdonk: *Centre for Plant Breeding Institute, P.O. Box 16, 6700 AA Wageningen, The Netherlands*
Keishiro Wada: *Department of Biology, Faculty of Science, Kanazawa University, Marunouchi 1-1, Kanazawa, Ishikawa 920, Japan*
Suzanne I. Warwick: *Biosystematics Research Centre, Agriculture Canada, Ottawa, Ontario, Canada K1A 0C6*
Andrew R. Watkinson: *School of Biological Sciences, University of East Anglia, Norwich NR4 7TJ, U.K.*
Robert Wyatt: *Department of Botany, University of Georgia, Athens, GA 30602, USA*
Elizabeth A. Zimmer: *Departments of Biochemistry and Botany, Louisiana State University and Louisiana Agricultural Experiment Station, Louisiana State University Agricultural Center, Baton Rouge, LA 70803, USA*

Preface

The Fourth International Symposium of Plant Biosystematics was held 10-14 July, 1989, in Kyoto on the special topic "Biological Approaches and Evolutionary Trends in Plants"; guest speakers included twenty-two of the world's leading plant biologists, representing ten different countries. The objectives of the Symposium were to bring these experts together, to present their significant new findings in evolutionary biology, and to discuss current issues in plant biosystematics in light of the new evidence and ideas brought forward at various levels of biological organization, from molecule to cell, individual, population, species and community.

In the Symposium's first session, "Biology and Evolution of Weeds and Weed-Crop Complexes", recent discoveries concerning parapatric differentiation of weed populations, including adaptive evolution in herbicide resistant biotypes, and complex evolutionary patterns in weed-crop complexes of various groups were introduced, and the implications of these new findings were discussed.

The second session, "Molecular Approaches in Plant Biosystematics", illuminated remarkable new evidence made possible by novel and powerful analytical tools in molecular biology: amino acid sequencing of proteins (e. g., Rubisco, ferredoxin), restriction-site variations of cpDNA, mitDNA, rDNA, etc., and chromosome-banding patterns revealed by differential staining. Recent rapid progress in these fields have moved plant biosystematics a great step forward in a new direction. There is no doubt that these new empirical discoveries at the molecular level will bring forth in the coming decade new insights concerning affinities or phylogenetic relationships among various members of the plant kingdom.

A new wave of research in plant population biology and evolutionary ecology since the 1970s has had a great impact on biology as well as biosystematics. In the third and fourth sessions, "Population Biology and Life History Evolution (1) Reproductive Biology of Plants and (2) Demography and Life History Evolution of Plants", numerous recent and fascinating findings in these fields and the results of theoretical examinations of the facts were presented. In current life-history and demographic studies, we first focus on various aspects of reproductive biology and evolutionary changes in significant reproductive parameters. We then attempt to demographically quantify these parameters. Through such studies, we are now gradually coming to understand the mechanisms of evolutionary changes in various significant life-history parameters of plant species or species group in relation to changes in environmental regimes, i. e., life history evolution of a particular taxon. In this connection, new information as well as theoretical frameworks introduced in the Kyoto Symposium no doubt will give great stimulation to future development of studies in plant population biology and evolutionary ecology. As was correctly predicted by Peter Raven in 1987, the vigorous and most effective hybridization of plant population biology with traditional plant biosystematics is bearing exciting and promising fruit today.

This volume includes all the invited papers presented at the IOPB Symposium held in Kyoto. The Symposium was organized and sponsored by the International Organization of Plant Biosystematists, the Society for the Study of Species Biology, the Weed Science Society of Japan, and the Japan Society of Plant Taxonomists. The generous financial support of the Commemorative Association for the Japan World Exposition, the Inamori Foundation, Kyoikusha Co., Ltd. and 50 other coorporations is gratefully acknowledged.

It is a pleasure to express my cordial gratitude to all invited speakers for their excellent contributions, as well as to all other participants in the Symposium, to whom we owe all the success of the Symposium. I also wish to thank the members of the Organizing Committee and Botany Department staff for providing all necessary assistance during the Symposium. Finally, but not the least of all, I extend my cordial gratitude to Junko Ueda for her excellent skill and patience editing this volume.

April 1990

Shoichi Kawano
Department of Botany,
Kyoto University, Kyoto;
Institute of Genetic Ecology,
Tohoku University, Sendai, Japan

Contents

Contributors vii
Preface ix

Part I Biology and Evolution of Weeds and Weed-Crop Complexes

1 Genetic variation in weeds — with particular reference to Canadian agricultural weeds 3
Suzanne I. Warwick
2 Differentiation and adaptation in the Genus *Capsella* (Brassicaceae) 19
Herbert Hurka
3 Parapatric differentiation of paraquat resistant biotypes in some Compositae species 33
Kazuyuki Itoh and Shooichi Matsunaka
4 Biosystematics of cultivated plants and their wild relatives 51
Leonard W. D. Van Raamsdonk
5 Weed-crop complex in cereal cultivation 67
Hisao Kobayashi and Sadao Sakamoto
6 Responses to flooding in weeds from river areas 81
Cornelis W. P. M. Blom

Part II Molecular Approaches in Plant Biosystematics

7 Chloroplast DNA and Nuclear rDNA variation: Insights into autopolyploid and allopolyploid evolution 97
Douglas E. Soltis and Pamela S. Soltis
8 Chloroplast DNA and phylogenetic studies in the Asteridae 119
Richard G. Olmstead, Robert K. Jansen, Helen J. Michaels, Stephen R. Downie and Jeffrey D. Palmer
9 Ribosomal DNA variation and its use in plant biosystematics 135
Christian Knaak, R. Keith Hamby, Michael L. Arnold, Monique D. LeBlanc, Russell L. Chapman and Elizabeth A. Zimmer
10 Molecular approach to plant systematics from protein sequence comparison 159
Keishiro Wada

11 A protein sequence study of the phylogeny and origin of the Dicotyledons 171
Peter G. Martin and Julie M. Dowd
12 Genome organisation and evolution in the genus *Vicia* 183
S. N. Raina

Part III Population Biology and Life History Evolution
(1) Reproductive Biology of Plants

13 Relationship between plant breeding systems and pollination 205
Mary T. Kalin Arroyo and Francisco Squeo
14 Variation and evolution of mating systems in seed plants 229
Spencer C. H. Barrett and Christopher G. Eckert
15 Reproductive biology of milkweeds (*Asclepias*): Recent advances 255
Robert Wyatt and Steven B. Broyles
16 Biology of asexually reproducing plants 273
Krystyna M. Urbanska
17 The demographic consequences of sexuality and apomixis in *Antennaria* 293
Paulette Bierzychudek
18 Mating systems and speciation in haplontic unicellular algae, desmids 309
Terunobu Ichimura and Fumie Kasai

Part IV Population Biology and Life History Evolution
(2) Demography and Life History Evolution of Plants

19 Optimal growth schedule of terrestrial plants 335
Yoh Iwasa
20 Annual plants: A life-history and population analysis 351
Andrew R. Watkinson
21 Evolution of size-dependent reproduction in biennial plants: A demographic approach 367
Naoki Kachi
22 The use of functional as opposed to phylogenetic systematics: A first step in predictive community ecology 387
Paul A. Keddy

Index 407

Part I

BIOLOGY AND EVOLUTION OF WEEDS AND WEED-CROP COMPLEXES

1 Genetic Variation in Weeds — with Particular Reference to Canadian Agricultural Weeds

SUZANNE I. WARWICK

Biosystematics Research Centre, Agriculture Canada, Ottawa, Ontario, Canada K1A 0C6

I. Introduction

Weeds can be defined as colonizing plant species that grow in habitats markedly disturbed by man and are commonly classified according to habitat-type, i.e. agrestals - agricultural land; ruderals - waste places, roadsides; grassland weeds - pastures, meadows, lawns; etc. (Baker, 1965; Holzner, 1982). In spite of their habitat and taxonomic diversity, weeds exhibit many similarities in growth patterns and the growth characteristics of an "ideal" weedy species are well documented (Baker, 1974; Holzner, 1982). The topic of genetic variation in weeds and the evolution of colonizing species, in general, has also received much attention recently and several excellent reviews are available (Brown and Marshall, 1981; Barrett, 1982; Oka, 1983; Clegg and Brown, 1983; Barrett and Richardson, 1986; Barrett and Shore, 1989; Warwick, 1989). Based on these studies, a number of shared genetic features are predicted for weedy species and are presumed to contribute to their success and spread in disturbed environments. These include: self-fertilization or clonal reproduction, high levels of enzyme multiplicity through polyploidy, multilocus associations providing a reduced number of genotypes resulting in genetically depauperate populations, substantial interpopulation differentiation, and high levels of phenotypic plasticity. A number of evolutionary and historical factors will be important in influencing patterns of genetic variation in weedy species. Primarily these include: founder effects and/or genetic drift; mating system; environmental homogeneity; polyploidy; domestication and crop/weed interactions; and direct selection pressure. The effects of each of these will be discussed briefly:

BIOLOGICAL APPROACHES AND EVOLUTIONARY TRENDS IN PLANTS ISBN 0-12-402960-4

Founder effects/mating system. Weedy species, particularly after long distance colonization events, are usually subject to genetic bottlenecks and drift, both potentially resulting in reduced variability of populations. The mating system of a plant species will influence the distribution of genetic variation within and among populations in the native range. Selfing is known to promote the development and maintenance of multilocus associations and reduce levels of heterozygosity, and as a result populations of selfers often tend to be genetically uniform but highly differentiated from one another (Loveless and Hamrick, 1984). Population differentiation in primarily outcrossing species is less apparent, with most of the variation characteristic of the species found in a given population. Therefore, depending on the number and size of introductions, the degree of selfing of the species will be correlated with the severity of genetic bottleneck effects. As a result, it is predicted that founder effects, or the loss of allelic variation, will be most apparent in predominantly selfing species, particularly if the source for the new introduction was limited. On the other hand genetic bottlenecks will be less evident in outbreeding species, provided that large population sizes are maintained following an introduction (Brown and Marshall, 1981; Barrett and Richardson, 1986).

Environmental homogeneity. Theoretical studies have suggested a positive relationship between the levels of genetic variation in populations and the degree of environmental heterogeneity (Hedrick *et al.*, 1976). Agricultural weedy habitats are generally considered less heterogeneous because of their simple structure and relatively low biotic diversity and the high level of predictability associated with land use patterns. As a result limited genetic variation is expected in agrestal or agricultural weedy species (Hamrick *et al.*, 1979). On this basis, Barrett (1982) predicted that populations of agrestal weeds would exhibit lower genetic variation than those of ruderal weeds. However, few studies have been conducted which test this hypothesis and conclusions to date are somewhat conflicting. Warwick and Black (1986a) studying ruderal and agrestal populations of two weed species, *Chenopodium album* L. and *Amaranthus retroflexus* L., from Ontario, Canada, found little evidence for such habitat-correlated differences in levels of genetic polymorphisms (with the exception of triazine-resistant agrestal populations which showed marked lower genetic diversity in both species). Bosbach and Hurka (1981) studying European populations of *Capsella bursa-pastoris* (L.) Md. found greater levels of genetic heterogeneity in "intensively disturbed" sites, i.e. recently cultivated or disturbed soil, compared with "less disturbed" sites, i.e. lawns and other habitats which have not been cultivated.

Polyploidy. Many weedy species are polyploid. As a result of polyploidy, particularly allopolyploidy and the addition of two different genomes, it is possible for a single individual to express stable multiple allelic forms of a duplicated gene locus. In terms of allozyme variation, considerable within-individual variation is often evident in the form of enzyme multiplicity (or "fixed" heterozygosity), even though populations themselves may be depauperate in terms of actual numbers of genotypes. Various proposals have suggested that such multi-enzyme phenotypes could potentially result in increased biochemical variability in an individual and permit adaptation to a wider range of environments (Roose and Gottlieb, 1976; Soltis and Rieseberg, 1986); although to date, few studies documenting the ecological or adaptive significance of such phenotypes exist for weedy species. This pattern of few, but potentially highly variable, multilocus "allozyme genotypes", which is characteristic of many of the polyploid weed species studied to date (Warwick, 1989), corresponds to the concept of the "general-purpose" genotype described by Baker (1965). The latter was described as being frequent in weedy species and important in conferring wide environmental tolerance, and responsible for explaining the large amounts of phenotypic plasticity characteristic of weedy species.

Domestication/Crop-weed interaction. Evolutionary bottlenecks due to selection for domestication will contribute to lower levels of genetic variation, particularly in the case of recently evolved weedy strains of a crop species. Exchange of genetic material between closely related weedy and cultivated taxa may compensate for reduced genetic variation and has been suggested as a factor contributing to the success of certain weedy taxa, such as the weed-crop complex in *Setaria* (Harlan, 1982; Darmency *et al.* 1987).

II. Studies of Genetic Variation

Studies on genetic variation in weeds can be conveniently divided into two types: I) those based on ecological genetic or genecological studies, and which normally measure adaptive life history variation in response to direct selection pressure; and II) those based on allozyme studies which measure genetic variation in the absence of direct selection. Examples of each will be provided below.

A Genecological or ecological genetic studies of weeds

Genecological weed studies generally document or investigate genetically-based intraspecific variation in key, presumably adaptive, life-history traits in response to particular and often single selective factors. The experimental design of such studies involves the collection of populations from contrasting habitats with obvious patterns of phenotypic differentiation and the comparative growth of these populations under standard cultivation. Given the discontinuous sampling strategy, the various adaptive forms (plants which maintain their differences under standard conditions) are usually described as biotypes or ecotypes. Typical life history traits measured include: patterns of phenology, seed germination and dormancy, seed and plant size and growth rates at different stages of maturity, allocation to reproductive, vegetative and underground growth, etc. Examples of ecotypic or adaptive differentiation in weedy species in relation to a number of selective factors including:

TABLE 1

Examples of genetic differentiation in morphological and physiological traits in weedy species in response to particular selective factors

Selective factors	References
Latitude/climate	
- life history variation	
- *Setaria lutescens*	Norris and Schoner, 1980
- *Avena fatua*	Miller *et al.*, 1982
- *Rumex crispus*	Hume and Cavers, 1983
- *Verbascum thapsus*	Reinartz, 1984
- *Convolvulus arvensis*	Oegennaro and Weller, 1984
- *Echinochloa crus-galli*	Potvin, 1986
Light regime/photoperiodic response	
- *Portulaca oleracea*	Singh, 1975
- *Chenopodium rubrum*	Tsuchiya and Ishiguri, 1981
- *Plantago lanceolata*	Teramura, 1983
Soil type	
- roadside salt tolerance	
- *Senecio vulgaris*	Briggs, 1978
- *Eclipta alba*	Choudhuri and Sharma, 1979
- *Anthoxanthum odoratum*	Kiang, 1982
- edaphic races (calciphile/calcicole)	
- *Gomphrena celosiodes*	Srivastava and Misra, 1970
- *Cynodon dactylon*	Gupta and Ramakrishnan, 1977
- *Gaillardia pulchella*	Heywood, 1986
- mineral nutrition/soil fertility	
- *Plantago* spp.	Kuiper, 1983
- *Echinochloa colunum*	Kapoor and Ramakrishnan, 1974
- *Stellaria media*	Sobey, 1987
Cutting/mowing	
- *Poa annua*	Warwick and Briggs, 1978 McNeilly, 1981
- *Plantago lanceolata*	Warwick and Briggs, 1979
- *Prunella vulgaris*	Warwick and Briggs, 1979
- *Plantago major*	Warwick and Briggs, 1979
- *Trifolium repens*	Horikawa, 1986

TABLE 1 (continued)

Selective factors	References
Grazing/Trampling	
- *Poa annua*	Law *et al.*, 1977
- *Plantago major*	Warwick and Briggs, 1979
Herbicides	
- triazine resistant weed species	Lebaron and Gressel, 1982
- triallate tolerance	Jana and Naylor, 1982
- *Avena fatua*	
Biotic factors	
- crop mimicry	
- *Echinochloa crus-galli*	Barrett, 1983
- competition	
- *Poa annua*	McNeilly, 1981
Cultivation practices	
- cropping/cultivation system	
- *Avena fatua*	Jana and Thai, 1987
- *Sonchus arvensis*	Pegtel, 1974
- *Elytrigia repens*	Neuteboom, 1980
- weeding pressure	
- *Senecio vulgaris*	Kadereit and Briggs, 1985

latitude/climate, light regime, soil type, cutting/mowing, grazing/trampling, herbicides, biotic factors, and cultivation practices, are given in Table 1. As is evident from the Table, studies investigating genotypic differentiation of agricultural weedy species in response to the selection pressures imposed by standard agricultural practices, such as cultivation methods, fertilizer applications, herbicides (with the exception of triazine-resistance), crop associations, etc. are limited.

Patterns of both genetic differentiation and phenotypic plasticity in response to such direct selection are common in weedy species. For example, in an investigation of adaptive variation in response to the selective factor, cutting/mowing, in six common lawn weeds (*Achillea millefolium* L., *Bellis perennis* L., *Plantago major* L., *Plantago lanceolata* L., *Poa annua* L., and *Prunella vulgaris* L.) collected from a range of habitats representing a cutting/mowing gradient (Warwick and Briggs, 1978, 1979), in two of the six species, *Poa annua* and *Plantago major*, the habitat-correlated differences in size and degree of erectness were due to genetic differentiation. In the other four species, populations from short turf habitats (lawn and pasture), were less homogeneous genetically, with both phenotypic plasticity and genotypic differentiation evident as adaptive strategies in such habitats.

B. Allozyme studies of weedy species

Studies of allozyme variation have become very important in documenting levels of genetic variation in all living species, weedy species receiving their share of attention. The data provide an objective estimation of genotypic diversity in the absence of direct selection pressure, as well as providing information on the mating systems and genetic structure of populations. Standard genetic parameters include number of polymorphic loci, average number of alleles per locus, Nei's (1973) index of gene diversity, levels of heterozygosity, and Wright's fixation index and allow a quantitative assessment of inter- and intra-populational differentiation. Allozyme variation is generally reflective and a good predictor of overall genetic variability, as independently measured by morphological and phenological variation. In general, levels of allozyme variation in weedy species are much lower than those for late successional plant species, 30 and 63% average percent polymorphic loci, respectively (Hamrick *et al.*, 1979). However, levels of genetic diversity in weedy species, as assessed from allozyme variability range from very low levels, one to a few genotypes, to species with extremely high levels of genetic variation (Table 2). The amount and organization of allozyme variation differs among species with contrasting life histories, but also within and among populations of a single species, as demonstrated for example in *Eichhornia paniculata* (Spreng.) Solms (Glover and Barrett, 1987).

Genetic Variation in Canadian Weeds. The examples provided below were drawn primarily from the author's research on genetic variation of agricultural weeds. The purposes of these studies were i) to characterize the genetic features for these weedy species and compare them with the genetic features predicted for successful colonizers; ii) to compare allozyme and life history patterns of variation as measures of genetic diversity, and iii) to assess the predictive value of each of these data sets as regards the genetic potential of these weedy species.

1. Genetic variation in five primarily selfing weeds of monocultures of soybean and maize in eastern Canada

A comparative study of allozyme and life history variation was carried out on five, introduced, weedy species that are undergoing rapid range expansion northward in eastern North America (Warwick, 1987, 1989; Warwick and Black, 1986b; Warwick *et al.*, 1984, 1987a; Weaver *et al.*, 1985). Occupying a narrow ecological setting, they are all primarily weeds of monocultures of soybean and maize, and their expansions northward in the last 15-30 years have paralleled the northern extension of the range of maize and soybean production in eastern North America.

TABLE 2
Examples of allozyme studies in weedy species

Taxa	Location	Reference
Low levels of allozyme variation		
Xanthium strumarium	Australia	Moran and Marshall, 1978
Chondrilla juncea	Australia	Burdon *et al.*, 1980
Bromus mollis	Australia	Brown and Marshall, 1981
Emex spinosa	Australia	Marshall and Weiss, 1982
Echinochloa microstachya	Australia	Barrett and Richardson, 1986
Sorghum halepense	Canada	Warwick *et al.*, 1984
Abutilon theophrasti	Canada	Warwick and Black, 1986a
Panicum miliaceum	Canada	Warwick, 1987
Setaria faberi	Canada	Warwick *et al.*, 1987b
Datura stramonium	Canada	Warwick, 1989
Polygonum lapthifolium	Canada,Europe	Consaul, 1988
Capsella bursa-pastoris	Europe	Bosbach and Hurka, 1981
Typha spp.	United States	Mashburn *et al.*, 1978
Avena barbata	United States	Brown and Marshall, 1981
Striga asiatic	United States	Werth *et al.*, 1984
Cyperus esculentus	United States	Horak and Holt, 1986
Hydrilla verticillata	World-wide	Verkleij *et al.*, 1983
High levels of allozyme variation		
Trifolium subterraneum	Australia	Brown and Marshall, 1981
Echium plantagineum	Australia	Brown and Burdon 1983
Helianthus annuus	Australia	Dry and Burdon, 1986
Trifolium hirtum	California	Jain and Martins, 1979
Avena barbata	California	Clegg and Allard, 1972
Apera spica-venti	Canada	Warwick *et al.*, 1987a
Carduus spp.	Canada	Warwick *et al.*, 1989

The species are self-compatible and presumed to be predominantly selfing. They include: *Abutilon theophrasti* Medic. (velvetleaf), *Datura stramonium* L. (jimson weed), *Panicum miliaceum* L. (proso millet), *Sorghum halepense* (L.) Pers. (johnsongrass), and *Setaria faberi* W. Herrmann (giant foxtail). Substantial inter-populational and intra-populational variation in morphological and phenological life history traits (including: seed size, percent germination and dormancy, seedling size and growth rates, time to flowering, flowering plant height and size of plant parts, percent allocation to reproductive, vegetative and underground growth) were observed for all five species in a series of standard greenhouse or garden cultivation trials (Table 3). The same populations exhibited low levels of allozyme variation (Table 4). From 0-3 enzyme loci within each species were polymorphic, each having only two or three alleles. True breeding multiple enzyme phenotypes (i.e.

TABLE 3
Life history variation in five self-fertilizing agricultural weeds of maize and soybean monocultures from eastern Canada. After Warwick *et al.* (1986-1989)

	Abutilon theophrasti	*Panicum miliaceum*	*Setaria faberi*	*Sorghum halepense*	*Datura stramonium*
No. populations	39	5	8	13	5
No. life history* features scored	51	7	16	46	21
- Significant population differences	33	7	16	46	21
Patterns of genetic variation	Continuous: latitudinal trends in some traits	6 biotypes: -Crop-like: large-seeded, nonshattering -Weedy: small seeded, shattering dormancy	Continuous	2 biotypes: -Large-seeded, annual; -Rhizomatous, small-seeded, overwintering perennial	Continuous: latitudinal trends in some traits

*Include: phenology, morphometric and growth features: height, size of plant parts, dry weight and resource allocations; seed dormancy and germination patterns, etc.

"fixed" heterozygotes) were present at 50% of the duplicated loci in three of the four polyploid species. A small number of allozyme genotypes were observed in each of the species; and within each species these genotypes were very similar to each other, differing by only one or two alleles. Almost all populations contained a single genotype. For example in *Abutilon theophrasti*, 33 of the 39 populations surveyed were homogeneous containing one of two genotypes. In contrast population differences were evident in 33 of 51 life history characters measured, many of which showed latitudinal trends along the north/south gradient. Similarly for *Panicum miliaceum*, all 39 weedy populations surveyed, which represented six biotypes with distinct life history differences, contained either one or the other of two allozyme genotypes. The most striking result was that for *Datura stramonium*, where all nine populations contained the same allozyme genotype, in contrast to significant population differences in several life-history traits.

In summary, the five weedy species examined shared a number of genetic features, many of which had been predicted for successful colonizing species. All were self-compatible, primarily selfing; four of the five were polyploid; high levels

TABLE 4
Isozyme variation in five self-fertilizing agricultural weeds of maize and soybean monocultures from eastern Canada. After Warwick *et al.* (1986-1989)

	Abutilon theophrasti	*Panicum miliaceum*	*Setaria faberi*	*Sorghum halepense*	*Datura stramonium*
Chromosome No.	$2n = 24$	$2n = 36$	$2n = 36$	$2n = 40$	$2n = 24$
No. of populations	39	39	8	13	9
No. of loci	27	19	24	21	22
No. (percent) of loci monomorphic	25 (93%)	18 (95%)	21 (88%)	18 (86%)	22 (100%)
No. (percent) of loci polymorphic	2 (7%)	1 (5%)	3 (12%)	3 (14%)	0
No. (percent) of duplicated loci with enzyme multiplicity	14 (52%)	8 (42%)	13 (54%)	3 (14%)	2 (9%)
No. of multilocus genotypes	4	2	9	10	1

of enzyme multiplicity were observed at duplicated loci in three of the four polyploid species; small numbers of genotypes were evident in each species; and in all species, populations were markedly depauperate containing one to a few allozyme genotypes. In all five species, a low level of allozyme variation contrasted sharply with substantial inter-populational and intra-populational variation in morphological and phenological life history traits.

2. Genetic variation in native versus introduced weedy populations

Since the limited allozyme variability seen in introduced North American weed populations may simply reflect founder effects, where possible it is important to compare levels of genetic variation in both native and introduced ranges. We have compared levels of genetic variation in populations of two weedy species from both their introduced and presumed native ranges: the outcrossing winter annual, *Apera spica-venti* L. (Warwick *et al.*, 1987b) and the predominantly selfing weed species, *Polygonum lapathifolium* L. (Consaul, 1988).

TABLE 5
A comparison of isozyme variation in nine introduced Canadian and six native European populations of the outcrossing weed species *Apera spica-venti*. After Warwick *et al.* (1987)

	Canada	Europe
No. of loci surveyed	17	17
Percent polymorphic loci	57	62
Mean no. of alleles	2.54	2.53
Mean heterozygosity	0.23	0.23
Total species diversity	0.211	0.208
Mean population diversity	0.209	0.203
Inter-populational gene diversity	0.010	0.024

Apera spica-venti, a widely distributed weed of winter cereals in Europe, has recently become established as a weed of winter cereals in Ontario, Canada. Levels of genetic variation were compared among and within nine Canadian and six European populations, representing the introduced and native ranges, respectively. Both the Canadian and European populations showed high levels of variation among populations with respect to 31 life history and morphological traits. Although greater differences were recorded among the European populations, this was primarily due to the distinctness of a single population from Poland. Levels of within-population variation of the same characters were similar in both the introduced and native populations. High levels of genetic variability in allozyme characters were also evident in both the Canadian and European populations (Table 5) (percentage of polymorphic loci = 57-62%, number of alleles per locus was 2.53). There was little or no divergence in allozyme characters among populations with only 1.0-2.4% of the total gene diversity (G_{ST}) allocated to the among-population component, which are similar to the average values described for other outcrossing, wind-pollinated grasses. The amounts and patterns of genetic variation were consistent with what one would expect in an outcrossing species and do not support the hypothesis of genetically depauperate introductions.

In contrast, a comparison of genetic diversity in the introduced and native ranges of the predominantly selfing weed species, *Polygonum lapathifolium* (Consaul,

1988) indicated very low levels of allozyme variation in populations from both North America and Europe. In a survey of 15 enzyme systems, representing 22 loci, only two (9%) of the loci were polymorphic in the North American populations, resulting in six multilocus genotypes (one genotype accounted for 85% of the individuals); while five (17%) of the enzyme loci were polymorphic in the European accessions, resulting in eight genotypes (one genotype accounted for 77% of the individuals). North American and European populations were distinguished by a single allele. This lack of allozyme variation contrasted sharply with the considerable intra-specific variation in morphological and physiological traits, observed in this species complex.

III. Conclusions

In summation the key factor affecting variation patterns in the agricultural weeds above appeared to be the mating system of the species. As indicated earlier, founder effects, or the loss of allelic variation, will be most apparent in predominantly selfing species, particularly if the source for the new introduction was limited and less evident in outbreeding species, provided that large population sizes are maintained following an introduction. The high levels of allozyme variation observed for *Apera spica-venti* and the other species listed in Table 2 are most clearly associated to their predominantly outcrossing mating system. Polyploidy, and the pattern of potentially highly variable, multiple enzyme genotypes, was also an important factor. However, it cannot account for the depauperate number of genotypes, since two of the diploid species *Datura stramonium* and *Polygonum lapathifolium* also contained few multilocus allozyme genotypes. Evolutionary bottlenecks due to selection for domestication, may be an important factor in explaining the low levels of variability in weedy biotypes of *Panicum miliaceum*, as it was shown in a further study by Warwick (1987) that variation in the crop accessions of proso millet also show relatively little allozyme variation, with a total of five multilocus allozyme genotypes among all weedy and nonweedy collections of the species. The genetic effects of the environmental homogeneity of agricultural monocultures was less obvious, in that reduced allozyme variability was observed in both the ruderal populations of *Polygonum lapthifolium* as well as the five weed species occupying soybean and maize monocultures.

Correlation between allozyme and life history variation. The relationship between allozyme variation and quantitative life history characters in weedy species is a complex one (Lewontin, 1984; Price *et al.*, 1984). In our own studies, the marked lack of allozyme variation in the six predominantly selfing weeds was a striking contrast to the inter- and intra-populational differences in life history features observed in these species. A similar lack of concordance between allozyme variation and that seen in life history traits has been described for other colonizing species (Barrett, 1982; Barrett and Richardson, 1986; Warwick, 1989). The above data would indicate that high levels of genetic diversity as estimated from allozymes are not a prerequisite for a successful weedy species. The differences in adaptive life history traits, evident in populations of weedy species, probably represent recent divergence in response to natural and human selection, a divergence which has not been paralleled in the allozyme characters. The latter are generally considered to be selectively neutral, although studies of population differentiation of wild barley (Nevo *et al.*, 1986) and *Plantago major* (Van Dijk, 1984) indicated close linkage of allozyme markers with life history differences, suggesting the possibility at least for indirect selection of allozyme characters. Lande (1976) provided a genetic rationale for the discrepancies between life history and allozyme characters, with the suggestion that large amounts of heritable variation could be maintained by mutation in polygenic quantitative traits, even when there is strong stabilizing selection for allozyme characters. In conclusion, allozyme variation would appear to reflect with reasonable accuracy the genetic and adaptive diversity of predominantly outcrossing weedy plants. However, the value of allozyme variation in predicting other types of genetic variation in predominantly selfing agricultural weed species is more limited.

References

Baker, H. G. (1965). Characteristics and modes of origin of weeds. *In* "The genetics of colonizing species" (H.G. Baker and G.L. Stebbins, eds), pp. 147-168. Academic Press, New York.

Baker, H. G. (1974). The evolution of weeds. *Annu. Rev. Ecol. Syst.* **5**, 1-24.

Barrett, S. C. H. (1982). Genetic variation in weeds. *In* "Biological control of weeds with plant pathogens" (R. Charudation and H. L. Walker, eds), pp. 73-98. John Wiley & Sons, New York.

Barrett, S. C. H. (1983). Crop mimicry in weeds. *Econ. Bot.* **37**, 255-282.

Barrett, S. C. H. and Richardson, B. J. (1986). Genetic attributes of invading species. *In* "The ecology of biological invasions: an Australian perspective" (R. H. Groves and J. J. Burdon, eds), pp. 21-33. Australian Academy of Science, Jacaranda Publishing, Canberra.

Barrett, S. C. H. and Shore, J. S. (1989). Isozyme variation in colonizing plants. *In* "Isozymes in plant biology" (D.E. Soltis and P.S. Soltis, eds). Discorides Press. (in press).

Bosbach, K. and Hurka, H. (1981). Biosystematic studies on *Capsella bursa-pastoris* (Brassicaceae): enzyme polymorphism in natural populations. *Plant. Syst. Evol.* **137**, 73-94.

Briggs, D. (1978). Genecological studies of salt tolerance in groundsel *Senecio vulgaris*, with particular reference to roadside habitats. *New Phytol.* **81**, 381-390.

Brown, A. H. D. and Burdon, J. J. (1983). Multilocus diversity in an outbreeding weed, *Echium plantagineum* L. *Aust. J. Biol. Sci.* **36**, 503-509.

Brown, A. H. D. and Marshall, D. R. (1981). Evolutionary changes accompanying colonization in plants. *In* "Evolution today" (G. G. Scudder and J. L. Reveal, eds), pp. 351-363. Proceedings of the Second International Congress of Systematic and Evolutionary Biology. Hunt Institute for Botanical Documentation, Carnegie-Mellon University, Pittsburgh.

Burdon, J. J., Marshall, D. R. and Groves, R. H. (1980). Isozyme variation in *Chondrilla juncea* L. in Australia. *Aust. J. Bot.* **28**, 193-198.

Choudhuri, G. N. and Sharma, B. D. (1979). Differential resource utilization by intraspecific variants of a compositaceous weed. *Indian J.Ecol.* **5**, 192-202.

Clegg, M. T. and Allard, R. W. (1972). Patterns of genetic differentiation in the slender wild oat species *Avena barbata*. *Proc. Natl. Acad. Sci.* U.S.A. **69**, 1820-1824.

Clegg, M. T. and Brown, A. H. D. (1983). The founding of plant populations. *In* "Genetics and conservation" (C. M. Schonewald-Cox, S. M. Chambers, B. MacBryde and L. Thomas, eds), pp. 216-228. Benjamin/Cummings, Menlo Park, California.

Consaul, L. (1988). A biosystematic study of the *Polygonum lapthifolium* L. complex in North America. Masters Thesis. Ottawa University.

Darmency, H., Zangre, G. R. and Pernes, J. (1987). The wild-weed-crop complex in *Setaria*: a hybridization study. *Genetica* **75**, 103-107.

Dry, P. J. and Burdon, J. J. (1986). Genetic structure of natural populations of wild sunflowers (*Helianthus annuus* L.) in Australia. *Aust. J. Biol. Sci.* **39**, 255-270.

Glover, D. E. and Barrett, S. C. H. (1987). Genetic variation in continental and island populations of *Eichhornia paniculata* (Pontederiaceae). *Heredity* **59**, 7-17.

Gupta, U. and Ramakrishnan, P. S. (1977). The effect of added salt on competition between two ecotypes of *Cynodon dactylon* (L.) *Pers. Proc. Ind. Acad. Sci.* **86**, 275-280.

Hamrick, J. L., Linhart, Y. B. and Mitton, J. B. (1979). Relationships between life history characteristics and electrophoretically detectable genetic variation in plants. *Annu. Rev. Ecol. Syst.* **10**, 173-200.

Harlan, J. F. (1982). Relationships between weeds and crops. *In* "Biology and ecology of weeds" (W. Holzner and M. Numata, eds.), pp. 91-96. W. Junk Publishers, The Hague, Netherlands.

Hedrick, P. W., Ginevan, M. E. and Ewing, E. P. (1976). Genetic polymorphism in heterogeneous environments. *Annu. Rev. Ecol. Syst.* **7**, 1-32.

Heywood, J. S. (1986). Edaphic races of *Gaillardia pulchella* in central Texas. *J. Heredity* **77**, 146-150.

Holzner, W. (1982). Concepts, categories and characteristics of weeds. *In* "Biology and ecology of weeds" (W. Holzner and M. Numata, eds), pp. 3-20. W. Junk Publishers. The Hague, Netherlands.

Horak, M. J. and Holt, J. S. (1986). Isozyme variability and breeding systems in populations of yellow nutsedge (*Cyperus esculentus*). *Weed Sci.* **34**, 538-543.

Horikawa, Y. (1986). Reproductive strategy in white clover, *Trifolium repens* L. of different habitats. *J. Jpn. Soc. Grassl. Sci.* **32**, 235-242.

Hume, L. and Cavers, P. B. (1983). Resource allocation and reproductive and life history strategies in widespread populations of *Rumex crispus*. *Can. J. Bot.* **61**, 1276-1282.

Jain, S. and Martins, P. S. (1979). Ecological genetics of the colonizing ability of rose clover (*Trifolium hirtum* All.). *Am. J. Bot.* **66**, 361-366.

Jana, S. and Naylor, J. M. (1982). Adaptation for herbicide tolerance in populations of *Avena fatua*. *Can. J. Bot.* **60**, 1611-1617.

Jana, S. and Thai, K. M. (1987). Patterns of changes of dormant genotypes in *Avena fatua* populations under different agricultural conditions. *Can. J. Bot.* **65**, 1741-1745.

Kadereit, J. W. and Briggs, D. (1985). Speed of development of radiate and non-radiate plants of *Senecio vulgaris* L. from habitats subject to different degrees of weeding pressure. *New Phytol.* **99**, 155-169.

Kapoor, P. and Ramakrishnan, P. S. (1974). Soil factors influencing the distribution of ecotype populations in *Echinochloa colonum* (L.) Link. (Gramineae). *Bot. J. Linn. Soc.* **69**, 65-78.

Kiang, Y. T. (1982). Local differentiation of *Anthoxanthum odoratum* populations on roadsides. *Am. Midl. Nat.* **107**, 340-350.

Kuiper, D. (1983). Genetic differentiation of various physiological parameters of *Plantago major* and their role in strategies of adaptation to different levels of mineral nutrition. *In* "Genetic Aspects of Plant Nutrition" (M. R. Saric and B. C. Loughman, eds), pp. 261-267. W. Junk Publishers, The Hague, Netherlands.

Lande, R. (1976). The maintenance of genetic variability by mutation in a polygenic character with linked loci. *Genet. Res.* **26**, 221-235.

Law, R., Bradshaw, A. D. and Putwain, P. D. (1977). Life history variation in *Poa annua*. *Evolution* **31**, 233-246.

Lebaron, H. M. and Gressel, J. (Eds) (1982). Herbicide resistance in plants. Wiley Co., New York.

Lewontin, R. C. (1984). Detecting population differences in quantitative characters as opposed to gene frequencies. *Am. Nat.* **123**, 115-124.

Loveless, M. D. and Hamrick, J. L. (1984). Ecological determinants of genetic structure in plant populations. *Annu. Rev. Ecol. Syst.* **15**, 65-96.

Marshall, D. R. and Weiss, P. W. (1982). Isozyme variation within and among Australian populations of *Emex spinosa* (L.) Campd. *Aust. J. Biol. Sci.* **35**, 327-322.

Mashburn, S. J., Sharitz, R. R. and Smith, M. H. (1978). Genetic variation among *Typha* populations of the southeastern United States. *Evolution* **32**, 681-685.

McNeilly, T. (1981). Ecotypic differentiation in *Poa annua* interpopulation differences in response to competition and cutting. *New Phytol.* **88**, 539-548.

Miller, S. D., Nalewaja, J. D. and Mulder, C. E. G. (1982). Morphological and physiological variation in wild oats. *Agron. J.* **74**, 771-775.

Moran, G. F. and Marshall, D. R. (1978). Allozyme uniformity within and variation between races of the colonizing species *Xanthium strumarium* L. (Noogora Burr.). *Aust. J. Biol. Sci.* **31**, 283-292.

Nei, M. (1973). Analysis of gene diversity in subdivided populations. *Proc. Natl. Acad. Sci. U.S.A.* **70**, 3321-3323.

Neuteboom, J. H. (1980). Variability of couch (*Elytrigia repens* (L.) Desv.) in grasslands and arable fields in two localities in the Netherlands. *Acta Bot. Neerl.* **29**, 407-417.

Nevo, E., Beiles, A., Kaplan, D., Golenberg, E. M., Olsvig-Whittaker, L. and Naveh, Z. (1986). Natural selection of allozyme polymorphisms: a microsite test revealing ecological genetic differentiation in wild barley. *Evolution* **40**, 13-20.

Norris, R. F. and Schoner, C. A. Jr., (1980). Yellow foxtail (*Setaria lutescens*) biotype studies: dormancy and germination. *Weed Sci.* **28**, 159-163.

Oegennaro, F. P. and Weller, S. C. (1984). Growth and reproductive characteristics of field bindweed (*Convolvulus arvensis*) biotypes. *Weed Sci.* **32**, 525-528.

Oka, H. I. (1983). Life-history characteristics and colonizing success in plants. *Am. Zool.* **23**, 99-109.

Pegtel, D. M. (1974). Effect of crop rotation on the distribution of two ecotypes of *Sonchus arvensis* L. in the Netherlands. *Acta. Bot. Neerl.* **23**, 349-350.

Potvin, C. (1986). Biomass allocation and phenological differences among southern and northern populations of the C_4 grass *Echinochloa crus-galli*. *J. Ecol.* **74**, 915-923.

Price, S. C., Shumaker, K. M., Kahler, A. L., Allard, R. W. and Hill, J. E. (1984). Estimates of population differentiation obtained from enzyme polymorphisms and quantitative characters. *J. Hered.* **75**, 141-142.

Reinartz, J. A. (1984). Life history variation of common mullein (*Verbascum thapsus*). I. Latitudinal differences in population dynamics and timing of reproduction. *J. Ecol.* **72**, 897-912.

Roose, M. L. and Gottlieb, L. D. (1976). Genetic and biochemical consequences of polyploidy in *Tragopogon*. *Evolution* **30**, 818-830.

Singh, K. P. (1975). Effect of different light intensities on the growth performance of two ecotypes of *Portulaca oleracea* L. *Trop. Ecol.* **16**, 163-169.

Sobey, D. G. (1987). Differences in seed production between *Stellaria media* populations from different habitat types. *Ann. Bot.* **59**, 543-549.

Soltis, D. E. and Rieseberg, L. H. (1986). Autopolyploidy in *Tolmiea menziesii* (Saxifragaceae): Genetic insights from enzyme electrophoresis. *Am. J. Bot.* **73**, 310-318.

Srivastava, A. K. and Misra, R. C. (1970). Ecotypic differentiation in *Gomphrena celosioides* Mart. *Indian J. Weed Sci.* **2**, 63-69.

Teramura, A. H. (1983). Experimental ecological genetics in *Plantago* IX. Differences in growth and vegetative reproduction in *Plantago lanceolata* L. (Plantaginaceae) from adjacent habitats. *Am. J. Bot.* **70**, 53-58.

Tsuchiya, T. and Ishiguri, Y. (1981). Role of the quality of light in the photoperiodic flowering response in four latitudinal ecotypes of *Chenopodium rubrum* L. *Plant Cell Physiol.* **22**, 525-532.

Van Dijk, H. (1984). Genetic variability in *Plantago* species in relation to their ecology. 2. Quantitative characters and allozyme loci in *P. major*. *Theor. Appl. Genet.* **68**, 43-52.

Verkleij, J. A. C., Pieterse, A. H., Horneman, G. T. A. and Torenbeek, M. (1983). A comparative study of the morphology and isoenzyme pattern of *Hydrilla verticillata* (L.f.) Royale. *Aquatica Bot.* **17**, 43-59.

Warwick, S. I. (1987). Isozyme variation in proso millet (*Panicum miliaceum* L.). *J. Hered.* **78**, 210-212.

Warwick, S. I. (1990). Allozyme and life history variation in five northwardly colonizing North American weedy species. Plant Syst. Evol. (in press).

Warwick, S. I. and Black, L. D. (1986a). Electrophoretic variation in triazime-resistant and susceptible populations of *Amaranthus retroflexus* L. *New Phytol.* **104**, 661-670.

Warwick, S. I. and Black, L. D. (1986b). Genecological variation in recently established populations of *Abutilon theophrasti* (velvetleaf). *Can. J. Bot.* **64**, 1632-1643.

Warwick, S. I. and Briggs, D. (1978). The genecology of lawn weeds. I. Population differentiation in *Poa annua* L. in a mosaic environment of bowling green lawns and flower beds. *New Phytol.* **81**, 711-723.

Warwick, S. I. and Briggs, D. (1979). The genecology of lawn weeds. III. Cultivation experiments with *Achillea millefolium* L., *Bellis perennis* L., *Plantago lanceolata* L., *Plantago major* L. and *Prunella vulgaris* L. collected from lawns and contrasting grassland habitats. *New Phytol.* **83**, 509-536.

Warwick, S. I., Bain, J. F., Wheatcroft, R. and Thompson, B. K. (1989). Hybridization and introgression in *Carduus nutans* and *C. acanthoides*. Reexamined. *Syst. Bot.* **14**, 476-494.

Warwick, S. I., Thompson, B. K. and Black, L. D. (1984). Population variation in *Sorghum halepense*, Johnson grass, at the northern limits of its range. *Can. J. Bot.* **62**, 1781-1790.

Warwick, S. I., Thompson, B. K. and Black, L. D. (1987a). Life history variation in populations of the weed species *Setaria faberi*. *Can. J. Bot.* **65**, 1396-1402.

Warwick, S. I., Thompson, B. K. and Black, L. D. (1987b). Genetic variation in Canadian and European populations of the colonizing weed species, *Apera spica-venti*. *New Phytol.* **106**, 301-317.

Weaver, S.E., Dirks, V. and Warwick, S. I. (1985). Variation among populations of *Datura stramonium* L. at the northern margin of its distribution. *Can. J. Bot.* **63**, 1303-1308.
Werth, C. R., Riopel, J. L. and Gillespie, N. W. (1984). Genetic uniformity in an introduced population of witchweed (*Striga asiatica*) in the United States. *Weed Sci.* **32**, 645-648.

2 Differentiation and Adaptation in the Genus *Capsella* (Brassicaceae)

HERBERT HURKA

Spezielle Botanik, University of Osnabrück, Osnabrück, Federal Republic of Germany

I. Introduction

Considerable interest exists in the characteristics of weeds and colonizing plants (Baker and Stebbins, 1965; Baker, 1974; Brown and Marshall, 1981; Barrett, 1982; Barrett and Richardson, 1986; Brown and Burdon, 1987; Warwick, this volume). Common features of weedy plants often include adaptation to a wide range of environments. This can be achieved by considerable phenotypic plasticity as well as ecotypic differentiation. It has even been postulated that genetic variation and phenotypic plasticity may be alternative strategies. However, there seems to be no reason to suspect that genetic variation and phenotypic plasticity will always be exclusive (Quinn, 1987). Another common feature of colonizing plants is marked local differentiation due to founder effects and restricted gene flow as a consequence of predominant self-pollination. Nevertheless, colonizers seem to display a wide range of possible evolutionary outcomes (Brown and Marshall, 1981).

To obtain more knowledge of the biology of weedy plants, we decided to concentrate on the genus *Capsella*. The shepherd's purse, *Capsella bursa-pastoris*, is one of the most frequent and most widespread flowering plants on earth (Coquillat, 1951). It is characterized by its colonizing ability and its wide ecological range, which suggests that the genus *Capsella* may be very well suited for the study of evolutionary biology. We will present the results of several lines of experiments which aimed at understanding the differentiation and adaptation of this very successful colonizing plant.

BIOLOGICAL APPROACHES AND EVOLUTIONARY TRENDS IN PLANTS ISBN 0-12-402960-4

II. The Genus *Capsella*

The genus *Capsella* comprises several species. However, due to the enormous morphological polymorphism detected within this genus, no clearcut species concept exists (Almquist, 1907, 1923, 1929; Shull, 1923, 1929; Svenson, 1983). The number of species recognized by taxonomists varies between one and ten. For our purposes we have adopted a conservative view and designated three species. The diploid *Capsella rubella* Reuter ($2n = 16 = 2x$) is predominantly self-fertilizing and is distributed in countries surrounding the Mediterranean Sea, rarely in other countries with Mediterranean climates. *Capsella grandiflora* (Fauché & Chamb.) Boiss., also diploid with $2n = 16 = 2x$, is self-incompatible. It is found only in Western Greece and, rarely, in Northern Italy. The tetraploid *Capsella bursa-pastoris* (L.) Med. ($2n = 32 = 4x$) is predominantly selfing with world-wide distribution except in the hot and wet tropics. In general, these species are annuals.

The genus *Capsella* almost certainly originated in the Old World (East Mediterranean area) and was presumably brought to Middle and Northern Europe with the spread of agriculture from the Middle East during the Bronze and Iron Ages some 4 to 5 thousand years B.C. From Europe it extended its range into the New World and Australasia with European colonists.

The two diploid species differ sharply in their breeding systems: self-pollinated (*C. rubella*) versus obligate outcrossing due to an incompatibility system (Riley, 1936). *Capsella rubella*, therefore, might be regarded as the derivative species. However, *Capsella grandiflora* ($2x$) and *Capsella bursa-pastoris* ($4x$) display twice as much DNA per cell as *Capsella rubella* ($2x$) and share common nucleotypic characteristics (Freundner and Hurka, in preparation). This would point to a derivative status for *Capsella grandiflora.*

The diploid species *Capsella rubella* and *Capsella grandiflora* have three AAT isozymes. One is located in the plastids. In the tetraploid *Capsella bursa-pastoris*, all three *Aat* isozyme loci are duplicated, resulting in multiple-banded patterns (interlocus hybrid-bands). The diversity of alleles at the duplicated *Aat* loci in *Capsella bursa-pastoris* differs from the sum of the diploid pattern. Some alleles were unique, which might indicate an ancient polyploid event. The inheritance of allozymes was disomic. This disomic inheritance and the true-breeding multiple-banded patterns ("fixed heterozygotes") would suggest an allotetraploid origin for *C. bursa-pastoris* (Hurka *et al.*, 1989).

Isoelectric focusing (IEF) of Rubisco (Ribulose-1, 5-bisphosphate carboxylase/oxygenase) has been used for analyzing the origin of allopolyploids and evolutionary relationships within plant genera (Gray, 1980; Wildman, 1983). Rubisco is composed of large subunits (LSU) coded by chloroplast DNA, and of small subunits (SSU) coded by nuclear DNA. IEF patterns of Rubisco and its subunits were studied in the genus *Capsella* (Mummenhoff and Hurka, in press). All three *Capsella* species share the same LSU banding pattern, indicating a close relationship. However, the species differ in their SSU patterns. The two diploid species have identical SSU patterns, whereas that of *Capsella bursa-pastoris* is different. The SSU pattern of the tetraploid may indicate an ancient allopolyploid, from which the original SSU multigene family has diverged with time.

III. Breeding System, Seed Dispersal, and the Soil Seed Bank

A. *Breeding system*

It is known that *Capsella bursa-pastoris* and *C. rubella* are self-fertile species whereas *Capsella grandiflora* is self-sterile. Shull (1929) assumes outcrossing rates for *C. bursa-pastoris* of 1-2% under field conditions.

Progeny analysis in *Capsella bursa-pastoris* revealed that overall outcrossing is rare, but higher than Shull (1929) would predict. Estimations of outcrossing rates based on allozymes were between 3% and 12% (Hurka *et al.*, 1989). Previous studies using morphological markers have also indicated that outcrossing rates in *Capsella bursa-pastoris* are variable (Hurka and Wöhrmann, 1977). The heterogeneity of outcrossing rates may be related to many factors. If, for instance, alleles within a population are patchily distributed, as will be shown below for a *Capsella* population, then individual plants will receive non-random samples of pollen, and estimates of the mating system will vary accordingly. Besides this spatial variation, temporal variation of the mating system of *Capsella bursa-pastoris* is almost certain. Such aspects of flower morphology as opening of the flower, exposure of the style, and position of the anthers depend on environmental factors. Cloudy and rainy weather seems to support self-pollination, whereas dry and sunny weather seems to favor outcrossing. At low temperatures (about 4-10°C) anthesis is prolonged up to five-fold, but allogamy is apparently reduced (Hurka *et al.*, 1976).

The predominantly autogamous yet flexible mating system combined with polyploidy may contribute to the colonizing ability of *Capsella bursa-pastoris.*

B. *Seed dispersal and the soil seed bank*

Gene flow in *Capsella* is either by pollen or by seed. Movement of genes via pollen is very limited in the predominantly self-fertilizing *Capsella bursa-pastoris*, but is variable.

Seeds of *Capsella bursa-pastoris* on average measure about 1 mm in length and weigh about 0.1 mg (Hurka and Benneweg, 1979). The number of seeds per fruit and the number of fruits per plant vary partly due to different genotypes but also to considerable phenotypic plasticity (Hurka and Wöhrmann, 1977; Neuffer and Hurka, 1986b). Estimations of seed production of several hundred plants grown in random block experiments gave minima of 5,000 seeds per plant and maxima of 90,000. Most individuals grown in those experiments produced between 30,000 and 60,000 seeds.

The seeds fall to the ground very close to the mother plant. All test plants grown in greenhouses scattered their seeds into an area 15-20 cm in diameter. (Hurka and

Haase, 1982). In the field, the distances travelled by the seeds are often of the same order of magnitude, which can be deduced from the the fact that distribution patterns of *Capsella bursa-pastoris* on an experimental field of 30 m^2 were accurately registered over a period of 4 years. After this time, the initial distribution pattern was still recognizable, in spite of removal of competing species and tilling of the ground. Taking into account the scale of the experimental field (30 m^2) and the heavy seed output, it seems reasonable to assume that under natural conditions dispersal distances of scattered seeds must be short (Bosbach *et al.*, 1982).

Seeds deposited on the soil surface can be incorporated into the soil seed bank. The size of the buried seed population is highly variable, ranging from almost zero to up to 30,000 buried seeds per m^2 (Hurka and Haase, 1982; and unpublished results).

The seed epidermis of *Capsella* produces a mucilage which, when moistened, becomes sticky and facilitates the attachment of the seeds to men, animals, transport vehicles etc. Repeated soaking and drying of the seed mucilage cover does not influence germinating capacity (Hurka and Haase, 1982). Thus, despite the fact that most seeds of *Capsella* display only short-distance transport, this plant is also provided with very effective long-distance transport mechanisms.

IV. Variation Within and Between Populations of *Capsella bursa-pastoris*

A. *Phenotypic level*

Variation of characters at the phenotypic level (polygenic characters) include such important life history traits as time to flowering, plant height, number of seeds, and germination behavior. The characters were measured from progeny of the mother plants collected in the field. Progeny were grown in random block experiments in an experimental field station either under Mid-European lowland conditions (Münster and Osnabrück, West Germany) or in alpine climates (Schynige Platte, near Interlaken/Switzerland, 2000 m). In another set of experiments, controlled growth chamber experiments under different temperature regimes were performed to analyze the degree of phenotypic plasticity.

Populations from different countries, continents, and climatic conditions have been analyzed. Based on parametric and non-parametric analysis of variance, the variation between families within populations is very often significant. This is especially true for time to flowering and to a lesser degree for plant height and rosette diameter. Only two populations out of a total of 65 grown in random block experiments were homogeneous with regard to flowering (Table 1). This high degree of polymorphism detected by the computer may be questionable. Raising the significance level at which the null-hypothesis is rejected will eventually reduce the

TABLE 1

Variation between families within populations. Random block experiments with 15-25 families per population and 8-15 offspring per family. Data from Neuffer and Hurka (1986a and b); Neuffer and Bartelheim (1989); and unpublished results

	Proportion of polymorphic populations		
Country of origin	Time to flowering	Plant height	Rosette diameter
N, Norway	5 from 5	4 from 4	2 from 3
SF, Finland	9 from 9	8 from 8	6 from 8
S, Sweden	1 from 1	1 from 1	1 from 1
IS, Iceland	1 from 1	-	-
D, West Germany	7 from 7	6 from 6	2 from 2
CH, Switzerland	22 from 23	12 from 20	14 from 23
A, Austria	0 from 1	-	-
AFG, Afghanistan	1 from 1	-	-
USA, United States of America	14 from 14	14 from 14	14 from 14

degree of polymorphism, and further statistical treatment of the data ("Least Square Differences", LSD) often condensed the family values to homogeneous subgroups (Neuffer and Bartelheim, in press). In addition, phenotypic plasticity may obscure the picture, as selection for different flowering strains out of a polymorphic population sometimes failed. It appears, therefore, that genetic polymorphism of quantitative characters may be overestimated if statistics are interpreted unreflectively. However, there is no doubt that the genetic variation of polygenic traits in natural populations of *Capsella bursa-pastoris* may be high and that the degree of polymorphism between populations may change on a microgeographical scale.

As far as we can judge from Table 1, *Capsella bursa-pastoris* populations introduced to North America do not exhibit lower genetic variation than European source populations. This is different from the general expectation that alien weeds often are genetically less variable than the source populations, as may be expected as a result of founder effects (Barrett, 1982; Clegg and Brown, 1983). However, there are also reports in the literature that introduced populations are not always genetically depauperate (e.g. Brown and Marshall, 1981, for *Bromus mollis* in Australia and Warwick *et al.*, 1987, for *Apera spica-venti* in Canada).

In all experiments where Scandinavian populations (high geographical latitude) were compared with populations from lower latitudes (e.g. the European Alps), populations from Scandinavia were early flowering (Neuffer and Hurka, 1986a) (Fig. 1). Even when planted in parallel experiments at "Schynige Platte" (46.38°N, 8.00°E; 2000 m; Switzerland) and Osnabrück (52.18°N, 8.00°E; 70 m; West Germany), Scandinavian populations proved to be early genotypes (Neuffer, 1986). However, there was a general delay in time to flowering at the alpine station for both Scandinavian and Swiss plants compared to their sister plants grown in

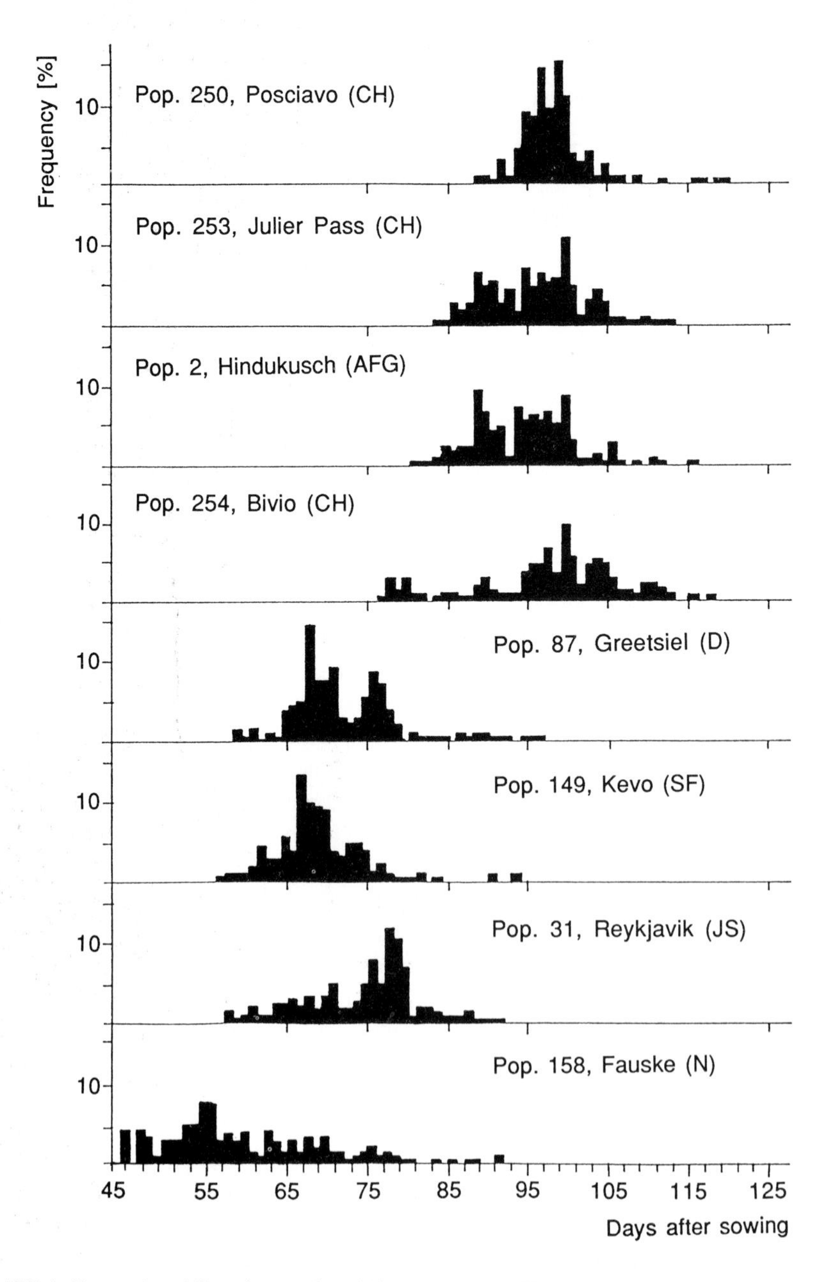

FIG. 1 Frequencies of flowering per day within and between different *Capsella bursa-pastoris* populations. 25 families per population with 10 progeny each. Experimental station: Münster, D (Middle European lowland). Countries of origin: AFG, Afghanistan; CH, Switzerland; D, Federal Republic of Germany; IS, Iceland; N, Norway; SF, Finland.

Osnabrück (Fig. 2). These latter findings point to the influence of the environment on time to flowering, whereas the differences between the population sets within an experiment must be attributed to genetic differences.

Growth chamber experiments with different temperatures were carried out. The progeny of five mother plants (= strains A to E) per population were divided into subgroups and treated with different temperature regimes. Strains are random samples of natural populations. Daytime and nighttime temperatures changed at 12:12 hour interval, synchronously with light and dark. Altogether, progeny of 18 populations were grown under five to seven controlled temperature regimes (Neuffer and Hurka, 1986a). Table 2 focuses on some main points.

Absolute values for days to flowering varied greatly, up to three- to fourfold within treatment groups and five- to sixfold between the groups. All strains (= treatment groups) were strongly delayed in flowering by the temperature regime 10:5°C. On average, in all temperature regimes strains from Scandinavian populations (Pop. 119 and 151) flowered earlier than the strains from Switzerland, demonstrating genetic differences between populations. Some strains have obligate vernalization requirements (cp. Pop. 257). Variation between strains (estimated by CV-values, Table 2) within populations was low in Pop. 119 and 151, but high in Pop. 280 and 281, indicating genetic polymorphism for temperature susceptibility in the latter two populations. The situation was further complicated by different variation patterns within the strains. Variation was low in some cases (low within CV-values -- temperature regime 10:5°C left apart -- for strains B to E in Pop. 119 and for strains A and E in Pop. 281) and high in others (strain D in Pop. 280 and strain C in Pop. 281). This means that there are differences not only between strains within the same temperature regime but also within strains between different temperature regimes.

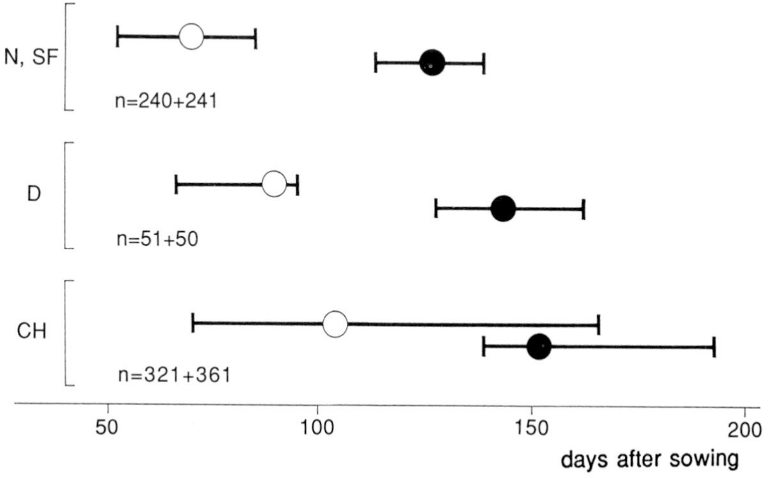

FIG. 2 Influence of different climates on flowering in *Capsella bursa-pastoris*. Sister plants grown in Osnabrück, D (Middle European lowland; open circles), and Schynige Platte, CH (Alpine climate; black circles). Sample size n: the first number refers to Osnabrück, the second to Schynige Platte. Countries of origin see FIG. 1. Data from Neuffer (1986).

TABLE 2

Effect of different temperature regimes on time to flowering (days after sowing) of *Capsella bursa-pastoris*. Mean values x (n = 5) for every treatment within each strain A to E are given. Countries of origin see Table 1. CV(1), coefficient of variation within strains (10:5°C excluded); CV(2) coefficient of variation between strains. Data from Neuffer and Hurka (1986a)

Country (Pop. No.)	Temp.	x_A	x_B	x_C	x_D	x_E	CV(2)
SF	25:15	41	47	46	47	56	0.114
(Pop. 119)	20:10		67	58	54	67	0.106
	25: 5	64	62	62	60	72	0.073
	10: 5	144	153	153	146	165	0.054
	30:10	46	51	50	48	55	0.067
	CV(1)	0.240	0.164	0.135	0.115	0.133	
N	25:15	46	43	48	48	46	0.044
(Pop. 151)	20:10	76	71	88	79	73	0.085
	25: 5	57	53	62	61	55	0.066
	10: 5	142	137	143	145	147	0.026
	30:10	53	51	54	52	53	0.021
	CV(1)	0.221	0.216	0.279	0.229	0.202	
CH	25:15	-	-	40	-	-	-
(Pop. 257)	20:10	-	-	56	-	-	-
	10: 5	191	186	120	169	186	0.172
	30:10	-	-	43	-	-	-
	CV(1)	-	-	0.183	-	-	
CH	25:15	70	189	190	272	135	0.436
(Pop. 280)	20:10	81	115	147	124	93	0.231
	25: 5	74	168	102	151	92	0.342
	10: 5	152	223	212	167	156	0.181
	30:10	89	149	123	188	89	0.330
	CV(1)	0.106	0.202	0.268	0.350	0.214	
CH	25:15	61	54	68		64	0.095
(Pop. 281)	20:10	68	71	149	105	80	0.356
	25: 5	67	55	78	88	68	0.174
	10: 5	186	196	271	259	186	0.190
	30:10	59	45	68	61	68	0.156
	CV(1)	0.069	0.192	0.431	0.262	0.098	

It appears from the open-field and growth chamber experiments that *Capsella bursa-pastoris* displays definite intraspecific genetic variation in the time required before flowering. Genetic adaptations of certain traits to local ecological conditions were evident. Populations also varied in the amount and pattern of plasticity, and it appeared that phenotypic plasticity may also be controlled by selection. There is pronounced ecotypic variation in time to flowering between "early " Scandinavian and "late" alpine populations. "Early" and "late" ecotypes could also be proved for North America. Populations from the Central Valley of California are early-flowering compared to those from Northern California and the Sierra Nevada (in preparation).

In addition to these marked geographical differences a strong correlation between time of flowering and elevation above sea-level was observed for populations from European Alpine regions. Alpine populations comprise an array of genotypes with different susceptibility to low temperatures. The higher the elevation, the more summer annual genotypes were replaced by winter annual genotypes. The later flowering genotypes within the whole set of Alpine ecotypes replaced the earlier genotypes along an altitudinal gradient. This topocline was paralleled by an ecocline expressed as shortening of vegetation period (Neuffer and Bartelheim, 1989).

Ecotypic patterns of flowering, growth form, and leaf form parameters (Steinmeyer *et al.*, 1985; Neuffer and Hurka, 1986b; Neuffer, 1989) contrast with variation in another important fitness character: germination behavior. Once dormancy was broken, seeds from all populations were able to germinate over the entire temperature range. Some populations revealed pronounced temperature optima for germination capacity; others germinated equally well over the entire temperature range (Neuffer and Hurka, 1988). This indicates genetic heterogeneity between and also within populations. However, no correlation between germinability and any environmental pattern was detected. Our data suggest that germination of *Capsella bursa-pastoris* in the field is mainly regulated by the factors contributing to the inception and breaking of dormancy, which depend on preharvest (maternal effects) and postharvest conditions. Since dormancy inception and release is highly influenced by environmental factors, this appears to be a satisfactory way to regulate seedling emergence, in time and space, for *Capsella*, which occupies a wide range of soil types and climatic conditions and which has to cope with the unpredictability of its habitats. Such a germination strategy comes close to an "general - purpose genotype".

To summarize the main points: phenotypic characters in *Capsella bursa-pastoris* may vary randomly within and between populations. However, it is obvious that some important life history traits like time to flowering and plant height are the product of the genotype and the environment, and consequently may also vary in an ecotypic manner. Germination strategy in *Capsella* comes close to an "general-purpose genotype".

B. Isozyme level

With regard to allozymes, homomorphic and polymorphic populations can be encountered (Hurka, 1983). The proportion of homomorphic populations is significantly higher than phenotypic characters would indicate. One reason for this discrepancy between estimations based on phenotypic and molecular markers may be that phenotypic characters as a rule are polygenic markers, whereas isozymes as defined here are coded by one locus and thus represent monogenic characters. Another reason may be the different adaptive significance of the two sets of characters. Allozyme variation pattern is often very patchy. One finds distinct and clear-cut changes of allozyme frequencies within a few kilometers or even meters. When analyzing subpopulation structure of a polymorphic *Capsella bursa-pastoris* populations, it turned out that allozymes were not evenly distributed over the population (Fig. 3). This might reflect restricted gene flow events and/or the influence of the soil-seed bank (see below).

On a macrogeographic scale, there are differences in allele frequencies between different parts of the world (Table 3, for instance, frequencies of *Aat 2B-1* in Scandinavia and Germany and of *Aat 3A-1* in Switzerland and North America). The genotype *Aat 3A-55/3B-55* was common in Scandinavia, rare in West Germany, locally in Switzerland and occasionally in New Zealand. It has not been found outside these areas so far.

The *Aat 1A-1* allele was nearly fixed (more than 7000 individuals from all over the world checked, frequency > 0.995). However, it was replaced in 35 individuals either by the allele *Aat 1A-2* (Finland, USA), by *Aat 1A-3* (USA), or by *Aat 1A-4* (Spain). It cannot be determined at the moment whether the occurrence of *Aat 1A-2* in Spain and North America goes back to independent mutations or rather should be interpreted by long-distance dispersal.

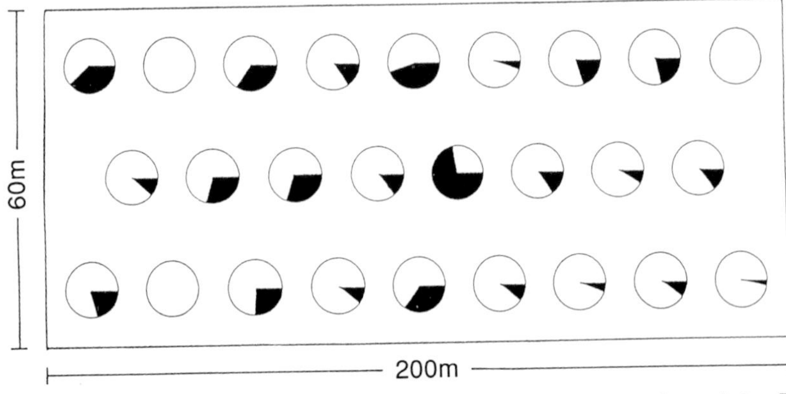

FIG. 3 Distribution pattern of the allele *Aat* 1B-2 within a *Capsella bursa-pastoris* population. Total population 60 x 200 m, subpopulations (indicated by circles) 7 m^2 and 20-25 m apart. Sample size of each subpopulation 20 plants. Allele frequencies are given by black sectors (white only: 0 %, black only: 100 %).

TABLE 3

Frequencies of *Aat* alleles in *Capsella bursa-pastoris* in different parts of the world. CAN, Canada; CH, Switzerland; D, Federal Republic of Germany; DK, Denmark; GB, Great Britain; GR, Greece; I, Italy; IS, Iceland; N, Norway; S, Sweden; SF, Finland; SP, Spain; USA, United States of America

	Scandinavia	IS	GB	D	CH	Med. Countr.	North America
Allele	(DK, N, S, SF)				(I, GR, SP)	(USA, CAN)	
Aat 1A-1	0.997	1.000	1.000	0.996	1.000	0.985	0.968
1A-2	0.002	-	-	0.004	-	0.013	0.029
1A-3	-	-	-	-	-	-	0.003
1A-4	-	-	-	-	-	0.002	-
Aat 1B-1	0.932	1.000	0.729	0.789	0.914	0.777	0.852
1B-2	0.033	-	-	0.136	0.029	0.084	0.012
1B-3	-	-	0.004	0.040	0.040	-	0.005
1B-4	0.035	-	0.266	0.015	0.015	0.139	0.130
Aat 2A-1	0.997	0.405	0.998	0.987	1.000	1.000	0.996
2A-2	0.002	0.595	0.002	0.013	-	-	0.004
Aat 2B-1	0.135	-	0.284	0.657	0.159	0.327	0.166
2B-2	-	-	0.003	0.014	-	0.050	0.004
2B-4	0.865	1.000	0.725	0.329	0.841	0.623	0.830
Aat 3A-1	0.031	1.000	0.763	0.608	0.156	0.602	0.902
3A-3	0.240	-	0.225	0.344	0.820	0.394	0.094
3A-5	0.728	-	0.011	0.046	0.323	0.004	0.004
3A-6	-	-	-	0.002	-	-	-
Aat 3B-2	0.002	-	0.064	0.002	0.323	0.036	0.007
3B-3	-	-	-	0.004	-	-	0.003
3B-5	0.998	1.000	0.935	0.994	0.677	0.964	0.990

The multilocus genotype *Aat 1A-11/1B-11; 2A-11/2B-44; 3A-11/3B-55* was nearly fixed in the Central Valley of California but very rare outside the valley. It was also detected in high frequencies in some Mediterranean populations and in Bolivia, and only occasionally outside these regions. Unfortunately, so far we do not have sufficient or any accessions of *Capsella bursa-pastoris* from Middle and South America and from the southern states of the USA. However, it seems that high frequencies of this multilocus genotype in America coincide with areas which were first colonized by Mediterranean people. Isozyme genotypes of *Capsella* may to some extent reflect colonization history.

V. Discussion

Capsella bursa-pastoris is characterized by a very complex variation pattern which is not easy to recognize, nor easy to describe. It is strongly influenced by the effective breeding system, seed dispersal mechanisms and properties of the soil-seed bank.

Most seeds fall to the ground near the mother plant and are incorporated into the soil-seed bank. Therefore, populations in already occupied sites are often recruited from autochthonous seed material. Thus, selection can operate over many years on the same gene pool, and adaptation to local environments may evolve. It seems important to stress that populations of colonizing species are not always true "colonizing" populations, but may rather be composed of gene complexes already adapted to a given locality.

Long-distance transport in *Capsella*, which is faciliated by the mucilaginous seed coat, may result in natural populations arising through the accidental transport of a small number of seeds to hitherto unoccupied sites. Therefore, such populations are often genetically rather uniform. If different populations were established by small groups of founders drawn from a polymorphic parental population, then variation between the founder populations could be expected. It can be assumed that the existence of genetic divergence between *Capsella* populations is partly due to such a mechanism.

The soil-seed bank can serve as a permanent source for the introduction of "new" genotypes into the actual plant population, depending on how well it is shaken up and depending on the degree of polymorphism stored in the soil. This introduction of genetic variation will be counteracted, however, by the competitive elimination of seedlings and plants under conditions of high successional displacement stress. Therefore, one would expect less variable populations in places where succession has already started. It is also to be expected that in freshly filled or newly broken soils more seeds are brought up to the surface than in bare, but otherwise nondisturbed, soils. This disturbance increases the probability of different genotypes in highly disturbed soils. And, indeed, we could establish differences in polymorphism between populations of highly and less disturbed soils (Bosbach *et al.*, 1982).

Patterns of variation may thus be patchy and random. They nevertheless might reflect adaptation to the local environment depending on the character under investigation. Some phenotypic characters (for instance flowering behavior and growth form parameters) often appear to be the product of the genotype and the environment and consequently may vary in an ecotypic manner. However, phenotypic plasticity may obscure genotypic differentiation. Germination, although displaying genetic polymorphism, seems to be regulated mainly by the factors contributing to the inception and breaking of seed dormancy, which depends on pre- and post-harvest conditions, and no correlation between the genotype and environmental parameters has been detected so far. Obviously, *Capsella bursa-pastoris* adopted different adaptive strategies associated with several life history traits.

Genetic variation at the isozyme level differs in quality from that at the phenotypic level. Allozyme variation does not seem to reflect adaptations to local environments, nor does it provide any image of the adaptive diversity of the species. Instead, allozyme variation is more likely to reflect genetic drift effects, gene flow events, and the effective mating system. Variation at the allozyme level may also reflect evolutionary history and colonization events.

Studies at the phenotypic and molecular level provide different information. Both are necessary for an understanding of evolution.

Acknowledgements

I wish to acknowledge the enthusiasm of all my co-workers. Especially, I thank Barbara Neuffer for helpful discussion. Portions of this research were supported by the German Research Foundation DFG.

References

Almquist, E. (1907). Studien über die *Capsella bursa-pastoris* (L.). *Acta Horti Bergiani* **4** (6), 1-92.

Almquist, E. (1923). Studien über *Capsella bursa-pastoris* (L.) II. *Acta Horti Bergiani* **7** (2), 41-95.

Almquist, E. (1929). Zur Artbildung in der freien Natur. *Acta Horti Bergiani* **9** (2), 37-76.

Baker, H.G. (1974). The evolution of weeds. *Annu. Rev. Ecol. Syst.* **5**, 1-24.

Baker, H.G. and Stebbins, G.L., eds (1965). "The Genetics of Colonizing Species". Acad. Press, London, New York.

Barrett, S.C.H. (1982). Genetic variation in weeds. *In* "Biological Control of Weeds with Plant Pathogens" (Ch. Raghavan and H.L.Walker, eds), pp. 73-98. John Wiley and Sons, New York.

Barrett, S.C.H. and Richardson, B.J. (1986). Genetic attributes of invading species. *In* "Ecology of Biological Invasions" (R.H. Groves and J.J. Burdon, eds), pp. 21-33. Cambridge University Press, Cambridge, London, New York, Melbourne, Sydney.

Bosbach, K., Hurka, H. and Haase, R. (1982). The soil seed bank of *Capsella bursa-pastoris* (Cruciferae): its influence on population variability. *Flora* **172**, 47-56.

Brown, A.H.D. and Burdon, J.J. (1987). Mating systems and colonizing success in plants. *Brit. Ecol. Soc. Symp.* **26**, 115-131.

Brown, A.H.D. and Marshall, D.R. (1981). Evolutionary changes accompanying colonization in plants. *In* "Evolution Today" (G.G.E. Scudder and J.L. Reveal, eds), pp. 351-363. *Proc. 2nd Internat. Congr. Syst. Evol. Biol.*, Hunt Instit. Bot. Document., Pittsburgh, PA.

Clegg, M.T. and Brown, A.H.D. (1983). The founding of plant populations. *In* "Genetics and Conservation" (C.M. Schonewald-Cox, S.M.Chambers, B. Mac Bryde and L. Thomas, eds), pp. 216-228. Benjamin and Cummings, Menlo Park, CA.

Coquillat, M. (1951). Sur les plantes les plus communes de la surface du globe. *Bull. Mens. Soc. Linn.. Lyon* **20**, 165-170.

Gray, J.C. (1980). Fraction I protein and plant phylogeny. *In* "Chemosystematics: Principles and Practice" (F.A. Bisby, J.G. Vaughan and C.A. Wright, eds), pp. 167-193. Academic Press, London.

Hurka, H. (1983). Enyzme profiles in the genus *Capsella. In* "Proteins and Nucleic Acids in Plant Systematics" (U. Jensen and D.E. Fairbrothers, eds), pp. 222-237. Springer-Verlag, Berlin, Heidelberg.

Hurka, H. and Benneweg, M. (1979). Patterns of seed size variation in populations of the common weed *Capsella bursa-pastoris* (Brassicaceae). *Biol. Zentralbl.* **98**, 699-709.

Hurka, H. and Haase, R. (1982). Seed ecology of *Capsella bursa-pastoris* (Cruciferae): dispersal mechanism and the soil seed bank. *Flora* **172**, 35-46.

Hurka, H. and Wöhrmann, K. (1977). Analyse der genetischen Variabilität natürlicher Populationen von *Capsella bursa-pastoris* (Brassicaceae). *Bot. Jahrb. Syst.* **98**, 120-132.

Hurka, H., Freundner, St., Brown, A.H.D. and Plantholt, U. (1989). Aspartate aminotransferase isozymes in the genus *Capsella* (Brassicaceae): subcellular location, gene duplication, and polymorphism. *Biochem. Genet.* **27**, 77-90.

Hurka, H., Krauss, R., Reiner, T and Wöhrmann, K. (1976). Das Blühverhalten von *Capsella bursa-pastoris* (Brassicaceae). *Plant Syst. Evol.* **125**, 87-95.

Mummenhoff, K. and Hurka, H. Evolution of the tetraploid *Capsella bursa-pastoris* (Brassicaceae): isoelectric focusing analysis of Rubisco. *Plant Syst. Evol.* (in press).

Neuffer, B. (1986). Transplantationsversuch Schynige Platte im Sommer 1985. Blühverhalten alpiner und skandinavischer Populationen von *Capsella bursa-pastoris* (Brassicaceae). *Beilage Jahresber.Alpengarten Schynige Platte (Berner Oberland, Schweiz)* **60**, 1-8.

Neuffer, B. (1989). Leaf morphology in *Capsella* (Cruciferae).Dependency on environments and biological parameters. *Beitr. Biol. Pflanzen* **64**, 39-54.

Neuffer, B. and Bartelheim, S. (1989). Genecology of *Capsella bursa-pastoris* from an altidudinal transsect in the Alps. *Oecologia* (Berl.) **81**, 521-527.

Neuffer, B. and Hurka, H. (1986a). Variation of development time until flowering in natural populations of *Capsella bursa-pastoris* (Cruciferae). *Plant Syst. Evol.* **152**, 277-296.

Neuffer, B. and Hurka, H (1986b). Variation of growth from parameters in *Capsella* (Cruciferae). *Plant Syst. Evol.* **153**, 265-279.

Neuffer, B. and Hurka, H. (1988). Germination behaviour in populations of *Capsella bursa-pastoris* (Cruciferae). *Plant Syst. Evol.* **161**, 35-47.

Quinn, J.A. (1987). Complex patterns of genetic differentiation and phenotypic plasticity versus an outmoded ecotype terminology. *In* "Differentation Patterns in Higher Plants" (K.M. Urbanska, ed.), pp. 95-113. Academic Press, London.

Riley, H.P. (1936). The genetics and physiology of self-sterility in the genus *Capsella. Genetics* **21**, 24-39.

Shull, G.H. (1923). The species concept from the point of view of a geneticist. *Am. J. Bot.* **10**, 221-228.

Shull, G.H. (1929). Species hybridization among old and new species of Shepherd's Purse. *Proc. Internat. Congr. Plant Sci I*, 837-888.

Steinmeyer, B., Wöhrmann, K. and Hurka, H. (1985). Phänotypenvariabilität und Umwelt bei *Capsella bursa-pastoris* (Cruciferae). *Flora* **177**, 323-334.

Svenson, S. (1983). Chromosome numbers and morphology in the *Capsella* complex (Brassicaceae). *Wildenowia* **13**, 267-276.

Warwick, S.J., Thompson, B.K. and Black, L.D. (1987). Genetic variation in Canadian and European populations of the colonizing weed species *Apera spica-venti. New Phytol.* **106**, 301-317.

Wildman, S.G. (1983). Polypeptide composition of Rubisco as an aid in studies of plant phylogeny. *In* "Proteins and Nucleic Acids in Plant Systematics" (U. Jensen and D.E. Fairbrothers, eds), pp. 182-190. Springer-Verlag, Berlin, Heidelberg, New York, Tokyo.

3 Parapatric Differentiation of Paraquat Resistant Biotypes in Some Compositae Species

KAZUYUKI ITOH[1] AND SHOOICHI MATSUNAKA[2]

[1]*Tropical Agriculture Research Center, MAFF, Tsukuba, Ibaraki 305, Japan*
[2]*Laboratory of Pesticide Science, Faculty of Agriculture, Kobe University, Kobe 657, Japan*

1. Introduction

Shortly after the introduction of herbicides for weed control, Harper (1956) predicted that the repeated use of the same herbicide would be almost inevitably followed by the development of a resistant biotype of weed, as often happened with fungicides and insecticides. In 1968, the first case of resistance development was reported by Ryan (1970) in Washington State in *Senecio vulgaris* grown in a nursery where atrazine and simazine had been used once or twice annually since 1958. It has been reported that the repeated use of *s*-triazine herbicides leads to the formation of resistant biotypes in 53 species (38 broad leaf, 15 grass) belonging to 32 genera due to new selection pressure in weedy plants (Holt and LeBaron,1989). Herbicide-resistant biotypes were observed not only for *s*-triazine herbicides but also for the major herbicides which are used on a high commercial scale internationally, like bipyridyl and dinitroaniline (Mudge *et al.*,1984) herbicides. These resistant biotypes are responsible for severe weed damage worldwide. This review considers the appearance, distribution, inheritance and physiology of weed biotypes in Compositae resistant to bipyridyl herbicides.

II. Discovery of Biotypes Resistant to Bipyridyl Herbicides

In 1980, one of the authors and his co-workers were informed that the control of *Erigeron philadelphicus* L. (synonym *Conyza philadelphicus*, Philadelphia

BIOLOGICAL APPROACHES AND EVOLUTIONARY TRENDS IN PLANTS ISBN 0-12-402960-4

fleabane) had become increasingly difficult in mulberry fields located along the Arakawa River at Fukiage in Saitama Prefecture. Replies to questionnaires distributed to mulberry farmers in that area revealed that paraquat had been applied 2 or 3 times annually during the preceding 8 to 11 years. It was thus assumed that new biotypes resistant to paraquat appeared approximately 5 or 6 years after paraquat was introduced to the mulberry fields (Watanabe *et al.*,1982).

In field experiments conducted at sites infested with *E. philadelphicus* plants resistant to paraquat, some of the plants were killed after the application of bentazon, glyphosate and MCPA at recommended dosages, suggesting that the plants were susceptible to these herbicides. The paraquat-resistant plants and normal susceptible plants originating from a site where paraquat had not been sprayed were compared with each other (Table 1).

A dose of 1.0 kg a.i./ha was sufficient to completely kill the green leaves of the susceptible plants. In sharp contrast, resistant plants treated with paraquat at a rate of 0.5 to 2.0 kg a.i./ha showed none of the symptoms observed in the susceptible plants. They retained a few green leaves even at a rate of 16.0 kg a.i./ha. The paraquat -resistant plants had also developed a resistance to diquat (Fig. 1).

Seedlings at the 1.5 leaf stage, derived from resistant or susceptible plants of *E. philadelphicus*, were sprayed with varying concentrations of paraquat. For the susceptible seedlings, the concentration ranged from 0.0039 to 0.5 kg a.i./ha, while it ranged from 0.031 to 16.0 kg a.i./ha for the resistant seedlings. The dose of paraquat required to kill the resistant seedlings was 250 times higher than that for the susceptible seedlings. The use of a paraquat solution at a concentration of 0.5 kg a.i./ha was sufficient to differentiate resistant from susceptible seedlings (Fig. 2).

TABLE 1

Control of paraquat-resistant *Erigeron philadelphicus* in a mulberry field with several herbicides applied at recommended doses

Herbicide	Rate (kg a.i./ha)	Top dry weight (g/m^2)
Paraquat-dichloride	0.96	93.9
Paraquat-dimethylsulphate	1.14	102.7
Bentazon	6.00	8.1
MCP-sodium	0.39	25.9
Glyphosate	4.10	0
Check	–	92.8

(Watanabe *et al.*, 1982)

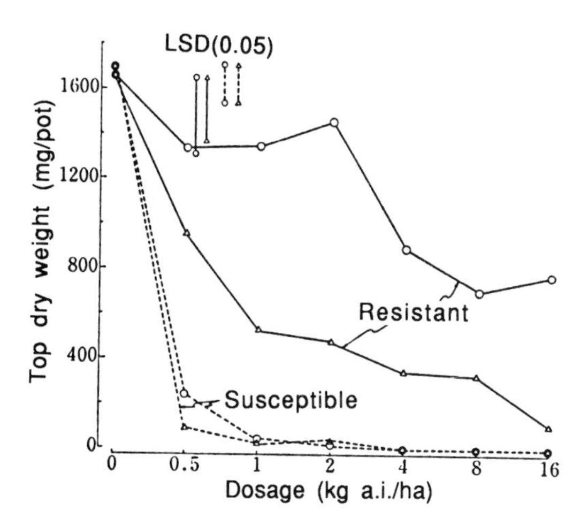

FIG. 1 Relationship between applied doses of paraquat (o) and disquat(Δ) and top dry weight of tested plants 10 days after application (Watanabe *et al.*, 1982).

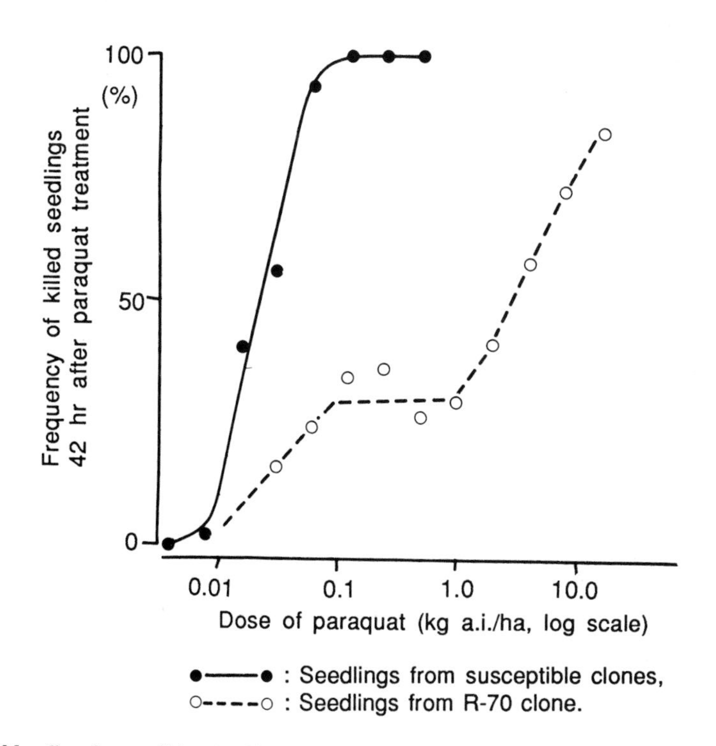

FIG. 2 Mortality of young *Erigeron philadelphicus* seedlings from a resistant clone and from susceptible clones following paraquat treatment (Itoh *et al.*, 1984).

An experiment was carried out to examine the response of leaf disks to various concentrations of a paraquat solution. The results clearly showed that the level of resistance to paraquat of the resistant plants of *E. philadelphicus* was 100 times higher than that of the susceptible plants (Fig. 3).

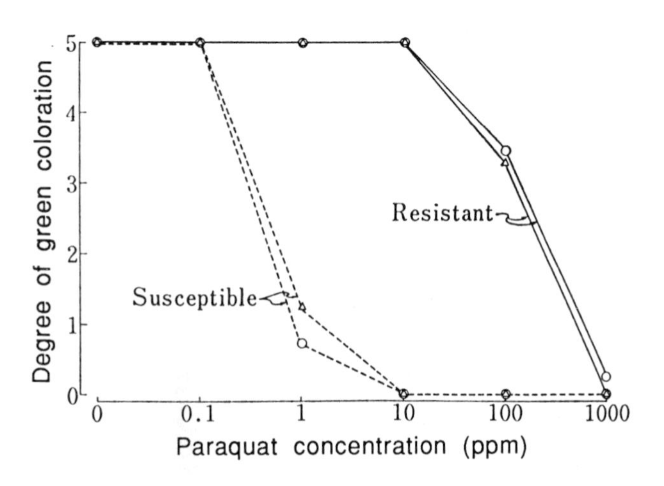

FIG. 3 Relationship between paraquat concentration and retention of green color of radical leaf (○) and stem leaf (△) disks in *Erigeron philadelphicus* 48 hours after dipping in herbicide solutions (Watanabe *et al.*, 1982).

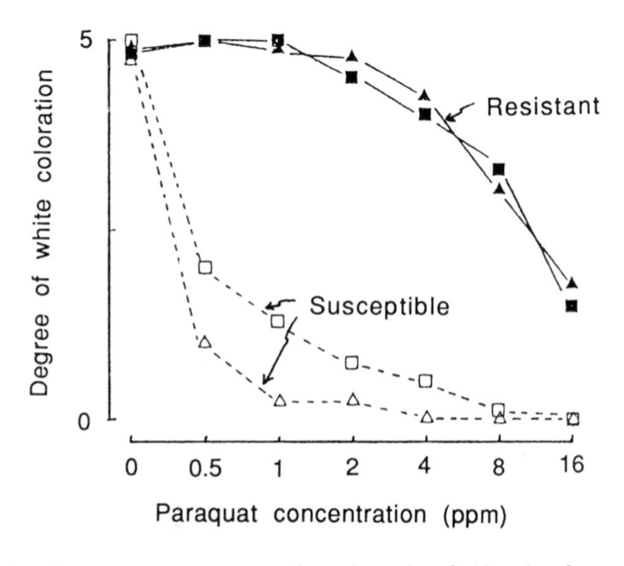

FIG. 4 Relationship between paraquat concentration and retention of white color of cut roots in *Erigeron philadelphicus* L. in the light (△,▲) and dark (□,■) 48h after dipping in the herbicide solution (Itoh and Miyahara, 1983).

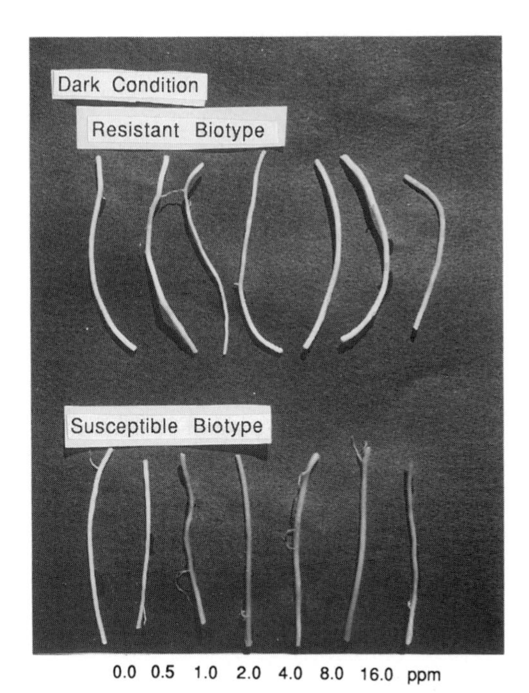

FIG. 5 Relationship between doses of paraquat and retention of white color of cut roots in *Erigeron philadelphicus* L. under dark conditions with dipping in the herbicide solution.

An examination of the response of cut roots 5cm long was also carried out with various concentrations of a paraquat solution. The results showed, not only under illumination but also under dark conditions, that the level of the resistance to paraquat of the resistant roots of *E. philadelphicus* was nearly 10 times higher than that of the susceptible ones (Figs. 4 and 5). This fact suggests that the mode of paraquat action was different in photosynthetic and non-photosynthetic systems.

Paraquat resistance of *E. philadelphicus* was also observed in fields cultivated with perennial crops such as mulberries, tea and various fruit trees (Hanioka,1986; Usami *et al.*, 1989). Kato *et al.* (1982) detected a biotype of *Erigeron canadensis* L. (*Conyza canadensis*, Canadian fleabane, horseweed) resistant to paraquat in vineyards in Osaka Prefecture, and Hanioka (1987) also found biotypes of *Conyza sumatrensis* (Retz.) Walker (*Erigeron sumatrensis* Retz.) and *Youngia japonica* (L.) DC. (Asiatic hawkbeard) resistant to paraquat in mulberry fields in Saitama Prefecture. The three resistant biotypes showed a similar pattern of resistance to that of *E. philadelphicus* (Figs. 6 and 7). Presently resistance has been observed in four

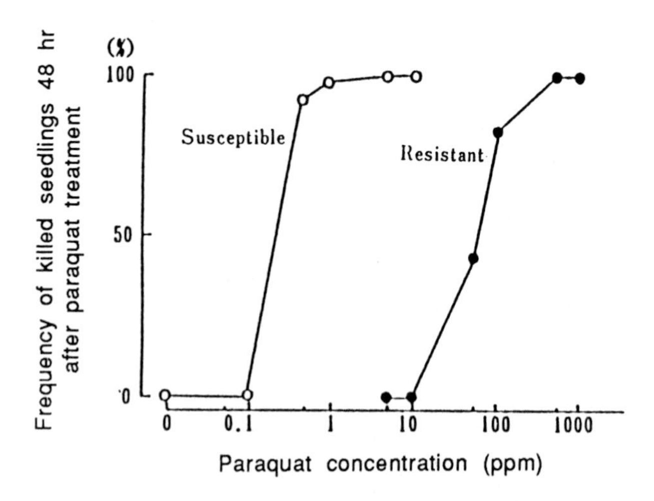

FIG. 6 Mortality of *Youngina japonica* seedlings (emerged after a month) derived from a resistant plant and from a susceptible plant after foliar spray of paraquat (Hanioka, 1987).

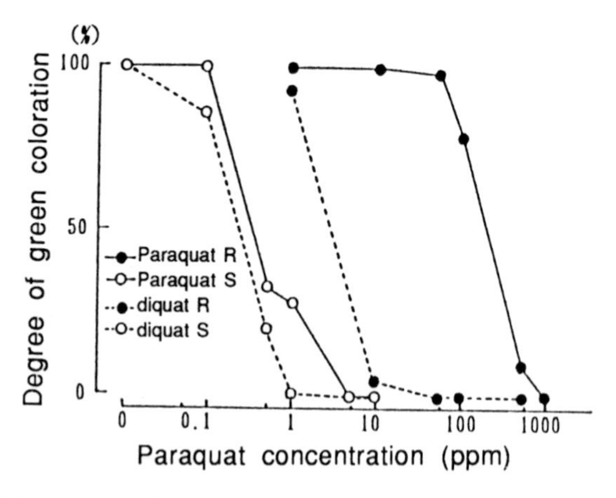

FIG. 7 Relationship between paraquat and diquat concentration and retention of green color of radical leaf disks, after dipping in herbicide solution for 48 hr in *Conyza sumatrensis* Retz (Hanioka, 1987).

biotypes of Compositae weeds in a single field or in a single levee of paddy field where paraquat had been applied 2 or 3 times annually during the preceding 15-20 years (Fig. 8). Biotypes resistant to paraquat have been detected in ten weed species (5 Compositae species, 3 grass) belonging to 7 genera worldwide (Table 2). Furthermore, a biotype resistant to both paraquat and atrazine was observed in *E. canadensis* (Pölös *et al.*, 1986). It is evident that the strong selection pressure

Fig. 8 Paraquat resistance has been observed in four biotypes of Compositae weeds such as *Erigeron philadelphicus* (in front of a mulberry tree), *E. canadensis* (center), *Conyza sumatrensis* (right) and *Youngia japonica* (lower left) in a single mulberry patch at Kumagaya in Saitama Prefecture in April 1989, where paraquat had been applied 2 or 3 times annually during the preceding 15 years.

TABLE 2
Distribution of paraquat-resistant weeds

	Genera Species	Common Name	Year Found	Location
1.	*Arctotheca calendula*	Capeweed	1986	Australia
2.	*Conyza bonariensis*	Hairy fleabane	1979	Egypt
	(*C. linifolius*)		1984	Hungary
	(*Erigeron bonariensis*)			
3.	*C. sumatrensis*		1987	Japan
	(*Erigeron sumatrensis*)			
4.	*Epilobium ciliatum*	American willowherb	1984	Belgium
5.	*Erigeron canadensis*	Canadian fleabane	1982	Japan
6.	*E. philadelphicus*	Philadelphia fleabane	1980	Japan
7.	*Hordeum glaucum*	Barley grass	1983	Australia
8.	*H. leporium*		1989	Australia
9.	*Poa annua*	Annual bluegrass	1978	United Kingdom
10.	*Youngia japonica*	Asiatic horseweed	1987	Japan

imposed by 2 or 3 yearly applications of paraquat and diquat led to the appearance of populations resistant to herbicides in four weed species infesting a field. If the occurrence of weed resistance to herbicide is to be minimized or delayed, weed control practices that rely on the use of one herbicide over a long period of time must clearly be avoided.

III. Distribution and Ecological Fitness of the Resistant Biotypes

These Compositae species are dominant in the early stage of secondary succession, and many species are plants originating in North and South America which become naturalized to Japan in modern times (Numata and Asano, 1969).

Figure 9 shows distribution of a paraquat-resistant biotype of *E. philadelphicus* in mulberry fields along the Arakawa River in Fukiage in April 1981. The resistance was only distributed in a circle whose radius was 2.0 km in that year. The presence of a starting point of the resistant biotype was inferred based on the figure.

The ratio of occurrence of paraquat-resistant biotypes was high in mulberry

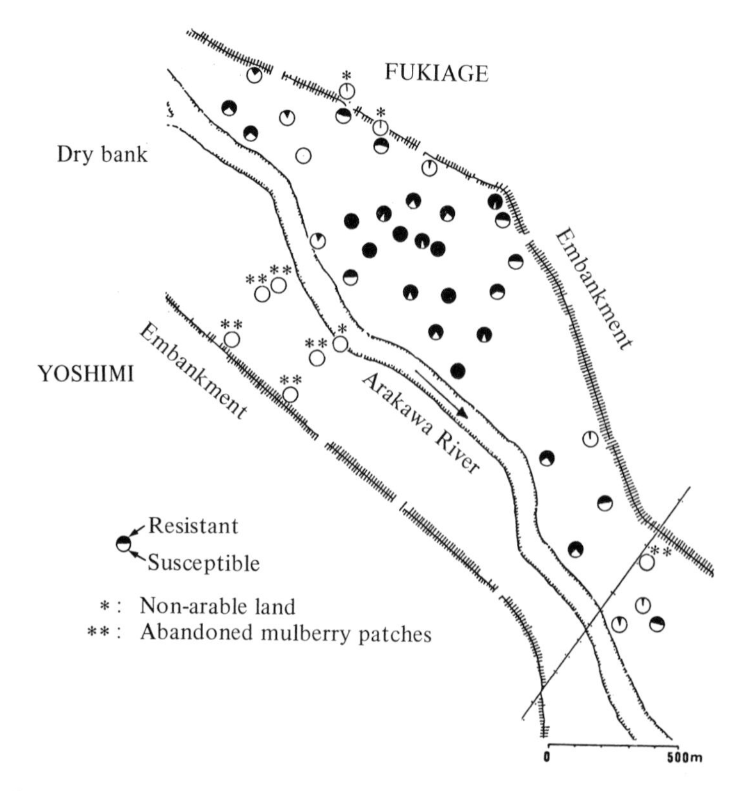

FIG. 9 Distribution of paraquat-resistant biotype of *Erigeron philadelphicus* in mulberry fields on the bank of the Arakawa River in April 1981 (Itoh and Miyahara, 1983).

patches with a higher population of *E. philadelphicus*, whereas in mulberry patches with smaller populations of the species the ratio was also correspondingly low. This fact indicates that the occurrence ratio of resistant biotypes is reflected by the frequency of paraquat application. Therefore questionnaires were submitted to sericultural farmers. It is interesting to note that the paraquat-resistant biotype was not detected in areas where mulberry cultivation had been discontinued only 2-3 years earlier (Table 3). Also in *E. canadensis*, we found that the growth of the paraquat-susceptible biotype was more vigorous than that of the resistant one in the absence of paraquat (Matsunaka and Moriyama, 1987). This fact seems to suggest that the resistant biotype is less competitive than the normal susceptible one in the absence of paraquat application. It implies that the ecological fitness of the resistant biotype is lower than that of the normal biotype, as in the case of the atrazine-resistant biotypes of *Senecio vulgaris* and *Amaranthus retroflexus* (Conard and Radosevich,1979). Recently it has been reported that the resistant *Hordeum glaucum* Steud. (wall barley) biotype grew less vigorously and produced a lower amount of dry matter and fewer tillers and heads than the susceptible biotype in the absence of paraquat (Tucker and Powles,1989).

About three months ago, we submitted a questionnaire on herbicide resistance in relation to kinds of species, herbicide, location and frequency of treatment per year to researchers engaged in weed control in agricultural experiment stations and related branches encompassing sericulture, horticulture, tea research and fruit tree research in nation-wide. The answers showed that *E. canadensis* was widely distributed in Southeast Japan including Honshu and Kyushu. *E. philadelphicus* showed a relatively narrow distribution in Central Japan. The distribution of the newly discovered species of *C. sumatrensis* and *Y. japonica* was narrower than that of *E. philadelphicus*, which may be found only in Saitama, Gunma and Ibaraki Prefectures (Satoh *et al.*, unpublished).

TABLE 3

Mean ratio of occurrence of paraquat-resistant biotype to the total number of plants examined of *Erigeron philadelphicus* under different frequencies of paraquat application

Paraquat application	Land utilization	No. of samples	Mean ratio of resistant biotype
Never applied	Vacant[a)]	10	2.1(%)
2-3 years ago	Abandoned mulberry patches	2	0.0
Sometimes[b)]	Vicinity of mulberry patches	14	47.5
2-3 times every year	Mulberry patches	24	80.5

Determined in April, 1982.

a): Embankment, unused land, etc.

b): Sometimes receiving paraquat due to drift or boundary application in adjacent mulberry patches.

TABLE 4
Germination rate of progenies derived from selfing and allogamy in *Erigeron philadelphicus*

Crossing treatments	(1) In a head	(2-a) No crossing	(2-b) In a plant	(2-c) In a clone	(2-d) Between clones	(2-e) Between biotypes
No. of heads used	18	20	13	7	26	15
Mean no. of florets/head	517*	309	329	278	323	306
Mean no. of germination achenia / head	0.1	0.6	2.2	1.9	112.2	119.6
Mean germination ratio**(%)	0.02	0.2	0.7	0.7	34.7	39.7

* Disk- and ray-florets. ** Mean number of achenia with germination/mean number of florets.

IV. Reproduction Mode and Inheritance of the Resistance

Experiments on the reproduction of *E. philadelphicus* showed that the percentage of self-pollination and apomixis was less than 1.0 %. In contrast, the values for the crosses between different clones of the same biotype and between different biotypes of *E. philadelphicus* were 34.7 and 39.7%, respectively. The percentage values were high, and the number of developed seedlings per head averaged 115 (Table 4). The resistant three species (except for *E. philadelphicus*) showed self- pollination, and *Erigeron annuus* (L.) Pers. (annual fleabane) showed a high percentage of apomixis (Tahara,1915).

The chromosome number of *C. sumatrensis* is $2n$=54, and the species may be hexaploid with the basic chromosome number of the genus n=9. For *Conyza bonariensis* (L.) Cronq. (*C. linefolia*, *Erigeron bonariensis*, hairy fleabane) the number is $2n$=52 or 54, for *E. canadensis* it is diploid ($2n$=18) and for *E. philadelphicus* it is n=9 or 16. For *E. annuus* it is $2n$=27 (Numata and Asano, 1969).

The order of detection of the paraquat-resistant biotype corresponded to the order of occurrence density (number of individuals) of the species in fields with paraquat application (Hanioka, 1987). In Japan, the next candidates for paraquat resistance are *C. bonariensis* biotypes. Resistant biotypes of *E. annuus* and *E. strigosus* have not been detected, although these species sometimes exhibited a high occurrence density in areas where paraquat had been applied. This fact and the difference in the sexual reproduction suggest that the new genus name *Stenactis* Cass. proposed by Kitamura in 1981 should be used for the two species.

Segregation of the phenotype in the F_1 population of the crosses, S x S, S x R, R x S and R x R, in 1982 and 1983 and the test crosses in 1983 provided evidence that a single dominant gene was responsible for the paraquat resistance in *E.*

TABLE 5

Segregation ratio in F_1 population of four kinds of crosses between paraquat-resistant (R) and susceptible (S) clones of *Erigeron philadelphicus* in 1982 and 1983

Parental clone Female Male	No. of F_1 heads	Total no. of seedlings	No. of seedlings which survived	No. of dead seedlings	Expected ratio*
					(R : S)
S x S	15	1882	1	1881	0 : 1
S x R hetero	3	433	211	222	1 : 1
S x R homo	2	242	238	4	1 : 0
R hetero x S	15	1791	893	898	1 : 1
R homo x S	1	138	138	0	1 : 0
R hetero x R hetero	17	1939	1449	490	3 : 1
R hetero x R home	11	1193	1179	14	1 : 0
R homo x R homo	9	1216	1211	5	1 : 0

*Single dominant gene.

S : Part or total of S-8, 15, 17, 22 and 45 clone.

R hetero : Part or total of R-31, 50, 70 and 74 clone.

R homo : Part or total of R-13 and 60 clone.

philadelphics (Table 5), although many species of weeds show a uniparental (maternal) inheritance of *s*-triazine resistance (Warwick *et al.*, 1980, Gressel *et al.*, 1981). Islam and Powles's (1988) data obtained from segregated populations indicated that paraquat resistance in *H. glaucum* is controlled by a single nuclear gene with incomplete dominance. Recently it has been reported that paraquat resistance in *C. bonariensis* which originated in Egypt is conferred by a single dominant gene (Shaaltiel *et al.*,1988). The majority of achenia sampled from wild resistant plants of *E. canadensis*, *C. sumatrensis* and *Y. japonica* showed resistance except for one in *C. sumatrensis*. In *E. philadelphicus* plants segregated to resistant and susceptible biotype. A large number of progenies of *C. sumatrensis* (S) x *C. sumatrensis* (R), *E. canadensis* (S) x *C. sumatrensis* (R) or *C. bonariensis* (S) x *C. sumatrensis* (R) were resistant (Hanioka, 1989). It is suggested that dominant major genes control the paraquat resistance in these Compositae species.

Thus, the occurrence of paraquat-resistant biotypes may not be associated with original intraspecific variation, but with mutation of a gene, as paraquat-resistant biotypes were detected in three genera and many areas at the same time.

It has been postulated that when a herbicide is used repeatedly, there is a very strong selection pressure against non-resistant individuals. Subsequently the non-resistant biotype can be easily eliminated by herbicide application (Table 6). In the case of *E. philadelphicus*, according to Macnair's (1981) estimation the complete manifestation of resistance in a biotype requires only eight generations since a single dominant gene involves paraquat resistance.

The models most commonly used for describing plant gene flow are island, stepping-stone and isolation-by-distance models (Jain and Bradshaw, 1966). Direct measurements of pollen or seed dispersal indicate that the isolation-by-distance model best fits the patterns observed. A microgeographic distribution of paraquat-resistant phenotypes in *E. philadelphicus* in Fukiage was determined using the model. Depression patterns of variation for correlograms obtained by spatial autocorrelation analysis based on morph (phenotype) frequency suggest the presence of parapatric differentiation, although information on quantitative gene flow was limited (Yamaguchi and Itoh, unpublished).

The present study showed clearly that the resistant plants developed not only from the achenia or ramets of resistant clones but also from pollination with pollen from resistant plants which is mediated by various insects (Tanaka, 1971) in the case of *E. philadelphicus*. Integrated weed control practices involving herbicides, cropping systems, mulching with organic matter, and appropriate tillage practices would, in combination, minimize the likelihood of weed species becoming resistant to herbicides.

TABLE 6

Distribution of paraquat resistance in *Erigeron philadelphicus* in a mulberry area in Fukaya City (quoted from Hanioka, 1986)

Observation time	Items	Mulberry patches	Vicinity of mulberry patches	Total
Oct. 1982	No. of sampling points	20	11	31
	Mean % of resistant plants	9.3	32.3	17.5
	Mean % of estimated resistant genotype*	6	21	12
Nov. 1985	No. of sampling points	16	13	29
	Mean % of resistant plants	60.8	78.7	68.8
	Mean % of estimated resistant genotype*	40	52	46

* : Single dominant gene, R hetero : R homo = 2 : 1 based on assumption

V. Physiological Aspects of the Resistance

Based on the studies of Tanaka *et al.* (1986), the leaves of the susceptible biotypes of *E. philadelphicus* and *E. canadensis* wilted when treated with more than 5 micro M paraquat at the cut end, whereas those from the resistant biotypes did not wilt even at 500 micro M. Autoradiographs indicated that ^{14}C-paraquat taken up through the cut ends was rapidly distributed through the vascular system in the leaves of the susceptible biotype, but was hardly translocated in the leaves of the resistant biotype. The amount of paraquat taken up during 48 hours by the resistant biotype was 0.5 % that of the susceptible biotype under illumination. The difference in the paraquat movement may be associated with paraquat resistance in the two species of *Erigeron.* This assumption is supported by Fuerst *et al.* (1985), who reported that a biotype of *C. bonariensis* was resistant to paraquat.

In our experiments using *E. canadensis*, the resistance to paraquat was observed at the protoplast level. Leaf disks and protoplasts showed a lower O_2-consumption (as determined by the Mehler reaction in the presence of paraquat in the resistant biotype) than the susceptible one, while in the intact chloroplasts there was no difference between the two biotypes (Fig. 10). The activity of the enzymes involved in the superoxide detoxification system, i.e., superoxide dismutase, ascorbate peroxidase, glutathion reductase and MDA reductase, was nearly twice as high in the resistant biotype as in that of the susceptible one (Table 7), as reported by Shaaletiel and Gressel in 1986. These results suggest that the resistance mechanism may be present at the protoplast level (Kawaguchi,1989).

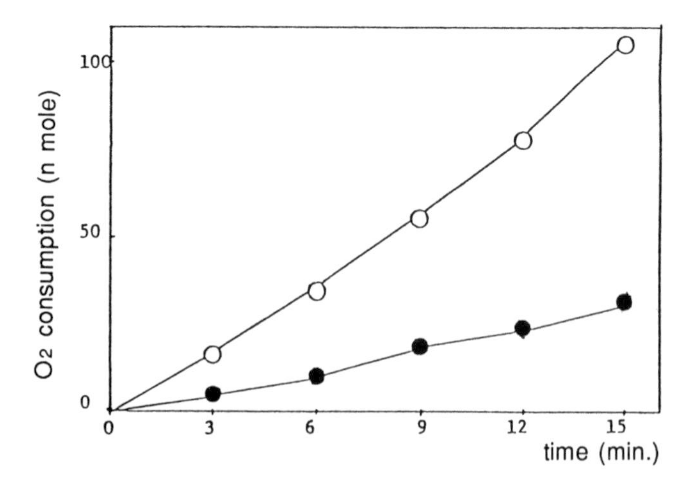

FIG. 10 O_2 consumption by protoplasts of resistant (●) susceptible (○) biotypes in *Erigeron canadensis* L. Chlorophyll concentration is 3 μ g/ml (Kawaguchi, 1989).

TABLE 7
Oxygen detoxifying enzyme activities in extracts of intact chloroplast of resistant (R) and susceptible (S) strains of *Erigeron canadensis* L.
(Kawaguchi, 1989)

Enzyme	Biotype	Activity	(% of susceptible biotype)
Superoxide dismutase	R	28.9[a]	(169)
	S	17.1	(100)
Ascorbate peroxidase	R	9.3[b]	(179)
	S	5.2	(100)
Glutathion reductase	R	2.6[c]	(161)
	S	1.6	(100)
MDA reductase	R	1.2[c]	(194)
	S	0.64	(100)

a : unit/mg protein, b : μ Mol/mg protein, c : m Mol NADPH/mg protein

VI. Conclusion

(1) Paraquat-resistant biotypes have been identified in *Erigeron philadelphicus*, *E. canadensis*, *Conyza bonariensis*, *C. sumatrensis* and *Youngia japonica* but not in *E. annuus*, in which **apomixis** may be present.
(2) A large difference in the degree of resistance to paraquat of *E. philadelphicus* was recognized among seedlings, mature whole plants, leaf disks, cut roots and protoplasts. These results suggest that the mode of paraquat action differed between photosynthetic and **non-photosynthetic** systems, and also between intact plants and leaf disks or **protoplasts.**
(3) The number of resistant individuals increased rapidly in the fields where paraquat had been applied, while such biotypes did not invade fields where paraquat had not been applied, as resistant plants are **less competitive** than the susceptible ones. Thus, colonies of resistant biotypes were scattered.
(4) Paraquat-resistance was found to be controlled by a **single dominant** gene in *E. philadelphicus* and *C. bonariensis*. It is probably the same in other species.
(5) The occurrence of paraquat-resistant biotypes may be due not to intraspecific variations, but to the **mutation** of a single gene, which is inherited and can be detected in 3 genera in some areas at the same time.

References

Conard, S. G. and Radosevich, S.R. (1979). Ecological fitness of *Senecio vulgaris* and *Amaranthus retroflexus* biotypes susceptible or resistant to atrazine. *J. Appl. Ecol.* **16**, 171-177.

Fuerst, E. P., Nakatani, H.Y., Dodge, A.D., Penner, D. and Arntzen, C.J. (1985). Paraquat resistance in *Conyza. Plant Physiol.* **77**, 984-989.

Gressel, J., Ezara, G. and Jain, S.M. (1981). Genetic and chemical manipulation of crops to confer tolerance to chemicals. *In* "Chemical Manipulation of Crop Growth and Development" (J. S. Mclarn, ed.), pp.79-91. Butterworth, London.

Hanioka, Y. (1986). Distribution of paraquat resistant *Erigeron philadelphicus* L. and the countermeasures to it in the mulberry field Saitama. *Shokuchou* **20 (7)**, 10-16. [in Japanese]

Hanioka, Y. (1987). Paraquat resistance in *Erigeron sumatrensis* Ritz. and *Youngia japonica* D.C. *Weed Res. Japan* **32** (Suppl.), 137-140. [in Japanese]

Harper, J. L. (1956). The evolution of weeds in relation to resistance to herbicides. Proc. 3rd Br. Weed Control Conf., 179-188.

Holt, J. S. and Lebaron, H.M. (1989). Significance and worldwide distribution of herbicide resistance. Abst. Weed Sci. Soc. Amer, 131.

Islama, A.K.M.R. and Powles, S.B. (1988). Inheritance of resistance to paraquat in barley grass *Hordeum glaucum* Steud. *Weed Res.* **28**, 393-397.

Itoh, K. (1988). Paraquat resistance in *Erigeron philadelphicus* L. *JARQ*. **22 (2)**, 85-90.

Itoh, K. and Miyahara, M. (1983). Distribution of *Erigeron philadelphicus* L. resistant to paraquat related to land use. *Weed Res. Japan* **28** (Suppl.), 187-188. [in Japanese]

Itoh, K. and Miyahara, M. (1984). Inheritance of paraquat resistance in *Erigeron philadelphicus* L. *Weed Res. Japan* **29**, 301-307.

Itoh, K. and Miyahara, M. (1985). A habitat of *Erigeron philadephicus* L. resistant to paraquat. Proc. 10th Conf. of Asian- Pacific Weed Sci. Soc., Chengmai, Thailand, 13-18.

Jain, S.K. and Bradshaw, A.D. (1966). Evolutionary divergence among adjacent plant populations. I. The evidence and theoretical analysis. *Heredity* **21**, 407-441.

Kato, A., Okuda, Y., Juri, T., Dan, M. and Uejyo, Y. (1982). Resistance to paraquat and diquat in *Erigeron canadensis* L. *Bull. Osaka Agr. Res. Center* **19**, 59-64. [in Japanese with English summary]

Kawaguchi, S. (1989). Utilization of paraquat resistance of *Erigeron canadensis* for the breeding of paraquat resistant crop and a study on the paraquat resistant mechanism. M. Sc. Thesis, Kobe Univ., pp.72. [in Japanese]

Kawano, S. (1974). "Speciation and Adaptation" Evolutionary Biology in Plants 2. pp. 407. Sanseido, Tokyo [in Japanese]

Kitamura, S. (1981). Compositae. *In* "Wild Flowers of Japan III" (Y. Satake *et al.*, eds), pp.156-235. Heibonsha, Tokyo.[in Japanese]

Macnair, M. (1981). Tolerance of higher plants to toxic materials. *In* "Genetic Consequences of Man Made Change"(A. Bishop and L. M. Cook, ed.) pp.177-207. Academic Press

Matsunaka, S. and Moriyama, A. (1987). A prosperity of paraquat resistant *Erigeron canadensis* L. *Weed Res. Japan* **32** (Suppl.), 141-142. [in Japanese]

Mudge, L. C., Gossett, J. and Murphy, T. R.(1984). Resistance of goosegrass (*Eleusine indica*) to dinitroaniline herbicides. *Weed Sci.* **32**, 591-594.

Numata, M. and Asano, S. (1969). "Biological Flora of Japan, Sympetalae 1", pp.78-89. Tsukiji Shokan Pub., Tokyo.

Pölös, E., Mikuras, J., Szigeti, Z., Matkovics, B., Quyhai, Do, Parducz, A. and Lehoczki, E. (1986). Paraquat and atrazine co- resistance in *Conyza canadensis*(L.) Conq. *Pestic. Biochem. Physiol.* **30**, 142-154.

Ryan, G. F. (1970). Resistance of common groundsel to simazine and atrazine. *Weed Sci.* **18**, 614-616.

Shaaltiel, Y. and Gressel, J. (1986). Multienzyme oxygen radical detoxifying system correlated with paraquat resistance in *conyza bonariensis. Pestic. Biochem. Physiol.* **26**, 22-28.

Shaaltiel, Y., Chua, N-H., Gepstein, S. and Gressel, J. (1988). Dominant pleiotropy controls enzymes co-segregating with paraquat resistance in *Conyza bonariensis. Theor. Appl. Genet.* **75**, 850- 856.

Tahara, M. (1915). Parthenogenesis in *Erigeron annuus* Pers. *Bot. Mag. Tokyo* **29** (344), 245-254. [in Japanese]

Tanaka, H. (1971). Pollination of some Compositae 3. *Saishyu-to-shiiku* **33**, 85-91. [in Japanese]

Tanaka, Y., Chisaka, H. and Saka, H. (1986). Movement of paraquat in resistant and susceptible biotypes of *Erigeron philadelphicus* and *E. canadensis*. *Physhiol. Plant.* **66**, 605- 608.

Tucker, E. S. and Powles, S. B. (1989). Competitiveness of paraquat-resistant wall barley (*Hordeum glaucum*). *Weed Sci.* **37** (in press).

Usami, Y., Koizumi, H., Saka, H. and Satoh, M. (1989). Distribution of *Erigeron philadelphicus* L. resistant to paraquat in Ibaraki Prefecture. *Weed Res. Japan* **34**, 63-67. [in Japanese with English summary]

Warwick, S. I. and Black, L. (1980). Uniparental inheritance of atrazine resistance in *Chenopodium album*. *Can. J. Plant Sci.* **60**, 751-753.

Watanabe, Y., Honma, T., Itoh, K. and Miyahara, M. (1982). Paraquat resistance in *Erigeron philadelphicus* L. *Weed Res. Japan* **27**, 49-54.

4 Biosystematics of Cultivated Plants and Their Wild Relatives

LEONARD W.D. VAN RAAMSDONK

Centre for Plant Breeding Institute,
P.O. Box 16, 6700 AA Wageningen, The Netherlands

I. History of Cultivated Plants

Assuming the observation of differences between plants as one of the basic activities of identification and taxonomy, the starting point of the taxonomy of cultivated plants can be traced back as early as to prehistoric times. The oldest cultivated plant, wheat, originally has a brittle spike which fell apart into spikelets at maturity. The first record of domestication concerns the selection of non-brittle races of *Triticum monococcum* in about 7000 B.C. Such an event was probably based on a conscious observation of the differences between the two traits. Such an observation does not imply that the resulting selection was also carried out consciously (Heiser, 1988). Similar cases concern the gradual replacement of 'hulled' grains by 'free-threshing' grains in wheat and two-rowed spikes by six-rowed spikes in barley.

Crop plants received special attention in the modern system of binary nomenclature. Linnaeus ascribed specific status to a great number of crops, mostly by using the specific epithet 'sativus' (garlic, *Allium sativum*; cucumber, *Cucumis sativus*; lettuce, *Lactuca sativa*) or the epithet 'domestica' (European plum, *Prunus domestica*). He indicated infraspecific varieties.

Experiments with a systematic scope, nowadays known as biosystematics, were also started by Linnaeus and his contemporaries. *Tragopogon porrifolius* and *T. pratensis* were crossed successfully in 1757. The hybrids showed an intermediate flower color, but the second generation sown in Leningrad segregated according to what are presently known as the Mendelian laws. Koelreuter saw these plants and called them 'half-hybrids'. Koelreuter himself produced hybrids, reciprocals and

BIOLOGICAL APPROACHES AND EVOLUTIONARY TRENDS IN PLANTS ISBN 0-12-402960-4

backcross hybrids, which is common practice in modern breeding, and emphasized the function of pollen and the role of insects in pollination.

II. Biosystematics

Modern biosystematic research (Vickery, 1984) consists of several stages (Fig. 1). The experimental approach focuses on building up and maintaining a living plant collection. In the next stage an identification of the plant material may be necessary prior to the actual gathering of data. A re-classification can result from the final classification scheme when the taxa appeared to be redefined. The multidisciplinary basis is reflected by using information from the fields of morphology, cytogenetics, content of chemicals, electrophoresis, embryo sac development, DNA analysis, plant geography, ecology, and crossing experiments (Stace, 1980; Poppendieck, 1985). The synthesis of these data may lead to the two main characteristics of biosystematic research. They result simultaneously from the research as two perpendicular but related dimensions. These are:

Vertical objective: This is the time dimension of evolutionary development. Biosystematics may give a description or a synthesis of evolution, relationships and isolating mechanisms leading to the present species of the studied group.

Horizontal objective: This is the variability dimension of classification. A grouping of species or other taxa in a classification scheme may be based on an evolutionary tree. Proper identification of the specimens used, well-defined descriptions of the taxa and prediction of crossability can be based on such a scheme.

When we focus on systematics of cultivated plants, the existence of isolation barriers between crops and wild relatives has three practical consequences.

One important technique of breeding crop plants is crossing and backcrossing, or 'introgressive hybridization' (Small, 1984; van Raamsdonk, 1986). The possibility of predicting crossability between cultivated plants and wild relatives is an important tool in plant breeding. Knowledge of isolation barriers gives the opportunity to avoid these barriers by artificial breeding methods.

Recently direct handling of DNA has become an important tool in plant breeding. Using bacteria as vector it is possible to transfer certain DNA sequences, which it is hoped will be or act as a specific gene, from a source to a cell culture. The cells of this culture can be regenerated to mature plants, from among which the tranformed ones can be selected. The result is a combination of characters which cannot be produced in a natural way and which is not found in the wild. Escapes from cultivation or hybridization with wild relatives can result in a plague or other undesired effect. Knowledge of isolation barriers is important in an opposite direction. The existence of strong isolation barriers alone, prevents the transfer of pollen from

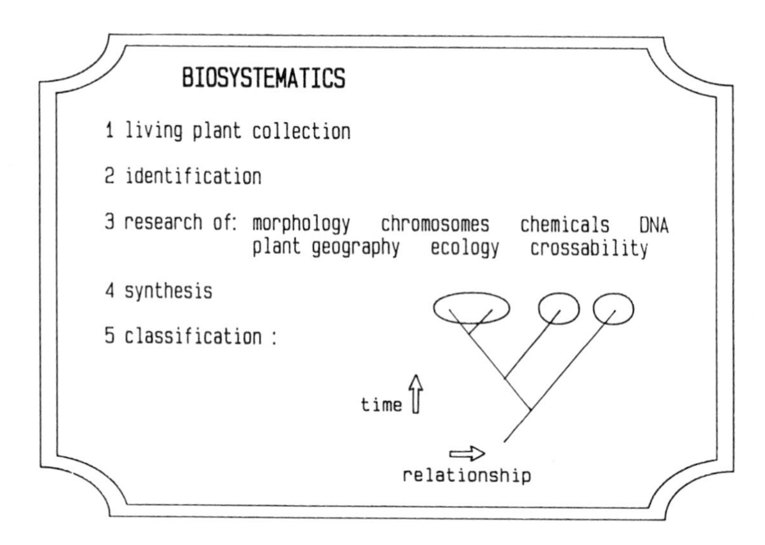

Fig. 1 A review of the different stages of biosystematic research, of the multidisciplinary approach and of a classification tree with four taxa in three groups.

these artificially enriched crop plants to wild relatives. Biosystematics will play an important role in risk assessment and legislation of biotechnological research.

Large germ plasm collections are usually documented by passport data like origin, source, species and cultivar names, classification and breeding system, and evaluation data concerning usage, yield, discriminative characters, etc. Especially for obtaining passport data, the results of biosystematic research are important.

The mutual relationship between biosystematics and plant breeding is bilateral. Attention was already paid to the role of knowledge of isolation barriers in plant breeding. The other way round, working out a crossing program as part of breeding a crop plant itself may contribute to biosystematic research.

Breeding by introgressive hybridization may corrupt proposed classifications of cultivars in higher categories (cultivar-groups, provars, convars; Jeffrey, 1968; Brandenburg, 1984) by a continuous production of cultivars and hybrid lines intermediate between groups.

III. Evolution and Domestication

Evolution of wild plants and domestication of cultivated plants have common aspects as well as deviating points. In the classical evolution theory of Darwin, selection and migration with isolation resulting from it are the main topics (Fig. 2). Darwin also noted the existence of major mutations, but these are not comparable to the mutations presently known, since Darwin's 'Origin of Species' (1859) was

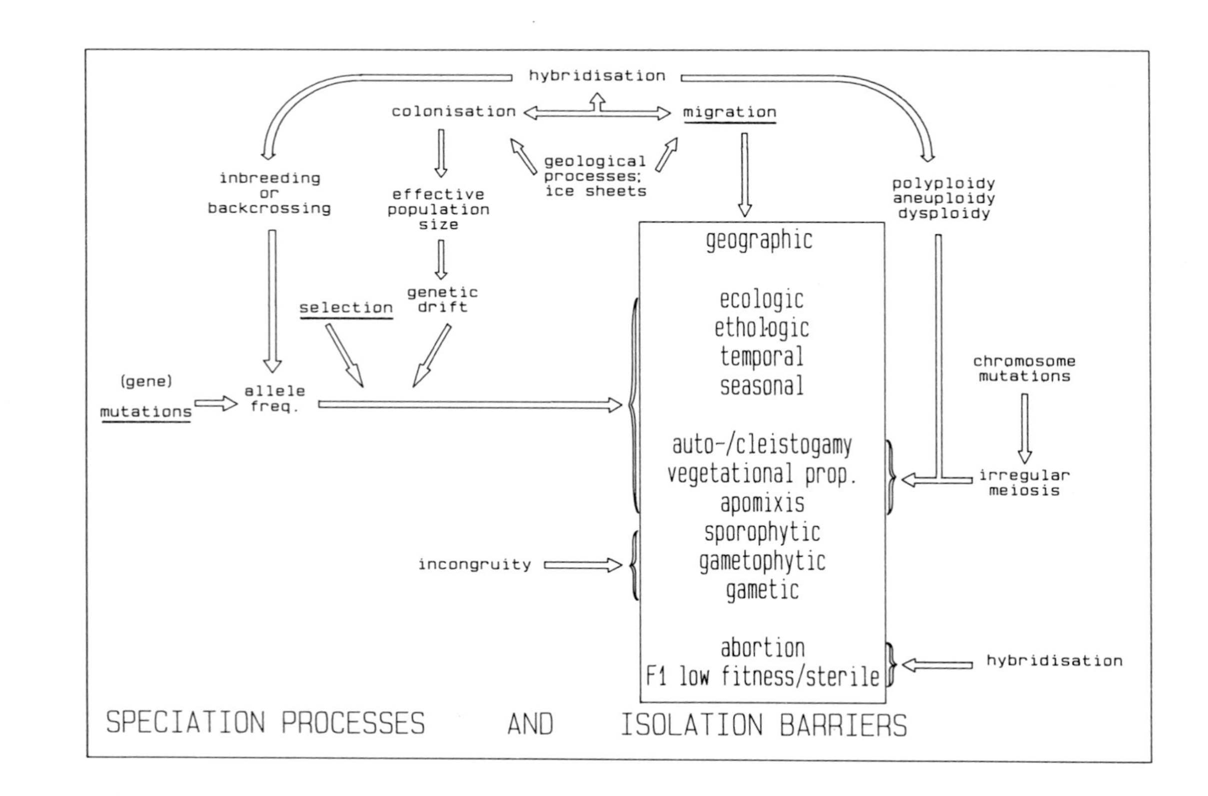

Fig. 2 Schematic representation of evolutionary processes and their relationship to isolation barriers. The basic principles of the evolution theory of Darwin are underlined.

published before Mendel's 'Versuche über Pflanzenhybriden' (1866) and far before the rediscovery of this basic work of Mendel by Bailey (1897) and De Vries, Correns and Tschermak (1900). Natural selection and artificial selection can be considered to be similar mechanisms. Selection pressures in breeding practice may be high, but from population genetical studies it can be concluded that under natural circumstances the same high pressures can exist (Warwick, this volume).

Besides the basic evolutionary mechanisms of selection, migration, and mutation, other mechanisms such as genetic drift (Ladizinsky, 1985), hybridization and introgression (Stace, 1975, 1987), polyploidization and other chromosomal rearrangements (Lewis, 1980; Grant, 1987; King, 1988; Soltis and Soltis, this volume), self-compatibility (Mulcahy, 1984) and vegetative propagation (Gadella, 1983) are discovered or recognised as important to evolution (Fig. 2). Some of these can be achieved artificially in a short time under controlled circumstances in plant breeding. This situation is advantageous to naturally occurring systems. In some cases crop-weed complexes may act as model systems to study evolutionary processes and mechanisms (Pickersgill, 1981). Knowledge about the specific systematic position of the crop under study will result as a by-product.

IV. Isolation Barriers

Species, and other taxa as well, exist and maintain their reproductive isolation by means of isolation barriers, which result from the evolutionary processes and mechanisms mentioned above. A list of isolation barriers is summed up in Fig. 3. They can be subdivided in several ways. Some of them act before gamete fusion (prezygotic) and others after gamete fusion (postzygotic). Simultaneously, isolation barriers may influence the external environment of the (mother) plant; others manifest themselves inside the plant (Stace, 1975, 1980). Some striking examples of isolation barriers are found and will be discussed with examples taken from crop plants.

External barriers are easily avoided by plant breeding methods. Related species of crop plants are transported from their original locations and habitats and included in plant collections, which means that geographical and ecological isolation barriers are withdrawn. Forcing late flowering plants to flower earlier or delaying early flowering plants is common practice in order to synchronize parents for crossing.

The opposite mechanisms of strict self-fertilizing or obligatory outbreeding and all the intermediate situations are important to evolution and plant breeding (Charlesworth, 1988; Barrett, 1988). Apomixis (Asker, 1984; pseudogamy excluded) and vegetative propagation seem to be preferable to sexual reproduction because there is no biomass and energy investment in male structures, but the absence of genetic recombination can be a disadvantage (Urbanska, this volume; Bierzychudek, this volume). In terms of classification both asexual reproduction

prezygotic	external	geographical	: allopatry (vs. sympatry)
		ecological	: allotopy (vs. syntopy)
		ethological	: different pollinators
		temporal	: flowering day-time differences
		seasonal	: flowering at different seasons
	internal	compatibility	: self-pollination and apomixis
		breeding	: vegetative propagation
		incongruity	: non-function of pollen-pistil system
postzygotic		abortion	
		hybrid inviability	
		non-fitness of F1 hybrids	
		F1 hybrid sterility	
		F2 hybrid inviability	
		non-fitness of F2 hybrids	
		F2 hybrid sterility	

Fig. 3 Isolation barriers.

and strict inbreeding may lead to highly isolated populations and the existence of numerous microspecies. Many ornamentals propagate vegetatively, i.e., by corms (gladioli), bulbils (tulip), bulbscales (lily), splitting (carnation) or cuttings (chrysanthemum). Breeding by introgressive hybridization of these ornamentals can be made difficult, for instance by long-lasting juvenile stages, and may require forcing plants to flower. On the other hand, many vegetables are propagated by seeds. This enables the production of homozygous cultivar lines and the use of heterosis effects in hybrids after crossing of inbred lines.

Incompatibility and incongruity are two principally different mechanisms. A pistil can be considered as a complex of barriers and promotors, pollen as a unit provided with genetic information for penetrating the barriers and cooperating with the promoting factors to reach successful fertilization. The relationship between pollen and pistil may be non-functional in two different ways. With incompatibility, a system exists which prevents or disturbs cooperation. This is usually based on multiple alleles. If in both pistil and pollen the same S-allele is acting, the pollen tube growth is inhibited or stopped. So incompatibility is an evolutionary solution to inbreeding, which may have negative effects in nature. With incongruity, the partners do not fully fit together because the matching of the genic systems of pistil and pollen is incomplete. This is a byproduct of evolutionary divergence. Incongruity may be bilateral or unilateral. In the latter case crossing of one parent with another may be successful, whereas the reciprocal crossing is not

(Hogenboom, 1975, 1978). A model system of incongruity is worked out in the tomato (*Lycopersicon, L. esculentum* and *L. pimpinellifolium* are self-compatible, *L. peruvianum* and some other species are not. After crossing *L. peruvianum* (female) with *L. esculentum* (male) unilateral incongruity was found; the reciprocal results in embryo abortion. After selection and inbreeding, both the self-incompatibility of *L. peruvianum* and the unilateral incongruity between *L. peruvianum* and *L. esculentum* were broken (Hogenboom, 1979).

Geographical isolation as well as polyploidization are involved in the evolution and domestication of coffee (*Coffea*), by far the most important crop plant. In particular, *C. canephora* (diploid, self-incompatible, Central Africa) and *C. arabica* (tetraploid, self-compatible, East Africa) are the basic parental species of the presently used crop. These species were used as a source of coffee beans over a number of years, in which breeding and selection took place separately. Originally these species were isolated geographically, but in spite of the different ploidy levels and obligate inbreeding of one of the species, hybrids were obtained after artificial hybridization. Triploid and tetraploid hybrids resulted from these crosses, while modern autotetraploid hybrids were produced after artificial chromosome doubling of the diploid *C. canephora* with colchicin (Ferwerda, 1976).

V. Methods

The use of advanced techniques is necessary to analyze the large amount of data in biosystematics obtained from quite different sources. Two types of data are available. Most data concern each specimen or object individually. This information can be organized in tables, with rows representing objects and columns representing variables. Distances between pairs of objects can be calculated from this table, and the resulting distance matrix is the basis of cluster analysis (CA) or principal coordinate analysis (PCoordA). Similarly, a matrix with correlations between pairs of variables is the basis of principal component analysis (PCA) (van Raamsdonk, 1988). A special group of data directly concerns the relation between pairs of objects, like the results of immunological techniques, DNA hybridization or crossing experiments. A crossability coefficient is developed which enables the numerical evaluation of crossing results by means of CA and PCoordA (van Raamsdonk, 1990). The process of crossing two plants and obtaining a hybrid population can be divided into three phases, indicated as follows by stage and parameter to be measured:

1. seed production, the number of produced seeds;
2. seed germination, the percentage of germinated seeds;
3. fertility of hybrids, the number of bivalents or the pollen fertility.

The crossability coefficient consists of a formula for each of these three stages, which are to be used independently of each other. The coefficient structure and the use of the formulae is presented in the form of a dichotomous key:

1a. No seed or seeds inviable:

$$D1_{ij} = 1 - .5N_{ij} / N_{max} \qquad (.5 < D1_{ij} < 1.0)$$

with: N_{max} = maximum number of ovulae of female parent
N_{ij} = number of seeds produced by combination i,j

1b. At least one germinable seed . . . . 2.

2a. Hybrids male sterile:

$$D2_{ij} = .75 - .5G_{ij} \qquad (.25 < D2_{ij} < .75)$$

with: G_{ij} = percentage of germination/100 obtained in combination i,j
$(0 <= G_{ij} <= 1)$

2b. Hybrids male fertile, at least partly:

$$D3_{ij} = .5 - .5F_{ij} \qquad (0.0 < D3_{ij} < .5)$$

with: F_{ij} = percentage of pollen fertility/100 obtained in combination i,j
$(0 <= F_{ij} <= 1)$

A formula based on bivalent production in meiosis is developed as an alternative to formula D3. A compound formula covering all three stages is also developed. The rationale of the coefficient is discussed by van Raamsdonk (1990). The distance matrix resulting from the coefficient is comparable to the crossability matrix between populations of the *Turnera ulmifolia* based on the number of seeds produced (phase 1; Shore and Barrett, 1985). Results of *Allium* and *Cucumis* analyzed with the aid of the set of formulae as shown will be discussed.

VI. Systematics of Crop Plants

At IVT, large crossing programs were carried out in *Allium L.* and *Cucumis L.* in order to include desired characteristics from wild species in the cultivated crop. The systematic positions of onion and cucumber are remarkably different.

A. *Cucumis*

The genus *Cucumis* consists of two subgenera which are slightly related. Cucumber (*C. sativus*) is the only representative of the Asiatic subgenus with

$2n$=14. Some wild species described from India and Nepal presumably belong to *C. sativus* and should be considered to be primitive landraces. The other (African) subgenus, Melo, consists of about 20 species with $2n$=24 belonging to four sections. Biosystematic research was based on morphology, cytogenetics, electrophoresis, RFLP mapping and geography. The multidisciplinary data were reviewed by van Raamsdonk *et al.* (1989).

Two different directions of evolution on the subgeneric level have been considered. The basic number x=12 could be derived from x=7 by fragmentation (Bhaduri and Bose, 1948; Ayyanger, 1967), or a derivation of x=7 from x=12 by fusion can be assumed (Trivedi and Roy, 1970). The latter hypothesis is most likely, since the basic number x=12 is most common in the Cucurbitaceae, and the group with the longest polyploid series can be considered to be the oldest one. The subgenus Melo is therefore most probably the oldest (Fig. 4). However, the subgenera are not closely related because of the large genetic differences, which result from long geographic isolation, accompanied by inbreeding, drastic genetic changes and migration as aspects of the quantum model of speciation (Grant, 1980).

Most African species belong to section Anguria (group 2 of Jeffrey, 1980) which are cross-compatible in a number of combinations. This group can be subdivided into two subgroups with the following species:

Myriocarpus subgroup: *C. africanus*, *C. heptadactylus* (4x), *C. myriocarpus* subsp. *leptodermis* and subsp. *myriocarpus*;

Anguria subgroup: *C. aculeatus* (4x), *C. anguria* subsp. *anguria*, *C. anguria* subsp. *longipes*, *C. diniae* (4x), *C. dipsaceus*, *C. ficifolius* (2x and 4x), *C. figarei* (4x and 6x), *C. prophetarum*, and *C. zeyheri* (2x and 4x).

Crossing results are shown in Fig. 5a and are obtained from Dane *et al.* (1980), Singh and Yadava (1984) and van Raamsdonk *et al.* (1989). The crossability between the species is correlated with the position in the subgroups. The hybrid between *C. myriocarpus* and *C. anguria* is not indicative of their relationship, since aberrant specimens of the species were used (Staub *et al.*, 1987). All polyploids except *C. heptadactylus* are crossable with diploid species of the Anguria subgroup. Differences between the subgroups are also found in geographical distributions (Fig. 4), contents of cucurbitacins (Enslin and Rehm, 1958), of flavonoids (Brown *et al.*, 1969), and in the occurrence of alleles of some enzyme systems (Esquinas-Alcazar, 1977). In some respects, *C. prophetarum* holds an intermediate position.

Crossing results between 12 species are represented in a dendrogram in Fig. 5b. Splitting levels indicate the grouping of the species in taxonomic groups and in crossability groups. The four groups determined on the taxonomic level agree with the division of Jeffrey (1980). Species linked together to the right of the crossability level are more or less interfertile. The tetraploid species show a low fertility level after crossing with their related diploids. In general, crossability appears to be positively correlated with relationship.

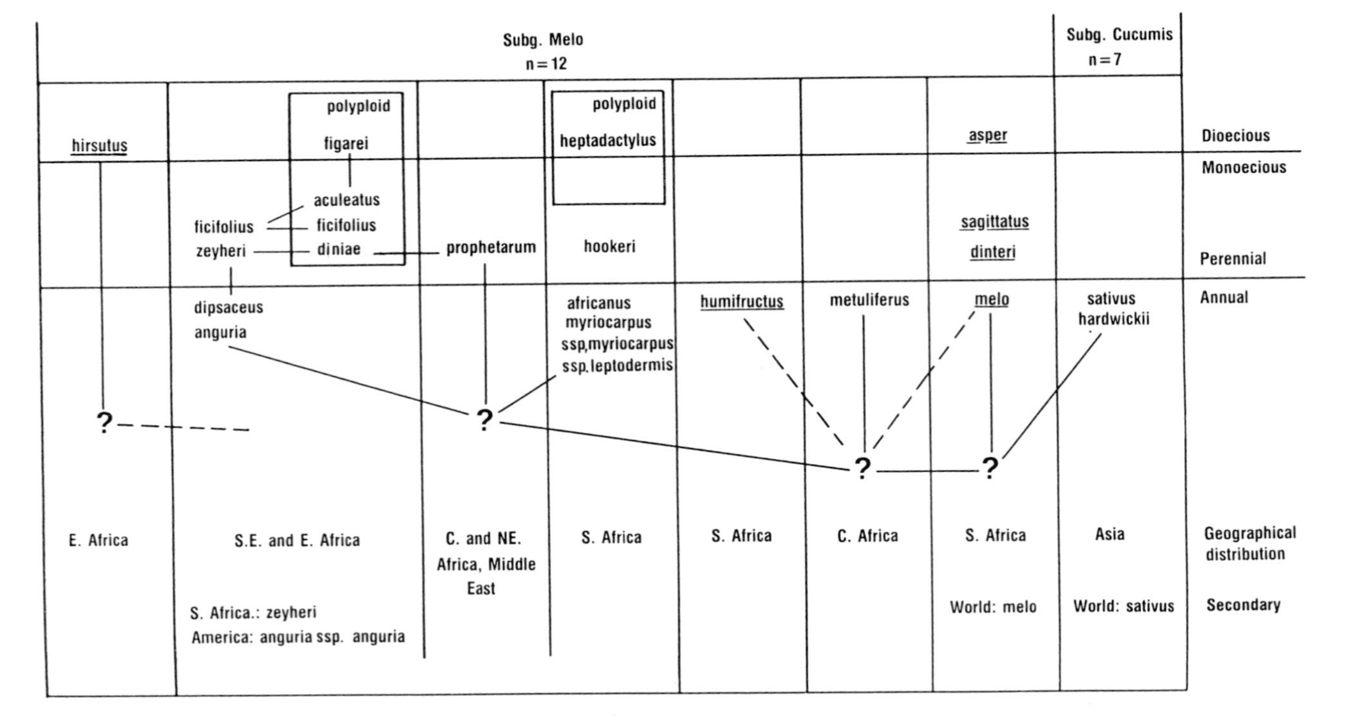

Fig. 4 Subgenera, groups and species of *Cucumis* with polyploidy, sex expression, growth form, geographical distribution and assumed phylogenetic tree indicated (the latter according to the chloroplastDNA-RFLP mapping of Perl-Treves and Galun, 1985).

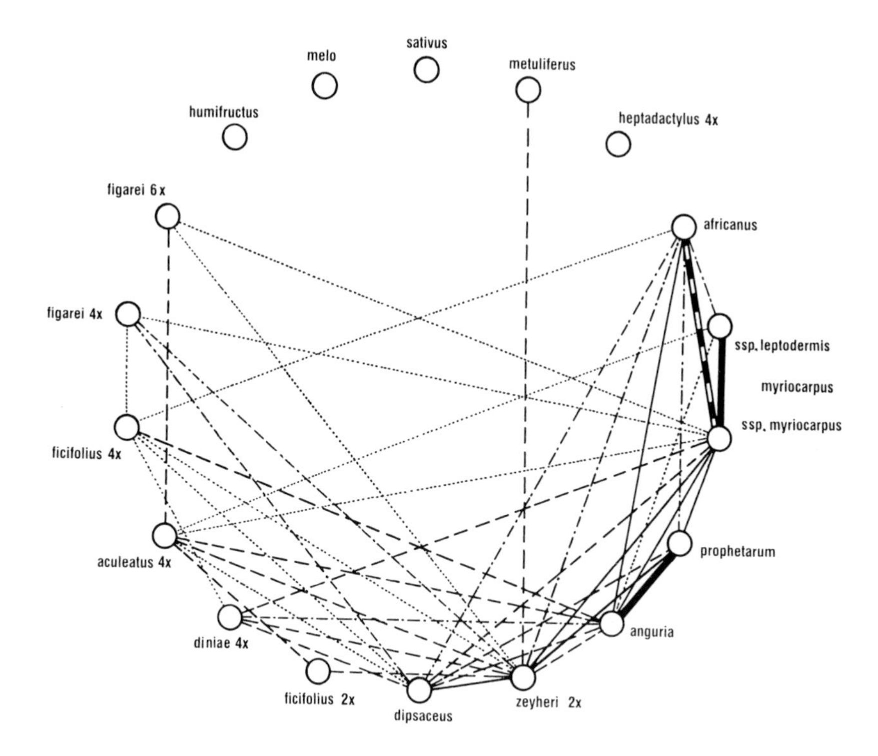

Fig. 5 a) Crossing polygon of *Cucumis* with results of meiotic analysis and fertility.
>11 bivalents and pollen fertility >80%,
number of bivalents unknown and pollen fertility 97%,
>11 bivalents and pollen fertility <46% (*C. anguria* x C. myriocarpus: 7 bivalents and pollen fertility 85%),
<11 bivalents and pollen fertility <54%,
F1 seed fertility reduced,
F1 seed sterile.
Lack of connecting line indicates cross not possible or not attempted.

B. *Allium*

Onion (*A. cepa*) belongs to a section with 7 species originating from Central Asia. This section can be subdivided in three groups. These groups are:

Basic group: *A. galanthum* and *A. pskemense*.

Derived group 1: *A. altaicum* and *A. fistulosum*.

Derived group 2: *A. oschaninii, A. vavilovii* and *A. cepa*.

A. roylei belongs to the related subgenus *Rhizirideum*.

Variation of 21 above-ground morphological characters is analyzed by PCA and is shown in Fig. 6, in which the three groups are recognizable. Hybrids hold an

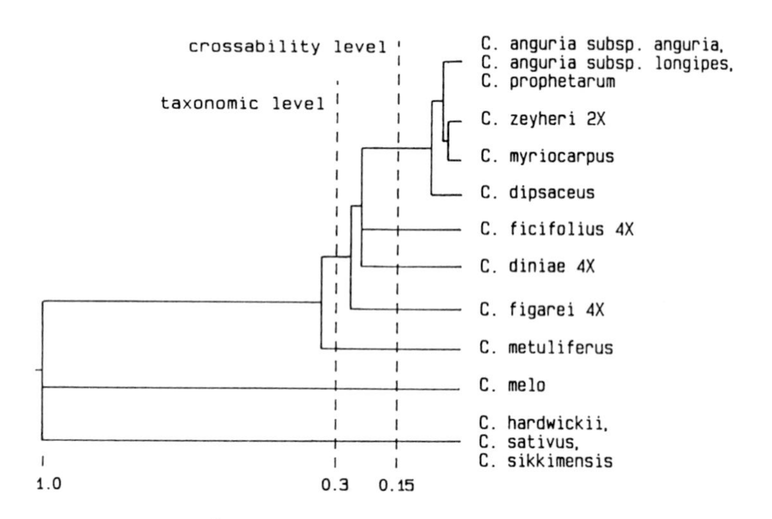

Fig. 5 b) Dendrogram of crossability results in *Cucumis*. Taxonomic and crossability levels are indicated.

intermediate position. The differences in morphology between reciprocals indicate a maternal effect in some cases. The basic group with *A. galanthum* and *A. pskemense* is distributed in Central Asia and shares some characteristics with both other groups of which the evolution can be characterized by changes in morphology and chromosome structure accompanied by migration from the primary gene centre.

A. oschaninii may have originated from this basic group by developing new morphological characteristics but maintaining a karyotype similar to *A. galanthum* and *A. pskemense*. This evolution was connected with migration westwards. *A. vavilovii* and *A. cepa* may have evolved from *A. oschaninii* by developing a different karyotype. *A. altaicum* and *A. fistulosum* may have originated from the basic group along a second evolutionary line by developing a deviating karyotype and new morphological characteristics. This evolution was connected with migration eastwards (Van Raamsdonk and De Vries, in prep.).

Species within groups are completely interfertile (*A. cepa* and *A. vavilovii*), or are completely isolated (*A. galanthum* and *A. pskemense*). Species of different groups can readily be crossed in some cases (*A. cepa* and *A. fistulosum*) (Fig. 7a). *A. roylei* of section *Rhizirideum* is able to produce hybrids together with *A. cepa*. Only a few seeds can be produced in the first generation from which moderately fertile hybrids were grown. This is an indication of the existence of some isolation mechanisms. However, backcross hybrids with *A. cepa* can easily be made (Van Raamsdonk *et al.*, in prep.).

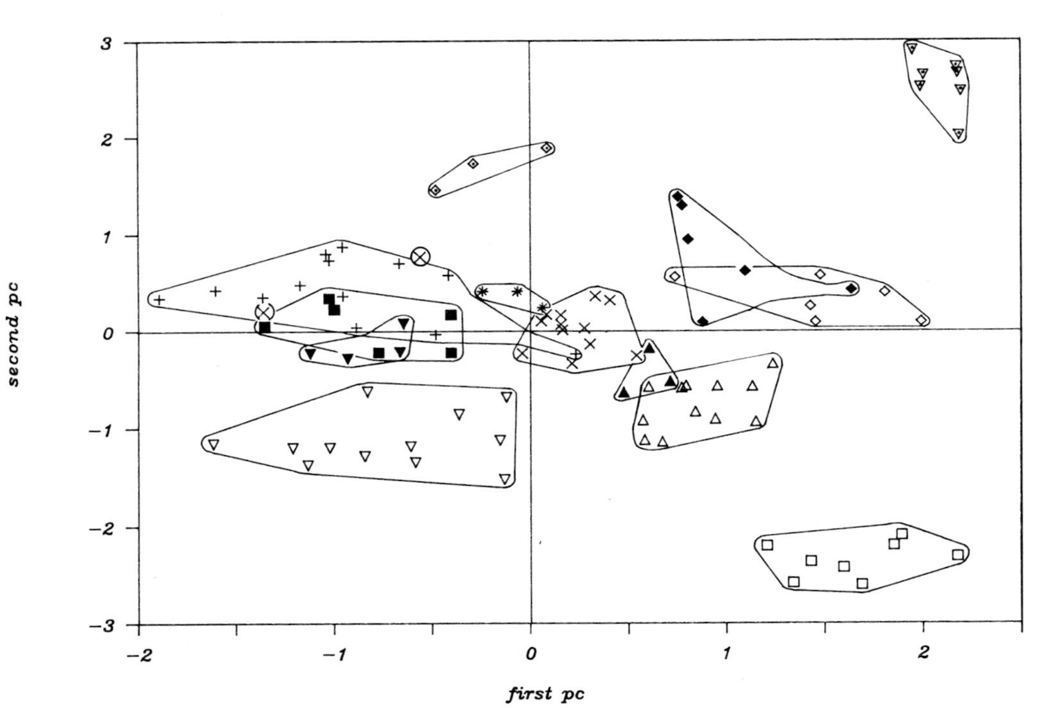

Fig. 6 Principal component diagram based on 21 above-ground characters with specimens of all species and some hybrids plotted against the first and second principal component on the x-axis and y-axis, respectively. The species and hybrids included are: closed squares, A. cepa; circled crosses, A. oschaninii; plus-signs, A. vavilovii; stars, A. pskemense; crosses, A. galanthum; open squares, A. fistulosum; dotted lower triangles, A. roylei; closed lower triangles, A. cepa x A. fistulosum; open lower triangles, A. fistulosum x A. cepa; closed upper triangles, A. galanthum x A. fistulosum; open upper triangles, A. fistulosum x A. galanthum; closed rhombs, A. roylei x A. fistulosum; open rhombs, A. fistulosum x A. roylei; dotted rhombs, A. cepa x A. roylei.

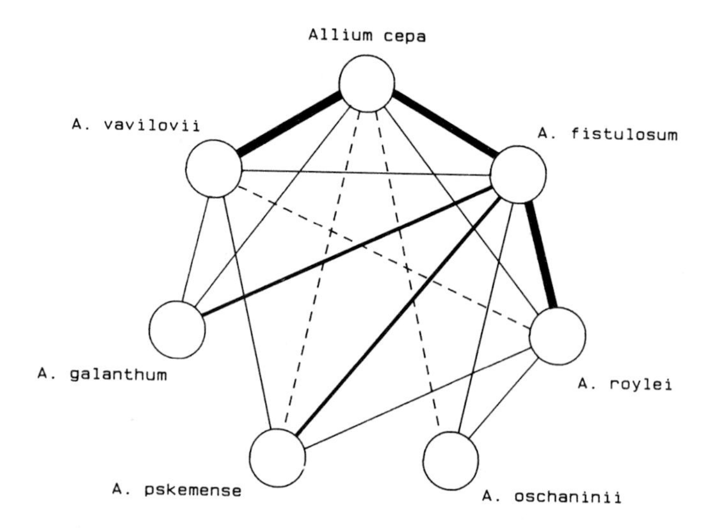

Fig. 7 a) Crossing polygon of *Allium* sect. *Cepa* with results of seed germination.
>50 seedlings / 100 flowers,
26-50 seedlings / 100 flowers,
1-25 seedlings / 100 flowers,
<1 seedling / 100 flowers.

Lack of connecting line indicates cross achieved, but no seed obtained.

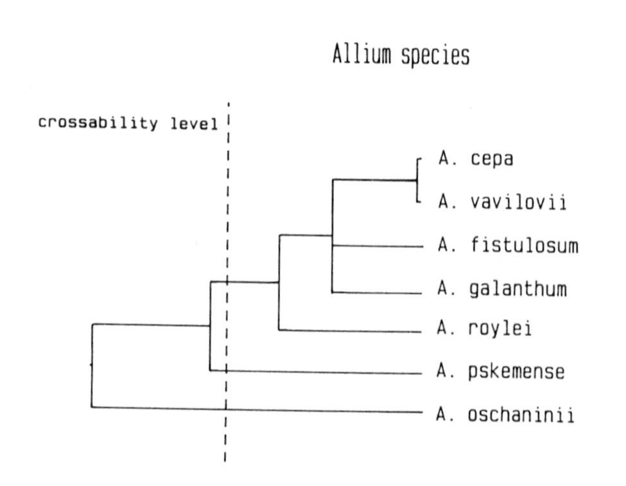

Fig. 7 b) Dendrogram of crossability results in *Allium* sect. *Cepa*. Crossability level is indicated.

Results of crossing experiments are represented in a dendrogram (Fig. 7b). Unrelated species like *A. cepa* and *A. roylei* are linked, while related species like *A. galanthum* and *A. pskemense* are separated on the crossability level. In *Allium*, crossability does not correlate with relationship.

VII. Conclusions

Biosystematic results of crop plants contribute to the optimalization of practical breeding and to the knowledge of isolating mechanisms, domestication and evolution. A considerable amount of information should be available to clarify deviations between relationship and crossability.

References

Ayyanger, K.R. (1967). Taxonomy of Cucurbitaceae. *Bull. Nat. Inst. Sci. India* **34**, 380-396.

Asker, S. (1984). Apomixis and biosystematics. *In* "Plant Biosystematics" (W.F. Grant, ed.), pp. 237-248. Academic Press, Toronto.

Barrett, S.C.H. (1988). The evolution, maintenance, and loss of self-incompatibility systems. *In* "Plant Reproductive Ecology" (J. & L. Lovett Doust, ed.), pp. 98-124. Oxford University Press, New York, Oxford.

Bhaduri, P.N. and Bose, P.C. (1948). Cytogenetic investigations in some Cucurbits with special reference to fragmentation of chromosomes as a physical basis of speciation. *J. Genet.* **48**, 237-256.

Brandenburg, W.A. (1984). Biosystematics and hybridization in horticultural plants. *In* "Plant Biosystematics" (W.F. Grant, ed.), pp. 617-632. Academic Press, Toronto.

Brown, G.B., Deakin, J.R. and Wood, M.B. (1969). Identification of *Cucumis* species by paper chromatography of flavonoids. *J. Am. Soc. Hort. Sci.* **94**, 231-234.

Charlesworth, D. (1988). Evolution of homomorphic sporophytic self-incompatibility. *Heredity* **61**, 445-453.

Dane, F., Denna, D.W. and Tsuchiya, T. (1980). Evolutionary studies of wild species in the genus *Cucumis*. *Z. Pflanzenzüchtung* **85**, 89-109.

Enslin, P.R. and Rehm, S. (1958). The distribution and biogenesis of the cucurbitacins in relation to the taxonomy of the Cucurbitaceae. *Proc. Linn. Soc. London* **169**, 230-238.

Esquinas-Alcazar, J.T. (1977). Alloenzyme variation and relation ships in the genus *Cucumis*. Ph.D. Diss., University of California, Davis, California.

Ferwerda, F.P. (1976). Coffees, *Coffea* spp. (Rubiaceae). *In* "Evolution of crop plants" (N.W. Simmonds, ed.), 257-260. Longman, London, New York.

Gadella, T.W.J. (1983). Some notes on the determination of the mode of reproduction in higher plants. *Proc. KNAW series C* **86(2)**, 155-166.

Grant, W.F. (1987). Genome differentiation in higher plants. *In* "Differentiation patterns in higher plants" (K.M. Urbanska, ed.), 9-32. Academic Press, London.

Grant, V. (1980). Plant Speciation. Columbia University Press, New York.

Heiser, C.B. (1988). Aspects of unconscious selection and the evolution of cultivated plants. *Euphytica* **37**, 77-81.

Hogenboom, N.G. (1975). Incompatibility and incongruity: two different mechanisms for the non-functioning of intimate partner relationships. *Proc. R. Soc. London B* **188**, 361-375.

Hogenboom, N.G. (1978). Exploitation of incongruity, a new toll for hybrid seed production. *In* "Proc. conf. on Broadening the genetic base of crops", 299-309. Wageningen.

Hogenboom, N.G. (1979). Incompatibility and incongruity in *Lycopersicon*. *In* "The Biology and Taxonomy of the Solanaceae" (J.G. Hawkes, R.N. Lester and A.D. Skelding, ed.), 435-444. Linn. Soc. Symp. series 7.

Jeffrey, C. (1968). Systematic categories for cultivated plants. *Taxon* **17**, 109-114.

Jeffrey, C. (1980). A review of the Cucurbitaceae. *Bot. J. Linn. Soc.* **81**, 233-247.

King, M. (1988). Chromosomal rearrangements, speciation and the theoretical approach. *Heredity* **59**, 1-6.

Ladizinsky, G. (1985). Founder effect in crop-plant evolution. *Econ. Bot.* **39**, 191-199.

Lewis, W.H. (1980). Polyploidy: its biological relevance. Plenum Press, New York.

Mulcahy, D.L. (1984). The relationship between self-incompatibility, pseudo-compatibility and self-compatibility. *In* "Plant Biosystematics" (W.F. Grant, ed.), pp. 229-237. Academic Press, Toronto.

Perl-Treves, R. and Galun, E. (1985). The *Cucumis* plastome: physical map, intrageneric variation and phylogenetic relationship. *Theor. Appl. Gen.* **71**, 417-429.

Pickersgill, B. (1981). Biosystematics of crop-weed complexes. *Die Kulturpflanze* **29**, 377-388.

Poppendieck, H.-H. (1985). Taxonomy. Evolution and classification of seed plants. Progress in Botany, vol. 47, 239-283. Springer-Verlag, Berlin, Heidelberg.

Raamsdonk, L.W.D. van (1986). The grey area between black and white: the choice between botanical code and cultivated code. *Acta Hort.* **182,** 153-158.

Raamsdonk, L.W.D. van (1988). IRIS: a program package for Information Regrouping to Identify Structures, version 3.1. IVT, Wageningen.

Raamsdonk, L.W.D. van, Nijs, A.P.M. den and Jongerius, M.C. (1989). Meiotic analysis of *Cucumis* hybrids and an evolutionary evaluation of the genus *Cucumis* (Cucurbitaceae). *Plant Syst. Evol.* **163**, 133-146.

Raamsdonk, L.W.D. van (1990). A crossability coefficient for the evaluation of crossing experiments. in preparation.

Raamsdonk, L.W.D. van and Vries, T. de, in prep. Biosystematic studies in *Allium* sect. *Cepa*.

Raamsdonk, L.W.D. van, Wietsma, W.A. and Vries, J.N. de, in prep. Crossing experiments in *Allium* sect. *Cepa*.

Shore, J.S. and Barret, S.C.H. (1985). Morphological differentiation and crossability among populations of the *Turnera ulmifolia* L. complex (Turneraceae). *Syst. Bot.* **10**, 308-321.

Singh, A.K. and Yadava, K.S. (1984). An analysis of interspecific hybrids and phylogenetic implications in *Cucumis* (Cucurbitaceae). *Plant Syst. Evol.* **147**, 237-252.

Small, E. (1984). Hybridization in the domesticated-weed-wild complex. *In* "Plant Biosystematics" (W.F. Grant, ed.), pp. 195-210. Academic Press, Toronto.

Stace, C.A. (ed.) (1975). Hybridization and the flora of the British Isles. Academic Press, London, New York.

Stace, C.A. (1980). Plant taxonomy and biosystematics. Edward Arnold, London.

Stace, C.A. (1987). Hybridization and the plant species. *In* "Differentiation Patterns in Higher Plants" (K.M. Urbanska, ed.), 115-130. Academic Press, London.

Staub, J.E., Frederick, L. and Marty, T. (1987). Electrophoretic variation in cross-compatible wild diploid species of *Cucumis*. *Can. J. Bot.* **65**, 792-798.

Trivedi, R.N. and Roy, R.P. (1970). Cytological studies in *Cucumis* and *Citrullus*. *Cytologia* **35**, 561-569.

Vickery, R.K. jr. (1984). Biosystematics 1983. *In* "Plant Biosystematics" (W.F. Grant, ed.), pp. 237-248. Academic Press, Toronto.

5 Weed-Crop Complex in Cereal Cultivation

HISAO KOBAYASHI[1] AND SADAO SAKAMOTO[2]

[1]*Department of Agronomy, Faculty of Agriculture, Yamaguchi University, Yamaguchi 753, Japan*
[2]*Plant Germ-plasm Institute, Faculty of Agriculture, Kyoto University, Kyoto 617, Japan*

I. Introduction

According to Harlan and de Wet (1965), many of our crops have weed forms. Most cultivated plants have one or more companion weed forms; generally the weed forms are not more primitive than the crops, but are sometime well adapted to their habitat in morphological as well as ecological characteristics. Weeds are not always progenitors of crop plants. The weeds we see in our fields today may be far removed from wild as their companion crops. The sequence, then, is more likely to be from wild plants adapted to naturally disturbed habitats to a crop-weed complex.

Some weeds always grow in particular cultivated plant fields. Those are called associated weeds. Most of them develop the characteristics of little or non-shattering seeds or fruits, this being one of the most typical characteristics of crops. Vavilov (1926) reported many kinds of weeds associated with bread wheat, such as *Cephalaria syriaca* (L.) Roem. et Schult., *Brassica* spp., *Sinapis* spp., *Pisum arvense* L. s. l., *Lathyrus sativus* L., *Linum usitatissimum* L. and *Fagopyrum tataricum* (L.) Gaertn., and with emmer wheat, such as *Avena* spp., and *Secale* spp. He proposed that among those associated weeds, some cultivated plants resulted as secondary crops, including *Brassica* spp., *Sinapis alba* L., *Pisum sativum* L., *Lathyrus sativus*, *Linum usitatissimum*. He and his colleagues also disclosed *plantae linicolae*, the weeds associated with flax fields, including *Camelina sativa* (L.) Crantz., *C. linicola* Shimp. et Smenn., *Eruca sativa* L. and others. Among them, *Camelina sativa* evolved as a secondary crop. These indicate that the study of associated weeds is a key to uncovering not only the weed-crop complex as a gene-exchangeable biological group but also the mechanism of the origin of cultivated plants.

BIOLOGICAL APPROACHES AND EVOLUTIONARY TRENDS IN PLANTS ISBN 0-12-402960-4

Crop evolution, as Harlan (1975) stated, is sometimes parallel with weed evolution; the same principles apply to both. Crop and weed races often have a common progenitor, and they sometimes belong to the same biological species, such as einkorn, barley, oats, sorghum, pearl millet and so on.

II. Association of Awnless Type *Lolium temulentum* with Wheat and of Awned Type with Barley and Oat

The present authors conducted field surveys on the weed-crop complex of wheat cultivation in Greece, Turkey and Romania in 1980 (Sakamoto and Kobayashi, 1982), and Greece and Turkey in 1982 (Kobayashi, 1984).

Lolium temulentum L. was frequently found in wheat and barley fields. *L. temulentum* is a famous mimic weed to wheat, as described in Mathew (13: 20-24). It produces non-shattering caryopses like crops.

In the reaped mounds of wheat, barley and oat, ear heads of *L. temulentum* and capsules of *Agrostemma githago* L. were frequently found. Such mounds, ready for threshing by machine, were observed everywhere in Greece and Turkey. In

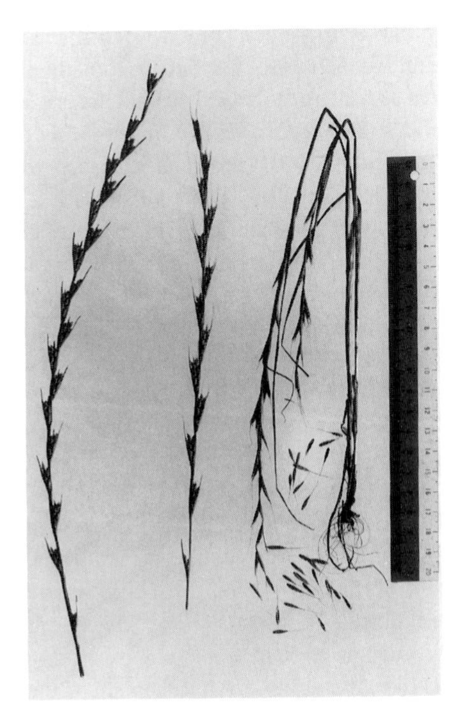

Fig. 1 Awnless (left) and awned types (center) of *Lolium temulentum* in the same durum wheat field, collection no. 6-15-6-16, Vrses near Neapoli, Cerete, Greece, and brittle type (right), collection no. 8-7-3-2B, 27km SW of Kars to Eruzurum, Turkey (Kobayashi, unpublished).

threshed grains, the seeds of *L. temulentum* and *A. githago*, both with less shattering seeds, were frequently found to contaminate grains found in markets and those in samples we collected.

Lolium temulentum observed in Greece, Turkey, and Romania was divided into three types, an awnless type usually with large head, an awned type usually small head, and a brittle type (Fig. 1).

Seventy eight samples of *L. temulentum* collected in 1982 in Greece and Turkey indicated this association more clearly. The awned type prevailed in both barley fields and reaped bundles (100 %), and in oats (87 %), while the awnless type was very rare. However, in both bread wheat and durum wheat fields, awnless type was found more frequently (about 55 %), than awned type. Such association was especially clear in the samples collected in Greece. Similar data were also obtained in threshed grain samples. The association of awnless type with wheat is clearer in the collections of Turkey than in those of Greece.

Both cereal fields and reaped bundles were contaminated with the other kinds of cereals, without exception. These situation are attributed to imperfect cleaning of grains for sowing, especially barley and oat. Barley and oats are usually used for feeding livestock; therefore, most farmers do not pay much attention to cleaning up such grains. This situation creates an association between awnless or awned types of weed and the kinds of cereals in which they are mixed.

Fifty eight families of *L. temulentum* were grown in Kyoto. Ear length correlated positively with plant height but negatively with number of tillers at higher nodes. The individual with tall plant height generally produced large heads but had fewer tillers at higher nodes.

The families from collection Nos. 82-7-9-4-10 and 82-7-9-4-11 showed a very interesting difference: awnless type was taller and produced longer ears, but produced fewer tillers than the awned type. Wheat and rye usually had taller plants than either barley or oats, based on field observations. The taller awnless type plants associated with wheat cultivation method are in harmony with the height of wheat plants. These data are clearly consistent with the observations obtained in the field.

The association of the awnless type of *L. temulentum* with wheat for bread and the awned type with barley and oats, especially for animals, seems to result from cleaning up of grains to remove weed seeds and other contaminators, mainly through winnowing, sifting, and washing, because weeding is done only once during the seedling stage in the field. Therefore, these three methods of cleaning grain seem to be the main selective agent for changing the characteristics of *L. temulentum*, such as mimicking.

Sifting is practiced only for wheat grains used in making bread and for sowing in the observed area. It is concluded that sifting seems to be a direct selective agent producing mimicry, especially in the form of grain size and shape, such as awned or awnless caryopsis.

III. Mimic Weeds of Japanese Barnyard Millet and Foxtail Millet

The progenitors of Japanese barnyard millet (*Echinochloa utilis* Ohwi et Yabuno) and foxtail millet (*Setaria italica* L.) are *Echinochloa crus-galli* L. and *Setaria virdis* (L.) P. Beauv., respectively. In Japan, millet is a relict crop, and its cultivation is confined to mountain villages or local islands. Figure 2 shows the locations of the villages surveyed since 1976 for weeds that mimic millets (Kobayashi, 1988).

A. Mimic weed of Japanese barnyard millet

A mimic weed that very closely resembled awned Japanese barnyard millet was found in Outou village in Nara Prefecture (Fig. 3). The mimic was called *orokabie*. *Hie* or *bie* in Japanese means barnyard millet, and *oroka* means volunteers, referring to the seedlings growing naturally without seeding after harvesting. Of course, the first time we thought those *orokabie* were *hie*, until we grasped their panicles and their caryopses dropped.

Mimic weed without or with awn also was found in cultivated fields of Japanese barnyard millet in Akiyamagou, Niigata Prefecture. This was called *zorobie*, meaning *hie* with self-shattering caryopses. Local cultivaters could not weed out *zorobie* throughly because of its almost complete mimicry. A local cultivator sometimes cut cultivated barnyard millet, and often overlooked this weed on the ridges where

Fig. 2 Japanese local villages investigated for mimic weeds to Japanese barnyard and foxtail millets.

some hundred *hie* individuals grew. His success rate in cutting this weed ranged from 14 % to 100 % varying with ridges in the field, and was mostly around 5O %. This indicates the high degree of mimicry of this weed.

The local name *onzoubie* in Shiramine village means "hateful *hie*" or a troublesome weed. *Zorobie* or *zurubie* in Akiyamagou and *sasarabi* or *sasarabe* in Ohnogun, Gifu Prefecture, all mean *hie* with shattering grains. Local farmers recognize those mimic weeds by their spread tillers and by their reddish color and as offtype of cultivated barnyard millet.

Caryopses of mimics are a bit smaller than the cultivated millet, especially in thickness. This supports the local farmers' recognition that mimic caryopses are smaller.

Barnyard millet and its mimics showed almost no difference in plant size, thickness of stem, or leaf shape, but mimics have many tillers. Degree of spread of mimics is higher than that of millet.

B. Mimic weed to foxtail millet

The weed that mimic foxtail millet was found in almost all areas where foxtail millet has been cultivated.

The local name of mimic weeds to foxtail millet, *yonoko*, *enogo*, *enoko* or *inno*, is derived from the meaning "dog's tail." *Oro*, *uro*, and *ororogusa* are from the concept of volunteers. Other names indicate the mimics to foxtail millets. In the mountain region in Nara Prefecture, the Kii Peninsula, many names including *hakora*, *hakoda*, etc., were used. Many kinds of meaning in local names were found.

Local farmers identify mimics by plant type and plant color. To weed mimics

Fig. 3 Mimic associated weed called *orokabie* (left two panicles) to awned Japanese barnyard millet (right) at Outou village, Nara Prefecture (Koboyashi, unpublished).

out, every year they change the cultivated fields or cultivated variety by varying the seedling color. In Japanese *waka* poetry of about three hundred years ago, mimic weeds are referred to; they also are written in *Chimin Yao Su* about 550 A.D. (Chia, 6C) and other ancient Chinese literary works.

In Japan, millet was grown formerly in sifting cultivation on mountain slopes as well as in fields near local farmers' houses. Around the field near farmers' houses, the weed species of *Echinochloa crus-galli* and *Setaria viridis* are found; they can cross with barnyard millet or foxtail millet freely. Most local farmers said they found no mimic weed on mountain slopes where sifting cultivation was practiced, and the above two weed species were not actually found in such fields on mountain slopes. Therefore, these mimic weeds are thought to have evolved in the fields near local farmers' houses.

IV. The Weed Associated with Millets in India

Millets and their weed relatives present many important problems concerning crop-weed complexes. Indian barnyard millet, *Echinochloa frumentacea* (Roxb.) Link., kodo millet, *Paspalum scrobiculatum* L., and little millet, *Panicum miliare* Lam., were originated and have been cultivated in the Indian subcontinent and its surrounding areas. Field surveys were conducted in South India, Karnataka, Tamil Nadu and Andhra Pradesh in 1985 (Kobayashi, 1987), and along the West Ghats Mountains in Maharashtra and the East Ghats Mountains in Orissa State in 1987 (Kobayashi, 1989). The millets that originated in India were sometimes observed to have associated weeds and to develop mimic types of those millets.

A. *Cultivated Paspalum scrobiculatum and its mimic weed to upland rice plants*

1. *Kodo millet*

Kodo millet, cultivated *Paspalum scrobiculatum*, was found and studied in South India in 1985 and in Orissa in 1987. According to de Wet *et al.* (1984), this crop is today grown from Kerala and Tamil Nadu, north Rajastan, to Uttar Pradesh, Bihar and West Bengal.

2. *Mimic Paspalum scrobiculatum weed to upland rice plants*

Mimic *Paspalum scrobiculatum* weed to upland rice plants was found in upland rice fields in the mountain region of Orissa (Fig. 4). This mimic weed was called *kodo* or *kodoghas*, meaning "grass of kodo millet," by the local farmers. This mimic has large spikelets, or grains.

Kodoghas resembles the tussock habit of upland rice. *Kodoghas* was sometimes very hard to recognize or find. This mimic weed was found in every upland rice field, but was not found in other crop or millet fields. With its large grains, it is completely associated with upland rice fields in this region.

The senior author encountered laborers of the Paraja tribe harvesting upland rice on the hill terrace of Orissa. They reaped upland rice plants along with *kodoghas*. They received large grains of weedy *P. scrobiculatum* along with their harvesting wages. They prepare these as *anna*, boiled grains, after dehulling them.

Ear head contamination of weedy *Paspalum* spp. in reaped bundles reached high levels: 8 heads per 1,110 upland rice panicles. In banded upland rice fields, more like paddy fields, *kodoghas* was not found in ten reaped bundles observed randomly, containing about four thousand five hundred rice panicles.

Grain threshed by being stamped on by bulls was cleaned off by winnowing. The chaff and other admixtures were removed. They were sometimes sifted by sieves made of bamboo or tin. *Kodoghas* was easily separated through winnowing and sifting.

Kodaghas is also called *mandia* or *kodo* in Orissa: the latter makes no distinction from cultivated kodo millet in name. However, this associated *P. scrobiculatum* was, according to local farmers, not sown. It is clearly weed, even though it has cultivated characteristics.

Fig. 4 Mimic *Paspalum scrobiculatum* weed (arrow) among upland rice plants at Badhitafar village (K87-10-14-4), Orissa (Kobayashi, unpublished).

B. Weed type and cultivated type of Setaria glauca

1. Weed type and mimic type of Setaria glauca

Weed of *Setaria glauca* was found everywhere in cultivated fields. It shattered matured grains.

Types of *Setaria glauca* that mimicked kodo millet were found in South India, along the boundary area of Karnataka, Andhra Pradesh and Tamil Nadu. They had grains which shattered scarcely, like cultivated plants.

The mimic type showed almost perfect mimicry of the red-colored plant body except for the stick-like spike with seta. It could not be distinguished from kodo millet before heading, as the local farmers pointed out.

The mimics were called *varagu-sakkalathi*, meaning kodo millet's "illegal wife", and also called *varagu-korali* and *ara-sama* (Fig. 5).

2. Cultivated type of Setaria glauca

The mixed field of little millet and cultivated *Setaria glauca* was found in the southern extremity of Andhra Pradesh. This cultivated type has slender stems and leaves like little millet and has non-shattering caryopses. In contrast, the weed

Fig. 5 "Illegal wives" of kodo and little millets (Kobayashi, unpublished).
From right to left: little millet (called *same*), its illegal wife (*sakkalathi same, pil same, koothi same* or *same melati*), kodo millet (*varagu*) and *Setaria glauca* (*varagu sakkalathi*).

Setaria glauca found in upland rice field and their surroundings has shattering caryopses.

Cultivated fields of *S. glauca* mixed with little millet were also encountered on the mountain region of Orissa (Fig. 6). Two types, reddish brown and whitish yellow in seta color, were observed in each field.

According to a local farmer, mixed seeds of little millet and *S. glauca* were sown in May, but the little millet was heavily suppressed by a severe drought that year, which delayed the rains. Cultivated *S. glauca* was called *lingudi*, and was eaten as *baht*, like boiled rice.

The summarized results of 1985 and 1987 are shown in Table 1. The cultivated type of *S. glauca* found in Orissa in 1987 had thick long stems with large wide leaves and large long heads with short seta, resembling the mimic type ("illegal wife") of kodo millet in South India. But the cultivated type of south India has slender leaves and stems, resembling little millet.

These characteristics contrast sharply to the cultivated *Setaria glauca* found in South India on the boundary regions of Karnataka, Tamil Nadu and Andhra Pradesh.

Cultivated and weed types of *S. glauca* in a village in the boundary area among Karnataka, Tamil Nadu and Andhra Predesh were called *lingudi* and *ghas-lingudi*, respectively. *Ghas-lingudi* means the "weed of *lingudi*." Only 4 km north of that village, the weed type was called *langri*. In another village, both types were called *kukuru-lange*, with no difference in the recognition or use of the local names. The cultivated types were called *kuku-lange* and *kukuru-lange*, meaning "dog's tail." The weed type of this species is called *lota*, whose meaning is unknown, and *bilai-lange*, meanimg "cat's tail."

These differences in plant characteristics and the local farmers' recognition,

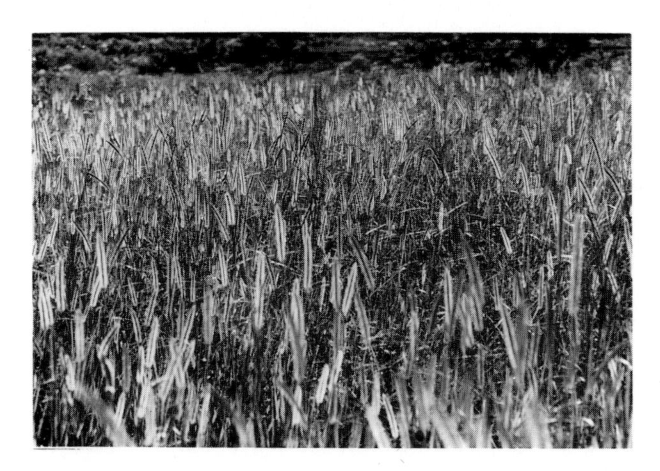

Fig. 6 Mixed field of cultivated *Setaria glauca* and little millet at Dakuta village (K87-10-12-8), Orissa (Kobayashi, unpublished).

TABLE 1
Comparison of three types of *Setaria glauca* in herbarium specimens collected in Orissa (1987) and in South India (1985)

Location and Type		Culm L	Spike L	Spike W	Seta L	Flag leaf Position	Flag leaf L	Flag leaf W	No. of tillers
Cultivated									
82-10-31-2	$\bar{x}$	54.5	4.6	6.0	8.0	31.8	11.2	3.6	5.5
(n=14)	σ	6.56	0.83	0	0.78	5.65	2.08	0.50	3.23
K87-10-12-5	$\bar{x}$	87.5	10.0	4.5	6.1	54.5	20.4	6.4	2.8
(n=16)	σ	21.44	2.40	0.53	0.20	17.87	2.96	0.84	1.84
Mimic type									
85-10-28-1	$\bar{x}$	65.5	5.8	6.3	7.3	38.3	10.9	4.7	10.8
(n=9)	σ	11.51	1.66	0.50	1.00	10.51	2.38	0.61	9.24
85-10-19-1	$\bar{x}$	48.3	7.8	6.3	6.4	29.5	11.0	6.5	12.4
(n=8)	σ	7.45	1.28	0.52	0.52	5.71	2.16	0.53	0.74
82-11-2-1	$\bar{x}$	56.8	7.2	7.2	3.8	41.6	18.3	7.0	6.6
(n=6)	σ	7.78	1.44	0.75	0.75	7.0	4.05	0.63	1.52
Weed									
85-10-28-1	$\bar{x}$	84.0	4.7	3.9	5.0	49.2	7.1	3.8	7.8
(n=9)	σ	10.30	1.14	0.25	0.82	8.50	1.14	0.50	4.0
K87-10-10-5	$\bar{x}$	81.0	10.6	4.5	6.3	72.5	7.0	5.1	6.1
(n=21)	σ	14.79	1.70	0.10	0.12	11.34	1.47	0.82	0.67
K87-10-10-5	$\bar{x}$	84.6	9.2	3.7	5.8	46.9	16.2	5.2	3.5
(n=9)	σ	10.56	2.84	0.69	0.63	12.44	6.39	1.50	1.60
K87-10-11-2	$\bar{x}$	72.5	7.0	5.1	6.1	47.5	10.7	5.0	3.8
(n=9)	σ	11.34	1.47	0.82	0.67	9.87	5.53	0.62	5.12

85- and K87- in location nos. indicate 1985 and 1987, respectively.
L: length; W: width. (Kobayashi, unpublished)

naming, and characteristic differentiation of *S. glauca* between Orissa and South India indicate the differences in their mode of origin and, therefore, their independent origins.

C. Cultivated type and associated type of *Brachiaria* species

1. Cultivated type of *Brachiaria* sp.

A cultivated field of *Brachiaria* sp. with brittle caryopses was found at about 100 km north of Bangalore in Karnataka (Fig. 7). This cultivated millet was called *korne*. It has brittle caryopses or grains that shattered only partly even when we grasped the panicles in our hands. However, its brittleness does not become a problem for harvesting because this millet is reaped in the morning before the dew on the panicles dries.

The open panicle type and closed panicle type of *korne* and a small weed species of *Brachiaria* were found in the cultivated field, as shown in Fig 7.

2. Weed of Brachiaria sp. associated with little millet

A weed of *korne*-like *Brachiarai* associated with little millet was found in northern Tamil Nadu of South India (Fig. 7). This associated weed was called *pill same*, *sakkalati same* or *koothi same*, meaning "illegal wife of little millet;" little millet is called *pani varagu*. Local farmers do not remove this weed, and reap it along with little millet. This weed shows shattering of grains and grows naturally without sowing by local farmers. Only when the population of this weed grows too largely, farmers uproot it and feed it to their livestock.

This weed is said to have high drought tolerance. Even when the yield of little millet is poor, local farmers can harvest its illegal wife, which they cultivate as insurance against failure of the little millet crop.

The local farmers can distinguish this weed from little millet because it has a round and thick stem and broad leaves. But its leaves are rolled and look like slender little millet leaves. This too can be called mimicry in some sense.

3. Weed of Braciaria sp. associated with kodo millet

In the kodo millet field of Orissa, *Brachiaria* sp. and *Echinochloa colona*, with caryopses which shattered scarcely, were found sporadically. *Brachiaria* sp. was called *ghusara-pata*. According to a local farmer, this means the weed is associated with kodo millet cultivation. *E. colona* was called *dhera*. These two weeds are threshed with kodo millet and eaten together. *Brachiaria* sp. was also found in kodo millet fields in other locations. This indicates that *Brachiaria* sp. is strongly associated with kodo millet cultivation and that it is just becoming a secondary crop.

Fig. 7 Open (right) and closed panicle types (center) of ***korne*** millet and weed ***Brachiaria*** sp. (left) in a ***korne*** field at Malleswarapua village, about 100km N from Bangalore, Karnataka, India (Kobayashi, unpublished).

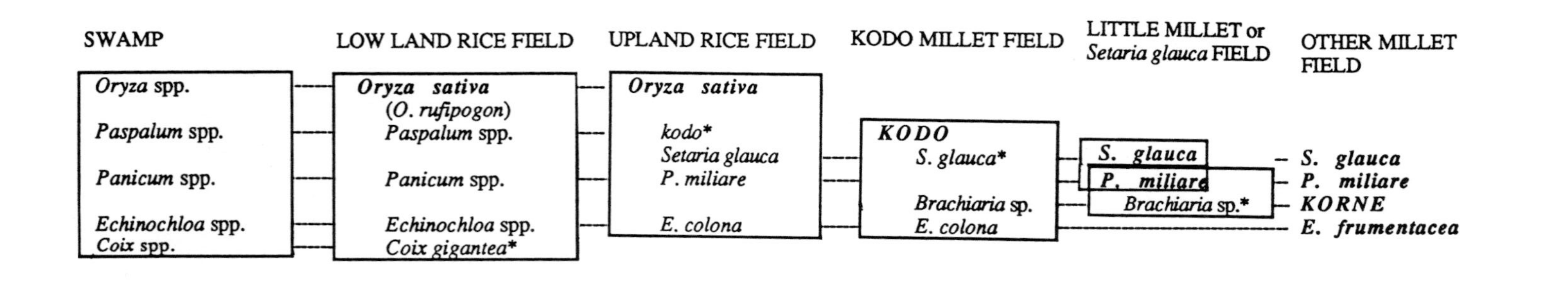

Fig. 8 Schematic diagram of the association of weeds, lowland or upland rice, and millet fields (Kobayashi, unpublished).
Bold letters indicate cultivated plants.
Border lines indicate fields.
* Mimic type to the main crop was developed.

V. Conclusion

The association relationship of weeds observed in India is summarized in Fig 8. What are the factors that cause weeds to be associated with or mimic weeds? The large and non-shattering grains of *Paspalum scrobiculatum* were assumed to be the result of the following three factors: First, winnnowing after threshing of rice contaminated with *Paspalum scrobiculatum*, as pointed out in the summarized discussion on *Camelina* evolution (Stebbins, 1950). Second, the collection of those grains by laborers to eat in addition to their wages for harvesting--in other words, the use of this weed by the farmers. Third, mimicking behavior must have been promoted in associated weeds. Mimicry is seems to be achieved mainly through hand weeding as well as the first two factors.

What is the agent promoting these weeds as a secondary crop? According to field observations, moisture conditions varied from fully moist to severe dryness. *Oryza rufipogon* Griff. grows in swamps where water is always available. The paddy field is kept full of water, at least in the growing period of lowland rice. In contrast, upland rice is cultivated on hilly slopes with a higher moisture content than other millet fields.

Brachiaria sp. associated with little millet in South India was given the special name "illegal wife of little millet." This illegal wife was fed as green fodder and also eaten as crop insurance in drought years; it might be said that it was becoming a secondary crop. The same situation was also observed in mimic *Setaria glauca* to kodo millet in South India. This mimic weed was also called "illegal wife." In the little millet field of Orissa, it was observed, becase of severe drought, that *Setaria glauca* suppressed the growth of little millet and had taken its place as the crop. Therefore, cultivated *Setaria glauca* and *Brachiaria*, called *korne*, must be capable of growing even in dry soil.

In summary, kodo millet, cultivated *S. glauca* and *Brachiaria* sp. called *korne* seem to have first evolved as associated or mimic weeds and thereafter to have been domesticated as secondary crops.

References

Chia Sus-Hisie 6(C). *Ch'imin Yao Su* (translated by B. Nishiyama and Y. Kumashiro) 1976. Asian Economic Press, Ltd. Vol. **1**, 347pp.

de Wet, J. M. J., Rao, K. E., Mengesha, M. H. and Brink, R. E. (1983). Diversity in kodo millet, *Paspalum scrobiculatum. Econ. Bot.* **37**, 159-163.

Harlan, J. R. and de Wet, J. M. J. (1965). Some thoughts about weeds. *Econ. Bot.* **19**, 16-24.

Harlan, J. R. (1975). Crops and Man. Amer. Soci. Agr. Crop Sci. Soci. Amer. pp. 295.

Kobayashi, H. (1987a). Close association of awnless type *Lolium temulentum* L. (darnel) with wheat and of awned type with barley or oat in Greece, Turkey and Romania. *In* "Domesticated Plants and Animals of the Southwest Eurasian Agro-pastoral Culture Complex", (Y. Tani and S. Sakamoto, eds), "I. Cereals" (S. Sakamoto, ed.), pp. 87-99. Res. Inst. Humanistic Studies, Kyoto Univ.

Kobayashi, H. (1987b). Mimic and associated weeds with millets and cultivation methods of millets in the southern Indian Subcontinent. *In* "A Preliminary Report of the Studies on Millet Cultivation and its Agro-pastoral Culture Complex in the Indian Subcontinent (1985)" (S. Sakamoto, ed.), pp. 15-40. Kyoto University.

Kobayashi, H. (1988). Associated mimic weeds of Japanese barnyard and foxtail millets (in Japanese). *In "Hatasakunohkou no Seiritsu"* (K. Sasaki and T. Matsuyama, eds), Japan Broadcasting Publ. Co. Ltd. pp. 165-187.

Kobayashi, H. (1989). Mimic and associated weeds with millet and rice cultivation in Orissa and Maharashtra in India. *In* "A Preliminary Report of the Studies on Millet Cultivation and its Agro-pastoral Culture Complex in the Indian Subcontinent (1987)" (S. Sakamoto, ed.), pp. 11-32. Kyoto University.

Sakamoto, S. and Kobayashi, H. (1982). Variation and geographical distribution of cultivated plants, their wild relatives and weeds native to Turkey, Greece and Romania. *In* "Preliminary Report of Comparative Studies on the Agrico-pastoral Peoples in Southwestern Eurasia II(1980)" (Y. Tani, ed.), pp. 41-104. Res. Inst. Humanistic Studies, Kyoto Univ.

Stebbins, G. L. (1950). "Variation and Evolution in Plants", 643pp. Columbia Univ. Press, N. Y.

Vavilov. N. I. (1926). Studies on the origin of cultivated plants. *Bull. Appl. Bot. Plant Breed.* (Leningrad) **16**, 1-248.

6 Responses to Flooding in Weeds from River Areas

CORNELIS W.P.M. BLOM

Department of Experimental Botany, Catholic University, Toernooiveld 6525 ED Nijmegen, The Netherlands

I. Introduction

In the northwestern part of Europe, as in many other parts of the world, the flooding of river areas occurs frequently. The research line of the Experimental Plant Ecology group at the University of Nijmegen is the investigation of flooding responses and adaptive characteristics in weeds occupying low- to high-situated habitats in river areas. The species under study - *Rumex, Chenopodium, Plantago* - are distributed along elevation gradients in the Dutch river ecosystems. These flood plains consist of river banks, wetlands, and former riverbeds. Dikes have been in existence since about 1200 A.D. to protect the inhabitants and their agricultural or industrial land from recurring flooding.

The flood plains are subject to regular inundation at times when the river is carrying large quantities of water. A considerable quantity of that water originates from snow melting in the mountains of Switzerland, southern Germany, and the northeastern part of France. The strongly fluctuating water levels that occur during the growing season are, however, mainly a consequence of unpredictable peak periods in precipitation. The main rivers, the Rhine, the Meuse, and their tributaries, drain large areas before reaching the Netherlands. The flooding of grassland found in river forelands used to be very rare in the growing season, but in recent decades extremely high water levels have brought about more frequent floodings of these forelands in the spring and summer. These floodings probably result from urbanization, an increase in the number and size of industrial areas, and the improved drainage of upstream agricultural areas. As a result, after heavy

BIOLOGICAL APPROACHES AND EVOLUTIONARY TRENDS IN PLANTS ISBN 0-12-402960-4

rainfall a surplus of water causes larger fluctuations in water levels in shorter lapses of time in both winter and summer.

The hypothesis in our studies is that plants growing in those river flood plains have to be adapted to survive periods of severe inundation. The severity of the flooding can range from waterlogging of the soil to submergence of the whole plant. While the responses of plants in waterlogged soils may differ from those of plants facing submergence, both conditions greatly influence the physiology, morphology, and population biology of the weeds that grow there (Van de Steeg, 1984; Blom, 1985). In our studies we have applied at least three approaches, with each approach being directed towards a different level of evolutionary relationships (Blom, 1987). The first approach has involved long-term observations in the field in order to study variation in plant characteristics and in environmental values. Under natural conditions the occurrence and performance of an individual plant reflect the effects of the interaction of many environmental factors. Consequently, our second approach has been the design of experiments that can be conducted under controlled conditions so that the separate and combined effects of the observed environmental factors on plant characteristics can be investigated. These experiments have been carried out in phytotrons, greenhouses, and experimental plots in the field. The third approach has been the re-evaluation of the experimental results in nature. Following these approaches, we first studied individual plants of different species from one or two genera to obtain insight into their morphological and physiological responses to flooding. Questions relating to population biology can be answered by comparing plants from the same species but different populations and by studying variations among plants within one population. The main question to be answered here is whether adaptive responses are caused by physiological plasticity or by genetic differentiation.

This paper reports on a study comparing various species of *Rumex* that have been shown to possess contrasting tolerances to flooding caused by increases in river-water level. *Rumex* species from sites at high, intermediate, and low elevations are experimentally subjected to various flooding regimes in order to identify and quantify any relevant adaptive features and to test whether their distribution might be caused by a differential response to flooding. We compare the adaptations of *Rumex* plants with the responses of *Chenopodium* species from over-wet soils. Life-history traits of *Plantago* plants from different populations and from plants of one subdivided population have also been studied. The adaptive values of the different responses upon flooding will be discussed.

II. Description of the Species

Studies on the adaptive responses to flooding were carried out with five *Rumex* species, *Chenopodium rubrum*, and *Plantago major* ssp. *pleiosperma*, all of which

TABLE 1
Characteristics of some plant species from river areas

	habitat	flooding intensity and duration	seed release and following germination period	longevity	survival strategy
R. thyrsiflorus	high-lying grassland	seldom very short	summer late summer	perennial	maintainance
R. acetosa	high-lying grassland	seldom very short	summer late summer	perennial	maintainance germination
R. crispus	low-lying grassland	infrequent short-long	autumn spring	perennial	maintainance germination
R. palustris	mud flat river bank	frequent long	autumn summer	biennial	germination maintainance
R. maritimus	mud flat river bank	frequent prolonged	autumn spring	annual-biennial	germination maintainance
Plantago major ssp. *pleiosperma*	river bank mud flat	frequent prolonged	autumn summer	perennial	germination maintainance
Chenopodium rubrum	river bank	frequent prolonged	autumn summer	annual	germination

occur in the Dutch river areas (Table 1). The distribution of the five *Rumex* species is correlated to a distinct gradient of flooded habitats in which flooding intensity and duration vary from seldom and short-term to frequent and prolonged, respectively. On the basis of preference for a common habitat, at least three groups can be distinguished within these species (Fig. 1). Within the *Rumex* genus, *R. thyrsiflorus* and *R. acetosa* are found on more highly elevated sites that are seldom flooded in the growing seasons. These sites, which are densely populated with tall grasses and perennials, mainly belong to communities of Arrhenatheretum and Lolio-Cynosuretum. The Arrhenatheretum sites containing *R. thyrsiflorus* are hayfields, and the Lolio-Cynosuretum type with *R. acetosa* are either grazed by cattle or managed as hayfields with an aftermath treatment. *Rumex crispus* occurs in more frequently flooded grassland having a relatively dense vegetation layer that belongs to the Ranunculo-Alopecuretum geniculatis. During dryer intervals between floods these areas are used as grazing areas by cattle. *Rumex maritimus* and *R. palustris* are mainly found on the mud flats of former river beds or on regularly flooded edges along the river. The pioneer vegetation type belongs to the Rumicetum maritimi community.

Plantago major ssp. *pleiosperma* can be also found in these open wet sites. The community in which this species occurs belongs to the Nanocyperion alliance. *Chenopodium rubrum* is mainly found on the sandy edges of rivers and forms the

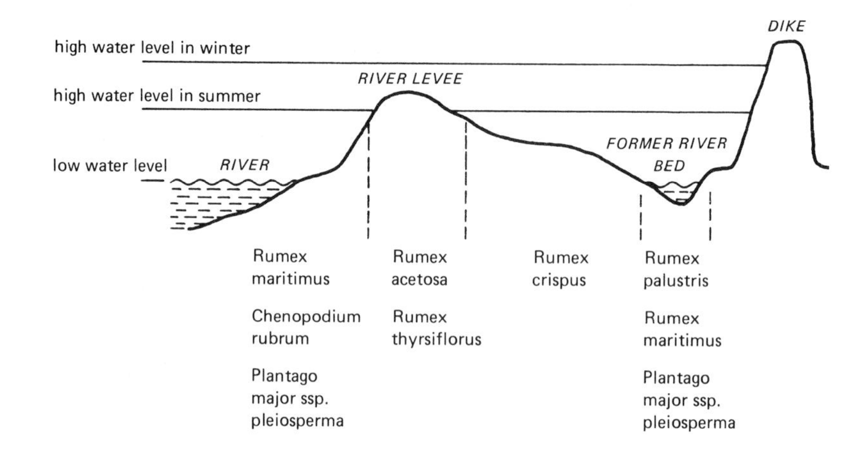

FIG. 1. The distribution of the species under study along a gradient of flooded areas in the river area.

characteristic Chenopodietum glauco-rubri community. Due to the unpredictability of the floodings, these sites are mostly unsuitable for any form of management or agricultural use. However, cattle and geese sometimes penetrate these areas during the short intervals between floodings. Plants surviving in these sites, therefore, also have to be adapted to grazing and trampling and to growing in compacted soils.

III. Strategies in *Rumex maritimus* and *Chenopodium rubrum*

Chenopodium rubrum was incorporated together with the *Rumex* species into our studies. We chose *Chenopodium* in order to compare its responses to flooding with those of *Rumex maritimus*: both are common pioneer species from the clay or sandy soil regions found along rivers in the Netherlands. An important question is whether these adjacent species have different strategies for surviving irregular flooding. Both species germinate from the end of May until late summer, after the water has receded from their habitat. All *Chenopodium* plants flower late in the summer under short-daylight conditions; *R. maritimus* plants may flower, but they may also winter as a rosette and flower the second year. Table 2 shows the results of some greenhouse experiments that were designed to study the effects of different waterlogging regimes on the growth of both species. It is very clear that *Chenopodium* plants suffer from waterlogging: their number of leaves, stem length, and biomass were significantly decreased. In contrast, only a small decrease in the number of leaves was found in the *Rumex* plants.

TABLE 2

Effects of flooding on the growth parameters of *Chenopodium rubrum* and *Rumex maritimus* (n=5; ±SE). Experiments were carried out in the greenhouse where 6-week-old plants were partially submerged for 6 weeks

	Chenopodium rubrum		*Rumex maritimus*	
	drained	flooded	drained	flooded
leaves (nb)	1100± 100	170± 50	220± 70	170± 30
stem length (cm)	51± 4	43± 6	100± 8	135± 6
dry weight (g)	23± 2	5± 1	24± 4	28± 4

Furthermore, plants under flooded conditions possessed more elongated petioles and higher biomasses than those under drained treatments. Table 3 demonstrates reproductive responses to flooding. Both species reacted most strongly when treatments were applied early in the life cycle. Seed production was dramatically reduced in *C. rubrum*; in contrast, a significant increase was found in *R. maritimus*. The results of these experiments are reported in greater detail in Van der Sman *et al.* (1988). The conclusions to be drawn are very clear: as far as growth and reproduction under greenhouse conditions are concerned, *R. maritimus* is much better adapted to flooding than *C. rubrum*. To bridge the distance between these fully conditioned circumstances and the real situation, large-scale experiments in an experimental garden were performed which mimicked the flooding situations in the river area. In these plots plants were kept under submerged conditions for varying periods of time. Provisional results are presented in Table 4. Under flooded conditions *C. rubrum* had a 65% survival rate; all surviving plants, however, developed flowers and produced seeds. Seed production was dramatically lower in plants subjected to the submerged treatments. *Rumex maritimus* reacted differently: all plants survived but only 35% flowered. In this species a reduction in seed production was also found, but the decrease was smaller than in *Chenopodium*.

TABLE 3

Effects of flooding on the reproductive parameters of *Chenopodium rubrum* and *Rumex maritimus*. Experiments were carried out in the greenhouse. Plants were subjected to partial submergence for 0, 2 and 6 weeks in an earlier (A) and later (B) stage of the life-cycle

		Chenopodium rubrum			*Rumex maritimus*		
weeks of flooding		0	2	6	0	2	6
seed production	A:	5.2	1.4	1.1	18.0	15.4	29.8
(g/plant)	B:	5.2	3.9	3.1	18.0	19.7	23.2

TABLE 4

Survival and flowering characteristics in *Chenopodium rubrum* and *Rumex maritimus* after submergence treatment in experimental plots. Dry: drained conditions. Wet: submerged conditions three times a year for ten days each

	Chenopodium rubrum		*Rumex maritimus*	
	dry	wet	dry	wet
% survival	100	65	100	100
% flowering	100	100	100	35
number of seeds/plant	44.000	990	13.000	950

Results obtained from plants grown in the experimental plots and in the greenhouse reveal the existence of different strategies in the two species. The *Rumex* strategy is maintenance, achieved by the survival of vegetative plants. However, if the plants were able to flower in spite of flooding, then seed production appeared to be relatively high. *Chenopodium* strategy is flowering, with production of relatively few seeds during the short intervals between the flooding periods. Both the experimental approaches used in these studies highlight the fact that *Rumex maritimus* is the most adaptive species. These results are in accordance with the behavior of these plants in nature (J. van der Sman and H. van de Steeg, personal communication).

IV. Morphological and Physiological Responses to Flooding

Since the first effects of flooding appear in the soil, special attention was given to changes in the morphology of root systems. A direct effect of flooding is a major decrease in the gas exchange between the atmosphere and the soil (Armstrong, 1979; Kozlowski, 1984). Any oxygen remaining in the soil is very soon exhausted as a result of the respiration of root and soil organisms. Waterlogged roots must, therefore, function in almost anaerobic soils.

We studied root architecture and morphology by means of the pin-board method (Blom, 1979), with the non-destructive method of the so-called perforated soil system (Van den Tweel and Schalk, 1981), and also with the line-intersect method (Voesenek & Blom, 1987). These methods enabled us to describe root turnover and growth under both experimental and field conditions. We found at least three responses in root morphologies upon flooding: (a) an increase in root branching; (b) the development of new adventitious roots; and (c) an altered vertical distribution of the laterals, with more roots concentrated in the upper layers of the soil (Laan *et al.*, 1989b; Voesenek *et al.*, 1989). In Table 5 the root lengths, total numbers, and dis-

TABLE 5
Root length, total numbers, and distribution of laterals (n=8; ±SE) after prolonged flooding in *Rumex* plants from habitats seldom to frequently flooded in river areas (data after Laan *et al.*, 1989b)

	length(m)	numbers	Distribution of new roots over the tap-root (% of total from base to apex)
R. thyrsiflorus	3.5±0.6	29	31-37-32
R. crispus	15.9±2.0	47	27-36-37
R. maritimus	24.1±2.6	115	59-22-19

tribution of new laterals of three *Rumex* species from different habitats after three weeks of flooding are given. These data prove that tolerant plants have very long roots which grow superficially.

Other responses to flooding are changes in root anatomy. Aerenchyma formation and a high porosity due to expansion of the intercellular spaces are among the most obvious adaptive responses. The formation of large channels in the root cortex enhances the diffusion of atmospheric or photosynthetic oxygen from the shoot to the roots so that aerobic respiration and growth can be maintained (Armstrong, 1979; Justin and Armstrong, 1987; Laan *et al.*, 1989a, b; Laan *et al.*, 1990). In an experiment in which three *Rumex* species were grown under aerobic and anaerobic conditions, clearcut differences in gas space development of the new formed laterals were found. *Rumex thyrsiflorus* was apparently unable to form aerenchyma under anaerobic conditions. When the cortex of the primary roots was examined 0.5 cm behind the root apex, the cross-sectional area occupied by the intercellular spaces was approximately 7%. In the newly formed roots that developed under anaerobic conditions we found a small increase to 11%.

Rumex crispus had only a small amount of aerenchyma under aerobic conditions, but a four-fold increase to 28% was observed under stagnant conditions. Aerenchyma in *R. maritimus* was formed in both primary and newly formed roots; this was reflected in a high cross-sectional area of intercellular spaces in the primary laterals (14%), which increased to around 36% in the new roots(for more details see Laan *et al.*, 1989b). Experiments investigating root architecture and anatomy clearly showed that aerenchyma formation is closely connected to an increase in root length upon flooding. Increased porosities of recovered root systems in flooding-resistant *Rumex* plants result in improved internal oxygen transport from the leaves to the root environment.

Differential flood-tolerances may be caused by differences in internal aeration and the use of aerial and photosynthetic oxygen in root respiration. Flood-intolerant *Rumex* plants did not show any internal aeration. Flood-tolerant plants from lower habitats possess internal aeration and oxygen loss around the roots (Laan *et al.* 1989a). Forty to fifty percent of the total root respiration in *R. crispus* and *R. maritimus* can be ascribed to the diffusion of aerial oxygen through the aerenchymatous tissues of the plants (for further details see Laan *et al.*, 1990). We may conclude, therefore, that the size of the root system, the distribution of the laterals over the tap-root, and the degree of internal aeration due to differences in porosity at least partly explain the differential responses towards flooding in *Rumex* species.

The next question is what we know about the phenotypic changes and related physiological processes that occur in the shoots of plants facing waterlogging or submergence. Experimental results strongly indicate that complicated physiological processes are involved: opposite effects were found in the upper parts of the *Rumex* species under study. For example, we measured shoot dry weights, leaf areas, leaf lengths, and number of leaves after 40 days of waterlogging (Table 6). The shoot biomass of *R. acetosa* decreased after waterlogging, but a significant increase in the shoot weights of *R. crispus* and *R. palustris* was found. These changes in biomass were mainly due to changes in leaf areas. Another response that was observed in the field as well as in greenhouses was that submerged plants were able to elongate their petioles until the leaf tips reached the water surface. This phenomenon appears to be a mechanism by which amphibious and some terrestrial plants can survive temporary floods.

In a field experiment, seedlings of *R. acetosa, R. crispus,* and *R. palustris* were planted in a low-lying, regularly flooded zone and in a higher dry region along the river. During the study period we compared petiole lengths before and after a flooding period of four days, in which only the lower zone was submerged: after the flooding period the petiole lengths of leaves of *R. acetosa* were shorter than at the start of the experimental period; in contrast, the petioles of both *R. crispus* and *R. palustris* were significantly longer after the flooding period -- approximately 100% and 120% respectively. The elongation of the non-flooded plants was approximately 20% and 5%, respectively (for more details see Voesenek and Blom,

TABLE 6

Changes in shoot characteristics in *Rumex* species after 6 weeks of waterlogging. d = drained treatments, w = waterlogged treatments (Data after Voesenek *et al.*, 1989b). Differences between treatments are significant in all species (*n*=3; ±SE)

	biomass (g)		leaf area (cm^2)		leaves (number)	
	d	w	d	w	d	w
R. acetosa	6.2±0.5	4.7±0.5	1100± 76	569± 54	55± 3	43± 5
R. crispus	3.3±0.3	6.8±1.3	851± 72	1116± 178	19± 2	22± 4
R. palustris	5.1±0.3	6.5±1.1	940± 84	1376± 313	37± 2	31± 1

1989b). In the greenhouse we subjected plants of *R. acetosa, R. crispus,* and *R. palustris* to various depths of submergence and consequently studied the capacity of their leaves to emerge above the water level. Both *R. crispus* and *R. palustris* were able to elongate their petioles under submerged conditions. *R. palustris* was the only species capable of overcoming deep inundations; within a few days its leaves could bridge heights of 40 cm and restore contact between the leaf tips and the atmosphere (see Voesenek and Blom, 1989a,b).

Our results also show that growth responses upon submergence were especially prevalent in the younger petioles of both *R. crispus* and *R. palustris*. Evidence from the literature (Osborne, 1984; Jackson 1985; Ridge, 1985) and from our own experiments indicates that the enhanced elongation of submerged shoots has to be attributed to an accumulation of ethylene and/or to its increased internal production in the shoot tissue. Petiole growth can be manipulated by the external application of ethylene. In an experiment in which the effects of both submergence and ethylene application were studied in the three above-mentioned *Rumex* species, we obtained similar results with both treatments. Relative to control plants the application of 0.5 Pa ethylene caused an increase in petiole length of approximately 60% in *R. crispus* and approximately 220% in *R. palustris*; under submerged conditions these increases were approximately 70% and 240% for *R. crispus* and *R. palustris*, respectively. Under both conditions no elongation of young petioles was observed in *R. acetosa*. Measurements of the endogenous ethylene level (Table 7) show that for all three species the ethylene concentrations in submerged plants exceed those in plants grown under drained conditions. These results clearly prove that variation in growth responses in petioles of the three *Rumex* species may be attributed to variation in sensitivity to enhanced ethylene in the plant tissue (for more details see Voesenek and Blom, 1989b).

Experiments investigating the hormonal regulations of adaptive responses to flooding were superimposed on the experiments described here. The purpose of one series of experiments was to determine continuously the course of ethylene production in a plant under waterlogged conditions. For these measurements we used a photoacoustic CO_2-wave guide laser (see Harren, 1988). With this photo-

TABLE 7

Minimum and maximum values of ethylene (concentrations given in Pscal) measured in the tissue of drained and submerged *Rumex* plants. Duration of treatments from 2-16 h

	drained	submerged
R. acetosa	0.015 - 0.028	0.019 - 0.122
R. crispus	0.012 - 0.034	0.032 - 0.103
R. palustris	0.014 - 0.026	0.030 - 0.088

acoustic laser system we succeeded in detecting the production of ethylene up to 6 ppt in waterlogged plants during a period of nearly 200 hours of growth. These experiments clearly demonstrate that there is a higher level of ethylene production in *R. crispus* and *R. palustris* than in *R. acetosa*.

V. Variation Between and Within Populations of *Plantago*

Up to this point this paper has reported on the adaptive responses to flooding shown by plants from different species of one or two genera. Other levels of investigation include the adaptive value of variation in characteristics of plants from different populations of one species and in characteristics of individual plants belonging to one single population. One of the aims of studies at these levels is to unravel the relationship between environmental variation and variation in the life-history characteristics of plants. An important question is just how much of the variation observed in plants is based on genetic differentiation and how much on phenotypic plasticity -- although differences in the means or degrees of plasticity are also genetically determined.

We have compared life-history traits such as growth and reproduction in different populations of *Plantago major* (Blom and Lotz, 1985; Lotz and Blom, 1986) as well as in different parts of one population occurring in a restricted area (Lotz, 1989). *Plantago major* is a self-compatible wind pollinator with a high self-fertilization rate, which suggests a low genetic variation within populations. Based on results from physiological experiments, we expected that differences in life-history characteristics between and within populations would be determined by environmental factors rather than by genetic differentiation.

To test this hypothesis we performed a reciprocal transplantation experiment. Two populations, differing in habitat and plant characteristics, were selected: one was located on an open flood plain on the Rhine river in the central part of the Netherlands; the other, in an embanked flood plain along the North Sea coast. The riverside site undergoes irregular flooding all year around, and the *Plantago* population there occurs adjacent to populations of *R. maritimus* and *C. rubrum*. During dry periods this area is grazed to a moderate degree by cattle. The site in the embanked flood plain is a former beach plain separated from tidal influences by a dike built in 1965. The short vegetation layer, grazed by rabbits, has an open structure and mainly consists of relatively short-living, slow-growing perennials. At the end of the first growing season the survival rate of plants from both populations was highest on the open flood plain. The river plants showed a significantly higher rate of survival in their own habitat than at other sites (for details see Lotz and Blom, 1986). During the study season no floods occurred, but during the following winter plants were subjected to flooding and subsequently died (Table 8). In the embanked flood plain nearly all of the plants were alive during the season after the year of

TABLE 8
Survival, growth, and reproduction of *Plantago major* ssp. *pleiosperma* in an open (1) and embanked (2) flood plain

population	site	survival %	biomass shoot(g)	biomass seed(mg)
1 open	1	75-->0*	17.8	1.5
	2	40-->30	0.5	0
2 embanked	1	70-->0	17.0	1.0
	2	50-->45	1.5	0.1

* Survival after the first and second season.

transplantation. Plants of both populations produced seeds at the riverside location. Shoot biomass and seed production of transplanted plants from both populations were significantly higher in the river habitat (Table 8). In *P. major* ssp. *pleiosperma* we found the same survival strategy as in *C. rubrum*; plants of both species produced as many seeds as possible within a relative short time. The transplantation experiments prove that this adaptive response to flooding is mainly due to phenotypic plasticity. In an accompanying greenhouse experiment (Lotz and Blom, 1986) genetic differences in the time of initiation of flowering and in other reproductive characteristics between the populations were found. In addition, plants from the river bank flowered earlier than those from the other population.

Different microhabitats were present within the embanked flood plain (see Blom, 1987; Lotz, 1989). The study site has a total area of about 4 ha and can be divided into a wetter part, where the soil is nearly permanently saturated by rain water, and a dryer part. Demographic research conducted over a four-year period revealed large differences between the sites: *P. major* ssp. *pleiosperma* from the wet part of the area showed significantly higher mortality and lower biomass and seed production than its cohorts from the dryer site (Blom, 1987). Within the small-scale mosaic environment of the dryer part of the area, Lotz(1989) distinguished three subpopulations that occur in clearly identifiable habitats situated only a few meters apart from each other. There are differences in the availability of macro-nutrients between the three subsites. Among subsites plants of *P. major* ssp. *pleiosperma* differ in their life-history traits.

The variation in plant characteristics was mainly due to phenotypic plasticity. Nevertheless, genetic differentiation between the subpopulations has also been found, particularly in leaf biomass and growth form (Lotz, 1989). Results of these field experiments as well as of greenhouse tests on the effects of varying levels of nutrients on growth and reproduction clearly proved that genetic differences do exist between and within populations of *P. major* ssp. *pleiosperma*. Different degrees of plasticity in life-history characteristics, e.g. in survival of transplants and in reproduction, were found. All experiments proved that variations in vegetative and generative development depended mainly on the environmental factors; only small population effects were found. This study demonstrate that phenotypic plasticity expressed as variation in vegetative and generative characteristics is an important response to the selective forces of irregular floodings.

VI. Conclusions

We distinguish at least six phases in the adaptive response of weeds from river areas to flooding.

(1) After the level of the river water increases, thereby causing waterlogging of the soil or submergence of the vegetation, plants accumulate ethylene in their tissues. The higher amounts of C_2H_4 can definitely be partly attributed to the lower diffusion rate of ethylene from the tissues to the submerged environment of the plant and probably partly to the higher production of this hormone upon waterlogging.

(2) Flood-tolerant weeds react differently to higher amounts of ethylene than intolerant ones. The enhanced ethylene concentration causes petiole and stem elongation in flood-tolerant plants. Depending on the height of the water level above the submerged plants and on the duration of the flooding period, the increased growth of the shoots may restore contact between the leaves and the open air.

(3) Tolerant plants possess or develop aerenchymatous tissue in the shoots and roots. Upon flooding, the primary root system ceases to grow or even succumbs; new porous laterals emerge. Conglomerates of new roots develop at the root-shoot junction.

(4) Increased porosity in the shoots and the newly formed root system result in improved internal oxygen transport from the leaves to the root environment. In this way, tolerant plants may use aerial oxygen for root respiration. Moreover, the radial oxygen losses around the roots partly re-establish aerobic conditions in the soil and reactivate oxygen-dependent soil microorganisms. This results in increased availability of nutrients for the tolerant plants and therefore possibly in increased growth and reproduction.

(5) Plants from different species growing in the same habitat and exposed to severe environmental stress factors may possess different life-history strategies to ensure survival. Some species behave as biennials or perennials and survive mainly in their vegetative phases. Others are able to germinate and produce seeds in the very short intervals between two successive floods.

(6) Between- and within-population variations exist among the individual plants. Some individuals succumb to flooding; other plants demonstrate plastic responses in physiological and morphological characteristics. Our results

indicate that most adaptive responses to flooding are environmentally induced; only a few adaptive characteristics are genetically determined.

Acknowledgements

I am very grateful to the members of the research group of Experimental Plant Ecology -- Gerard Bögemann, Willem Engelaar, Peter Laan, Carlo van der Rijt, Jeannette van der Sman, Harry van de Steeg, and Rens Voesenek -- for fruitful discussions on earlier versions of this manuscript. They all worked together on the scientific questions described in this paper, and many of their results are incorporated.

References

Armstrong, W. (1979). Aeration in higher plants. *Adv. Bot. Res.* **7**, 225-332.

Blom, C.W.P.M. (1979). Effects of trampling and soil compaction on the occurrence of some *Plantago* species in coastal sand dunes. Thesis University of Nijmegen.

Blom, C.W.P.M. and Lotz, L.A.P. (1985). Phenotypic plasticity and genetic differentiation of demographic characteristics in some Plantago species. *In* "Structure and Functioning of Plant Populations 2" (J. Haeck and J.W. Woldendorp, eds). North Holland Publ., Amsterdam.

Blom, C.W.P.M. (1985). "Observation and Explanation" a consideration of Plant Ecology. Inaugural speech Catholic University of Nijmegen (in Dutch, summary in English available).

Blom, C.W.P.M. (1987). Experimental plant ecology as an approach in coastal population biology. *In* "Vegetation between Land and Sea". (A.H.L. Huiskes, C.W.P.M. Blom and J. Rozema, eds). Junk Publ., Dordrecht.

Harren, F. (1988). The photoacoustic effect, refined and applied to biological problems. Thesis University of Nijmegen.

Jackson, M.B.(1985). Ethylene and the responses of plants to soil waterlogging and submergence. *Annu. Rev. Plant Physiol.* **36**, 145- 174.

Justin, S.H.F.W. and Armstrong, W. (1987). The anatomical characteristics of roots and plant response to soil flooding. *New Phytol.* **106**, 465-495.

Kozlowski, T.T. (1984). "Flooding and Plant Growth". Acad. Press, London.

Laan, P., Smolders, A., Blom, C.W.P.M. and Armstrong, W. (1989a). The relative roles of internal aeration, radial oxygen losses, iron exclusion and nutrient balances in flood-tolerance of *Rumex* species. *Acta Bot. Neerl.* **38**, 131-145.

Laan, P., Berrevoets, M.J., Lythe, S., Armstrong, W. and Blom, C.W.P.M. (1989b). Root morphology and aerenchyma formation as indicators for the flood-tolerance of *Rumex* species. *J. Ecol.* **77**, 693-703.

Laan, P., Tosserams, M., Blom, C.W.P.M. and Veen, B.W. (1990). Internal oxygen transport in *Rumex* species and its significance for respiration under hypoxic conditions. *Plant & Soil* **122**, 39-46.

Lotz, L.A.P. and Blom, C.W.P.M. (1986). Plasticity in life-history traits of *Plantago major* L. ssp. *pleiosperma* Pilger. *Oecologia* (Berl.) **69**, 25-30.

Lotz, L.A.P. (1989). Variation in life-history characteristics between and within populations of *Plantago major* L. Thesis University of Nijmegen.

Osborne, D.J.(1984). Ethylene and plants of aquatic and semi-aquatic environments: a review. *Plant Growth Regulation* **2**, 167- 185.

Ridge, J. (1985). Ethylene and petiole development in amphibious plants. *In* "Ethylene and Plant Development" (J.A. Roberts and G.A. Tucker), pp. 229-239. Butterworths, London.

Van de Steeg, H.M. (1984). Effects of summer inundation on flora and vegetation of river foreland in the Rhine area. *Acta Bot. Neerl.* **33**, 365-366.

Van der Sman, A.J.M., Van Tongeren, O.F.R. and Blom, C.W.P.M. (1988). Growth and reproduction of *Rumex maritimus* and *Chenopodium rubrum* under different waterlogging regimes. *Acta Bot. Neerl.* **37**, 439-450.

Van den Tweel, P.A. and Schalk, B. (1981). The horizontally perforated soil system: a new observation method. *Plant & Soil* **59**, 163-165.

Voesenek, L.A.C.J. and Blom, C.W.P.M. (1987). Rooting patterns of *Rumex* species under drained conditions. *Can. J. Bot.* **65**, 1638-1642.

Voesenek, L.A.C.J., Blom, C.WP.M. and Pouwels, R.H.W. (1989). Root and shoot development of *Rumex* species under waterlogged conditions. *Can. J. Bot.* **67**, 1865-1869.

Voesenek, L.A.C.J. and Blom, C.W.P.M. (1989a). Ethylene and flooding responses of *Rumex* species. *In* "Biochemical and Physiological Aspects of Ethylene Production in Lower and Higher Plants" (H. Clijsters *et al.*, eds), pp. 245-253. Kluwer Academic Press, Dordrecht.

Voesenek, L.A.C.J. and Blom, C.W.P.M. (1989b). Growth responses of *Rumex* species in relation to submergence and ethylene. *Plant Cell and Environ.* **12**, 433-439.

Part II

MOLECULAR APPROACHES IN PLANT BIOSYSTEMATICS

7 Chloroplast DNA and Nuclear rDNA Variation: Insights into Autopolyploid and Allopolyploid Evolution

DOUGLAS E. SOLTIS AND PAMELA S. SOLTIS

Department of Botany, Washington State University, Pullman, WA 99164, USA

I. Introduction

Restriction fragment analysis of chloroplast DNA (cpDNA) has proven to be of tremendous utility in phylogenetic reconstruction (see reviews by Giannasi and Crawford, 1986; Palmer, 1987; Palmer *et al.*, 1988). The basic structure of the chloroplast genome, as well as the numerous advantages of cpDNA in phylogenetic reconstruction, have been well-reviewed (Curtis and Clegg, 1984; Giannasi and Crawford 1986; Gillham *et al.*, 1985; Palmer 1985a, 1985b, 1987; Palmer *et al.*, 1988). In most angiosperms, maternal inheritance of cpDNA has been reported, although biparental inheritance has been reported in some taxa; conifers, in contrast, exhibit paternal inheritance of cpDNA (Corriveau and Coleman, 1988; Kirk and Tilney-Basset, 1978; Neale *et al.*, 1986; Sears 1980, 1983; Szmidt *et al.*, 1987; Whatley, 1982). Thus, in most studies of cpDNA variation, a maternally based phylogeny is generated. Although cpDNA has been used most extensively in the comparison of congeneric species, its value in phylogenetic analysis has also been demonstrated at other taxonomic levels, such as the subtribal level in Compositae (Jansen and Palmer, 1987, 1988), the generic level (D. Soltis *et al.*, 1990a; Sytsma and Gottlieb, 1986), and even in the comparison of conspecific populations (Neale *et al.*, 1988; Soltis *et al.*, 1989a, b).

Although it has not had the impact in phylogenetic reconstruction realized by cpDNA, restriction fragment analysis of nuclear ribosomal RNA genes (rDNA) has similarly proven to be of phylogenetic importance (see reviews by Doyle *et al.*, 1984; Giannasi and Crawford, 1986; Schaal and Learn, 1988). Nuclear ribosomal

BIOLOGICAL APPROACHES AND
EVOLUTIONARY TRENDS IN PLANTS ISBN 0-12-402960-4

RNA genes occur as several families, and lucid reviews of the structure and evolution of these gene families have been published (Doyle *et al.*, 1984; Flavell, 1986; Gerbi, 1986; Giannasi and Crawford, 1986; Jorgensen and Cluster, 1988; Long and Dawid, 1980; Rodgers and Bendich, 1987; Schaal and Learn, 1988). Unlike the chloroplast genome, which typically is uniparentally inherited, nuclear rDNA is biparentally inherited. The coding sequences of rDNA are conservative in their evolution and have proven valuable in phylogenetic reconstruction, as illustrated by Sytsma and Schaal (1985) on *Lisianthius*. In contrast, the intergenic spacer region is highly variable and has permitted population-level analyses (reviewed in Schaal and Learn, 1988).

In addition to their tremendous value in phylogenetic reconstruction, cpDNA and nuclear rDNA variation also provide valuable genetic markers for the analysis of polyploids. Several landmark papers provided an important impetus for a series of DNA-based analyses of polyploids upon which the present review is based. Early papers demonstrated the potential value of cpDNA analysis in elucidating the origins of allopolyploids (Berthou *et al.*, 1983; Bowman *et al.*, 1983; Erickson *et al.*, 1983; Ogihara and Tsunewaki, 1982; and Palmer *et al.*, 1983). The papers by Erickson *et al.* (1983) and Palmer *et al.* (1983) were particularly important to students of evolution because they independently demonstrated the insights that cpDNA could provide in *Brassica,* an already well-studied allopolyploid complex. Chloroplast DNA data not only revealed the maternal parents of three allotetraploid Brassicas, but also implicated specific populations of diploids that contributed the cytoplasm to each allotetraploid, and in the case of one allotetraploid, *B. napus*, suggested the possibility that introgressive hybridization had occurred (Palmer *et al.*, 1983).

Several early studies also demonstrated the potential value of nuclear rDNA in elucidating the origin and evolution of polyploids. Doyle *et al.* (1985) illustrated that the different repeat lengths for nuclear ribosomal RNA genes of *Tolmiea menziesii* and *Tellima grandiflora* (Saxifragaceae) were combined in their naturally occurring intergeneric hybrid. More recently, Zimmer *et al.* (1988) demonstrated additivity of rDNA profiles in maize hybrids. Palmer *et al.* (1983) reported that the allotetraploid *Brassica napus* combined the rDNA profiles of its diploid parents. Doyle *et al.* (1984) successfully employed restriction fragment analysis of nuclear rDNA in the study of polyploids in *Claytonia virginica.* These studies demonstrated that analysis of nuclear rDNA could potentially provide important markers for ascertaining the parentage of allopolyploids.

In this paper we will illustrate the important information that restriction fragment analysis of cpDNA and nuclear rDNA can provide in the study of both autopolyploidy and allopolyploidy. Specifically, we will emphasize that cpDNA and rDNA are powerful tools for: 1) determining the parentage of allopolyploids; 2) demonstrating multiple origins of both allopolyploids and autopolyploids, and 3) providing novel insights into polyploid evolution. Although cpDNA and nuclear rDNA variation can provide important phylogenetic and evolutionary information for polyploids when used individually, another theme of this review will be to stress the value of employing a uniparentally inherited marker (cpDNA) and a biparentally inherited maker (nuclear rDNA and/or allozymes) in the analysis of polyploids. When used in concert, these genetic markers can provide new information regarding polyploid evolution. To accomplish the three goals noted above, we will concentrate on several examples with which our laboratory has been involved. The taxa we will use to exemplify the role that DNA analysis can play in the study of polyploids are the allopolyploids *Tragopogon mirus* and *T. miscellus*, *Polystichum californicum*, and *Draba lactea*, and the autopolyploids *Heuchera micrantha*, *H. grossulariifolia*, and *Tolmiea menziesii*.

II. Examples

A. *Tragopogon mirus* and *T. miscellus*

Tragopogon mirus and *T. miscellus* (Compositae) are classic examples of recent allopolyploid speciation (Ownbey, 1950). They are important evolutionary models because, unlike most polyploids, the time of their formation is known with a high degree of certainty. These allotetraploids arose very recently, probably within the last 80 years, in the Palouse region of eastern Washington and adjacent Idaho in the Northwestern U.S. Although an Old-World genus, the diploid progenitors of the polyploids, *T. dubius*, *T. porrifolius*, and *T. pratensis*, were introduced into North America during the early years of this century and are now widely naturalized. Whereas *T. porrifolius* and *T. pratensis* are confined to waste places and lawns in towns and are relatively rare, *T. dubius* occurs commonly in waste places, fields, and along roadsides.

The parents of *Tragopogon mirus* are *T. dubius* and *T. porrifolius*; those of *T. miscellus* are *T. dubius* and *T. pratensis* (Ownbey, 1950). Since their recent formation, the two allotetraploids have become conspicuous in some towns of the Palouse region.The tetraploids typically are more abundant than either of the rare diploids, *T. porrifolius* and *T. pratensis*, and in some habitats the tetraploids are even more common than the widespread diploid, *T. dubius*.

Subsequent cytological, flavonoid, genetic, and morphological studies (Ownbey and McCollum, 1953, 1954; Brehm and Ownbey, 1965) led to the hypothesis that *T. mirus* arose independently at least three times and that *T. miscellus* arose independently at least two times just within the Palouse. In their classic electrophoretic investigation of allopolyploidy, Roose and Gottlieb (1976) obtained electrophoretic data that supported the hypothesis that *T. mirus* has had at least three separate origins within the Palouse. However, their electrophoretic data were equivocal in this regard. Their results could also be explained via a single polyploid event involving an unreduced gamete from *T. dubius* that was heterozygous at the critical electrophoretic loci; subsequent segregation and seed dispersal could yield the genotypic distributions observed. Electrophoretic data were inconclusive with regard to the possibility of multiple origins of *T. miscellus*. *Tragopogon mirus* and *T. miscellus* have also been reported from Arizona (Brown and Schaak, 1972), suggesting additional separate origins of the two tetraploids, but on a much larger geographic scale.

Using cpDNA data in conjunction with nuclear rDNA data, we obtained clear evidence for multiple origins of both allotetraploids, just within the Palouse region, and provided novel insights into allopolyploid evolution.The cpDNA data are discussed in greater detail in Soltis and Soltis (1989a). In our analysis of cpDNA variation, we identified six restriction-site mutations and three length mutations.The chloroplast genomes of the the three parental diploids were readily distinguished by these mutations. All populations of the tetraploid *T. mirus* have the chloroplast genome of *T. porrifolius*, indicating that *T. porrifolius* is the maternal parent of *T. mirus*. Because of this and also the fact that no cpDNA polymorphisms were found within *T. porrifolius*, cpDNA data were inconclusive with regard to the possibility of multiple origins of *T. mirus*.

Analysis of the 18S-25S nuclear rRNA genes did, however, demonstrate the presence of polymorphisms within *T. porrifolius*, with different populations characterized by different 18S-25S restriction profiles (Fig. 1). Different populations of *T. mirus* each exhibit one of the several different *T. porrifolius* profiles, as well as the 18S-25S repeat of the other diploid parent, *T. dubius*. Therefore, nuclear rDNA data not only confirm the parentage of allotetraploid *T. mirus*, but also provide convincing evidence for separate origins of *T. mirus* from the Palouse. Evidence for two separate origins is provided in Fig. 1.

Nuclear rDNA data for *T. mirus* also provided additional evolutionary information. In several locations, plants of *T. mirus* were growing with plants of *T. dubius* and *T. porrifolius*, suggesting that these represented possible sites of *in situ* formation of the allotetraploid. This hypothesis seemed reasonable given the very recent origin of *T. mirus*. One of these localities in fact appears to be the site, or at least close to the site, where Ownbey first detected *T. mirus*. However, in both instances in which we observed *T. mirus* growing with its diploid progenitors, rDNA data indicated that the tetraploid plants could not have been formed from the diploids present at that locality. Co-occurence of an allotetraploid (even one of recent origin) with its diploid progenitor species does not, therefore, imply that the tetraploid was formed *in situ*.

This analysis of *T. mirus* also demonstrates the value of employing cpDNA and nuclear rDNA in concert when analyzing allopolyploids. Nuclear rDNA data clearly demonstrate multiple origins of *T. mirus* in a small geographic area. Chloroplast DNA data indicate that the rare diploid parent, *T. porrifolius*, has consistently been the maternal parent of the tetraploid, whereas the much more common *T. dubius* has consistently been the paternal parent. This scenario is further strengthened by our recent analysis of a natural hybrid between *T. dubius* and *T. porrifolius*. The hybrid plant also has *T. porrifolius* as the maternal parent.

Our analysis of the tetraploid *T. miscellus* provides an interesting contrast to *T. mirus*, again demonstrating the importance of using both cpDNA and nuclear rDNA in the analysis of allopolyploids. Analysis of 18S-25S rDNA confirmed that the parents of *T. miscellus* are *T. dubius* and *T. pratensis*. However, because we observed no polymorphism for 18S-25S rDNA in either diploid parent, the question of multiple origins of *T. miscellus* remained unanswered. However, cpDNA provided unambiguous evidence for multiple origins of *T. miscellus* (Soltis and Soltis, 1989a). Two populations of *T. miscellus*, both from Pullman, Washington, exhibit the chloroplast genome of *T. dubius*, whereas all remaining populations of *T. miscellus* have the chloroplast genome of *T. pratensis*. Thus, cpDNA data clearly indicate that there have been at least two separate origins of *T. miscellus*, one with *T. dubius* as the female parent and one with *T. pratensis* as the female parent.

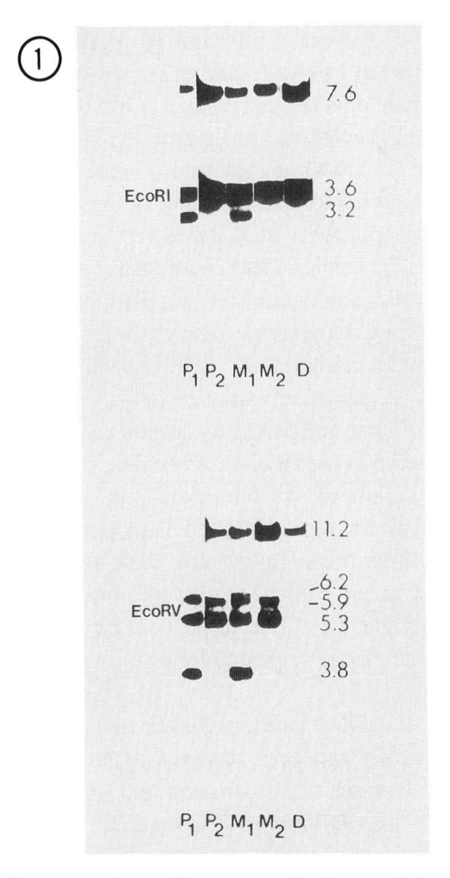

FIG. 1 18S-25S ribosomal gene variation in *Tragopogon*. Numbers to right of photograph indicate sizes of fragments in kb. D = *T. dubius*; P_1 and P_2 = different populations of *T. porrifolius* having different rDNA profiles; M_1 and M_2 = different populations of *T. mirus* possessing the two different *T. porrifolius* rDNA profiles.

Based on a cytoplasmic factor governing ligule length, Ownbey and McCollum (1953) hypothesized that *T. miscellus* probably had at least two independent origins: once with *T. dubius* as the maternal parent (resulting in an allotetraploid with long ligules) and once with *T. pratensis* as the maternal parent (resulting in an allotetraploid with short ligules). Ownbey and McCollum based their hypothesis on the fact that long- and short-liguled populations of *T. miscellus* are present in nature and that in crosses between *T. pratensis* and *T. dubius*, with *T. pratensis* as the maternal parent, all F_1 progeny were short-liguled. However, Ownbey and McCollum (1953) were unable to generate F_1 progeny from the reciprocal cross. cpDNA data corroborate the Ownbey and McCollum (1953) hypothesis; populations of *T. miscellus* with short ligules all have *T. pratensis* as the maternal parent, whereas the long-liguled Pullman populations have *T. dubius* as the maternal parent.

Consideration of several cpDNA restriction-site polymorphisms within *T. dubius* provided additional evolutionary information regarding the origin of the two populations of *T. miscellus* having *T. dubius* as the maternal parent (see Soltis and Soltis, 1989a, for a more detailed discussion). Of the 11 populations of *T. dubius* that were examined, five possessed two restriction-site mutations not observed in the other populations. The two populations of *T. miscellus* having *T. dubius* as the maternal parent exhibit the chloroplast genome of these same five populations of *T. dubius*. Thus certain populations of *T. dubius* are implicated in the formation of these two populations of *T. miscellus*. These results are similar to the findings of Palmer *et al.* (1983) who pinpointed a few populations of diploid *Brassica* species as the contributors of the chloroplast genome to allotetraploids in the genus.

Chloroplast DNA in conjunction with nuclear rDNA also provide novel evolutionary insights into polyploid evolution in *Tragopogon*. It is now clear that *T. miscellus* and *T. mirus* arose independently several times within the relatively small geographic region known as the Palouse. A scenario of island biogeography can be envisioned for the origin of the tetraploids. In several of the small towns characteristic of the Palouse region, diploid Tragopogons were brought into close proximity. Surrounding these towns are vast agricultural tracts in which Tragopogons do not occur. Only *T. dubius* is found on rare occasions along roadsides outside of towns. DNA data argue convincingly that on several separate occasions in towns sometimes separated by as little as 10 km, the allotetraploids formed.

It is also apparent from DNA data that the two rare diploid species, *T. porrifolius* and *T. pratensis*, typically appear as maternal parents of the two allotetraploids. Although *T. dubius* is much more common and widespread, it is the maternal parent for only two populations of *T. miscellus*. The rare *T. porrifolius* is consistently the maternal parent of *T. mirus* and also of a naturally occurring F_1 hybrid with *T. dubius*. Although successful crosses were made between *T. dubius* and *T. pratensis* in both directions (Ownbey and McCollum, 1953), in nature the more common *T. dubius* appears to be favored as a paternal parent. These findings suggest that pollen load, carried by insect pollinators, may be an important factor in determining the male and female parents of allopolyploid, as well as hybrid, angiosperms. The more abundant species may be more likely to serve as a paternal parent and the rarer species as a maternal parent. This hypothesis should be tested in other angiosperms using cpDNA and a biparentally inherited marker such as nuclear rDNA and/or allozymes.

B. *Polystichum californicum*

The tetraploid fern *Polystichum californicum* (Dryopteridaceae) is part of a species complex from western North American that comprises diploids, tetraploids, and a hexaploid (D. Wagner, 1979). The parentage of several polyploids, including *P. californicum*, is still in dispute. To ascertain the parentage of polyploids within this complex, allozymes, cpDNA, and nuclear rDNA have been employed (P. Soltis *et al.*, 1990). Herein, we discuss only *P. californicum* because it is most relevant to the objectives described in the introduction.

Two hypotheses have been suggested for the origin of the tetraploid *P. californicum*. W. Wagner (1973) proposed that the diploids *P. munitum* and *P. dudleyi* were parents of *P. californicum*. In contrast, D. Wagner (1979) suggested that *P. californicum* is polyphyletic. He suggested that northern populations of *P. californicum* had *P. dudleyi* and *P. imbricans* as parental diploids, whereas southern populations of the tetraploid had *P. dudleyi* and *P. munitum* as parents.

Allozyme data clearly demonstrate the role of *P. dudleyi* in the formation of the allotetraploid *P. californicum* (P. Soltis *et al.*, 1990). However, allozyme data could not unambiguously discriminate between the two competing hypotheses due to a high degree of allozymic similarity between two of the diploids, *P. munitum* and *P. imbricans*. Nuclear rDNA data also were not of value because both *P. munitum* and *P. imbricans* possess identical restriction profiles for 18S-25S rDNA. However, despite a high degree of allozymic similarity, *P. munitum* and *P. imbricans* are easily distinguished by restriction fragment analysis of cpDNA (Fig. 2). All populations of *P. californicum* examined, representing both northern and southern populations of the tetraploid, have the chloroplast genome of *P. imbricans*. The parentage of *P. californicum* is therefore *P. dudleyi* x *P. imbricans*. There is no evidence that *P. munitum* has been involved in the formation of some or all populations of *P. californicum* as D. Wagner and W. Wagner suggested, respectively.

The analysis of *P. californicum* therefore presents an interesting contrast to our results for allotetraploid Tragopogons, which indicated multiple origins. Although the tetraploid *P. californicum* was thought by some to be polyphyletic in origin (*P. dudleyi* x *P. munitum* and *P. dudleyi* x *P. imbricans*), cpDNA data in conjunction with allozyme data suggest only the involvement of *P. dudleyi* and *P. imbricans*. It is possible that multiple origins involving only these two diploids have occurred, but no convincing evidence for this scenario has been obtained.

C. *Draba lactea*

Draba lactea (Cruciferae) is a circumpolar hexaploid that is morphologically similar to the diploid *D. fladnizensis*. The parentage of this hexaploid is not known with certainty, but *Draba fladnizensis* is considered a likely diploid parent. Populations of *D. lactea*, as well as the diploid *D. fladnizensis*, occur disjunctly in southern Norway, northern Norway, and Spitsbergen.

Data from enzyme electrophoresis (Brochmann *et al.*, 1989b, in prep.) revealed fixed heterozygous enzyme phenotypes at several loci. Thus, isozyme data demonstrated that *D. lactea* is a genetic allopolyploid and also indicated the

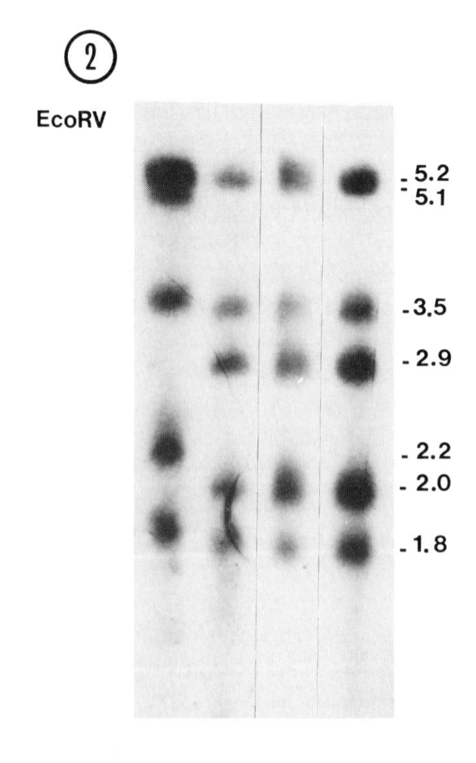

FIG. 2 Chloroplast DNA variation in *Polystichum*. Total DNAs were digested with the restriction enzyme indicated and transferred to a nylon filter. Chloroplast DNA probes from petunia representing a portion of the large single copy region were used. M = *P. munitum*; I = *P. imbricans*; C_N = *P. californicum* (northern population); C_S = *P. californicum* (southern population). Both northern and southern populations of the tetraploid *P. californicum* have the chloroplast genome of *P. imbricans*.

involvement of diploid *D. fladnizensis*, but provided no definitive evidence regarding the tetraploid parent. The high degree of allozymic variation among the hexaploid populations examined gave, however, clear evidence of multiple origins.

In contrast, restriction fragment analysis of nuclear rDNA not only indicated that *D. fladnizensis* is the likely diploid parent, but also demonstrated that *D. lactea* has had at least two separate origins. For the endonucleases *Eco*RI and *Eco*RV, populations of *D. lactea* and *D. fladnizensis* from southern Norway share unique restriction profiles for the 18S-25S gene family that differentiate them from populations of these same two species from northern Norway and Spitsbergen (Fig. 3). Thus, rDNA data indicate separate origins of the hexaploid in the two geographic regions.

The nuclear rDNA data are particularly intriguing when compared with the crossing data of Brochmann *et al.* (1989a). These investigators observed that artificial crosses between south Norwegian and Spitsbergen populations of the diploid *D. fladnizensis* produced completely sterile F_1 hybrids. Brochmann *et al.* (1989a) hypothesize that this sterility may be due to chromosomal rearrangements that differentiate the diploid populations from these two geographic areas. This suggests that there are incompatible races or "sibling species" within the single taxonomic species *D. fladnizensis*. The rDNA data also suggest that populations from these two areas are genetically differentiated. *Draba lactea* is morphologically very similar to *D. fladnizensis* and it is possible therefore that the hexaploid originated several times from different combinations of incompatible "sibling species" that are all part of one diploid taxonomic species. Recent work indicates that multiple origins of allopolyploids in *Draba* may be the rule rather than the exception. Brochmann *et al.* (in prep.) have genetic evidence for multiple origins in several polyploid *Draba* species, including *D. cacuminum*, *D. norvegica*, and *D. corymbosa*.

Nuclear rDNA data do not, however, completely resolve the parentage of hexaploid *D. lactea*. For example, most of the related species of *Draba* that we

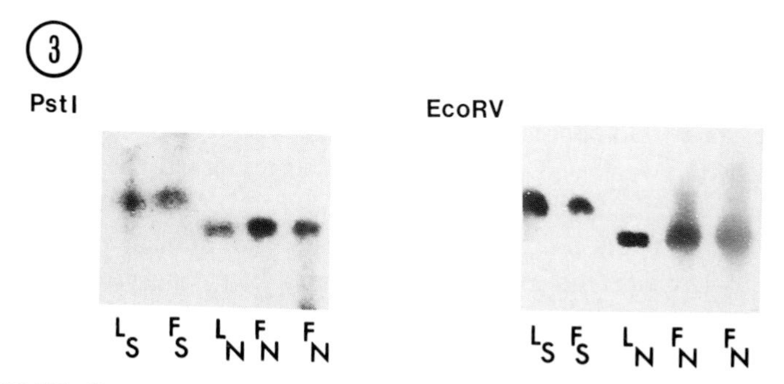

FIG. 3 18S-25S ribosomal RNA gene repeat length variation in *Draba*. L_S = *D. lactea* (southern population); L_N = *D. lactea* (northern population); F_S = *D. fladnizensis* (southern population); F_N = *D. fladnizensis* (northern population).

examined have the same rDNA profile observed in both *D. fladnizensis* and *D. lactea* from northern Norway and Spitsbergen. If another "race" of *D. fladnizensis* or even another species of *Draba* were involved in the formation of southern Norwegian populations of *D. lactea*, why then are two different rDNA profiles not combined in at least some hexaploid populations? Although the parentage of *D. lactea* was not completely elucidated, this example nonetheless illustrates how a relatively small amount of rDNA data can provide intriguing evolutionary information. These rDNA data, coupled with allozymic and crossability data, indicate a more complex evolutionary history for *Draba lactea* than previously envisioned.

D. *Heuchera micrantha*

Heuchera micrantha (Saxifragaceae) is a morphologically variable species from western North America. Five varieties have been recognized: *diversifolia*, *erubescens*, *hartwegii*, *micrantha*, and *pacifica*. Ness *et al.* (1989) documented the presence of both diploid ($2n = 14$) and tetraploid ($2n = 28$) populations within *diversifolia*, *hartwegii*, and *pacifica*; all populations examined of *erubescens* and *micrantha* are diploid and tetraploid, respectively. The tetraploids appear to be good examples of autotetraploidy because: 1) enzyme electrophoresis revealed a very high degree of genetic similarity between diploids and tetraploids (Ness *et al.*, 1989), 2) electrophoretic analyses of progeny arrays demonstrated tetrasomic segregation in tetraploid plants (Soltis and Soltis, 1989b), 3) the two cytotypes are morphologically indistinguishable, although tetraploids are larger than diploids when grown under similar conditions in the greenhouse (Ness *et al.*, 1989).

Restriction site analysis of cpDNA provided additional evolutionary insights into autopolyploidy in *H. micrantha* (Soltis *et al.*, 1989a). Levels of cpDNA variation were detected within *H. micrantha* that are much higher than those reported in earlier investigations of other plant species (e.g., Banks and Birky, 1985; Rieseberg *et al.*, 1988). The high levels of cpDNA variation detected within *H. micrantha* presented the unusual opportunity to study phylogenetic relationships among populations. A single most parsimonious tree was generated by PAUP (Fig. 4) which revealed that diploid and tetraploid populations appear together on several of the branches. Chloroplast DNA data actually suggest three separate origins of autoploids within variety *diversifolia* alone. This result was unexpected because diploid and tetraploid populations of var. *diversifolia* are morphologically indistinguishable (Ness *et al.*, 1989). In fact, statistical evaluation of morphology in *Heuchera micrantha* did not differentiate among these populations and could not provide strong evidence for multiple origins of the polyploids. Significantly, allozyme data also failed to provide evidence for multiple origins of autotetraploids in *H. micrantha* (Ness *et al.*, 1989).

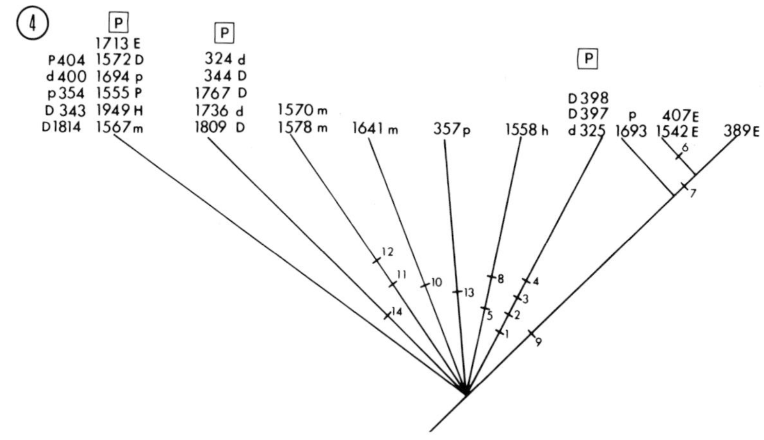

FIG. 4 Chloroplast DNA phylogeny for *Heuchera micrantha* based on 14 restriction site mutations (modified from Soltis *et al*., 1989a). Numbers beside slash marks designate restriction site mutations. Remaining numbers (324-1949) are collection numbers. Upper- and lowercase letters designate diploid and tetraploid populations, respectively (D/d = var. *diversifolia*; E = var. *erubescens*; H/h = var. *hartwegii*; m = var. *micrantha*; P/p = var. *pacifica*). Each P indicates a separate origin of autotetraploid populations.

E. *Heuchera grossulariifolia*

Results of a cpDNA investigation of *H. grossulariifolia* closely parallel those obtained for *H. micrantha* (Wolf *et al*., 1990). Like *H. micrantha*, *H. grossulariifolia* is native to the Pacific Northwest of North America. It too comprises diploid and tetraploid cytotypes ($2n$ = 14 and 28, respectively), and cytological, electrophoretic and inheritance data similarly indicate that the tetraploid is of autopolyploid origin (Wolf *et al*., 1989, 1990). Tetraploid populations are largely restricted to the dry confines of the Salmon River canyon, whereas diploids occupy more mesic sites and are more widespread. Diploids are found throughout central Idaho and adjacent western Montana. Disjunct diploid populations also occur in the Columbia River Gorge of Oregon and Washington.

A high degree of intraspecific cpDNA variation permitted the phylogenetic analysis of populations of *H. grossulariifolia*. A single most parsimonious tree was generated by PAUP (Fig. 5). As was observed in *H. micrantha*, diploid and tetraploid populations occur together on different branches of the tree. Just as in *H. micrantha*, this analysis indicated that there had been at least three separate origins of autotetraploids. This result could not have been predicted based on morphology because the tetraploid populations analyzed are so homogeneous. Allozyme data did, however, suggest the possibility of multiple origins because of the large number of alleles shared by the diploid and tetraploid cytotypes.

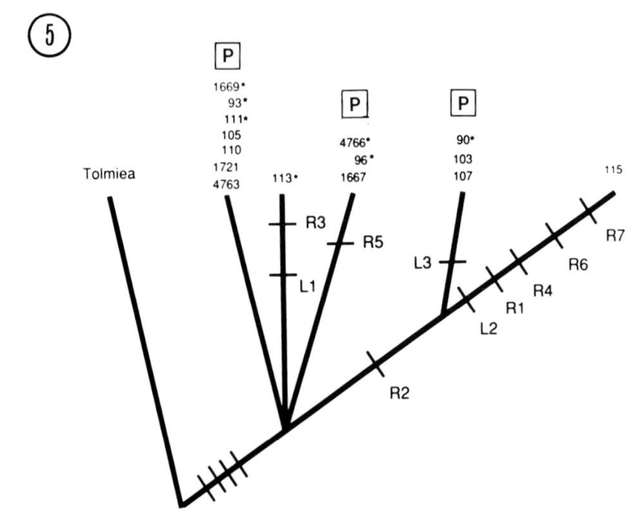

FIG. 5 Chloroplast DNA phylogeny for *Heuchera grossulariifolia* (modified from Wolf *et al.*, 1990). R1-R7 are restriction site mutations; L1-L3 are length mutations. *Tolmiea* is an outgroup; more than four mutations separate *H. grossulariifolia* from *Tolmiea*. The numbers 90-4766 are collection numbers of populations. Collection numbers followed by an asterisk are tetraploid populations, others are diploid populations. Each P indicates a separate origin of autotetraploid populations.

The fact that autotetraploids in both *H. micrantha* and *H. grossulariifolia* have had multiple origins is of considerable evolutionary significance because autopolyploidy has traditionally been considered to be rare. Although multiple origins of alloploids had been demonstrated previously, such was not the case for autoploids. Indeed, these findings for *H. micrantha* and *H. grossulariifolia* indicate that traditional views of autopolyploidy should be reevaluated. Successful autopolyploid events may be much more frequent, at least in some groups such as Saxifragaceae, than previous workers have maintained (see reviews by Stebbins, 1950; Lewis, 1980; Levin, 1983; Soltis and Rieseberg, 1986).

F. *Tolmiea menziesii*

Tolmiea menziesii (Saxifragaceae) aptly demonstrates some of the profound novel evolutionary information that cpDNA analysis can provide into polyploid evolution (Soltis *et al.*, 1989b). *Tolmiea* is monotypic; the single recognized species, *T. menziesii*, occurs west of the Cascades of western North America from California to Alaska. *Tolmiea menziesii* comprises two cytotypes (Soltis 1984); diploids ($2n = 14$) occupy the southern portion of the range, whereas tetraploids ($2n = 28$) occur in the northern portion of the range. Evidence from numerous sources, including morphology, cytology, enzyme electrophoresis, and flavonoid chemistry (reviewed in Soltis *et al.*, 1989b; Soltis and Soltis, 1989c) indicate that the tetraploid is of autopolyploid origin. Significantly, the tetraploid exhibits tetrasomic inheritance

(Soltis and Soltis, 1988). Restriction fragment analysis of the 5S and 18S-25S ribosomal RNA genes also suggests autopolyploidy (Soltis and Doyle, 1987).

Although diploid and tetraploid *Tolmiea* had been studied intensively using nuclear markers, cpDNA variation provided additional evolutionary information not revealed in earlier analyses (Soltis *et al.*, 1989b). All diploid populations are separated from all tetraploid populations by a minimum of three length mutations and three restriction site mutations; the mean sequence divergence between the cytotypes is high (0.08%). Thus, the chloroplast genomes of diploid and tetraploid *Tolmiea* are as distinct as those of many pairs of congeneric species of angiosperms.

Phylogenetic analysis of the cpDNA data surprisingly indicated that the primitive chloroplast genome is present in tetraploid rather than diploid *Tolmiea* (Fig. 6). This result was unexpected because one would predict that a diploid and its autotetraploid derivative would be identical, or nearly so, for numerous characteristics, provided the autotetraploid were of relatively recent origin. This is, in fact, the case for morphological and cytological characters, as well as for flavonoids, anthocyanins, allozymes, and restriction fragment patterns of nuclear ribosomal RNA genes. The chloroplast DNA findings therefore suggest that either: 1) diploid and tetraploid *Tolmiea* have diverged since the origin of the autotetraploid, 2) the original diploid donor of the cytoplasm present in the tetraploid subsequently became extinct, 3) the diploid was actually derived from the tetraploid via polyhaploidy, or 4) some diploid populations do have a chloroplast genome identical to that found in tetraploids, but by chance were not sampled. These possibilities are critically evaluated in Soltis *et al.* (1989); they conclude that the most likely scenarios are 1 and 2 above.

The cpDNA data for *Tolmiea* also have important implications regarding gene flow. Although diploid and tetraploid populations of *Tolmiea menziesii* occur within a few meters of each other in central Oregon, no diploid or tetraploid population examined exhibited a chloroplast genome characteristic of the other cytotype. Hence, despite close geographic proximity, diploid and tetraploid *Tolmiea* do not experience cytoplasmic gene flow. In contrast, cpDNA investigations have provided evidence suggestive of hybridization or introgression between species in other groups (e.g., Palmer *et al.*, 1983, 1985). Thus, the cytotypes of *Tolmiea* may be well-isolated reproductively in nature. The high degree of chloroplast DNA divergence between diploid and tetraploid *Tolmiea* and absence of cytoplasmic gene flow suggest, in fact, that diploid and tetraploid *Tolmiea* may best be considered distinct species. It is also important to note, however, that the chloroplast genome is maternally inherited in Saxifragaceae (D. Soltis *et al.*, 1990b), so pollen movement will not transfer the chloroplast genome in *Tolmiea*. Because seed dispersal is probably very limited in *Tolmiea* (Savile, 1975), cytoplasmic gene flow would be more likely if the chloroplast genome were paternally or biparentally inherited. It

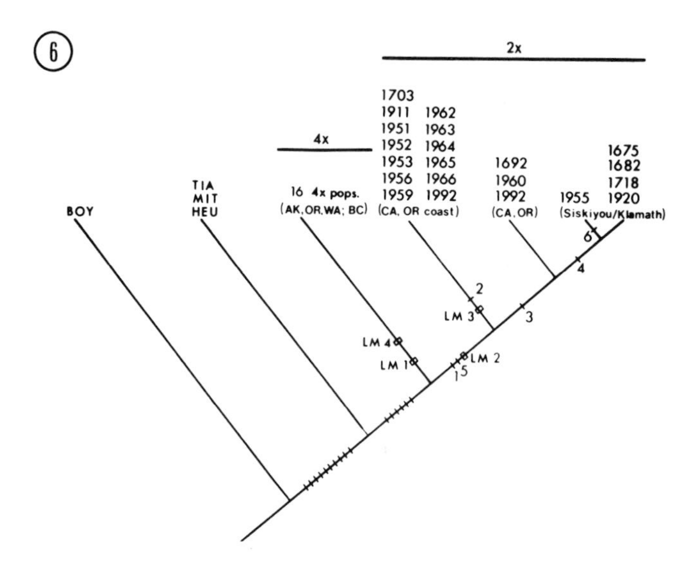

FIG. 6 Chloroplast DNA phylogeny for *Tolmiea menziesii* (from Soltis *et al.*, 1989). Numbers 1-7 refer to restriction site mutations; LM 1, 2, 3, 4, and 5 refer to length mutations. Slash marks designate restriction site mutations, open rectangles designate length mutations. The remaining numbers (1675-1992) represent the collection numbers of populations. BOY, TIA, MIT, and HEU are abbreviations for outgroups (see Soltis *et al.*, 1989b for more details).

also is possible that some nuclear gene flow between diploid and tetraploid *Tolmiea* could occur. Electrophoretic and crossability data suggest, however, that the possibility of significant nuclear gene flow between the cytotypes is slight (see discussion in Soltis *et al.*, 1989b).

The cpDNA data for *Tolmiea* also provide an interesting contrast to the results reviewed above for autotetraploids in *Heuchera micrantha* and *H. grossulariifolia*. In the two *Heuchera* examples, cpDNA data provided unequivocal evidence for multiple origins of the autotetraploids. Allozyme data also indicated the possibility of multiple origins of *H. grossulariifolia* (Wolf *et al.*, 1990), but were inconclusive for *H. micrantha* (Ness *et al.*, 1989). In *Tolmiea*, however, allozyme data suggested multiple origins of the tetraploid (Soltis and Soltis, 1989c), whereas cpDNA data were uninformative in this regard. Again, the lesson of these examples is that it is important to examine polyploids exhaustively, using both uniparentally inherited and biparentally inherited markers. This dual approach is most likely to yield significant evolutionary insights into polyploidy.

III. Conclusions

Restriction fragment analysis of cpDNA and nuclear rDNA is an invaluable tool for analyzing polyploids and investigating polyploid evolution. Because the chloroplast genome is maternally inherited and nuclear ribosomal RNA genes are biparentally inherited, they can provide complementary data for investigating the parentage of polyploids and documenting multiple origins of polyploids. DNA data can also provide novel evolutionary insights into polyploid evolution, even in polyploids that have been thoroughly studied with other methodologies. This is exemplified by the cpDNA analyses of allopolyploids in both *Brassica* (Palmer *et al.*, 1983) and *Tragopogon* (Soltis and Soltis, 1989a), as well as autopolyploids in *Heuchera* and *Tolmiea* (Soltis *et al.*, 1989a, b). Furthermore, when cpDNA and nuclear rDNA are employed in concert in the analysis of polyploids, they can provide evolutionary information that could not be obtained through the analysis of cpDNA or nuclear rDNA alone. This last point is aptly demonstrated by data presented herein for *Tragopogon.*

Use of these relatively new molecular methodologies should not, however, preclude further analyses of polyploid complexes with enzyme electrophoresis. Allozymes provide additional nuclear markers and should continue to be employed routinely in the analysis of polyploids. Allozyme data may, for example, provide evidence for the parentage of an allopolyploid when cpDNA and nuclear rDNA are inconclusive because the putative progenitors exhibit identical restriction profiles. Furthermore, in some instances, allozymes can provide evidence for multiple origins of polyploids when cpDNA and/or rDNA are uninformative, such as in *Tolmiea menziesii.* Allozymes are also invaluable markers in analyzing the nature of polyploidy (autopolyploidy vs. allopolyploidy). Large progeny arrays can be scored easily to determine whether a polyploid has polysomic or disomic inheritance (critical characteristics of autopolyploids and allopolyploids, respectively).

The important role that rDNA and particularly cpDNA can play in identifying the parents of allopolyploids has now been well demonstrated (Berthou *et al.*, 1983; Bowman *et al.*, 1983; Brochmann *et al.*, 1986b, in prep.; Doyle *et al.*, 1990; Erickson *et al.*, 1983; Hosaka, 1986; Hilu, 1988; Ogihara and Tsunewaki, 1982; Palmer *et al.*, 1983; Ranker *et al.*, 1989; Soltis and Soltis, 1989a, unpubl.; P. Soltis *et al.*, 1990). In several of these examples (e.g., *Brassica* and *Tragopogon*), previous studies had unequivocally documented the diploid parents of the allopolyploids, but cpDNA data provided additional information by identifying the maternal parents of these allopolyploids and pinpointing certain populations as the contributors of the cytoplasm to allotetraploids. Molecular genetic data have also elucidated the parentage of problematic allopolyploids. For example, competing hypotheses had been proposed based on morphology for the origin of the allotetraploid fern *Polystichum californicum* (D. Wagner, 1979; W. Wagner, 1973). Chloroplast DNA data in conjunction with allozyme data demonstrated that neither of these previous hypotheses was entirely accurate and also correctly identified the diploid parents. The recent work of Wendel (1989) provides a particularly exciting example of the role cpDNA can play in elucidating the parentage of allopolyploids. He has demonstrated that New World tetraploid cottons (*Gossypium*) actually possess the cytoplasm of Old World diploids. Chloroplast DNA analyses have also helped to elucidate the origins of allopolyploid crops, such as finger millet (*Eleusine corocana* subsp. *coracana*; Hilu, 1988), *Coffea arabica* (Berthou *et al.*, 1983), polyploid species of *Aegilops* and *Triticum* (including wheat; Bowman *et al.*, 1983; Ogihara and Tsunewaki, 1982), and *Brassica* (Erickson *et al.*, 1983; Palmer *et al.*, 1983).

Few previous studies have employed cpDNA or rDNA in the analysis of autopolyploids. Using cpDNA, Doebley *et al.* (1987) found that *Zea perennis* and *Z. diploperennis* are distinguished by only a single restriction site mutation, in agreement with the hypothesis that the former is a relatively recent autopolyploid derivative of the latter. Similarly, nuclear rDNA data supported the hypothesis of autopolyploidy for *Tolmiea menziesii* (Soltis and Doyle, 1987). The cpDNA data reviewed herein for three autopolyploids in Saxifragaceae not only further illustrate the role that cpDNA analyses can play in evaluating hypothesized examples of autopolyploidy, but also demonstrate the profound new insights that this approach can provide into autoploid evolution.

Analyses of cpDNA and nuclear rDNA variation continue to reveal multiple origins of allopolyploids, as well as autopolyploids. Documentation of multiple origins of autopolyploids in *Heuchera micrantha* and *H. grossulariifolia* (Soltis *et al.*, 1989a; Wolf *et al.*, 1990) is particularly significant because successful autopolyploid events have traditionally been considered to be extremely rare. A rapidly growing molecular data base suggests that multiple origins of allopolyploids may be the rule, rather than the exception. Multiple origins of allopolyploids have now been documented in a large number of taxa, including *Tragopogon mirus* and *T. miscellus* (Soltis and Soltis, 1989a, unpubl.; Roose and Gottlieb, 1976), *Draba lactea*, as well as several other Scandinavian *Draba* allopolyploids (Brochmann *et al.*, 1989b), allopolyploids in the *Glycine tabacina* complex (Doyle *et al.*, 1990), several allopolyploid ferns, including allopolyploid species of *Asplenium* (Werth *et al.*, 1985), *Hemionitis palmata* (Ranker *et al.*, 1989), and *Pteris cretica* (Suzuki and Iwatsuki, 1990), and an allotetraploid moss, *Plagiomnium medium* (Wyatt *et al.*, 1988). In several of these studies, multiple origins of an allopolyploid have been demonstrated on a small geographic scale. In *Tragopogon mirus* and *T. miscellus*, for example, independent origins of tetraploids have occurred in several different towns separated by only 10 to 100 km. Brochmann *et al.* (1989b) have similarly demonstrated multiple origins of *Draba* allopolyploids on a local geographic scale in Scandinavia.

Either cpDNA or nuclear rDNA can potentially reveal multiple origins of tetraploids. For example, cpDNA data demonstrated multiple origins of the tetraploid *Tragopogon miscellus*, but rDNA was uninformative in this regard. In contrast, nuclear rDNA data documented multiple origins of allotetraploid *Tragopogon mirus*. These examples again illustrate the value of employing both markers in analyzing polyploids. Chloroplast DNA and nuclear rDNA also can provide novel evolutionary insights, particularly when used in concert. For example, because nuclear rDNA data and cpDNA data were not concordant, Palmer *et al.* (1983) concluded that introgressive hybridization may have occurred in some populations of the allotetraploid *Brassica napus*. Chloroplast DNA and rDNA also provided novel evolutionary insights into the allopolyploids *Tragopogon mirus* and *T. miscellus*. Chloroplast DNA data indicated that the rare diploid parent typically has been the maternal parent of both of these allotetraploids whereas the common diploid parent has been the paternal parent. These data, when taken together, suggest that pollen load transported by insects may be an important factor in determining the parentage of allopolyploid, as well as hybrid, angiosperms.

The methodologies now available offer the opportunity to elucidate parentage and relationships in even the most morphologically problematic polyploid complexes. Through the combined application of biparentally inherited markers, such as allozymes and rDNA, and uniparentally inherited markers, such as cpDNA, the identification of the parents of alloploids should, in most cases, become fairly routine. Distinguishing between examples of auto- and allopolyploidy also should be less problematic. More significantly, these methods offer the student of polyploidy the opportunity to address questions that could not be adequately approached in the past. By using cpDNA, rDNA, and even allozymes in concert, we can, for example, ask: 1) are multiple origins typical of polyploids and do they occur frequently on a local geographic scale?; 2) how many times do multiple origins actually occur within single autopolyploids and allopolyploids?; 3) are some allopolyploids actually polyphyletic, involving different progenitors in different geographic areas?; 4) in insect-pollinated taxa, do rare diploid species consistently act as the maternal parent and common diploid species as the paternal parent in the formation of allopolyploid and hybrid angiosperms?; 5) how frequently do hybridization and introgression occur between different cytotypes within an autoploid complex, or between alloploids and their diploid progenitors?; 6) in cases of hybridization and introgression, how extensive is cytoplasmic gene flow vs. nuclear gene flow? Utilization of cpDNA and rDNA should initiate a new era of investigation of polyploid evolution.

Acknowledgements

We thank Paul Wolf and Jeff Doyle for reading an early draft of the manuscript. Portions of this research were supported by National Science Foundation Grants No. BSR-8717471 and BSR-8620444.

References

Banks, J. A.and Birky,W., Jr. (1985). Chloroplast DNA diversity is low in a wild plant, *Lupinus texensis*. *Proc. Natl. Acad. Sci. USA* **82**, 6950-6954

Berthou, F., Mathieu, C. and Vedel, F. (1983). Chloroplast and mitochondrial DNA variation as indicator of phylogenetic relationships in the genus *Coffea* L. *Theor. Appl. Genet.* **65**, 77-84.

Brehm, B. and Ownbey, M. (1965). Variation in chromatographic patterns in the *Tragopogon dubius-pratensis-porrifolius* complex (Compositae). *Am. J. Bot.* **52**, 811-818.

Bowman, C. M., Bonnard, G. and Dyer, T. A. (1983). Chloroplast DNA variation between species of *Triticum* and *Aegilops*. Location of the variation on the chloroplast genome and its relevance to the inheritance and classification of the cytoplasm. *Theor. Appl. Genet.* **65**, 247-262.

Brochmann, C., Borgen, L. and Stedje, B. (1989a). Chromosome numbers and crossing experiments in Nordic populations of *Draba* (Brassicaceae). Biological Approaches and Evolutionary Trends in Plants. 4th International Symposium of Plant Biosystematics, pp. 39 (Abstract).

Brochmann, C., Soltis, P. S. and Soltis, D. E. (1989b). Evolutionary trends in Nordic populations of *Draba* (Brassicaceae). Biological Approaches and Evolutionary Trends in Plants. 4th International Symposium of Plant Biosystematics, pp. 39 (Abstract).

Brochmann, C., Soltis, P. S. and Soltis, D. E. (In prep.). Multiple origins of allopolyploids in *Draba*.

Brown, R. K., and Schaak, C. G. (1972). Two new species of *Tragopogon* for Arizona. *Madroño* **21**, 304.

Corriveau, J. L. and Coleman, A. W. (1989). Rapid screening method to detect potential biparental inheritance of plastid DNA and results for over 200 angiosperm species. *Am. J. Bot.* **75**, 1443-1458.

Doebley, J., Renfroe, W. and Blanton, A. (1987). Restriction site variation in the *Zea* chloroplast genome. *Genetics* **117**, 139-147.

Curtis, S. E. and Clegg, M. T. (1984). Molecular evolution of chloroplast DNA sequences. *Mol. Biol. Evol.* **1**, 291-301.

Doyle, J. J., Beachy, R. N. and Lewis, W. H. (1984). Evolution of rDNA in *Claytonia* polyploid complexes *In* "Plant Biosystematics" (W. F. Grant, ed.), pp. 321-341. Academic Press, Toronto.

Doyle, J. J., Grace, J. G. and Brown, A. H. D. (1990). Multiple origins of polyploids in the *Glycine tabacina* complex inferred from chloroplast DNA polymorphism. *Evolution* (in press).

Doyle, J. J., Soltis, D. E. and Soltis, P. S. (1985). An intergeneric hybrid in the Saxifragaceae: evidence from the ribosomal RNA genes. *Am. J. Bot.* **72**, 1388-1391.

Erickson, L. R., Straus, N. A. and Beversdorf, W. B. (1983). Restriction patterns reveal origins of chloroplast genomes in *Brassica* amphidiploids. *Theor. Appl. Genet.* **65**, 201-206.

Flavell, R. B. (1986). Ribosomal RNA genes and control of their expression. *In* "Oxford Surveys of Plant Molecular and Cell Biology", Volume 3 (B. J. Miflin ed.), pp. 251-275. Oxford Univ. Press, Oxford, England.

Gerbi, S. A. (1986). The evolution of eukaryotic ribosomal DNA. *Biosystems* **19**, 247-258.

Giannasi, D. E. and Crawford, D. J. (1986). Biochemical systematics II. A reprise. *In* "Evolutionary Biology", Volume 20 (M. K. Hecht, B. Wallace, and G. T. Prance, eds), pp. 25-248. Plenum, New York.

Gillham, N. W., Boynton, J. E. and Harris, E. H. (1985). Evolution of plastid DNA. *In* "DNA and Evolution: Natural Selection and Genome Size" (T. Cavalier-Smith, ed.), pp. 299-351. Wiley, N.Y.

Hilu, K. W. (1988). Identification of the "A" genome of finger millet using chloroplast DNA. *Genetics* **118**, 163-167.

Hosaka, K. (1986). Who is the mother of the potato? - restriction endonuclease analysis of chloroplast DNA of cultivated potatoes. *Theor Appl. Genet.* **72**, 606-618.

Jansen, R. K. and Palmer, J. D. (1987). A chloroplast DNA inversion marks an ancient evolutionary split in the sunflower family (Asteraceae). *Proc. Natl. Acad. Sci. USA* **84**, 5818-5822.

Jansen, R. K. and Palmer, J. D. (1988). Phylogenetic implications of chloroplast DNA restriction site variation in the Mutisieae (Asteraceae). *Am. J. Bot.* **75**, 753-766.

Jorgensen, R. A. and Cluster, P. D. (1988). Modes and tempos in the evolution of nuclear ribosomal RNA: New characters for evolutionary studies and new markers for genetic and population studies. *Ann. Mo. Bot. Gard.* **75**, 1238-1247.

Kirk, J. T. O. and Tilney-Bassett, R. A. E. (1978). The plastids: their chemistry, structure, growth, and inheritance, Ed. 2. Elsevier, Amsterdam.

Levin, D. A. (1983). Polyploidy and novelty in flowering plants. *Am. Nat.* **122**, 1-25.

Lewis, W. H. (1980). Polyploidy in species populations. *In* "Polyploidy" (W. H. Lewis, ed), pp. 103-144. Plenum, New York.

Long, E. O. and Dawid, I. B. (1980). Repeated genes in eukaryotes. *Annu. Rev. Biochem.* **49**, 727-764.

Neale, D. B., Saghai-Maroof, M. A., Allard, R. W., Zhang, Q. and Jorgensen, R. A. (1988). Chloroplast DNA diversity in populations of wild and cultivated barley. *Genetics* **120**, 1105-1110.

Neale, D. B., Wheller, N. C. and Allard, R. W. (1986). Paternal inheritance of chloroplast DNA in Douglas-fir. *Can. J. For. Res.* **16**, 1152-1154.

Ness, B. D., Soltis, D. E. and Soltis, P. S. (1989). Autopolyploidy in *Heuchera micrantha* Dougl. (Saxifragaceae). *Am. J. Bot.* **76**, 614-626.

Ogihara, Y. and Tsunekawi, K. (1982). Molecular basis of the genetic diversity of the cytoplasm in *Triticum* and *Aegilops*. I. Diversity of the chloroplast genome and its lineage revealed by the restriction pattern of ct-DNAs. *Jpn. J. Genet.* **57**, 371-396.

Ownbey, M. (1950). Natural hybridization and amphiploidy in the genus *Tragopogon*. *Am. J. Bot.* **37**, 487-499.

Ownbey, M. and McCollum, G. (1953). Cytoplasmic inheritance and reciprocal amphiploidy in *Tragopogon*. *Am. J. Bot.* **40**, 788-796.

Ownbey, M. and McCollum, G. (1954). The chromosomes of *Tragopogon*. *Am. J. Bot.* **56**, 7-21.

Palmer, J. D. (1985a). Evolution of chloroplast and mitochondrial DNA in plants and algae. *In* "Monographs in Evolutionary Biology: Molecular Evolutionary Genetics" (R. J. MacIntyre, ed.), pp. 131-240. Plenum, N.Y.

Palmer, J. D. (1985b). Comparative anatomy of chloroplast genomes. *Annu. Rev. Genet.* **19**, 325-354.

Palmer, J. D. (1987). Chloroplast DNA evolution and biosystematic uses of chloroplast DNA variation. *Am. Nat.* **130**, S6-S29.

Palmer, J. D., Jorgensen, R. A. and Thompson, W. F. (1985). Chloroplast DNA variation and evolution in *Pisum*: patterns of change and phylogenetic analysis. *Genetics* **109**, 195-213.

Palmer, J. D., Shields, C. R., Cohen, D. B. and Orton, T. J. (1983). Chloroplast DNA evolution and the origin of amphidiploid *Brassica*. *Theor. Appl. Genet.* **65**, 181-189.

Palmer, J. D., Jansen, R. K., Michaels, H. J., Chase, M. W. and Manhart., J. R. (1988). Chloroplast DNA variation and plant phylogeny. *Ann. Mo. Bot. Gard.* **75**, 1180-1206.

Ranker, T. A., Haufler, C. H., Soltis, P. S. and Soltis, D. E. (1989). Genetic evidence for allopolyploidy in the neotropical fern *Hemionitis pinnatifida* (Adiantaceae) and the reconstruction of an ancestral genome. *Syst. Bot.* **14**, 439-447.

Rogers, S. O. and Bendich, A, J. (1987). Ribosomal RNA genes in plants: variability in copy number and in the intergenic spacer. *Plant Molec. Biol.* **9**, 509-520.

Rieseberg, L. H., Soltis, D. E. and Palmer, J. D. (1988). A molecular reexamination of introgression between *Helianthus annuus* and *H. bolanderi* (Compositae). *Evolution* **42**, 227-238.

Roose, M. L. and Gottlieb, L. D. (1976). Genetic consequences of polyploidy in *Tragopogon*. *Evolution* **30**, 818-830.

Savile, D. B. O. (1975). Evolution and biogeography of Saxifragaceae with guidance from their rust parasites. *Ann. Mo. Bot. Gard.* **62**, 354-361.

Schaal, B. A. and Learn, G. H., Jr. (1988). Ribosomal DNA variation within and among plant populations. *Ann. Mo. Bot. Gard.* **75**, 1207-1216.

Sears, B. B. (1980). Elimination of plastids during spermatogenesis and fertilization in the plant kingdom. *Plasmid* **4**, 233-255.

Sears, B. B. (1983). Genetics and evolution of the chloroplast. *Stadler Symp.* **15**, 119-139.

Soltis, D. E. (1984). Autopolyploidy in *Tolmiea menziesii* (Saxifragaceae). *Am. J. Bot.* **71**, 1171-1174.

Soltis, D. E. and Doyle, J. J. (1987). Ribosomal RNA gene variation in diploid and tetraploid *Tolmiea menziesii* (Saxifragaceae). *Plant Syst. Evol.* **15**, 75-78.

Soltis, D. E. and Rieseberg, L. H. (1986). Autopolyploidy in *Tolmiea menziesii* (Saxifragaceae): genetic insights from enzyme electrophoresis. *Am. J. Bot.* **73**, 310-318.

Soltis, D. E. and Soltis, P. S. (1988). Electrophoretic evidence for tetrasomic inheritance in *Tolmiea menziesii* (Saxifragaceae). *Heredity* **60**, 375-382.

Soltis, D. E. and Soltis, P. S. (1989a). Allopolyploid speciation in *Tragopogon*: insights from chloroplast DNA. *Am. J. Bot.* **76**, 1119-1124.

Soltis, D. E. and Soltis, P. S. (1989b). Tetrasomic inheritance in *Heuchera micrantha* (Saxifragaceae). *J. Hered.* **80**, 123-126.

Soltis, D. E. and Soltis, P. S. (1989c). Genetic consequences of autopolyploidy in *Tolmiea* (Saxifragaceae). *Evolution* **43**, 586-594.

Soltis, D. E., Soltis, P. S. and Bothel, K.D. (1990a). Chloroplast DNA evidence for the origins of the monotypic *Bensoniella* and *Conimitella* (Saxifragaceae). *Syst. Bot.* (in press).

Soltis, D. E., Soltis, P. S. and Ness, B. D. (1989a). Chloroplast DNA variation and multiple origins of autopolyploidy in *Heuchera micrantha* (Saxifragaceae). *Evolution* **43**, 650-656.

Soltis, D. E., Soltis, P. S. and Ness, B. D. (1990b). Maternal inheritance of the chloroplast genome in *Heuchera* and *Tolmiea* (Saxifragaceae). *J. Heredity* (in press).

Soltis, D. E., Soltis, P. S., Ranker, T. A. and Ness, B. D. (1989b). Chloroplast DNA variation in a wild plant, *Tolmiea menziesii*. *Genetics* **121**, 819-826.

Soltis, P. S., Soltis, D. E. and Wolf, P. G. (1990). Allozymic and chloroplast DNA analyses of polyploidy in *Polystichum*. I. The origins of *P. californicum* and *P. scopulinum*. *Syst. Bot.* (in press).

Stebbins, G. L. (1950). "Variation and Evolution in plants." Columbia University Press, New York.

Systma, K. J. and Schaal, B. A. (1985). Phylogenetics of the *Lisianthius skinneri* (Gentianaceae) species complex in Panama utilizing DNA restriction fragment analysis. *Evolution* **39**, 594-608.

Systma, K. J. and Gottlieb, L. D. (1986). Chloroplast DNA evidence for the origin of the genus *Heterogaura* from a species of *Clarkia* (Onagraceae). *Proc. Natl. Acad. Sci. USA* **83**, 5554-5557.

Suzuki, T. and Iwatsuki, K. (1990). Recurrent hybrid origins of agamosporous triploids in a fern, *Pteris cretica* L. in Japan. *Heredity* (in press).

Szmidt, A. E., Alden, T. and Hallgren, J.-E. (1987). Paternal inheritance of chloroplast DNA in *Larix*. *Plant Mol. Biol.* **9**, 59-64.

Wagner, D. H. (1979). Systematics of *Polystichum* in western North America north of Mexico. *Pteridoliga* **1**, 64.

Wagner, W. H., Jr. (1973). Reticulation of holly ferns (*Polystichum*) in the western United States and adjacent Canada. *Am. Fern J.* **63**, 99-115.

Wendel, J. F. (1989). New world tetraploid cottons contain Old World cytoplasm. *Proc. Natl. Acad. Sci. USA* **86**, 4132-4136.

Whatley, J. M. (1982). Ultrastructure of plastid inheritance: green plants to angiosperms. *Biol. Rev.* **57**, 527-569.

Wolf, P. G., Soltis, P. S. and Soltis, D. E. (1989). Tetrasomic inheritance and chromosome pairing behavior in the naturally occuring autotetraploid *Heuchera grossulariifolia* (Saxifragaceae). *Genome* **35**, 655-659

Wolf, P. G., Soltis, D. E. and Soltis, P. S. (1990). Chloroplast-DNA and allozymic variation in diploid and autotetraploid *Heuchera grossulariifolia*. *Am. J. Bot.* **72**, 232-244.

Wyatt. R., Odrzykoski, I. K. and Stoneburner, A. (1988). Allopolyploidy in bryophytes: Recurring origins of *Plagiomnium medium*. *Proc. Natl. Acad. Sci. USA* **85**, 5601-5604.
Zimmer, E. A., Jupe, E. R. and Walbot., V. (1988). Ribosomal gene structure, variation and inheritance in maize and its ancestors. *Genetics* **120**, 1125-1136.

8 Chloroplast DNA and Phylogenetic Studies in the Asteridae

RICHARD G. OLMSTEAD[1], ROBERT K. JANSEN[2], HELEN J. MICHAELS[3], STEPHEN R. DOWNIE[1], AND JEFFREY D. PALMER[1]

[1]*Department of Biology, Indiana University, Bloomington, IN 47405, USA*
[2]*Department of Ecology and Evolutionary Biology, University of Connecticut, Storrs, CT 06268, USA*
[3]*Department of Biology, University of Michigan, Ann Arbor, MI 48109, USA*

I. Introduction

The quest for understanding angiosperm phylogeny has led plant biosystematists to pursue many avenues of research. An abbreviated chronology of plant systematics since Linnaeus indicates that the study of floral structure and plant anatomy predominated through the 18th, 19th, and early 20th centuries, followed by the introduction of cytology and palynology in the mid-twentieth century, secondary plant chemicals in the 1960's, and protein variation in the 70's. This diverse array of approaches has in common one important link; all involve the study of phenotypes arising from the underlying genetic material through various biochemical and developmental pathways. Only in this decade have advances in molecular biology enabled plant systematists to readily examine the genetic material itself, DNA, to investigate phylogenetic relationships.

Of the three genomes in plants (nuclear, chloroplast, and mitochondrial), the chloroplast genome has proved to be the most useful for phylogenetic analyses to date. Its presence in high copy number, often 5,000 genomes per cell, makes it relatively easy to extract, and even total DNA extracts from 2-3 grams of fresh leaf material are rich enough in chloroplast DNA (cpDNA) for most systematic purposes. The chloroplast genome is small and varies little in size in green land plants, 120-217 kb, with much of the size variation accounted for by difference in the size of a large inverted repeat (Fig. 1). The chloroplast reproduces clonally with little or no recombination in most plants and is inherited maternally in most flowering

BIOLOGICAL APPROACHES AND
EVOLUTIONARY TRENDS IN PLANTS ISBN 0-12-402960-4

plants. The structure and function of the chloroplast genome have been the subject of several recent reviews (Whitfield and Bottomley, 1983; Palmer, 1985; Zurawski and Clegg, 1987). The low rate of nucleotide substitution in cpDNA relative to nuclear DNA and animal mtDNA (Wolfe *et al.*, 1987), combined with a highly conserved gene content and arrangement, makes both restriction site mapping (using chloroplast probes from even distantly related species) and nucleotide sequencing feasible approaches for comparative studies. However, the much larger size and greater gene content of the nuclear genome and the apparently slower rate of nucleotide substitution (Wolfe *et al.*, 1987) in the plant mitochondrial genome suggest that each will have valuable phylogenetic utility in the future.

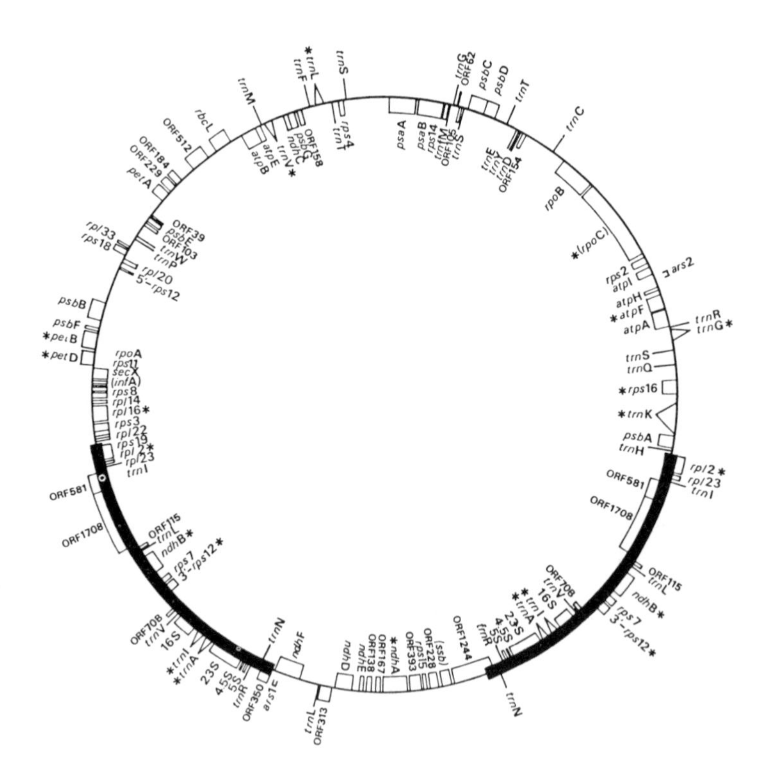

FIG. 1 Circular map of the cpDNA chromosome in tobacco (*Nicotiana tabacum*) showing the densely packed arrangement of genes and open reading frames. The total genome size is 155,844 base pairs and is organized into large (87 Kb) and small (18 Kb) single copy regions separated by two copies of the inverted repeat (25 Kb) indicated by the dark bars. Genes transcribed clockwise are shown on the inside of the circle; those transcribed in the reverse direction are on the outside.

Evolutionary change in cpDNA can be categorized into two distinct classes, nucleotide substitution (i.e. point mutation) and structural rearrangements (i.e. insertions, deletions, and inversions). Both classes can be exploited for phylogenetic inference. In our research on the systematics of the subclass Asteridae, we are assessing variation from both sources to address questions concerning phylogenetic relationships at different levels.

II. Nucleotide Substitutions

Point mutations resulting in nucleotide substitutions are the most common source of DNA variation among species. Nucleotide substitutions can be detected by restriction site analysis when mutations occur in restriction endonuclease recognition sites and by direct sequence comparison of homologous sequences such as genes, introns, or conserved spacer regions. Most mutations observed in restriction sites will be at sites in non-coding regions of the genome and at "silent"sites within protein genes, where nucleotide substitution rates are highest, making restriction site analysis most appropriate for phylogenetically closely related organisms. Rates of nucleotide substitution in genes and some introns are lower than in non-coding regions due to functional constraints on sequence evolution, making DNA sequencing of specific genes more appropriate for comparisons at greater phylogenetic distance.

A. *Restriction site analysis*

The analysis of restriction site variation involves the digestion of the cpDNA by restriction enzymes followed by the comparison of DNA fragments separated by gel electrophoresis. Restriction site mutations are detected by the presence or absence of fragments on the gel. Two small fragments indicate the presence of a restriction site in one sample, whereas a single larger fragment in another sample comprised of the two smaller fragments indicates its absence. Early studies involved the visual inspection of the gels themselves, however multiple mutational differences and the presence of length mutations (insertions/deletions) between samples create problems for interpreting homologous fragments. This approach is sufficient only for closely related taxa cut with restriction enzymes yielding few (5-20) fragments and requires purified cpDNA. The preferred method today is the filter hybridization approach, in which the DNA is transferred from the gel to nylon filters, which are then probed successively with cloned cpDNA fragments from a completely mapped reference genome. By this method more divergent taxa can be examined and more frequently cutting enzymes used (this reduces the number of enzymes needed to sample the same number of bases)with little problem in determining homology among fragments, because only a small portion of the genome is examined with each successive round of hybridization. For our current cpDNA analysis of the Solanaceae we have constructed a set of 40 clones, ranging in size from 2-5 kb, of tobacco to use as probes in hybridization, from larger clones kindly provided by M. Sugiura. In selecting restriction enzymes for a systematic survey, it is worth noting that enzymes with A/T-rich recognition sites (eg. DraI, TTTAAA) are more likely to provide a greater proportion of variable sites, because the non-coding regions of land plant cpDNA are A/T rich.

Carrying out systematic research using cpDNA requires that attention be paid to many of the same considerations important in studies using "conventional" data. In studies of large groups it will not be possible to sample all taxa, so representative specimens must be chosen carefully. It is important to include more than one out-group, so that character polarity can be assessed and the hypothesis of monophyly for the study group can be tested adequately. The conservative nature of cpDNA evolution suggests that single individuals will adequately represent most taxa at the generic-level and above (unless doubts exist concerning the monophyly of the group). Chloroplast DNA polymorphism within species is often very low (Banks and Birky, 1985; Michaels, unpublished data), but can be considerable (Soltis *et al.*, 1989) and may extend beyond species limits among very closely related species, thereby creating conflicting phylogenetic inferences (Doyle *et al.*, 1989). Reconstructing phylogeny from any data source requires that homologs be identified for the variable characters observed. The best way to assess homology of restriction sites is to construct restriction maps for each taxon for each enzyme used. Invariant sites provide reference points for aligning maps even though they do not provide phylogenetic information. Likewise, sites which differ in only a single taxon are phylogenetically uninformative. Only those sites present in two or more taxa and missing in at least two other taxa are phylogenetically informative. The phylogenetic reconstruction is then carried out using a cladistic analysis based on either Wagner parsimony (site gains and losses treated equally) or Dollo parsimony (single gain permitted for each site, but multiple losses allowed). For studies at the intrageneric level, little conflict is expected in the data and either method of parsimony is likely to produce the same tree (e.g. Palmer and Zamir, 1982; Sytsma and Gottlieb, 1986). However, as more divergent taxa are surveyed, more conflict in the data will arise and Dollo parsimony will be preferred based on the premise that parallel restriction site gains are expected to be much less likely than parallel site losses (DeBry and Slade, 1985; Jansen and Palmer, 1988).

Two examples of interspecific studies in the Solanaceae will serve to illustrate the advantages and limitations of cpDNA analysis among congeneric species. *Lycopersicon* was the subject of the first cladistic analysis of cpDNA restriction site variation (Palmer and Zamir, 1982). Included were eight species of *Lycopersicon* and two species of *Solanum* as outgroups. The cpDNAs were cut with 25 restriction enzymes and 39 variable sites were observed, of which 14 were phylogenetically informative. The resulting cladistic analysis (Fig. 2) suggests that all of the restriction site variation is consistent with the most parsimonious tree except for a single site, which is implied by the analysis to have been lost in parallel in two lineages, thus giving a consistency index (Kluge and Farris, 1969) of 0.93. This extremely high level of consistency within the cpDNA data suggests that the resulting tree provides a reliable estimate of phylogenetic relationships within *Lycopersicon*. The cpDNA phylogeny confirms that *Solanum pennellii* belongs in *Lycopersicon* and that red fruits are a derived feature within *Lycopersicon*, since the three species bearing red fruits included in this study, *L. esculentum, L. pimpinellifolium*, and *L. cheesmanii*, form a monophyletic group. At the same time, however, the data are insufficient to completely resolve relationships within *Lycopersicon* as evidenced by two internal trichotomies and identical cpDNA for *L. chilense* and three accessions of *L. peruvianum* for the restriction sites surveyed. Also, cpDNA polymorphism is observed within the widely distributed and morphologically variable species, *L. peruvianum*, for which six accessions were examined.

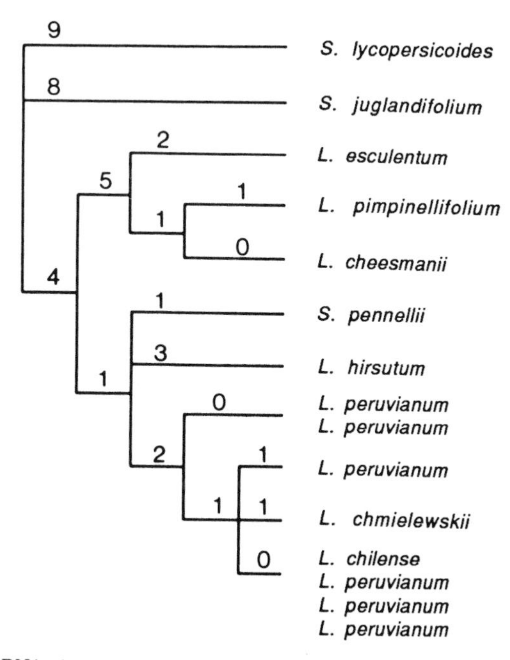

FIG. 2 Chloroplast DNA phylogeny of eight species of *Lycopersicon* (including *Solanum pennellii*). Numbers indicate the number of restriction site mutations defining each branch.

A more recent study of the genus *Nicotiana* (Olmstead and Palmer, unpublished data) investigates 21 species of *Nicotiana*, with one species of *Lycopersicon* and three species of *Petunia* included as outgroups. The analysis of restriction site variation in the large single copy region of the chloroplast genome yielded 187 variable sites, of which 108 were informative. A consensus tree of the twelve most parsimonious trees is shown in Fig.3. This analysis differs from that of *Lycopersicon* in having a greater number of taxa and informative restriction sites, but is similar in having highly consistent data for a study of its size (CI = 0.80) and in yielding an incompletely resolved tree. Note that in both of these studies Wagner and Dollo parsimony produce identical trees. In *Nicotiana* the incomplete resolution of the cpDNA tree carries an interesting implication for the evolution of that genus in Australia. Relationships among the six species of *Nicotiana* from Australia, *N. velutina, N. rotundifolium, N. megalosiphon, N. excelsior, N. gossei,* and *N. exigua,* cannot be resolved on the basis of cpDNA, because so little variation exists, suggesting a very recent radiation there. Also, the polyploid hybrid species, *N. tabacum,* has identical cpDNA with that of one of its putative progenitors, *N. sylvestris,* confirming the identity of the latter as the maternal hybrid parent. The association of highly consistent data with incomplete resolution of phylogeny emerges as a pattern in many studies of cpDNA of congeneric species (Palmer and Zamir, 1982; Palmer *et al.*, 1983; Hosaka *et al.*, 1984; Sytsma and Schaal 1985; Perl-Treves and Galun, 1985; Coates and Cullis, 1987; Doebley *et al.*, 1987).

The conclusions drawn from interspecific studies suggest that much more divergent chloroplast genomes may be mapped and that their restriction site variation may be used effectively for phylogeny reconstruction. With increasing evolution-

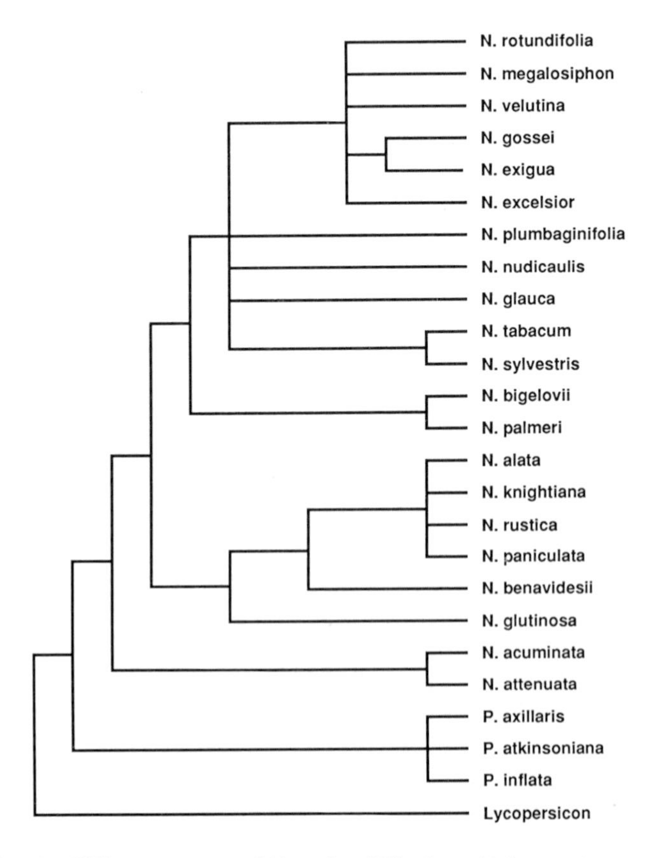

FIG. 3 Chloroplast DNA consensus tree of 21 species of *Nicotiana* with *Lycopersicon esculentum* and three species of *Petunia* as outgroups.

ary divergence among taxa in a cpDNA survey, two predictions can be made: 1) restriction site mutations will be more numerous and will enable complete resolution of phylogenetic relationships and 2) a greater number of inconsistencies will arise in the cpDNA data, particularly parallel losses of restriction sites. The first study undertaken to examine intrafamilial relationships focused on the family Asteraceae (Jansen and Palmer, 1988; Jansen *et al.*, submitted, a). A restriction site analysis using 11 enzymes was carried out on 57 genera representing 5 tribes. The bootstrap method of Felsenstein (1985) was used in conjunction with Dollo parsimony to analyze the large (927 variable sites and 328 informative sites) and less consistent (CI=0.39) data set produced by this analysis (Fig. 4). The bootstrap analysis provides a means of assessing the support for groups within the tree as indicated by the number of bootstrap replicates which support each group on the tree. As predicted on the basis of studies at the species-level, there are a sufficient number of informative restriction site mutations to provide a completely resolved estimate of phylogeny. However, also as predicted, the diminished consistency of the data, relative to data from studies at the species-level, results in conflicting estimates of phylogeny for some groups as reflected in the low bootstrap figures on some branches of the tree. The consistency index of 0.39 (note that Dollo parsimony will always produce lower CI values than Wagner parsimony) is low relative to studies among congeneric species cited above. However, other considerations, including the distribution of inconsistent characters (Jansen, et al., submitted, a) and an inverse correlation between consistency index and number of taxa in phylogenetic analyses in general (Sanderson and Donoghue, 1989; Archie, 1989), suggest that the cpDNA analysis is still a strong estimate of phylogenetic relationships in the Asteraceae.

The cpDNA phylogeny of the Asteraceae provides a test of the several hypotheses of tribal relationships in the Asteraceae that have been advanced by systematists studying the family, along with the predictions of those hypotheses regarding which tribe retains ancestral elements of the family. The results identify the tribe Mutisieae as the ancestral tribe in the family, finally settling a long-standing debate. However, the results go further to show that the Mutisieae is not a monophyletic tribe and that the subtribe Barnedesiinae is the sister group to the rest of the family (Jansen and Palmer, 1988; Jansen et al., submitted, b). In addition, a rigorous cladistic analysis of the Asteraceae using non-molecular characters (Bremer, 1987) is available for comparison with the cpDNA analysis (Jansen et al., submitted, b). The correct tribal placement of problem genera, whose tribal affinities have been obscure due to unique morphological attributes, often can be made readily based on cpDNA relationships (Keeley and Jansen, 1989; Jansen et al., submitted, b).

A second important Asteridae family, the Solanaceae, is currently the subject of an extensive restriction site survey (Olmstead and Palmer, unpublished data) including 133 species and 11 restriction enzymes. Unlike the Asteraceae, there has been no rigorous analysis of phylogenetic relationships in the Solanaceae. The cpDNA analysis will provide the first reliable estimate of tribal relationships in the family and, with a broad representation of the large genus *Solanum* (31 species included), will furnish information concerning the relationship of that important genus to many other allied genera.

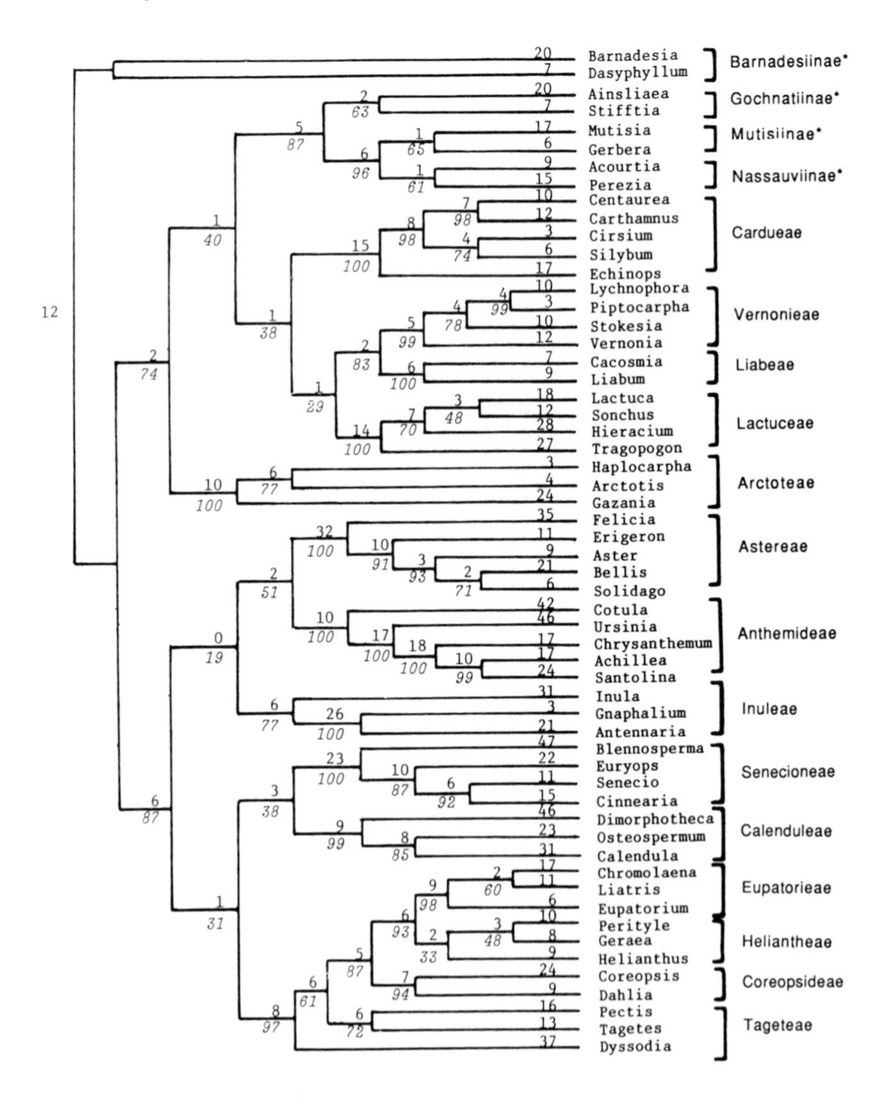

FIG. 4 Chloroplast DNA majority rule bootstrap tree of 57 genera of Asteraceae constructed using Dollo parisimony. Numbers above the line represent the number of restriction site mutations and the numbers below the line represent the percent of bootstrap replicates supporting each branch. Genera and tribal designations are at right.

B. DNA sequencing

The selection of a sequence for comparison should take into consideration the degree of phylogenetic divergence between the taxa under study. The length of the sequence and degree of divergence are important variables to consider when choosing an appropriate sequence. Published compilations of lengths and percent sequence divergence comparisons for the full complement of chloroplast genes between *Nicotiana tabacum* and *Marchantia polymorpha* (Wolfe and Sharp, 1988) and *N. tabacum* and *Oryza sativa* (Sugiura, 1989), the only three completely sequenced chloroplast genomes, provide an excellent starting point for the choice of a sequence for a phylogenetic study. As with restriction site analysis, identification of homologs when using sequence data for phylogenetic analysis is critical. All chloroplast genes, including those on the inverted repeat, evolve as single copy genes, thereby avoiding problems of gene homology inherent in the study of many nuclear genes. In sequence comparisons, individual nucleotides are the characters and correct identification of homologs depends on proper sequence alignment. If all sequences are the same length then alignment usually is trivial, however insertions, deletions, and highly divergent sequences present problems. As with mapping restriction sites, the invariant nucleotides are important for proper alignment and only those sites in which at least two nucleotides are shared by two or more taxa are phylogenetically informative in a parsimony analysis. For very distantly related organisms, in which silent nucleotide positions (e.g. most third codon positions in protein coding genes) may approach saturation with substitutions, the translated amino acid sequences of chloroplast protein genes may provide a more reliable indication of relationships. The simplest and most common approach to analysing sequence data entails weighting all sequence substitutions equally, however, some justification exists for applying weights to different substitutions. Two proposed systems of weighting include weighting transversions (purine-pyrimidine) over transitions (purine-purine or pyrimidine-pyrimidine) and applying a weighting scheme in which any substitution is weighted as the inverse of the number of other character states (nucleotides or amino acids) at that position, thereby reducing the impact of substitutions at those positions most likely to exhibit convergent evolution (Felsenstein, 1981). Alternative methods of phylogenetic reconstruction using sequence data are available that take all nucleotide positions into account, including distance matrix and maximum likelihood methods (Felsenstein, 1988).

Most of the interest in sequencing chloroplast genes for phylogenetic analysis of angiosperms has centered on the large subunit of ribulose bisphosphate carboxylase (*rbc*L), although no specific phylogenetic studies using *rbc*L sequences have been published (but see Palmer *et al.*, 1988 for a phenetic comparison of published sequences). The *rbc*L sequence similarity at the amino acid level for comparisons of tobacco with *Marchantia* and tobacco with rice are 91% and 93%, respectively (Wolfe and Sharp, 1988; Sugiura, 1989). A slowly evolving gene such as *rbc*L may be expected to exhibit insufficient nucleotide substitution to be useful for inferring phylogeny among closely related species, but should be useful at higher taxonomic levels. In the Asteridae we are using *rbc*L sequence data for phylogenetic studies at the interfamilial level, however enough sequences are available within two families, the Asteraceae and Solanaceae, to examine the effectiveness of *rbc*L sequence data for inferring intrafamilial relationships as well. This work-in-progress has focused on the families most closely related to the Asteraceae, but will be expanded to include 30-40 species representing all of the major families in the Asteridae, along with outgroups chosen from the Rosidae. To date 19 sequences have been determined for *rbc*L within the Asteridae. A phylogenetic analysis of these sequences was conducted with spinach as the outgroup using the bootstrap and Wagner parsimony (Fig. 5). Of the 1437 nucleotide positions in the typical *rbc*L sequence in the Asteridae, 384 were variable and the resulting tree has a CI of 0.56. The results show that numerous nucleotide substitutions distinguish representatives of different families whereas relatively few substitutions distinguish genera within each of the two families represented by more than one species. The greater interfamilial resolution is reflected in the greater number of bootstrap replicates supporting family-level clades, thus lending confidence to the use of *rbc*L sequences for phylogenetic reconstruction in the Asteridae. The interfamilial relationships depicted in Fig. 5 should be viewed as

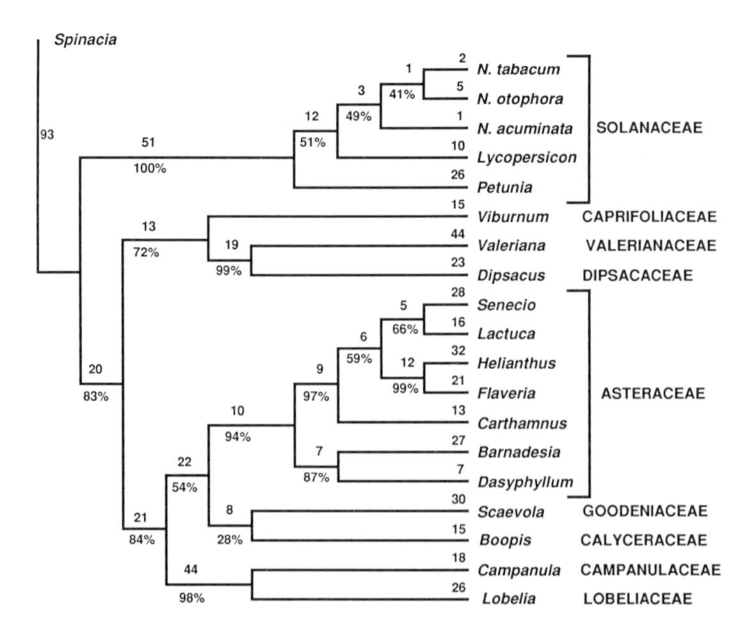

FIG. 5 Chloroplast DNA majority rule bootstrap tree using Wagner parsimony of 17 species representing 7 families of the subclass Asteridae based on *rbc*L sequence data. Numbers above the line represent the number of nucleotide substitutions along the branch and the numbers below the line are the percent of bootstrap replicates supporting each branch.

preliminary until more families have been sampled, but these preliminary results identify the Goodeniaceae, or a clade comprised of the Goodeniaceae and Calyceraceae, as the best candidate for the sister group to the Asteraceae of the families represented thus far.

III. Structural Rearrangements

Structural rearrangements are the result of several kinds of mutations including insertions and deletions in either coding or non-coding DNA and sequence inversions. Most of the previously documented rearrangements of the chloroplast genome have been discovered through research aimed at understanding its structure and function. The importance of rearrangements for understanding phylogeny has been recognized only recently and the first efforts aimed at surveying chloroplast genome structure specifically for systematic ends are now underway, with little published data yet available. Insertions and deletions of a small number of bases (1-20) may be very common, though difficult to detect, and the assessment of homology is difficult when they occur in non-coding regions. Larger insertions and deletions (100 + bp) occur less frequently and are readily detected by restriction site mapping, making the assessment of homology easier. However, drawing phylogenetic inference from insertions and deletions in non-coding sequences must proceed with caution, because these mutations tend to occur most frequently in "hotspots" (Tassopulu and Kung, 1984; Palmer, 1985; Palmer *et al.*, 1988). Exact size estimates and map locations must be confirmed using several restriction enzymes in order for confidence to be placed in the homology of such insertions and deletions. An example of a deletion in a non-coding region is found in the Solanaceae, where a 650 base pair deletion distinguishes *Nicotiana* from the rest of the family (Fig. 6). The size of the deletion is confirmed by mapping 10 restriction enzymes and the site of the deletion can be localized to within a 285 base pair region of the tobacco sequence.

The loss of genes or introns from the chloroplast genome is a special class of deletions, which occur rarely, thereby making the shared absence of a given gene or intron a powerful phylogenetic statement. The presence of a particular gene or intron can be detected by hybridization using cloned fragments of DNA specific to that particular gene or intron from another species of plant. Recently documented cases of chloroplast gene loss include the *rpl*22 gene, which is absent in all members of the legume family surveyed (Palmer *et al.*, 1988; Palmer and Doyle, unpublished data), and the *tuf*A gene, which is missing from the chloroplasts of all land plants (Baldauf and Palmer, submitted). Introns, although non-coding, also are conserved evolutionarily and the absence of an intron can furnish a useful phylogenetic marker. The intron in the *rpl*2 gene is missing from all sampled members of the Caryophyllales (Palmer and Zurawski, unpublished data), providing additional

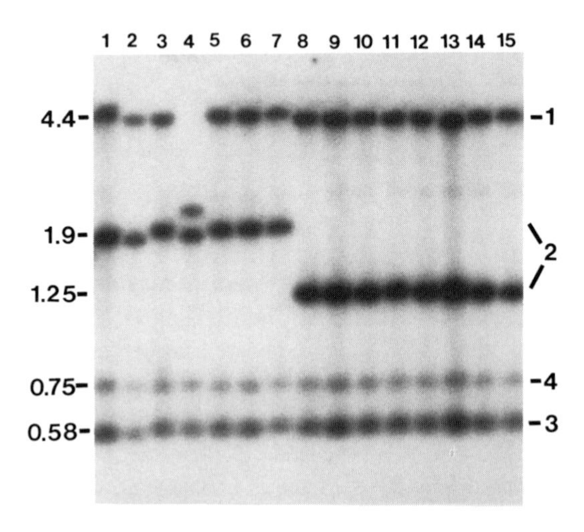

FIG. 6 Evidence of an approximately 650 bp deletion in *Nicotiana*. Numbers on the right indicate the sequence of fragments on the cpDNA molecule, numbers on the left indicate size of fragments in kb. Lanes 1-4 are various genera of Solanaceae, lanes 5-7 *Petunia*, and lanes 8-15 *Nicotiana*. All DNAs were digested with EcoRI. The approximately 1.9 kb fragment in *Petunia* is conserved with only minor size variation throughout the Solanaceae, implying that the smaller fragment in *Nicotiana* is the result of a ca. 650 bp deletion.

molecular evidence for that well-supported clade. In addition, the *rpl*2 intron apparently has been lost independently in other dicot lineages and perhaps in some monocots as well (Downie and Olmstead, unpublished data). A survey for missing chloroplast genes, introns, and conserved open reading frames (ORFs) in 88 species representing 36 families of the Asteridae (Table 1, Downie, unpublished data) indicates a non-random distribution of this class of deletions, suggesting that some yet unknown factors may trigger rearrangements and gene losses in certain chloroplast lineages.

TABLE 1

Genes, introns, and open reading frames (ORFs) absent from members of the Asteridae as represented by 94 species and 35 families (Downie, unpublished data)

Gene/intron/ORF	Missing from (species missing/species sampled):
*rpl*2 intron	Convolvulaceae/Cuscutaceae (4/4)
5' ORF 512	Oleaceae (3/4); Campanulaceae/Lobeliaceae (8/8)
3' ORF 512	Oleaceae (1/4); Campanulaceae/Lobeliaceae (7/8)
ORF 1244	Campanulaceae/Lobeliaceae (8/8)
ORF 581	Campanulaceae/Lobeliaceae (5/8)
ORF 75-ORF350	Convolvulaceae/Cuscutaceae (4/4); Campanulaceae (2/8)

The loss of one copy of the inverted repeat is another special case of a deletion which has been documented in three plant families, Pinaceae, Leguminosae, and Geraniaceae (Palmer *et al.*, 1987; Strauss *et al.*, 1988; Lavin 1990; Calie and Palmer, unpublished data). Whereas the loss of one copy of the inverted repeat is an extremely rare event, variation in size of the inverted repeat is common and usually consists of an expansion or contraction of the inverted repeat into, or out of, a single copy region, thereby making homology of inverted repeat size variants difficult to assess. Large insursions into what is normally single copy DNA by the inverted repeat have been documented in *Pelargonium* and *Geranium* (Palmer *et al.*, 1987; Calie and Palmer, unpublished data) and in *Nicotiana acuminata* (Shen *et al.*, 1982), where a sister species, *N. attenuata*, which differs in only two of over 600 restriction sites, has a typical size inverted repeat (Olmstead and Palmer, unpublished data).

Inversions are rare, but like gene losses make excellent systematic markers and can be detected by hybridization experiments. A fragment of DNA containing the end point of an inversion will hybridize to two widely separated fragments in a species without the inversion. For a more complete discussion of the methods for detecting cpDNA rearrangements see Palmer *et al.* (1988). Inversions have been documented in several plant families and are usually recognizable through restriction mapping as a unique event delineating all or part of a family as a monophyletic unit (Palmer, 1985; Jansen and Palmer, 1987; Palmer *et al.*, 1988). One such well-characterized inversion occurs in the Asteraceae, where all members of the family except those in the subtribe Barnedesiinae carry a 22 kb inversion relative to all other known land plant cpDNAs (Jansen and Palmer, 1987).

Deletions, insertions, and inversions represent a class of phylogenetic data distinct from that derived from the analysis of nucleotide substitutions. No generally accepted means exists of incorporating these two disparate forms of molecular evidence into a single phylogenetic reconstruction. Unusual rearrangements, for which a strong argument can be made for their uniqueness, may best be treated as undisputed evidence of monophyly for a group (Jansen and Palmer, 1987), whereas more commonly occurring rearrangements, such as small insertions and deletions, or "length mutations" in restriction site analyses, may be variously weighted (Morden and Golden, 1989), treated equivalent to mucleotide substitutions (Coates and Cullis, 1987; Doebley *et al.*, 1987), or considered unreliable for phylogenetic inference (Perl-Treves and Galun, 1985; Sytsma and Gottlieb, 1986; Jansen and Palmer, 1988; Soltis *et al.*, 1989). The fact that no firm theoretical justification exists for treating disparate forms of molecular data means that judgements inevitably must be made on a case-by-case basis. In this respect, systematists incorporating disparate sorts of conventional data, such as cytology and morphology, are confronted with similar problems.

IV. Conclusions

Chloroplast DNA systematics is a multi-faceted research program which is suited to all levels of taxonomic hierarchy in flowering plants, although limitations in the amount of useful variation may be experienced in within-species studies. Species-level phylogeny is best studied using restriction site analysis, whereas higher-level relationships are best studied by direct gene sequencing. Structural rearrangements, because of their infrequent occurrence, provide excellent systematic markers and can be assayed easily. As with any source of data used for phylogenetic analysis, proper attention must be paid to phylogenetic principles when it comes to selecting taxa, sampling within taxa, and data analysis. Many of the problems that plague analyses relying on other sources of data also affect cpDNA, including the correct assessment of homology, inconsistent data arising from parallel evolution, and incomplete resolution of phylogenetic trees. However, cpDNA offers many advantages over conventional sources of data, including the ready accessibility of numerous independent characters, characters that are independent of morphology (enabling tests of hypotheses concerning phenotypic evolution), and highly consistent data sets, especially at the species-level. The use of cpDNA in systematics is emerging as a mature research program and while not the panacea of early predictions, it will be an important tool for many years and will contribute to the resolution of many outstanding systematic problems.

Acknowledgements

This research was supported by NSF Grants BSR-8415934 to J.D.P. and R.K.J. and BSR-8717600 to J.D.P.

References

Archie, J. W. (1989). Homoplasy excess ratios: New indices for measuring levels of homoplasy in phylogenetic systematics and a critique of the consistency index. *Syst. Zool.* **38**, 253-269.

Baldauf, S. L. and Palmer, J. D. (1990). Evolutionary transfer of the chloroplast *tuf*A gene to the nucleus. *Nature* (in press).

Banks, J. A. and Birky, C. W. (1985). Chloroplast DNA diversity is low in a wild plant, *Lupinus texensis*. *Proc. Natl. Acad. Sci. U.S.A.* **82**, 6950-6954.

Bremer, K. (1987). Tribal interrelationships of the Asteraceae. *Cladistics* **3**, 210- 253.

Coates, D. and Cullis, C. A. (1987). Chloroplast DNA variability among *Linum* species. *Am. J. Bot.* **74**, 260-268.

DeBry, R. W. and Slade, N. A. (1985). Cladistic analysis of restriction endonuclease cleavage maps within a maximun-likelihood framework. *Syst. Zool.* **34**, 21-34.

Doebley, J., Renfroe, W. and Blanton, A. (1987). Restriction site variation in the *Zea* chloroplast genome. *Genetics* **117**, 139-147.

Doyle, J. J., Doyle, J., and Brown, A. H. D. (1989). The limits of chloroplast DNA in phylogeny reconstruction: polymorphism and phylogeny in the B genome of *Glycine*. *Am. J. Bot.* **76**(suppl.), 239.

Felsenstein, J. (1981). A likelihood approach to character weighting and what it tells us about parsimony and compatibility. *Biol. J. Linn. Soc.* **16**, 183-196.
Felsenstein, J. (1985). Confidence limits on phylogenies: an approach using the bootstrap. *Evolution* **39**, 783-791.
Felsenstein, J. (1988). Phylogenies from molecular sequences: inference and reliability. *Annu. Rev. Genet.* **22**, 521-565.
Hosaka, K., Ogihara, Y., Matsubayashi, M. and Tsunewaki, K. (1984). Phylogenetic relationship between the tuberous *Solanum* species as revealed by restriction endonuclease analysis of chloroplast DNA. *Jpn. J. Genet.* **59**, 349-369.
Jansen, R. K. and Palmer, J. D. (1987). A chloroplast inversion marks an ancient evolutionary split in the sunflower family (Asteraceae). *Proc. Natl. Sci. U.S.A.* **84,** 5818-5822.
Jansen, R. K. and Palmer, J. D. (1988). Phylogenetic implications of chloroplast DNA restriction site variation in the Mutisieae (Asteraceae). *Am. J. Bot.* **75**, 751-764.
Jansen, R. K., Holsinger, K. E., Michaels, H. J. and Palmer, J. D. Phylogenetic analysis of chloroplast DNA restriction site data at higher taxonomic levels: an example from the Asteraceae. *Evolution* (submitted, a).
Jansen, R. K., Michaels, H. J. and Palmer, J. D. Phylogeny and character evolution in the Asteraceae based on chloroplast DNA restriction site mapping. *Evolution* (submitted, b).
Keeley, S. C. and Jansen, R. K. (1989). Chloroplast DNA evidence for tribal placement of *Brachylaena, Tarchonanthus,* and *Pluchea* (Asteraceae). *Am. J. Bot.* **76**(suppl.), 250.
Kluge A. G. and Farris, J.S. (1969). Quantitative phyletics and the evolution of anurans. *Syst. Zool.* **18**, 1-32.
Lavin, M. *et al.* (1990). Evolutionary significance of the loss of the chloroplast DNA inverted repeat in the Leguminosae subfamily Papillionidae. *Evolution* **44(2)**, 390-402.
Morden, C. W. and Golden, S. S. (1989). *psb*A genes indicate common ancestry of prochlorophytes and chloroplasts. *Nature* **337**, 382-385.
Palmer, J. D. (1985). Comparative organization of chloroplast genomes. *Annu. Rev. Genet.* **19**, 325-354.
Palmer, J. D. and Zamir, D. (1982). Chloroplast DNA evolution and phylogenetic relationships in *Lycopersicon*. *Proc. Natl. Acad. Sci. U.S.A.* **79**, 5006-5010.
Palmer, J. D., Nugent, J. M. and Herbon, L. A. (1987). Unusual structure of geranium chloroplast DNA: a triple-sized inverted repeat, extensive gene duplications, multiple inversions, and two repeat families. *Proc. Natl. Acad. Sci. U.S.A.* **84,** 769-773.
Palmer, J. D., Shields, C. R., Cohen, D. B. and Orton, T. J. (1983). Chloroplast DNA evolution and the origin of amphidiploid *Brassica*. *Theor. Appl. Genet.* **65**, 181-169.
Palmer, J. D., Jansen, R. K., Michaels, H. J., Chase, M. W. and Manhart, J. R. (1988). Chloroplast DNA variation and phylogeny. *Ann. Mo. Bot. Gard.* **75**, 1180-1206.
Perl-Treves, R. and Galun, E. (1985). The *Cucumis* plastome: physical map, intrageneric variation and phylogenetic relationships. *Theor. Appl. Genet.* **71**, 417-429.
Sanderson, M. J. and Donoghue, M. J. (1989). Patterns of variation and levels of homoplasy. *Evolution* **43**, 1781-1795.
Shen, G. F., Chen, K., Wu, M. and Kung, S. D. (1982). *Nicotiana* chloroplast genome IV. *N. acuminata* has larger inverted repeats and genome size. *Mol. Gen. Genet.* **187**, 12-18.
Soltis, D. E., Soltis, P. S. and Ness, B. D. (1989). Chloroplast DNA variation and multiple origins of autopolyploidy in *Heuchera micrantha* (Saxifragaceae). *Evolution* **43**, 650-656.
Strauss, S. H., Palmer, J. D., Howe, G. T. and Doerksen, A. H. (1988). Chloroplast genomes of two conifers lack a large inverted repeat and are extensively rearranged. *Proc. Natl. Acad. Sci. U.S.A.* **85**, 3898-3902.
Sugiura, M. (1989). The chloroplast chromosomes in land plants. *Annu. Rev. Cell Biol.* **5**, 51-70.
Sytsma, K. J. and Gottlieb, L. D. (1986). Chloroplast DNA evolution and phylogenetic relationships in *Clarkia* sect. *Peripetasma* (Onagraceae). *Evolution* **40**, 1248-1261.

Sytsma, K. J. and Schaal, B. A. (1985). Phylogenetics of the *Lisianthius skinneri* (Gentianiaceae) species complex in Panama utilizing DNA restriction fragment analysis. *Evolution* **39**, 594-608.

Tassopulu, D. and Kung, S. D. (1984). *Nicotiana* chloroplast genome 6. Deletion and hot spot - a proposed origin of the inverted repeats. *Theor. Appl. Genet.* **67**, 185-193.

Whitfield, P. R. and Bottomley, W. (1983). Organization and structure of chloroplast genes. *Annu. Rev. Plant Physiol.* **34**, 279-310.

Wolfe, K. H. and Sharp, P. M. (1988). Identification of functional open reading frames in chloroplast genomes. *Gene* **66**, 215-222.

Wolfe, K. H., Li, W. H. and Sharp, P. M. (1987). Rates of nucleotide substitutions vary greatly among plant mitochondrial, chloroplast, and nuclear DNAs. *Proc. Natl. Acad. Sci. U.S.A.* **84**, 9054-9058.

Zurawski, G. and Clegg, M. T. (1987). Evolution of higher plant chloroplast DNA-encoded genes: implication for structure-function and phylogenetic studies. *Annu. Rev. Plant Physiol.* **38**, 391-418.

9 Ribosomal DNA Variation and its Use in Plant Biosystematics

CHRISTIAN KNAAK[1], R. KEITH HAMBY[1], MICHAEL L. ARNOLD[1], MONIQUE D. LEBLANC[1], RUSSELL L. CHAPMAN[2] AND ELIZABETH A. ZIMMER[1,2]

Departments of Biochemistry [1]and Botany[2], Louisiana State University and Louisiana Agricultural Experiment Station, Louisiana State University Agricultural Center, Baton Rouge, LA 70803, USA

I. Introduction

Ribosomal RNA (rRNA) genes, or rDNA, are the set of DNA sequences that code for the synthesis of ribosomal RNA. These genes are particularly amenable for use as macromolecular yardsticks. Essential in protein synthesis, rRNA molecules and rDNA are ubiquitous in all cells. The rRNA genes are present as multiple copies in most organisms, ranging in number from 5 to 10 repeats in bacteria to more than 40,000 in some higher plant species (Long and Dawid, 1980; Rogers and Bendich, 1987a). These genes are found in all three plant genomes: nuclear, chloroplast, and mitochondrial. This review will deal primarily with **nuclear** rDNA repeat units, but we will also discuss general aspects of **organellar** (chloroplast and mitochondrial) rRNA genes in the **Background** section.

Plants typically contain 1,000 to 10,000 copies of the nuclear repeat per cell (Ingle *et al.*, 1975). Copies of nuclear rRNA genes exist in long tandem arrays, at one or several chromosomal loci, forming the nucleolar organizing region (Long and Dawid, 1980). Within a single species, the number of copies may vary as much as four-fold (Cullis and Davies, 1975) . The highly reiterated nature of these arrays allows very small amounts of DNA to be used in molecular characterizations.

BIOLOGICAL APPROACHES AND
EVOLUTIONARY TRENDS IN PLANTS ISBN 0-12-402960-4

Although there is variability among rDNAs between species, the rDNA repeat units are highly homogeneous within an array (Brown *et al.*, 1972). This phenomenon is termed "concerted evolution" (Zimmer *et al.*, 1980; Arnheim, 1983). It is thought that gene conversion, unequal crossing over, replicative transposition, gene amplification, or a combination of these mechanisms is responsible for the concerted evolution of ribosomal gene families (Arnheim, 1983). The individual rRNA gene repeat units are composed of a number of regions, which vary in functional constraint and evolutionary rates; therefore, any proposed mechanism(s) for concerted evolution must take this regional heterogeneity into account.

The abundance of rRNA (up to 90% of the extant cellular RNA's, cytoplasmic and chloroplast included), facilitates both rRNA isolation and characterization. Sequencing and restriction site analyses of the rDNA arrays have revealed that these arrays are phylogenetically informative at different levels. Sequence variation within the rapidly evolving nuclear nontranscribed spacers, which separate the coding regions in the arrays, can be used to differentiate closely related species and even lineages within the same species. Variation within the more conserved transcribed spacers can be used to differentiate genera and families, while the rRNA coding sequences are most informative at higher levels of classification. These features of nuclear ribosomal genes have led to rDNA comparative sequence analysis in studies of a wide variety of organisms.

II. Background

A. *General structure of plant nuclear rDNA repeats*

The structure of the nuclear rDNA repeat, on which this article focusses, is given in Figure 1. Each repeat unit contains a transcription unit from which the precursor rRNA is transcribed, and a nontranscribed, or intergenic, spacer between the transcription units (Reeder, 1974; Delgarno and Shine, 1977). The precursor rRNA is cleaved and enzymatically modified after transcription to yield the mature products; the 18S, 5.8S, and 26S rRNAs (Jordan *et al.*, 1976; Delgarno and Shine, 1977). For plants and animals, 5S rRNA genes are similarly organized as clusters of tandem repeats, but are located elsewhere in the genome (Monier, 1974; Reeder, 1974; Zimmer *et al.*, 1988); in yeasts and slime molds, the 5S genes are linked with the 18S and 26S genes (Rubin and Sulston, 1973; Maizels, 1976; Cockburn *et al.*, 1976; Bell *et al.*, 1977).

The rDNA array shows heterogeneity with respect to three basic modes of variation: length, nucleotide sequence, and base modification (Appels and Dvorak, 1982a; Siegel and Kolacz, 1983; Waldron *et al.*, 1983; Jorgensen *et al.*, 1987). A fourth mode of rDNA variation is variation in the copy number of rDNA repeats per haploid genome (Rogers and Bendich, 1987b). However, since this variation is a quantitative character, it is not always measured and is of marginal use in phylogenetic analyses. In fact, rDNA copy number is unlikely to be informative taxonomically because it is highly variable within species; it is, however, useful in genetic analyses.

The regions of the rDNA that correspond to the mature sequence are the coding regions, which are separated by the internal transcribed spacers (ITS). The 5' leader sequence is known as the external transcribed spacer (ETS). The two ITSs consist of several hundred base pairs which separate the 18S cistron from the 5.8S cistron and the 5.8S cistron from the 26S cistron. The region separating the transcription units is called the nontranscribed, or intergenic spacer (IGS). The IGS ranges in length from one to eight kilobases in most plants (Jorgensen and Cluster, 1988). The IGS region contains tandem subrepeat sequences which vary interspecifically in length (usually from 100 to 200 base pairs). Within species, subrepeat length varies only slightly. Overall, natural selection appears to act to conserve functionally important RNA secondary structure (Wheeler and Honeycutt, 1988). Higher amounts of variation at the sequence level are seen among closely related taxa for the portions of rRNAs (in this case, 5S rRNAs) constituting the helical stem portions of the molecule. Changes may occur by compensating substitutions, in which base paired nucleotides in opposite strands of the stem change in reponse to one another. A degree of mismatch is apparently tolerated; short bulges do not prevent helical structure formation as long as they are flanked by base paired regions (Wheeler and Honeycutt, 1988).

For large rRNAs, a range of functional constraints along the RNA product is also observed. There are some stretches of rRNA sequence that do not vary in any known organism. Conversely, there are portions of the large 26S rRNA gene that are either variable or lacking altogether in some species of the same genus (Hadjilov, 1980). This heterogeneity of rRNA evolution, then, makes comparisons possible across a broad phylogenetic spectrum (Zimmer *et al.*, 1989). We recently

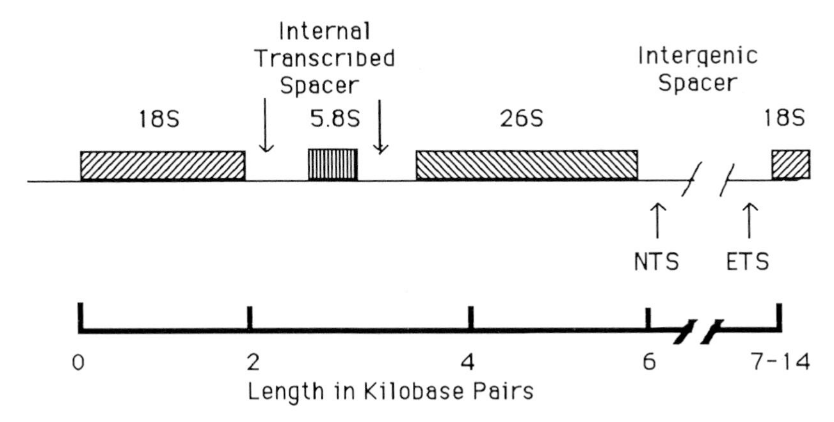

FIG 1 A typical nuclear rRNA repeat unit in plants.

TABLE 1

Base Pairing Differences[a,b] between Maize and Rice and Maize and Soybean

	Maize vs Rice	Maize vs Soybean
Loop	50 (4.1)	94 (7.8)
Stem	6 (0.8)	19 (2.8)

a) Numbers in parentheses are percent changes out of the total number of bases in that class.

b) In maize, 680 bases are involved in pairing interactions (Stem), and 1206 bases are involved in non-pairing interactions (Loop).

completed a comparative secondary structure analysis of maize, rice, and soybean 18S rRNA (Knaak *et al.*, unpublished), based on the proposed secondary structure of maize 18S rRNA (Gutell *et al.*, 1985). We found that proposed base paired positions are conserved at least five times as strongly as single stranded loop positions between corn and rice, and almost three times as much between corn and soybean (Table 1).

If we consider the rest of the rDNA repeat unit, the transcribed spacer regions show intermediate levels of variability in interspecific studies, consistent with some degree of functional constraint existing for these sequences (Appels and Dvorak, 1982b; Sytsma and Schaal, 1985). Substantial sequence conservation is observed in the ITS's of closely related species, reflecting the presumptive presence of processing signals, for which a degree of conservation is expected (Appels and Dvorak, 1982a; Schaal and Learn, 1988).

The intergenic spacer (IGS) is the most rapidly evolving portion of the rDNA. Since it shows the greatest amount of variation within and among populations, it is the region most useful for microevolutionary studies. It appears that the IGS consists of at least three regions that may differ in function and evolve at different rates (Appels and Dvorak, 1982a,b; Yakura *et al.*, 1984; Lassner and Dvorak, 1986; McMullen *et al.*, 1986; Toloczyki and Felix, 1986; Rogers *et al.*, 1986; Appels *et al.*, 1986). While there are no constraints on the primary sequences in this region, there may be some functional constraints on their overall structure (Federoff, 1979; Reeder, 1984; Dover and Flavell, 1984). It has been suggested that the length variable region of the spacer has a role in the recombination and evolution of the rDNA gene family (Federoff, 1979). Recent work in *Xenopus* has demonstrated that the subrepeats within the length-variable region also may have an enhancer effect on RNA polymerase I transcription (Reeder, 1984). Similar observations have been made in studies of wheat (Martini *et al.*, 1982; Flavell, 1986; Thompson and Flavell, 1988).

B. General structure of organelle rDNA repeats

With respect to rRNA genes, the plant mitochondrial and chloroplast genomes are more closely related to bacterial genomes. They contain 5S rRNA, but lack 5.8S rRNA (Leaver and Gray, 1982). Higher plant chloroplast rRNA genes exist, with few exceptions, in large rRNA encoding inverted repeats (Palmer, 1986). It is the only large repeated gene sequence commonly found in these genomes and includes the 16S, 23S, and 5S rRNA coding sequences (Figure 2). In addition, chloroplast ribosomes contain a component designated 4.5S rRNA. Thus far, 4.5S has been found only in higher plants (Bowman and Dyer, 1979).

Among plants, only a few angiosperm mitochondrial rDNAs have been well characterized. They have been shown to encode 26S, 18S, and 5S rRNA genes (Figure 3). The 18S and 26S rRNA genes are always found unlinked, at least 16 kilobases (Kb) apart, while the 5S gene is always found linked to the 18S gene (reviewed in Leaver *et al.*, 1983; Sederoff, 1984). The mitochondrial-specific 5S rRNA appears to be unique to angiosperms, and has yet to be found in animal or fungal mitochondrial genomes. There has been no indication of variation in the coding content of the mitochondrial rRNA genes examined among different plants (Bonen and Gray, 1980; Stern *et al.*, 1982; Iams and Sinclair, 1982; Huh and Gray, 1982), although overall plant mitochondrial genome size varies widely, even within species.

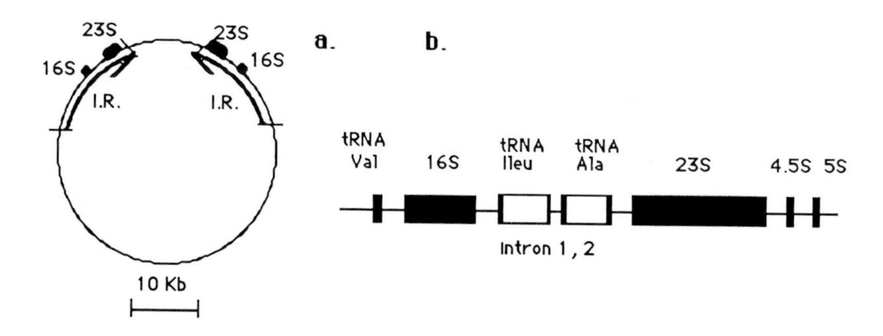

FIG. 2 Chloroplast r RNA transcription unit in a higher plant, tobacco. (a) Location of the rRNA gene in chloroplast DNA. The position of the inverted repeat is denoted by arrows. (b) Organization of the rRNA transcription unit (from Palmer, 1986. Filled in regions denote actual coding sequences.

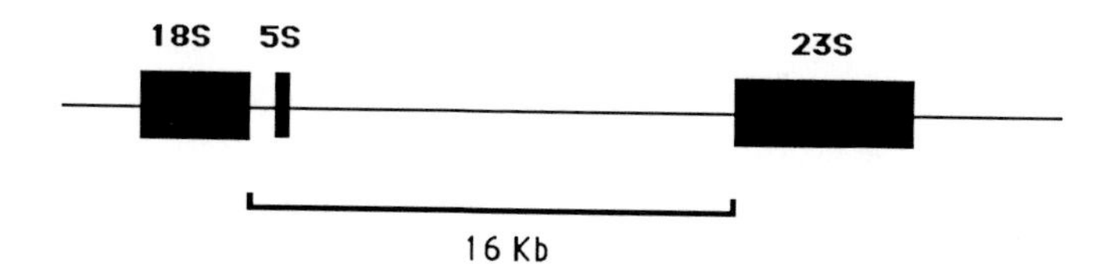

FIG. 3 Arrangement of the rRNA genes in maize mitrochondria (from Stern *et al.*, 1982).

C. Comparison of molecular approaches

A number of different approaches for biosystematic studies in plants are available at the molecular level. These include hybridization and hybrid melting point (Tm) analyses (including determination of rRNA gene copy number), restriction site analyses, and sequencing studies. The latter techniques are now enhanced by the advent of polymerase chain reaction (PCR) technology, which greatly facilitates the ability to obtain usable sequences. Below, we discuss each of these categories.

1. Hybridization and hybrid melting point analyses

While the analysis of genome relatedness by hybrid Tm (heteroduplex denaturation) analysis has traditionally been considered of limited value in producing well-resolved phylogenies, this technique has proven valuable in examining groupings of bacteria (Klipper-Bälz and Schleifer, 1981; Mordarski *et al.*, 1981). Hybrid Tm studies using probes cloned from the rDNA region allow many taxa to be analysed within relatively short periods of time.

DNA-DNA hybridization is also useful for estimating the number of rRNA genes in a species, and their quantitative variation in gene expression during different phases of plant development. This technique has its shortcomings, as it provides only distance measures (i.e., degree of relatedness). It can be a useful tool, however, in conjunction with restriction endonuclease and sequencing studies. For example, Appels and Dvorak (1982a), showed, by using probes of ~100 bp subcloned from rDNA spacer subregions, that a profile of differential sequence divergence could be observed across the spacer region of several *Triticum* species with Tm analysis. DNA-DNA hybridization and melting experiments also have been useful in determining relative numbers of rRNA genes in different lineages. Carrozza *et al.* (1980) found that greater than 100% variation in copy number exists in different cultivars of wheat; similar results have been obtained for maize (Rivin *et al.*, 1986). These studies reinforce the utility of hybridization studies in surveying quantitative variation in evolutionary studies.

2. Restriction site analysis

a) Methodology and initial characterization

Restriction endonucleases (RE's) cleave the DNA at specific nucleotide sequences and generate fragments of reproducible size. These fragments are then electrophoresed, hybridized to radioactively labeled RNA or DNA probes, and autoradiographed. The observed band fragments are then compared among individuals or species. While the resulting fragment patterns do not always discriminate among species, the distinctive restriction fragment length polymorphism (RFLP) patterns produced may be useful in a diagnostic role (see, e.g., Springer *et al.*, 1989). One problem with inferring a tree from raw fragment data is that lengths of DNA segments are measured. A single mutation can produce the loss of a restriction site leading to a dramatic change in the length of the DNA segment that is observed.

Real sequence divergences of 3 to 10% may yield much higher divergence estimates (from 40 to 80%) with fragment pattern based metrics (Rae *et al.*, 1981). One other pitfall in RE analysis in plants is that a high degree of methylation occurs in plant DNA, particularly on CG and CXG residues, thereby obscuring many potential restriction sites (Gerlach and Bedbrook,1979; Goldsbrough and Cullis, 1981; Jorgensen *et al.*, 1982; Siegel and Kolacz, 1983; Jorgensen and Cluster, 1988). Thus, models in which rDNA evolution is based solely upon RE length measurements must be carefully examined.

The first step in characterizing the rDNA array is the determination of the unit repeat length. This can be determined by a partial digest with an enzyme which has a unique restriction site within the repeat unit.. For ribosomal genes in particular, the hexanucleotide RE's *Xba*I, *Eco*RV and *Eco*RI are most often useful. The size interval between the resulting fragments gives an estimate of the length of the repeat unit (Doyle *et al.*, 1985). This technique, however, may not be feasible with species where the rDNA repeat units are very large (e.g., from gymnosperms). In that situation, double digestions with RE's which have a unique site within the repeat unit, combined with differential gene probing, should be used (Zimmer *et al.*, 1988).

RE studies of the nuclear rDNA unit in plants have also provided much information about the physical organization of these genes. A listing of plant rDNA units analyzed by RE mapping is presented in Table 2.

b) Characterization and analysis of variation in plant rRNA genes

The ultimate goal in RE analysis is to create a map detailing the location of specific recognition sites along the rDNA array. RE maps can be determined either by using partial digestion techniques, or by analysis of fragments created by single and double digestions with a set of RE's.

Since the coding regions are relatively invariant, many RE studies have focussed on characterization of the ribosomal gene's IGS and other highly variable regions (e.g., Long and Dawid, 1980; Appels and Honeycutt, 1986). Within the grasses closely related to wheat, for example, a central spacer region defined by the RE *TaqI* contained rapidly evolving sequences (Appels and Dvorak, 1982b). The 133 bp repetitive sequence within this *TaqI*-defined spacer region showed little cross hybridization with the analogous region of a related cereal (rye). Published maize and wheat IGS sequences cannot even be aligned using computers (Tolocyzki and Feix, 1986; Rogers and Bendich, 1987b). Finally, it has been observed that for the IGS region, **intra**specific length variation may be just as extensive as **inter**specific. This has been best documented in studies of wheat cultivars (Dvorak and Chen, 1984; Appels and Dvorak, 1982b). Such variation also has been documented in barley (Saghai-Maroof *et al.*, 1984) and in broad bean (Yakura *et al.*, 1984).

Species possessing a characteristic length fragment are not necessarily more

TABLE 2
Nuclear rDNA Restriction Enzyme Analyses in Plants

Genus*	RFLP Maps +/–	# Enzymes	Reference
Avena (Oat)	+	1	Cluster *et al.*, 1984
Brassica (Cabbage)	–	1	Quiros *et al.*, 1987
	+	8	Delseny *et al.*, in press
Claytonia	+	2	Doyle *et al.*, 1984
Clematis sp.	+	1	Learn & Schaal, 1987
Cucurbis (Pumpkin)	+	11	Siegel & Kolacz, 1983
Cytisus	+	7	Jorgensen & Cluster, 1988
Glycine (Soybean)	+	6	Doyle & Beachy, 1985
Hordeum (Barley)	+	5	Gerlach & Bedbrook, 1979
Lathyrus	+	7	Jorgensen & Cluster, 1988
Lilium	–	4	Von Kalm & Smyth, 1983
	+	16	Von Kalm *et al.*, 1986
Linum (Flax)	+	5	Goldsbrough & Cullis,1981
	–	7	Saghai-Maroof *et al.*, 1984
Lisianthus sp.	–	11	Sytsma & Schaal, 1985
Lupinus	+	7	Jorgensen *et al.*, 1988
Lycopersicon (Tomato)	+	8	Vallejo *et al.*, 1986
Medicago	+	7	Jorgensen & Cluster, 1988
Petunia	+	4	Waldron *et al.*, 1983
Phaseolus (Dwarf bean)	+	7	Jorgensen & Cluster, 1988
Phlox	+	6	Schaal *et al.*, 1987
Pisum (Pea)	+	2	Ellis *et al*, 1984
	+	31	Jorgensen *et al.*, 1987
	+	7	Jorgensen & Cluster, 1988
Raphanus (Radish)	–	4	Delseny *et al.*, 1979; 1983
	+	7	Jorgensen & Cluster, 1988
Tolmiea, Tellema	–	3	Doyle *et al*, 1985
Trifolium	+	7	Jorgensen & Cluster, 1988
Triticum (Wheat)	+	7	Appels & Dvorak, 1982a
	+	5	Appels *et al.*, 1986
Vicia (Broad bean)	–	1	Rogers & Bendich, 1987a
	+	2	Yakura & Tanifuji, 1983
Wisteria	+	7	Jorgensen *et al.*, 1988
Zea (Maize)	+	15	Zimmer *et al.*, 1988

* Common names are given in parentheses.

closely related than those that do not, since extensive spacer length variation may-also be found among individuals in natural populations. In *Phlox,* individual plants may contain as many as five types of rDNA length variants (Schaal *et al.*, 1987). Clear differentiation of rDNA repeat units among populations of a *Phlox* subspecies occurs, and there are also differences between two different subspecies in both the number and the lengths of rDNA repeats. Learn and Schaal (1987) found that within a population of *Clematis fremontii* sampled along a geographical transect, rDNA spacer variation was due, at least partly, to the numbers of subrepeats in the IGS in an individual. Increases in overall spacer length, however, were not correlated with position along the transect. In a RE mapping study of rDNA variation in the polyploid complex *Claytonia*, Doyle *et al.* (1984) observed that different races within the complex showed distinctive fragment patterns due to both restriction site changes and repeat length variation. These patterns were found to be consistent with conventional systematics of the group.

Applications of RE analysis of the rDNA repeat in relation to phylogeny have been carried out in *Brassica* (Quiros *et al.*, 1987; Delseny *et al.*, 1979), *Zea*, and *Sorghum* (see next section). For all three studies, the data did not establish a relationship between spacer length variants and taxonomic relatedness. Genetic studies with *Brassica* RFLP's for rDNA and other loci did, however, yield information on amphidiploid species origins and on rDNA array localization (Quiros *et al.*, 1987). Likewise, Doyle and Beachy (1985), in studies of rDNA repeat length variation and RFLP's within subgenera of *Glycine,* have found that their data are consistent with that of previous biosystematic studies, and may even be indicative of divergences not observed with other methods.

3. Sequencing studies

For a molecular evolutionary study, the complete sequencing of the region of interest must be the ultimate characterization and comparative tool. The use of primary DNA sequences in phylogenetic studies is still in its early stages, but holds great potential for answering difficult evolutionary questions. Sequencing is particularly useful in phylogenetic studies where traditional methods of classification such as morphological and paleontological analysis have not proven conclusive. Unresolved trees may be due to the lack of consistent characters (synapomorphies) in all the taxa to be examined, or to a patchy fossil record. Primary sequencing makes possible the examination of the pattern of mutation that is of the numbers of transitions and transversions that occurred, which may be useful in character weighting, and in assessing evolution over long time spans (Lake, 1987).

The development of "universal" rRNA-specific primers (Lane *et al.*, 1985; Qu *et al.*, 1983; Hamby *et al.*, 1988) has now made it possible to selectively sequence many portions of the rDNA repeat unit. These primers can be used for the complete sequencing of cloned rDNA genes, for the partial sequencing of rRNA directly through the use of reverse transcriptase (Hamby *et al.*, 1988), or for the amplification of rDNA via the polymerase chain reaction prior to direct DNA sequencing (Kocher and White, 1989).

A list of ribosomal gene sequences for which the primary sequence is known is presented in Table 3 below. Surprisingly, considering the number of plant rRNA sequences now available, apart from studies in our laboratory (see next section), there has been little work done on **phylogenetic** analysis of plants based on the larger rRNAs. Phylogenies from 5S and 5.8S rRNAs have been made (Hori *et*

TABLE 3
Listing of Plant rRNA Sequences That Have Been Completely Sequenced

rRNA	Species/Genus	Reference
5S	*Acorus calumus*	Bobrova *et al.*, 1987
	Chlamydomonas reinhardii	Darlix & Rochaix, 1981
	Chlorella	Luehrsen & Fox, 1981
	Dryopteris	Hori *et al.*, 1984
	Equisetum	Hori *et al.*, 1984
	Ginkgo	Hori *et al.*, 1985
	Helianthus	Vandenberghe *et al.*, 1984
	Lemna minor	Vandenberghe *et al.*, 1984
	Ligularia cathifolia	Bobrova *et al.*, 1987
	Lophocolea	Katoh *et al.*, 1983
	Lupinus	Rafalski *et al.*, 1982
	Lycopersicon	Vandenberghe *et al.*, 1984
	Lycopodium	Hori *et al.*, 1985
	Metasequoia	Hori *et al.*, 1985
	Nitella	Katoh *et al.*, 1983
	Secale	Vandenberghe *et al.*, 1984
	Scenedesmus	Luehrsen & Fox, 1981
	Spinacia	Delihas *et al.*, 1981
	Spirogyra	Hori *et al.*, 1985
	Psilotum	Hori *et al.*, 1985
	Phaseolus	Vandenberghe *et al.*, 1984
	Triticum aestivum	Barber & Nichols, 1978
	Ulva	Lim *et al.*, 1984
	Vicia faba	Vandenberghe *et al.*, 1984
	Zamia	Hori *et al.*, 1985
5.8S	*Lupinus luteus*	Rafalski *et al.*, 1983
	Oryza sativa	Takaiwa *et al.*, 1985
	Triticum vulgare	McKay *et al.*, 1980
	Vicia faba	Tanaka *et al.*, 1980
	Chlamydomonas reinhardii	Darlix & Rochaix, 1981
18S	*Glycine max*	Eckenrode *et al.*, 1985
	Oryza sativa	Takaiwa *et al.*, 1984
	Zea mays	Messing *et al.*, 1984
	Zamia	Nairn & Ferl, 1988
26S	*Oryza sativa*	Takaiwa *et al* ., 1985
Cp 16S	*Chlamydomonas reinhardtii*	Dron *et al.*, 1982
	Nicotiana tobaccum	Shinozaki *et al.*, 1986
	Zea mays	Schwartz & Kössel, 1980
Cp 23S	*Marchantia*	Ohyama *et al.*, 1983
	Nicotiana tobaccum	Shinozaki *et al.*, 1986
	Zea mays	Edwards & Kössel, 1981

rRNA	Species/Genus	Reference
Cp 5S	*Marchantia*	Yamano *et al.*, 1984
	Nicotiana tobaccum	Shinozaki *et al.*, 1986
Cp 4.5S	*Acorus*	Bobrova *et al.*, 1987
	Dryopteris	Takaiwa *et al.*, 1982
	Ligularia	Bobrova *et al.*, 1987
	Marchantia	Ohyama *et al.*, 1983
	Mnium rugicum	Troitsky *et al.*, 1984
	Nicotiana tobaccum	Takaiwa & Sugiara, 1980
	Spinacia	Kumagai *et al.*, 1982
	Spirodela	Keus *et al.*, 1983
	Triticum aestivum	Wildeman & Nazar, 1980
	Zea mays	Edwards & Kössel, 1981
Mt 26S	*Zea mays*	Dale *et al.*, 1984
Mt 18S	*Oenothera*	Brennicke *et al.*, 1985
	Zea mays	Chao *et al.*, 1984
Mt 5S	*Glycine max*	Morgens *et al.*, 1984
	Oenothera	Brennicke *et al.*, 1985
	Triticum aestivum	Spencer *et al.*, 1981
	Zea mays	Erdman *et al.*, 1983

al., 1985) for green plants. However, results from these studies are not reliable due to the limited information present in these genes (Steele *et al.*, 1989). Studies involving the 18S and/or 26S sequence data (Takaiwa *et al.*,1984,1985; Messing *et al.*,1984; Nairn and Ferl, 1988) have been restricted primarily to comparisons with non-plant taxa (e.g. *Xenopus,* yeast, rat and *Eschericia coli*) and are not very informative with respect to plant systematics. Bobrova *et al.* (1987) have constructed trees based upon sequences from chloroplast 4.5S and plant cytoplasmic 5S rRNA's. Only three taxa are available for which both sequences are known, and it appears that placement of these taxa on the trees is similar to traditional assignments.

An issue that has been addressed in part with published rRNA sequence data is the time of divergence of angiosperms. This is a major event in higher plant evolutionary history, with a somewhat confusing fossil record. Fossil record-based dates for the first appearance of angiosperms range from the late Triassic through the mid-Cretaceous (Crane *et al.*, 1989). Recently, this divergence time has been estimated (using molecular clock assumptions) by reconstructing phylogenetic trees from chloroplast DNA sequences (including chloroplast rDNA), in conjunction with sequences of the 18S and 26S rRNAs from rice, maize, soybean, and cycad (18S) and rice and lemon (26S) (Wolfe *et al.*,1989). Two independent phylogenetic approaches have led to an estimate of the angiosperm divergence as 200 million years ago. This contrasts significantly from the date proposed by Martin *et al.*

(1989), from consideration of the sequence divergence of the gene encoding glyceraldehyde-3 phosphate dehydrogenase (G3PDH) of higher plants, of 320 million years.

III. Work in Our Laboratories

A. *Restriction endonuclease studies*

1. Within the Andropogonae

Work in our laboratory and that of our collaborator, Jeffrey Bennetzen, has examined the structure of nuclear rRNA genes in the tribe Andropogonae of the family Poaceae. Initially we studied maize and its wild relatives, the teosintes (genus *Zea*) and *Tripsacum* (Zimmer *et al.*, 1988). Restriction digests of the 18S, 26S, and 5.8S genes have yielded essentially a single map for the approximately 10,000 repeat units per individual. Both length and site variation were detected among species and were concentrated in the intergenic spacer region of the rDNA repeat unit. Digestion of these nuclear DNAs with *Bam*H1 and subsequent hybridization with a 5S RNA gene specific probe allowed determination of the size of the 5S repeat unit in these species. Different lineages of *Zea* and different species in *Tripsacum* were characterized by distinct repeat unit types. The degree and nature of ribosomal gene variation, as discussed in the review of other RE work above, was insufficient to develop phylogenies within *Zea* and between *Zea* and *Tripsacum*. The rDNA and 5S RNA restriction site variation among the species can be interpreted, however, in a phylogenetic context and agrees with biochemical, karyotypic, and morphological evidence that places maize closest to the Mexican teosintes.

We detected in this initial work on *Zea* a restriction site polymorphism for *Eco*R1 in the 26S coding region of some inbred maize lines. This polymorphism has proven useful for probing the degree and localization of heterogeneity in the maize ribosomal array for DNA sequence, DNA methylation, chromatin structure and gene expression (Jupe, 1988).

We have also, using fewer restriction endonucleases, determined the rDNA repeat unit structures for two other genera in the tribe Andropogonae, *Sorghum* and *Saccharum* (Springer *et al.*, 1989). Unlike the situation in *Zea*, no site variation was observed in *Sorghum*, and each line or species usually contained a single repeat unit length type. Key restriction site polymorphisms can be used to differentiate among *Zea, Sorghum*, and *Saccharum*, while repeat unit length can discriminate among certain races and lines of *Sorghum* and detect *Sorghum* hybrid lines.

2. The Louisiana irises

We have used restriction fragment length polymorphism (RFLP) analysis of diagnostic rDNA markers to examine the process of introgressive hybridization

(Anderson, 1949) in two Louisiana irises, *Iris hexagona* and *I. fulva*. The two species are characterized on the basis of different fixed length RFLPs in the nuclear rDNA, while hybrids show a combination of both RFLPs in varying proportions, as shown by densitometry scans. The pattern of these RFLP markers along the geographical transects where populations were sampled indicates that bidirectional introgression is occurring in parapatric associations of *I. fulva* and *I. hexagona* (Arnold *et al.*, 1989; Bennett *et al.*, 1989).

B. Sequencing studies

1. Ribosomal RNA studies within the seed plants

In our labs, we have directed a major effort to reconstructing the phylogeny of green plants based on comparisons of the primary structure of the major rRNAs. Our technique involves the direct sequencing of rRNA through dideoxy (Sanger) sequencing, using oligonucleotide primers complementary to conserved regions of each molecule in the presence of deoxynucleotides, dideoxynucleotides, and reverse transcriptase. The phylogenetically informative positions are identified from the aligned sequences and analyzed using the phylogenetic inference program PAUP (Swofford, 1985). Alternate topologies are then explored with PAUP.

In particular, the resolution of distant phylogenetic relationships among higher plants is one of the most challenging problems in systematic biology. Even when similar features are shared by different taxa, homology is often uncertain. There is no fossil record establishing continuity of structure for most characters, due to a weak preservation of plant remains. The embryological data suggesting relationships are hard to interpret because the genetic processes underlying these features are incompletely understood. In addition, numerous cases of parallel and convergent evolution of morphological characters leave particular branches difficult to resolve on morphologically based phylogenetic trees.

We have sequenced five regions of the 18S rDNA and two or three regions of the 26S rDNA gene for 46 taxa. These eight primers used are referred to as 18E, 18G, 18H, 18J, 18L, 26C, 26D, and 26F (Hamby *et al.*, 1988). Each primer sequenced gave about 210 nucleotides, for a total of over 1700 nucleotides sequenced and compared. The ratio of transitions to transversions was approximately 2 for each primer sequenced.

Our results have indicated a unique origin for seed plants, and, within them, for the flowering plants, as well as coherence of the Gnetales (Zimmer *et al.*, 1989). The placement of the Gnetales relative to the angiosperms is equivocal. For seed plants, coherence of many certifiable families and/or orders such as the grasses (Poaceae; Hamby and Zimmer, 1988), Alismatales, Arales, Coniferales, and Cycadales are observed. Our trees differ from traditional interpretations and from recent cladistic hypotheses in that we find a group consisting of the Piperales, Nymphaeales, and monocots (Donoghue and Doyle's Paleoherb group, 1989), rather than the magnoliids, at the base of the angiosperm radiation (Figure 4). Movement of the magnoliid group (which has traditionally thought to be basal in the angiosperm radiation) to the base of the tree resulted in an increase of 27 steps over the most parsimonious tree length (1684 steps), indicating support for our non-traditional hypothesis.

The placement of the aquatic taxa *Sagittaria, Potamogeton,* and *Echinodorus* indicates that the monocots may be a paraphyletic assemblage, although clustering of the monocots only adds two steps to the tree. Likewise, the placement of *Ceratophyllum* and *Nelumbo* with *Magnolia* suggests that the classical "Nymphaeales" are not a natural group. Clustering of all the so-called water lilies adds 16 steps to our tree. Donoghue and Doyle (1989a,b), with morphological characters, have also placed *Nelumbo* away from *Nymphaea* and *Cabomba.*

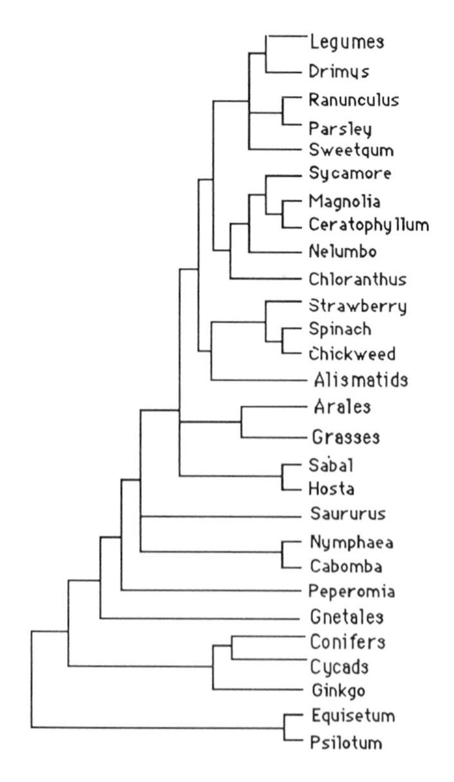

FIG 4 This is the general structure for the most parsimonious tree for angiosperm phylogeny[a,b], derived from a preliminary PAUP run with 46 taxa (C.I. = 0.492).
a. Genera included in each category: 1) Legumes- *Glycine, Pisum*; 2) Alismatids- *Sagittaria, Potamogeton, Echinodorus;* 3) Arales- *Colocasia, Pistia*; 4) Grasses- *Zea, Tripsacum, Sorghum, Saccharum, Oryza, Avena, Hordeum, Triticum, Arundinaria;* 5) Gnetales- *Gnetum, Welwitschia, Ephedra* (2 spp.); 6) Conifers- *Pinus, Juniperus, Cryptomeria.*
b. Branch lengths are not drawn to scale.

2. Ribosomal RNA studies within the green algae

A widely cited current classification of the green algae (Mattox and Stewart, 1984) is based on comparative cytology with an emphasis on ultrastructural features. Five classes of algae (Chlorophyceae, Pleurastrophyceae, Ulvophyceae, Charophyceae, and Micromonadophyceae) were described, and several older concepts of green algal phylogeny and taxonomy were modified or abandoned in the new system. The proposed classification provides a nested series of phylogenetic hypotheses that can be tested with cladistic analysis of additional, independent characters like those provided by cytoplasmic rRNA sequence data from both the 18S and 26S subunits. Our preliminary results from such studies support some aspects of the five-class proposal but challenge others. The monophyly of the Charophyceae and land plants is supported in our analysis, but the Ulvophyceae, Pleurastrophyceae, and Micromonadophyceae do not appear to be monophyletic assemblages. Although our studies have not focussed on the Chlorophyceae *per se*, there is evidence supporting the derived nature of the group as suggested by Mattox and Stewart (1984). Table 4 summarizes our preliminary results relative to the original Mattox and Stewart scheme.

Although ultrastructural features are widely accepted as important taxonomic characters, in some cases the interpretation of relationships inferred from ultrastructural characters differs among investigators. For example, ultrastructural features of the flagellar apparatus provide a basis for placing the unicellular flagellate *Tetraselmis* among the Micromonadophyceae (also known as the Prasinophyceae; Moestrup, 1978). However, ultrastructural features of cyto- and karyokinesis led Mattox and Stewart (1984) to place *Tetraselmis* in the Pleurastrophyceae. Analysis

TABLE 4
Five Classes of Green Algae sensu Mattox and Stewart 1984 ???

Algal Group*	rRNA Results
Chlorophyceae	[still unclear]
Pleurastrophyceae	[not a monophyletic assemblage]
Ulvophyceae	[not a monophyletic assemblage]
Charophyceae	[monophyletic with land plants]
Micromonadophyceae	[not a monophyletic assemblage]

* Groups are listed from bottom to top in order of most primitive to most derived according to Mattox and Stewart, 1984.

of the sequence data supports the latter concept. *Pedinomonas* is one of the smallest, naked (i.e., non scaly) micromonads, and differs in rRNA sequences from all other green algae we have studied thus far (more than 50 taxa). Also, it appears to be basal in all analyses we have performed. Sequencing of additional *Pedinomonas* species is underway.

The Trentepohliales (subaerial, branched, filamentous green algae) have been classified as Ulvophyceae, Charophyceae, or Pleurastrophyceae (see e.g., Chapman, 1984; Chapman and Henk, 1986; and Raven, 1987). The uncertainty about the phylogenetic position of this order reflects the lack of resolution provided by analysis of cytological, ultrastructural, and biochemical characters and the subjectivity inherent in previous assessments (which have not included cladistic analysis.). The analysis of rRNA sequences supports a clear affinity between the Trentepohliales and species in the Ulvophyceae. If our initial inference is upheld as we complete our survey of the green algae, then the evolutionary history of the phragmoplast-type cytokinesis (which is present in only three or four taxa in the Charophyceae and in two Trentepohliales in the Ulvophyceae) will be an interesting puzzle that may involve convergent or very early evolution of the phragmoplast.

Although the Trentepohliales clearly appear in a clade with some ulvophycean taxa, Ulvophyceae themselves are apparently not a monophyletic group. The initial analyses support a more traditional treatment of "ulotrichalean" greens as a distinct lineage not within the Ulvophyceae as proposed by Mattox and Stewart (1984). It must be noted, however, that we are at a very early stage of this study, and the inferences are preliminary.

Our study of the Chlorophyceae has focussed largely on the taxonomically difficult genus *Chlamydomonas*, which contains hundreds of species. There appear to be at least three lineages within the genus. The level of sequence divergence between *C. reinhardtii* and *C. moewusii*, purportedly intrageneric lineages, is (for some regions) comparable to that between angiosperms and gymnosperms (Jupe *et al.*, 1988). The preliminary data support the hypothesis that a third lineage, the *C. humicola* lineage, includes the monads *Polytoma* and *Haematococcus*. The genealogy we infer from the nuclear-encoded rRNA sequences is, thus far, congruent with that being developed by Lemieux and coworkers (see e.g., Lemieux *et al.*, 1985; Turmel *et al.*, 1987), with whom we are collaborating.

Overall, our preliminary results demonstrate that rRNA sequence data are a useful supplement to other types of taxonomic character data sets and that they will help to provide a better understanding of green algal phylogeny. This clearer understanding will lead to further revision of the taxonomy of Chlorophyta and bring us closer to a natural classification.

3. *Future prospects*

Future work in our laboratory will continue with the addition of taxa to the angiosperm sequencing project data set, particularly of other magnoliids and the Aristolochiales, and of the aquatic monocots, which traditionally have been difficult to classify. The latter will be used to help resolve the exact relationships among the monocots and could give a more stable arrangement of the basal taxa in angiosperm radiation.

In the future, the addition of polymerase chain reaction (Kocher and White, 1989) and automated sequencing technologies (Prober *et al.*, 1987) will greatly increase the efficiency of data procurement. For rDNAs, primers have been developed that selectively amplify nuclear or chloroplast rRNA coding regions from total DNA preparations (Hamby, unpublished data).

Other areas we are pursuing are studies of chloroplast ribosomal RNA gene evolution, as well as the influence of secondary structure on ribosomal RNA evolution. Unlike nuclear genomes, most chloroplast genomes contain rRNA genes in two copies only. Thus, using cp rRNAs for phylogenetic comparisons makes it possible to avoid any potential ambiguities arising from heterogeneity in the nuclear repeat units. An ideal approach is to utilize both of these methods in building a comprehensive data set. Likewise, the addition of secondary structure information to our data set will enable us to develop weighting schemes for our phylogenetic reconstruction methods.

Acknowledgements

This work was supported in part by NSF Grant DEB-BSR-8615212 and LEQSF Contract 86-LBR-(048)-08 to E.A.Z., by NSF Grant DEB-BSR-8722739 to R.L.C. and E.A.Z., by NSF Grant BSR-8308420 to R.L.C., and by an American Iris Society Grant to B.D. Bennett and M.L.A.

References

Anderson, E. (1949). *Introgressive Hybridization.* John Wiley and Sons, Inc. N.Y.

Appels, R. and Dvorak, J. (1982a). Relative rates of divergence of spacer and gene sequences within the rDNA region of the species in the Triticeae: Implications for the maintenance of homogeneity of a repeated gene family. *Theor. Appl. Genet.* **63**, 361-365.

Appels, R. and Dvorak, J. (1982b). The wheat ribosomal DNA spacer region: Its structure and variation in populations and among species. *Theor. Appl. Genet.* **63**, 337-348.

Appels, R. and Honeycutt, (1986). rDNA: Evolution over a billion years. *In* "DNA Systematics", Vol. II, Plant DNA. (S.K. Dutta, ed.), pp. 81-136. CRC Press, Boca Raton, Fla.

Appels, R., McIntyre, C.L., Clark, B.C. and May, C.E. (1986). Alien chromatin in wheat: Ribosomal DNA spacer probes for etecting specific nucleolar organizing region loci introduced into wheat. *Can. J. Genet. Cytol.* **28**, 665-672.

Arnold, M.L., Bennett, B.D. and Zimmer, E.A. (1989). Natural hybridization between *Iris fulva* and *I. hexagona*: Pattern of ribosomal DNA variation. *Evolution* (in press).

Arnheim, N. (1983). Concerted evolution of multigene families. *In* "Evolution of Genes and Proteins" (M. Nei and R.K. Kohn, eds), pp. 38-61. Sinauer, Sunderland, Mass.

Barber, J.C. and Nichols, M. (1978). The characterization of 5S rRNA from *Triticum aestivum*. *Can. J. Biochem.* **56**, 357-364.

Bell, G.I., DeGennaro, L.J., Gelfand, D.H., Bishop, R.J., Valenzuela, P. and Rutter, W.J. (1977). Ribosomal RNA genes of *Saccharomyces cerevisiae*. I. Physical map of the repeating unit and location of the regions coding for 5S, 5.8S, 18S, and 25S ribosomal RNAs. *J. Biol. Chem.* **252**, 8118-8125.

Bennett, B.D., Arnold, M.L., Grace, J.B. and Zimmer, E.A. (1989). An ecological and molecular genetic analysis of two parapatric species of *Iris*. (Mss. submitted.).

Bobrova, V.K., Troitsky, A.V., Ponomarev, A.G. and Antonov, A.S. (1987). Low-molecular-weight rRNA sequences and plant phylogeny reconstruction: Nucleotide sequences of chloroplast 4.5S rRNAs from *Acorus calamus* (Araceae) and *Ligularia calthifolia* (Asteraceae). *Plant Syst. Evol.* **156**, 13-27.

Bonen, L. and Gray, M.W. (1980). Organization and expression of the mitochondrial genome of plants. I. The genes for wheat mitochondrial ribosomal and transfer RNA: Evidence for an unusual arrangement. *Nucleic Acids Res.* **8**, 319-335.

Bowman, C.M. and Dyer, T.A. (1979). 4.5S ribonucleic acid, a novel ribosome component in the chloroplasts of flowering plants. *Biochem. J.* **183**, 605-621.

Brennicke, A., Moller, S. and Blanz, P.A. (1985). The 18S and 5S ribosomal RNA genes in *Oenothera* mitochondria: Sequence rearrangements in the 18S and 5S rRNA genes of higher plants. *Mol. Gen.Genet.* **198**, 404-410.

Brown, D.C., Wensink, P.C. and Jordan, E. (1972). A comparison of the ribosomal DNAs of *Xenopus laevis* and *Xenopus mulleri*: The evolution of tandem genes. *J. Molec. Biol.* **63**, 57-73.

Carrozza, M.L., Giorgi, L. and Cremonini. (1980). Nuclear DNA content and number of ribosomal RNA genes in cultivars and selected lines of *Durum* wheat. *Z. Pflanzenzücht.* **84**, 284-293.

Chao, S., Sederoff, R. and Levings III, C. S. (1984). Nucleotide sequence and evolution of the 18S ribosomal RNA gene in maize mitochondria. *Nucleic Acids Res.* **12**, 6629-6645.

Chapman, R.L. (1984). An assessment of the current state of our knowledge of the Trentepohliaceae. *In* "Systematics of the Green Algae" (D.E.G. Irving and D.M. John, eds), pp. 230-250. Academic Press, London.

Chapman. R.L. and Henk, M.C. (1986). Phragmoplasts in cytokinesis of *Cephaleuros parasiticus* (Chlorophyta) vegetative cells. *J. Phycol.* **22**, 83-88.

Cluster, P.D., Jorgensen, R.A., Bernatsky, R., Hakim-Elahi, A. and Allard, R.W. (1984). The genetics and geographical distribution of ribosomal DNA spacer-length variation in the wild oat *Avena barbata*. *Genetics* **107**, s21 [Abstract].

Cockburn, A.J., Newkirk, M.J., and Firtel, R.A. (1976). Organization of the ribosomal RNA genes of *Dictyostelium discoideum*: Mapping of the nontranscribed spacer regions. *Cell* **9**, 605-613.

Crane, P.R., Donoghue, M.J. Doyle, J.A. and Friis, E.M. (1989). Response to Martin *et al.*, *Nature* **342**, 131-132.

Cullis, C. and Davies, D.R. (1975). Ribosomal DNA amounts in *Pisum sativum*. *Genetics* **81**, 485-492.

Dale, R.M.K., Mendu,, N, Ginsburg, H. and Kridl, J.C. (1984). Sequence analysis of the maize mitochondrial 26S rRNA gene and flanking regions. *Plasmid* **11**, 141-150.

Darlix, J-L. and Rochaix, J-D. (1981). Nucleotide sequence and structure of cytoplasmic 5S RNA and 5.8S RNA of *Chlamydomonas reinhardii*. *Nucleic Acids Res.* **9**, 1291-1300.

Delgarno, L. and Shine, J. (1977). Ribosomal RNA. *In* "The Ribonucleic Acids" (P.R. Stewart and D.S. Letham, eds), pp. 195. Springer-Verlag, New York.

Delihas, N., Anderson, J., Sprouse, H.M., Kashdan, M. and Dudock, B. (1981). The nucleotide sequence of spinach cytoplasmic 5S ribosomal RNA *J. Biol. Chem.* **256**, 7515-7517.

Delseny, M., Aspart, L., Cooke, R., Grellet, F. and Penon, P. (1979). Restriction analysis of radish nuclear genes coding for rRNA: evidence for heterogeneity. *Biochem. Biophys. Res. Commun.* **91**, 540-547.

Delseny, M., Laroche, M. and Penon, P. (1983). Sequence heterogeneity in radish nuclear ribosomal genes. *Plant Sci. Lett.* **30**, 107-119

Donoghue, M.J. and Doyle, J.A. (1989a). Phylogenetic studies of seed plants and angiosperms based on morphological characters. *In* "The Hierarchy of Life" (B. Fernholm, K. Bremer and H. Jornvall, eds) Elsevier Science Publishers B.V.

Donoghue. M.J. and Doyle, J.A. (1989b). Phylogenetic analysis of angiosperms and the relationships of Hamamelidae. *In* "Evolution, Sytematics, and Fossil History of the Hamamelidae" (P.R. Crane and S. Blackmore, eds), Clarendon Press, Oxford.

Dover, G.A. and Flavell, R.B. (1984). Molecular coevolution: DNA divergence and the maintenance of function. *Cell* **38**, 622-623

Doyle, J.J. and Beachy, R.N. (1985). Ribosomal gene variation in soybean (*Glycine*) and its relatives. *Theor. Appl. Genet.* **70**, 369-376.

Doyle, J.J., Beachy, R.N. and Lewis, W.H. (1984). Evolution of rDNA in *Claytonia* polyploid complexes. *In* "Plant Biosystematics" (W.F. Grant, ed.), pp. 321-341. Academic Press, London.

Doyle, J.J., Soltis, D.E. and Soltis, P.S. (1985). An intergeneric hybrid in the Saxifragaceae: Evidence from ribosomal RNA genes. *Am. J. Bot.* **72**, 1388-1391.

Dron, M., Rahire, M. and Rochaix, J.-D. (1982). Sequence of the chloroplast 16S rRNA gene and its surrounding regions of *Chlamydomonas reinhardtii. Nucleic Acids Res.* **10**, 7609-7618.

Dvorak, J. and Chen, K.C. (1984). Distribution of nonstructural variation between wheat cultivars along chromosome arm 6Bp: evidence from the linkage map and physical map of the arm. *Genetics* **106**, 325.

Eckenrode, V.K., Arnold, J. and Meagher, R.B. (1985). Comparison of the nucleotide sequence of soybean 18S rRNA with the sequences of other small-subunit rRNAs. *J. Mol. Evol.* **21**, 259-269.

Edwards, K., Bedbrook, J., Dyer, T. and Kossel, H. (1981). 4.5S rRNA from *Zea mays* chloroplasts shows structural homology with the 3' end of prokaryotic 23S rRNA. *Biochem. Intern.* **2**, 533-538

Ellis, T.H.N., Davies, D.R., Castleton, J.A. and Bedford, I.D. (1984). The organization and genetics of rDNA length variants in peas. *Chromosoma* (Berl.) **91**, 74-81

Erdmann, V.E., Huysmans, E. Vandenberghe and DeWachter, R. (1983). Collection of published 5S and 5.8S ribosomal RNA sequences. *Nucleic Acids Res.* **11**, r105-r131.

Federoff, N.V. (1979). On spacers. *Cell* **16**, 697-710.

Flavell, R.B. (1986). The structure and control of expression of ribosomal RNA genes. *Oxf. Surv. Plant Mol. Cell Biol.* pp. 251-274.

Flavell, R.B., O'Dell, M., Sharp, P. and Nevo, E. (1986). Variation in the intergenic spacer of ribosomal DNA of wild wheat, *Triticum dicoccoides,* in Israel. *Mol. Biol. Evol.* **3**, 547-558.

Gerlach, W.L. and Bedbrook, J.R. (1979). Cloning and characterization of ribosomal RNA genes from wheat and barley. *Nucleic Acids Res.* **7**, 1869-1889

Goldsbrough, P.B. and Cullis, C.A. (1981). Characterization of the genes for ribosomal RNA in flax. *Nucleic Acids Res.* **9**, 1301-1309.

Gutell, R.R., Weiser, B., Woese, C.R. and Noller, H.F. (1985). Comparative anatomy of 16-S-like ribosomal RNA. *Prog. Nucleic Acid Res. Mol. Biol.* **32**, 155-215.

Hadjilov, A.A. (1980). Biogenesis of ribosomes in eukaryotes. *In* "Subcellular Biochemistry" Vol 7, (D.B. Roodyn, ed.). Plenum Press, New York.

Hamby, R.K. and Zimmer, E.A. (1988). Ribosomal RNA sequences for inferring phylogeny within the grass family (Poaceae). *Plant Syst. Evol.* **160**, 29-37.

Hamby, R.K., Sims, L.E., Issel, L.E. and Zimmer, E.A. (1988). Direct ribosomal RNA sequencing: optimization of extraction and sequencing techniques for work with higher plants. *Plant Mol. Biol. Rep.* **6**, 179-197.

Hori, H., Lim, B. and Osawa, S. (1985). Evolution of green plants as deduced from 5S rRNA sequences. *Proc. Natl. Acad. Sci. USA* **82**, 820-823.

Hori, H., Osawa, S., Takaiwa, F. and Sugiura, M. (1984). The nucleotide sequences of 5S rRNA from a fern *Dryopteris acuminata* and a horsetail *Equisetum arvense. Nucleic Acids Res.* **12**, 1573-1576.

Huh, T.Y. and Gray, M.W. (1982). Conservation of ribosomal RNA gene arrangement in the mitochondrial DNA of angiosperms. *Plant Mol. Biol.* **1**, 245-249.

Iams, K.P. and Sinclair, J.H. (1982). Mapping the mitochondrial DNA of *Zea mays*: Ribosomal gene localization. *Proc. Natl. Acad. Sci. USA.* **79**, 402-409.

Ingle, I., Timmis, I.N. and Sinclair, J. (1975). The relationship between satellite deoxyribonucleic acid, ribosomal ribonucleic acid gene redundancy, and genome size in plants. *Plant Physiol.* **55**, 496-501.

Jordan, B.R., Jourdan, R. and Jacq, B. (1976). Late steps in the maturation of *Drosophila* 26S ribosomal RNA: Generation of 5.8S and 2S RNAs by cleavages occurring in the cytoplasm. *J. Mol. Biol.* **101**, 85-105.

Jorgensen, R.A. and Cluster, P.D. (1988). Modes and tempos in the evolution of nuclear ribosomal DNA: New characters for evolutionary studies and new markers for genetic and population studies. *Ann. Mo. Bot. Gard.* **75**, 1238-1247.

Jorgensen, R.A., Cuellar, R.E. and Thompson, W.F. (1982). Modes and tempos in the evolution of nuclear encoded ribosomal RNA genes in legumes. *Carnegie Inst. Washington Year Book.* **81**, 98-101

Jorgensen, R.A., Cuellar, R.E., Thompson, W.F. and Kavanagh, T.A. (1987). Structure and variation in ribosomal RNA genes of pea. *Plant Mol. Biol.* **8**, 3-12.

Jupe, E.A. (1988). DNA methylation, chromatin structure and expression of maize ribosomal genes. Ph.D. Dissertation, Louisiana State University, Baton Rouge, Louisiana.

Jupe, E.R., Chapman, R.L. and Zimmer, E.A. (1988). Nuclear ribosomal RNA genes and algal phylogeny - the *Chlamydomonas* example. *Biosystems* **21**, 223-230.

Katoh, K., Hori, H. and Osawa, S. (1983). The nucleotide sequences of 5S rRNAs from four Bryophyta species. *Nucleic Acids Res.* **11**, 5671-5674.

Katoh, A., Yakura, K. and Tanafuji, S. (1985). Repeated DNA sequences found in the large spacer of *Vicia faba* rDNA. *Biochem. Biophys. Acta* **825**, 411-415.

Keus, R.J.A., Roovers, D.J., Dekker, A.F. and Groot, G.S.P. (1983). The nucleotide sequence of the 4.5S and 5S rRNA genes and flanking regions from *Spirodela oligorhiza* chloroplasts. *Nucleic Acids Res.* **11**, 3405-3410.

Klipper-Bälz, R. and Schleifer, K.H. (1981). DNA-rRNA hybridization studies among staphylococci and some other gram positive bacteria. *FEBS Microbiol. Lett.* **10**, 357-361.

Kocher, T.D. and White, T.J. (1989). Evolutionary analysis via PCR. *In* "PCR Technology: Principles and Applications for DNA Amplification" (H. Erlich, ed.). Stockton Press, New York, N.Y.

Kolosha, V.O., Kryukov, V.M. and Fodor, I. (1986). Sequence analysis of *Citrus limon* DNA coding for 26S rRNA. *FEBS* **197**, 89-92.

Kumagai, I., Pieler,T., Subramanian, A.R. and Erdman, V.A. (1982). Nucleotide sequence and secondary structure analysis of spinach chloroplast 4.5S rRNA. *J. Biol. Chem.* **257**, 12924-12928.

Lane, D.J., Pace, B., Olsen, G.J., Stahl, D.A., Sogin, M.L. and Pace, N.R. (1985). Rapid determination of 16S ribosomal RNA sequences for phylogenetic analyses. *Proc. Natl. Acad. Sci. USA.* **82**, 6955-6959.

Lake, J.A. (1987). A rate-independent technique for analysis of nucleic acid sequences: Evolutionary parsimony. *J. Mol. Evol.* **4**, 167-191.

Lassner, M. and Dvorak, J. (1986). Preferential homogenization between adjacent and alternate subrepeats in wheat rDNA. *Nucleic Acids Res.* **14**, 5499-5512.

Learn, G.H. and Schaal, B.A. (1987). Population subdivision for ribosomal DNA repeat variants in *Clematis fremontii*. *Evolution* **41**, 433-438.

Leaver, C.J. and Gray, M.W. (1982). Mitochondrial genome organization and expression in higher plants. *Annu. Rev. Plant Physiol.* **33**, 373-402

Leaver, C.J., Dixon, L.K., Hack, E., Fox, T.D. and Dawson, A.J. (1983). Mitochondrial genes and their expression in higher plants. *In* "Structure and Function of Plant Genomes" (O. Ciferri and L. Dure, eds), pp. 347-361. Plenum Press, NY.

Lemieux, B., Turmel, M. and Lemieux, C. (1985). Chloroplast DNA variation in *Chlamydomonas* and its potential application to the systematics of this genus. *Biosytems* **18**, 293- 298.

Lim, B.L., Kubota, M., Katoh, K., Hori, H. and Osawa, S. (1984). Phylogeny of land plants deduced from 5S rRNA sequences. *Proc. Japan Acad.* **60**, 178-182.

Long, E.O. and Dawid, I.B. (1980). Repeated genes in eukaryotes. *Annu. Rev. Biochem.* **49**, 727-764.

Luehrsen, K.R. and Fox, G.E. (1981). Secondary structure of eukaryotic cytoplasmic 5S ribosomal RNA. *Proc. Natl. Acad. Sci. USA*. **78**, 2150-2154.

McKay, R.M. (1981). The origin of plant chloroplast 4.5S RNA. *FEBS Lett.* **123**, 17-23.

McKay, R.M., Spencer, D.I., Doolittle, W.F. and Gray, W.M. (1980). Nucleotide sequences of wheat embryo cytosol 5S + 5.8S ribosomal ribonucleic acids. *Eur. J. Biochem.* **112**, 561-576.

Maizels, N. (1976). *Dictyostelium* 17S, 25S, and 5S rDNAs lie within a 38,000 base pair repeated unit. *Cell* **9**, 431-438.

Martin, W., Gierl, A. and Saedler, H. (1989). Molecular evidence for pre-cretaceous angiosperm origins. *Nature* **339**, 46-48.

Martini, G., O'Dell, M. and Flavell, R.B. (1982). Partial inactivation of the wheat nucleolus organizers by the nucleolus organizer chromosomes of *Aegilops umbellata*. *Chromosoma* (Berl.) **84**, 687-692.

Mattox, K.R. and Stewart, K.D. (1984). Classification of the green algae: A concept based on comparative cytology. *In* "Systematics of the Green Algae" (D.E.G. Irving and D.M. John, eds), pp. 29-72. Academic Press, London.

Messing, J., Carlson, J., Hagen, G., Rubenstein, I. and Oleson, A. (1984). Cloning and sequencing of the ribosomal RNA genes in maize; the 17S region. *DNA* **3**, 31-40.

McMullen, M.D., Hunter, B., Phillips, R.L. and Rubenstein, I. (1986). The structure of the maize ribosomal DNA spacer region. *Nucleic Acids Res.* **14**, 4953-4968.

Moestrop, O. (1978). On the phylogenetic validity of the flagellar apparatus in green algae and other chlorophyll a+b containing organisms. *Biosystems* **10**, 117-144.

Monier, R. (1974). 5S RNA. *In* "Ribosomes" (M. Nomura, A. Tissieres and P. Lingyel, eds), pp. 141-178. Cold Spring Harbor Press, New York.

Mordarski, M., Kacz, A., Goodfellow, M., Pulverer, G., Peters, G. and Schumacher-Perdreau, F. (1981). Ribosomal RNA similarities in the classification of *Staphylococcus*. *FEMS Microbiol. Lett.* **11**, 159-169.

Morgens, P.H., Grabau, E.A., and Gesteland, R.F. (1984). A novel soybean mitochondrial transcript resulting from a DNA rearrangement involving the 5S rRNA gene. *Nucleic Acids Res.* **12**, 5665-5684.

Nairn, C.J. and Ferl, R.J. (1988). The complete nucleotide sequence of the small-subunit ribosomal RNA coding region for the cycad *Zamia pumila*: Phylogenetic implications. *J. Mol. Evol.* **27**, 133-141.

Ohyama, K., Yamano, Y., Fukuzawa, H., Komano, T., Yamagishi, H., Fujimoto, S. and Sugiura, M. (1983). Physical mappings of chloroplast DNA from liverwort Marchantia polymorpha L. suspension cultures. *Mol. & Gen. Genet.* **189**, 1-9.

Palmer, J.D. (1986). Evolution of chloroplast and mitochondrial DNA in plants and algae. *In* "Molecular Evolutionary Genetics" (R.J. MacIntyre, ed.), pp. 131-240. Plenum Publishing Co.

Prober, J.M, Trainor, G.L., Dam,R.J., Hobbs, F.W, Robertson, C.W., Zagursky, R.J., Cocuzza, A.J. Jensen, M.A. and Baumeister, K. (1987). A system for rapid DNA sequencing with fluorescent chain-terminating dideoxynucleotides. *Science* **238**, 336-341.

Qu, L.H., Michot, B. and Bachellerie, J.P. (1983). Improved methods for structure probing in large RNAs: A rapid 'heterologous' sequencing approach is coupled to the direct mapping of nuclease accessible sites. *Nucleic Acids Res.* **11**, 5903-5920.

Quiros, C.F., Ochoa, S.F., Kianian, S.F. and Douches, D. (1987). Analysis of the *Brassica oleracea* genome by the generation of *B. campestris-oleracea* chromosome addition lines: Characterization by isozymes and rDNA genes. *Theor. Appl. Genet.* **74**, 758-766.

Rae, P.M.M., Barnett, T. and Murtiff, V.L. (1981). Nontranscribed spacers in *Drosophila* ribosomal RNA. *Chromosoma* (Berl.) **82**, 637-642.

Rafalsky, J.A., Wiewiorowski, M. and Soll, D. (1983). Organisation of ribosomal DNA in yellow lupine (*Lupinus lutens*) and sequence of the 4.8S gene. *FEBS Lett.* **152**, 241-246.

Raven, J.A. (1987). Biochemistry, biophysics and physiology of chlorophyll b-containing algae: Implications for taxonomy and phylogeny. *In* "Progress in Phycology" (F.E. Round and D.J. Chapman, eds), **5**, 1-122. Biopress Ltd., Bristol.

Reeder, R.H. (1974). Ribosomes from eucaryotes: Genetics. *In* "Ribosomes" (M. Nomura, A. Tissieres and P. Lengyel, eds), pp. 489. Cold Spring Harbor Press, New York.

Reeder, R.H. (1984). Enhancers and ribosomal gene spacers. *Cell* **38**, 349-351

Rivin, C.J., Cullis, C.A. and Walbot, V. (1986). Evaluating quantitative variation in the genome of *Zea mays* . *Genetics* **113**, 1009-1019.

Rogers, S.O. and Bendich, A.J. (1987a). Heritability and variability in ribosomal RNA genes of *Vicia faba*. *Genetics* **117**, 285-295.

Rogers, S.O. and Bendich, A.J., (1987b). Ribosomal RNA genes in plants: Variability in copy number and in the intergenic spacer. *Plant Mol. Biol.* **9**, 509-520.

Rogers, S.O., Honda, S. and Bendich, A.J. (1986). Variation of the ribosomal RNA genes among individuals of *Vicia faba*. *Plant Mol. Biol.* **6**, 339-345.

Rubin, G.M. and Sulston, J.E. (1973). Physical linkage of the 5S cistrons to the 18S and 28S ribosomal RNA cistrons in *Saccharomyces cerevisiae*. *J. Mol. Biol.* **79**, 521-530.

Saghai-Maroof, M.A., Soliman, K.M., Jorgensen, R.A. and Allard, R.W. (1984). Ribosomal DNA spacer length polymorphisms in barley: Mendelian inheritance, chromosomal location, and population dynamics. *Proc. Natl. Acad. U.S.A.* **81**, 8014-8018.

Schaal, B.A. and Learn, G.H., Jr. (1988). Ribosomal DNA variation within and among plant populations. *Ann. Mo. Bot. Gard.* **75**, 1207-1216.

Schaal, B.A., Leverich, W.J. and Nieto-Sotelo, J. (1987) Ribosomal DNA variation in the native plant *Phlox divaricata*. *Mol. Biol. Evol.* **4**, 611-621.

Schwartz, Z. and Kössel, H. (1980). The primary structure of 16S rRNA from *Zea mays* chloroplast is homologous to *E.Coli*. 16S rRNA. *Nature* **283**, 739-742

Sederoff, R.R. (1984). Structural variation in mitochondrial DNA. *Adv. Genet.* **22**, 1-108.

Shinozaki, K. Ohme, M., Tanaka, M.,Wakasugi, T., Hayashida, N., Matsubayashi, T., Zaita, N., Chunwongse, J., Obokota, J., Yamaguchi-shinozaki, K., Ohto, C., Torazawa, K., Meng, B.Y., Sugita, M., Deno, H., Kamogashira, T., Yamada, K., Kuasda, J., Takaiwa, F., Kato, A., Tohdoh, N., Shimada, H. and Sugiura, M. (1986). The complete nucleotide sequence of the tobacco chloroplast genome: its gene organization and expression. *EMBO J.* **5**, 2043-2049.

Siegel, A. and Kolacz, K., (1983). Heterogeneity of pumpkin ribosomal DNA. *Plant Physiol.* **72**, 166-171.

Southern, E.M. (1975). Detection of specific sequences among DNA fragments separated by gel electrophoresis. *J. Molec. Biol.* **98**, 503-517.

Spencer, D.F, Bonen, L. and Gray, M.W. (1981). Primary sequence of wheat mitochondrial 5S ribosomal ribonucleic acid: Functional and evolutionary implications. *Biochemistry* **20**, 4022-4029.

Springer, P., Zimmer, E.A. and Bennetzen, J.L. (1989). Genomic organization of the ribosomal DNA of *Sorghum* and its close relatives. *Theor. Appl. Genet.* **77**, 844-850.

Steele, K.P., Holsinger, K.E., Janen. R.K. and Taylor, R.W. (1989). Estimating the reliability of sequence data for phylogenetic analysis- An example using the 5S rRNA from green plants (in press).

Stern, D.B., Dyer, T.A. and Lonsdale, D.M. (1982). Organization of the mitochondrial ribosomal RNA genes of maize. *Nucleic Acids Res.* **10**, 3330-3340.

Swofford, D.L. (1985). PAUP: Phylogenetic analysis using parsimony, Version 2.4. Champaign: Illinois Natural History Survey.

Sytsma, K.J. and Schaal, B.A. (1985). Phylogenetics of the *Lisianthius skinneri* (Gentianaceae) species complex in Panama utilizing DNA restriction fragment analysis. *Evolution* **39**, 594-608.

Takaiwa, F. and Sugiura, M. (1980). The nucleotide sequence of 4.5S rRNA from tobacco chloroplasts. *Nucleic Acids Res.* **8**, 4125-4129.

Takaiwa, F., Kusuda, M., and Sugiura, M. (1982). The nucleotide sequence of chloroplast 4.5S rRNA from a fern, *Dryopteris acuminata. Nucleic Acids Res.* **10**, 2257-2260.

Takaiwa, F., Oono, K. and Sugiura, M. (1984). The complete nucleotide sequence of a rice 17S ribosomal RNA gene. *Nucleic Acids Res.* **12**, 5441-5448.

Takaiwa, F., Oono, K., Iida, Y. and Sugiura, M. (1985). The complete nucleotide sequence of a rice 26S ribosomal RNA gene. *Gene* **37**, 255-289.

Tanaka, Y., Dyer, T.A. and Brownlee, G.G. (1980). An improved direct RNA sequence method: Its application to *Vicia faba* 5.8S ribosomal RNA. *Nucleic Acids Res.* **4**, 2495-2502.

Thompson, W.F. and Flavell, R.B. (1988). DNase I sensitivity of rRNA genes in chromatin and nucleolar dominance in wheat. *J. Mol. Biol.* **204**, 535-548

Toloczyki, C. and Feix, G. (1986). Occurence of 9 homologous repeat units in the external spacer region of a nuclear maize rRNA gene unit. *Nucleic Acids Res.* **14**, 4969-4986.

Troitsky, A.V., Bobrova, V.K., Ponomarev, A.G. and Antonov, A.S. (1984). The nucleotide sequence of chloroplast 4.5S rRNA from *Mnium rugicum (Bryophyta)*: Mosses also possess this type of RNA. *FEBS Lett.* **176**, 105-109.

Turmel, M., Bellemare, G. and Lemieux, C. (1987). Physical mapping of differences between the chloroplast DNAs of the interfertile algae *Chlamydomonas eugametos* and *Chlamydomonas moewusii. Curr. Genet.* **11**, 543-552.

Vandenberghe, A., Chen, M-W., Dams, E., deBaere, R., deRoeck, E., Huysmans, E. and deWachter, R. (1984). The corrected nucleotide sequences of 5S RNAs from six angiosperms. *FEBS* **171**, 17-22.

Vallejos, C.E., Tanksley, S.D. and Bernatzky, R. (1986). Localization in the tomato genome of DNA restriction fragments containing sequences homologous to the rRNA (45S), the major chlorophyll a/b binding polypeptide and the ribulose bisphosphate carboxylase genes. *Genetics* **112**, 93-105.

Von Kalm, L. and Smyth, D.R. (1983). Variation in the ribosomal RNA genes of *Lilium henryi. In* "Manipulation and Expression of Genes in Eukaryotes", pp. 239-240. Academic Press, Australia.

Von Kalm, L. Vize, P.D. and Smyth, D.R. (1986). An undermethylated region in the spacer of ribosomal RNA genes of *Lilium henryi. Plant Mol. Biol.* **6**, 33-39.

Waldron, J., Dunsmuir, P. and Bedbrook, J. (1983). Characterization of the rDNA repeat units in the Mitchell *Petunia* genome. *Plant Mol. Biol.* **2**, 57-65.

Wheeler, W.C. and Honeycutt, R.L. (1988). Paired sequence differences in ribosomal RNAs: Evolutionary and phylogenetic implications. *Mol. Biol. Evol.* **5**, 90-96.

Wildeman, A. G. and Nazar, R.N. (1980). Nucleotide sequence of wheat chloroplastid 4.5S ribonucleic acid. Sequence homologies in 4.5S rRNA species. *J. Biol. Chem.* **255**, 11896-11900.

Williams, S.W., DeBry, R.W. and Feder, J.L. (1988). A commentary on the use of ribosomal DNA in systematic studies. *Syst. Zool.* **37**, 60-62.

Wolfe, K. H., Gouy, M., Yang, Y-W., Sharp, P.M. and Li, W-H. (1989). Date of the monocot-dicot divergence estimated from chloroplast DNA sequence data. *Proc. Natl. Acad. Sci. USA.* **84**, 9045-9058.

Yakura, K., Kato, A. and Tanifuji, S. (1984). Length heterogeneity in the large spacer of *Vicia faba* rDNA is due to the differing number of a 325 bp repetitive sequence element. *Mol. & Gen. Genet.* **193**, 400-405.

Yakura, K. and Tanifuji, S. (1983). Molecular cloning and restriction analysis of *Eco* RI-fragments of *Vicia faba* rDNA. *Plant Cell Physiol.* **24**, 1327-1330.

Yamano, Y., Ohyawa, K. and Komano, T. (1984). Nucleotide sequences of chloroplast 5S rRNA from cell suspension cultures of the liverworts *Marchantia polymorpha* and *Jungermannia subulata. Nucleic Acids Res.* **12**, 4621-4624.

Zimmer, E. A., Martin, S.L., Beverley, S.M., Kan, Y.W. and Wilson, A.C. (1980). Rapid duplication and loss of genes coding for the alpha chains of hemoglobin. *Proc. Natl. Acad. Sci. U.S.A.* **77**, 2158-2162.

Zimmer, E.A., Jupe, E.R. and Walbot, V. (1988). Ribosomal gene structure, variation and inheritance in maize and its ancestors. *Genetics* **120**, 1125-1136

Zimmer, E.A., Hamby, R.K., Arnold, M.L., LeBlanc, D.A. and Theriot, E.L. (1989). Ribosomal RNA phylogenies and flowering plant evolution. *In* "The Hierarchy of Life" (B. Fernholm, K. Bremer and H. Jornvall, eds), pp. 205-214. Elsevier Science Publishers B.V.

10 Molecular Approach to Plant Systematics from Protein Sequence Comparison

KEISHIRO WADA

Department of Biology, Faculty of Science, Kanazawa University, Marunouchi 1-1, Kanazawa, Ishikawa 920, Japan

I. Introduction

Since prehistory people have used edible and medical plants that grew in their environment. It was inevitable that they would come to classify plants as useful or harmful. This may have been the beginning of plant systematics. Today we can recognize great number of plant species in their characters, which are obtained from morphology, ecology, biochemistry and others. It is of critical importance for plant scientists to have a good understanding of plant systematics and the living components in the environment in which we live.

When amino acid sequences of insulins from domestic animals were determined in the early 1950s by F. Sanger, he suggested that the difference between amino acid sequences of insulin molecules from different animals was a reflection of species specificity. Since then, the sequences of many proteins have been reported and the comparison of a protein sequence from various species has led to the study of biological evolution and systematics at a molecular level. In the 1980s Dickerson introduced three-dimensional structural studies to the investigation of molecular evolution of proteins. Moreover, evolutionary study has advanced to a new stage with the recent remarkable progress of DNA sequence analyses. In combination with conventional studies, these investigations at a molecular level will make it possible to resolve problems in plant systematics that we have faced in herbarium and field work.

We investigated the molecular evolution of plant ferredoxins, because ferredoxins are widely distributed in plants, ranging from primitive prokaryotic algae to

BIOLOGICAL APPROACHES AND
EVOLUTIONARY TRENDS IN PLANTS ISBN 0-12-402960-4

higher plants, easily extracted and purified from plant tissues and their small molecular sizes facilitate structural analysis.

This paper focuses on the chloroplast-type 2Fe-2S ferredoxins. Amino acid sequences of the 62 ferredoxins so far established are compared to study the phylogenetic relationship of plants. Previous reports have been given in Symposia that were held at Leningrad (1975), Bayreuth (1982) (Matsubara and Hase, 1983) and Berlin (1987).

II. What is Ferredoxin?

Ferredoxin is one of the iron-sulfur proteins indispensable to all living cells. It has one or two redox centers consisting of equimolar amounts of non-heme iron and acid labile sulfur atoms. Ferredoxin functions in diverse redox systems as important electron carriers (Arnon, 1988).

In photosynthetic cells of green plants, light energy is converted to reducing power of ferredoxin, in addition to the chemical energy of ATP. The activated electrons in reduced ferredoxin are delivered to some important metabolic pathways, such as NADP reduction, nitrogen and sulfur assimilations (Bowsher *et al.*, 1988; Aketagawa and Tamura, 1980) and the thioredoxin system for enzyme regulation (Buchanan, 1980). It was recently realized that this 2Fe-2S type of ferredoxin was present in non-photosynthetic cells (Wada *et al.*, 1986). A new electron transfer system from NADPH supplied from the pentose phosphate pathway to the electron requiring enzyme reactions was proposed (Wada *et al.*, 1989).

In addition to sequence data, we have established the three-dimensional structure of a ferredoxin from cyanobacterium, *Spirulina platensis* (Fukuyama *et al.*, 1980).

Fig. 1 is a sequence alignment of 62 chloroplast-type ferredoxins. Some gaps were placed to get the maximal homology among them. Four cysteine residues, which contribute to the chelation of the iron-sulfur cluster, are all conserved and located at positions 43, 48, 51 and 82 from the amino terminus. Apart from these cysteine residues, only 6 residues (Pro-40, Ser-42, Gly-46, Gly-58, Thr-94 and Glu-97) are completely conserved. It is apparent that sequence homology is higher between the closely related species than between the distantly related ones. In the light of the three-dimensional structure, few amino acid substitutions were found in the important portion of the molecule, which contributes to holding the iron-sulfur

Fig. 1 Comparison of sequences of 2Fe-2S ferredoxins from higher plants to cyanobacteria. Amino acids are written by one letter abbreviations. Several gaps are inserted to geve higher homology among ferredoxin sequences. References are: Nos. 1, 2, 5-8, 12, 14-19, 26-30, 32, 34, 35, 37, 38, 41, 45-49 and 55-62, listed in Matsubara and Hase (1983). No. 3, Takahashi *et al.* (1983); No. 4, Sakai *et al.* (Unpublished); Nos. 9-11, Wada, *et al.* (1989); No. 13, Smeekens *et al.* (1985); Nos. 20-23, Sakai *et al.* (Unpublished); Nos. 24 and 25, Masui *et al.* (Unpublished); No. 31, Minami *et al.*(1985b); No. 33, Schmitter *et al.* (1988); No. 36, Minami *et al.* (1985a); No. 39, Nagashima, H., (Personal communication, 1989); No. 40, Inoue *et al.* (1984); No. 42, Uchida *et al.* (1988); No. 43, Masui *et al.* (1988a); No. 44, Ambler, R. P. (Personal communication); Nos. 50 and 51, Wada *et al.* (1988); No. 52, Masui *et al.* (1988b); No. 53, Alam *et al.* (1987); and No. 54, Chen *et al.* (1983). * Data from nucleotide sequences.

```
                                                                  10        20        30        40        50        60        70        80        90        100       110
                                                        ....|....|....|....|....|....|....|....|....|....|....|....|....|....|....|....|....|....|....|....|....|....|
No.  1 : Leucaena glauca                  1 --A-FKVKLLT-PDG-PKEFECPDDVYILDQAEELGIDLPYSCRAGSCSSCAGKLVEGDL-DQSDQSFLDDEQIEEGWVLTCAAYPRSDVVIETHKEEELTG
No.  2 : Spinacia oleracea I              2 -AA-YKVTLVT-PTG-NVEFQCPDDVYILDAAEEEGIDLPYSCRAGSCSSCAGKLKTGSL-NQDDQSFLDDDQIDEGWVLTCAAYPVSDVTIETHKEEELTA
No.  3 : Spinacia oleracea II             3 -AT-YKVTLVT-PSG-SQVIECGDDEYILDAAEEKGMDLPYSCRAGACSSCAGKVTSGSY-DQSDQSFLEDGQMEEGWVLTCIAYPTGDVTIETHKEEELTA
No.  4 : Fagopyrum cymosum                4 -AV-HKVKLVT-PEG-EKEFECPDDVYILDQAEELGIDLPYSCRAGSCSSCAGKVVAGAV-DQSDQSFLEDGQMEEGWVLTCVAYPTGPVVIETHKEEELTA
No.  5 : Medicago sativa                  5 -AS-YKVKLVT-PEG-TQEFECPDDVYILDHAEEEGIVLPYSCRAGSCSSCAGKVAAGEV-NQSDGSFLDDDQIEEGWVLTCVAYAKSDVTIETHKEEELTA
No.  6 : Sambucus nigra                   6 -AS-YKVKLIT-PDG-PQEFECPDDVYILDHAEELGIDIPYSCRAGSCSSCAGKLVAGSV-DQSDQSFLDDEQIEEGWVLTCVAYPKSDVTIETHKEEELTA
No.  7 : Petrocelinum crispum             7 -AT-YNVKLIT-PDG-EVEFKCDDDVYVLDQAEEEGIDIPYSCRAGSCSSCAGKVVSGSI-DQSDQSFLDDEQMDAGYVLTCHAYPTSDVVIETHKEEEIV-
No.  8 : Brassica napus                   8 -AT-YKVKFIT-PEG-EQEVECDDDVYVLDAAEEAGIDLPYSCRAGSCSSCAGKVVSGFV-DQSDESFLDDDQIAEGFVLTCAAYPTSDVTIETHKEEELV-
No.  9 : Raphanus sativus (leaf I)        9 -AT-YKVKFIT-PEG-EQEVECDDDVYVLDAAEEAGIDLPYSCRAGSCSSCAGKVVSGSV-DQSDQSFLDDDQIAEGFVLTCAAYPTSDVTIETHREEDMV-
No. 10 : Raphanus sativus (root I)       10 SAV-YKVKLIG-PDGQENEFDVPDDQYILDAAEEAGVDLPYSCRAGACSTCAGKIEKGQV-DQSDGSFLEDHHFEKGYVLTCVAYPQSDLVIHTHKEEELF-
No. 11 : Raphanus sativus (root II)      11 SAV-YKVKLIG-PEGEENEFEVQDDQFILDAAEEAGVDLPYSCRAGACSTCAGQIVKGQV-DQSEGSFLEDDHFEKGFVLTCVAYPQSDCVIHTHKETELF-
No. 12 : Arctium lappa                   12 -AT-YKVTLIT-PEG-KQEFEVPDDVYILDGAAEEVGDLPYSCRAGSCSSCAGKVTAGSV-DQSDGSFLDDDQMEAGWVLTCVAYPTSDVTIETHKEEELTA
No. 13 : Silene platensis                13 -AT-YKVTLITKESG-TVTFDCPDDVYVLDEAEEEGIDLPYSCRAGSCSSCAGCVVAGSV-DQSDQSFLDDDQIEAGWVLTCAAYPSADVTIETHKEEELTA
No. 14 : Phytolacca americana I          14 -AT-YKVTLVT-PSG-TQTIDCPDDTYVLDAAEEAGLDLPYSCRAGSCSSCTGKVTAGTV-DQEDQSFLDDDQIEAGFVLTCVAFPKGDVTIETHKEEDIV-
No. 15 : Phytolacca americana II         15 AAS-YKVTFVT-PSG-TNTITCPADTYVLDAAEESGLDLPYSCRAGACSSCAGKVTAGAV-NQEDGSFLEEEQMEAGWVLTCVAYPTSDVTIETHKEEDLTA
No. 16 : Phytolacca esculenta I          16 -AT-YKVTLVT-PSG-TQTIDCPDDTYVLDAAEEAGLDLPYSCRAGSCSSCTGKVTAGTV-DQEDQSFLDDDQIEAGFVLTCVAYPKGDVTIETHKEEDIA-
No. 17 : Phytolacca esculenta II         17 AAS-YKVTFVT-PSG-TKTITCPADTYVLDAAEDTGLDLPYSCRAGACSSCAGKVTAGSV-NQEDGSFLDEEQMEAGWVLTCVAYPTSDVTIETHKEEDLSA
No. 18 : Triticum aestivum               18 -AT-YKVKLVT-PEG-EVELEVPDDVYILDQAEEEGIDLPYSCRAGSCSSCAGKLVSGEI-DQSDQSFLDDDQMEAGWVLTCHAYPKSDIVIETHKEEELTA
No. 19 : Colocasia esculenta             19 -AT-YKVKLVT-PSG-QQEFQCPDDVYILDQAEEVGIDLPYSCRAGSCSSCAGKVKVGDV-DQSDGSFLDDEQIGEGWVLTCVAYPVSDGTIETHKEEELTA
No. 20 : Colocasia antiquorum I          20 -AA-YKVKLVT-PEG-EQEFECPDDVYILDQAEEVGIDLPYSCRAGSCSSCAGKVKGGDV-DQSDGSFLDDEQIEQGWVLTCVAYPTSDVVIETHKEEELTA
No. 21 : Colocasia antiquorum II         21 -AT-YKVKLVT-PSG-EQEFECPDDVYILDQAEEVGIDLPYSCRAGSCSSCAGKVKVGDV-DQSDGSFLDDDQIGEGWVLTCVAYPTGDVVIETHKEEELTA
No. 22 : Alocasia macrorhyza A           22 -AT-YKVKLVT-PQG-QQEFDCPDDVYILDQAEEEGIDLPYSCRAGSCSSCAGKVKQGEV-DQSDGSFLDDEQMEQGWVLTCVAFPTSDVVIETHKEEELTA
No. 23 : Alocasia macrorhyza B           23 -AT-YKVKLVT-PSGQPLEFECPDDVYILDQAEEEGIDLPYSCRAGSCSSCAGKVKNGNV-DQSDGSFLDDDQIGEGWVLTCVAYPTSDVVIETHKEEELTA
No. 24 : Trillium kamtschaticum          24 -AA-YKVKLIT-PEG-PVEFDCPDDVYILDQAEEEGIDLPFSCRAGSCSSCAGKIVAGEV-DQSDASFLDDDQLAEGWVLTCHAYPTSNVTIETHKEEELV-
No. 25 : Trillium smallii                25 -AA-YKVKLIT-PEG-PVEFDCPDDVYILDQAEEEGIDLPFSCRAGSCSSCAGKIVAGEV-DQSDASFLDDDQLAEGWVLTCHAYPTSNVTIETHKEEELV-
No. 26 : Gleichenia japonica             26 -AI-FKVKFLT-PDG-ERTIEVPDDKFILDAGEEAGLDLPYSCRAGACSSCTGKLLDGRV-DQSEQSFLDDDQMAEGFVLTCVAYPAGDITIETHAEEKL--
No. 27 : Equisetum telmateia I           27 --A-YKTVLKT-PSG-EFTLDVPEGTTILDAAEEAGYDLPFSCRAGACSSCLGKVVSGSV-DQSEGSFLDDGQMEEGFVLTCIAIPESDLVIETHKEEELF-
No. 28 : Equisetum telmateia II          28 --A-YKVTLKT-PDG-DITFDVEPGERLIDIASEKA-DLPLSCQAGACSTCLGKIVSGTV-DQSEGSFLDDEQIEQGYVLTCIAIPESDVVIETHKEDEL--
No. 29 : Equisetum arvense I             29 --A-YKTVLKT-PSG-EFTLDVPEGTTILDAAEEAGYDLPFSCRAGACSSCLGKVVSGSV-DESEGSFLDDGQMEEGFVLTCIAIPESDLVIETHKEEELF-
No. 30 : Equisetum arvense II            30 --A-YKVTLKT-PDG-DITFDVEPGERLIDIGSEKA-DLPLSCQAGACSTCLGKIVSGTV-DQSEGSFLDDEQIEQGYVLTCIAIPESDVVIETHKEDEL--
No. 31 : Marchantia polymorpha           31 --T-FKVTLNT-PTG-QSVIDVEDDEYILDAAEEAGLSLPYSCRAGACSSCAGKVTAGEV-DQSDESFLDDDQMDEGYVLTCIAYPTSDLTIDTHQEEALI-
No. 32 : Scenedesmus quadricauda         32 -AT-YKVTLKT-PSG-DQTIECPDDTYILDAAEEAGLDLPYSCRAGACSSCAGKVEAGTV-DQSDQSFLDDSQMEGGFVLTCVAYPTSDCTIATHKEEDLF-
No. 33 : Chlamydomonas reinhardtii       33 ----YKVTLKT-PSG-DKTIECPADTYILDAAEEAGLDLPYSCRAGACSSCAGKVAAGTV-DQSDQSFLDDAQMGNGFVLTCVAYPTSDCTIQTHQEEALY-
No. 34 : Dunaliella salina I             34 --S-YMVTLKT-PSG-EQKVEVSPDSYILDAAEEAGVDLPYSCRAGSCSSCAGKVESGTV-DQSDQSFLDDDQMDSGFVLTCVAYATSDCTIVTHQEENLY-
No. 35 : Dunaliella salina II            35 --A-YKVTLKT-PSG-DQTIEVSPDAYILDAAEEAGLDLPYSCRAGACSSCAGKVEAGTI-DQSDQSFLDDDQQGRGFVLTCVAYATSDCTISTHQEESLY-
No. 36 : Bryopsis maxima                 36 -AS-YKVTLKL-DDGSEAVIDCDEDSFILDVAEEEGIDIPFSCRSGSCSTCAGKIEGGTV-DQSEQTFLDDDQMEEGYVLTCVAYPTSDCTILTHQEEEMIG
No. 37 : Porphyra umbilicalis            37 -AD-YKIHLVSKEEGIDVTFDCSEDTYILDAAEEEGIELPYSCRAGACSTCAGKVTEGTV-DQSDQSFLDDEQMLKGYVLTCIAYPESDCTILTHVEQELY-
No. 38 : Cyanidium caldarium M           38 -AS-YKIHLVNKDQGIDETIECPDDQYILDAAEEQGLDLPYSCRAGACSTCAGKLLEGEV-DQSDQSFLDDDQVKAGFVLTCVAYPTSNATILTHQEESLY-
No. 39 : Cyanidium caldarium RK-1        39 --M-YKIQLVNQKEGVDVTINCPGDQYILDAAEEQGVDLPYSCRAGACSTCAGKLVKGSV-DQSDQSFLDEEQINNGFILTCVAYPTSDCVIQTHQEEALY-
No. 40 : Phodymenia palmata              40 -AVKYTVTLST-PGG-VEEIEGDETTYVLDSAEDQGIDLPYSCRAGACSTCAGIVELGTV-DQSDQSFLDDDQLNDSFVLTCVAYPTSDCQIKTHQEEKLY-
No. 41 : Bumilleriopsis filiformis       41 -AT-YSVTLVNEEKNINAVIKCPDDQFILDAAEEQGIELPYSCRAGACSTCAGKVLSGTI-DQSEQSFLDDDQMGAGFLLTCVAYPTSDCKVQTHAEDDLY-
No. 42 : Peridinium bipes                42 ----FKVTLDT-PDG-KKSFECPGDSYILDKAEEEGLELPYSCRAGSCSSCAGKVLTGSI-DQSDQAFLDDDQGGDGYCLTCVTYPTSDVTIKTHCESEL--
No. 43 : Ochromonas danica               43 -AK-YKVRLLNNSHNLDTIIDCPDDKFILEAAEDNNIELPYSCRAGSCSTCLGKITKGTV-DQSDQSFLDDEQIQEGFVLTCVAYPTSDVTISTHEEENLY-
No. 44 : Euglenoid strain LJ-1           44 -AT-YSVKLIN-PDG-EVTIECGEDQYILDAAEDAGIDLPYSCRAGACSSCTGIVKEGTV-DQSDQSFLDDDQMAKGFCLTCTTYPTSNCTIETHKEDDLF-
No. 45 : Synechocystis 6714              45 -AS-YTVKLIT-PDG-ENSIECSDDTYILDAAEEAGLDLPYSCRAGACSTCAGKITAGSV-DQSDQSFLDDDQIEAGYVLTCVAYPTSDCTIETHKEEDLY-
No. 46 : Aphanothece sacrum I            46 -AS-YKVTLKT-PDG-DNVITVPDDEYILDVAEEEGLDLPYSCRAGACSTCAGKLVSGPA-PDEDQSFLDDDQIQAGYILTCVAYPTGDCVIETHKEEALY-
No. 47 : Aphanothece sacrum II           47 -AT-YKVTLINEEEGINAILEVADDQTILDAGEEAGLDLPSSCRAGSCSTCAGKLVSGAAPNQDDQAFLDDDQLAAGWVMTCVAYPTGDCTIMTHQESEVL-
No. 48 : Aphanothece halophitica         48 -AS-YKVTLINEEMGLNETIEVPDDEYILDVAEEEGIDLPYSCRAGACSTCAGKIKEGEI-DQSDQSFLDDDQIEAGYVLTCVAYPASDCTIITHQEEELY-
No. 49 : Synechococcus sp.               49 -AT-YKVTLVR-PDGSETTIDVPEDEYILDVAEEQGLDLPFSCRAGACSTCAGKLLEGEV-DQSDQSFLDDDQIEKGFVLTCVAYPRSDCKILTNQEEELY-
No. 50 : Synechococcus 6301              50 -AT-YKVTLVNAAEGLNTTIDVADDTYILDAAEEQGIDLPYSCRAGACSTCAGKVVSGTV-DQSDQSFLDDDQIAAGFVLTCVAYPTSDVTIETHKEEDLY-
No. 51 : Synechococcus 6301*             51 -AT-YQVEVIY--QGQSQTFTADSDQSVLDSAQAAGVDLPASCLTGVCTTCAARILSGEV-DQPDAMGVGPEPAKQGYTLLCVAYPRSDLKIETHKEDELYALQFGQPG
No. 52 : Synechococcus 6307              52 -AS-YKVTLVNESEGLNKTIEVPDDQYILDAAEEQGIDLPYSCRAGACSTCAGKLTSGTV-DQSDQSFLDDDQIEAGFVLTCVAYPTSDCTIKTHTEEELY-
No. 53 : Anabaena 7120*(variabilis)      53 -AT-FKVTLINEAEGTKHEIEVPDDEYILDAAEEQGYDLPFSCRAGACSTCAGKLVSGTV-DQSDQSFLDDDQIEAGYVLTCVAYPTSDVVIQTHKEEDLY-
No. 54 : Anabaena variabilis 29413       54 -AT-FKVTLINEAEGTSNTIDVPDDEYILDAAEEEGYDLPFSCRAGACSTCAGKLVSGTV-DQSDQSFLDDDQVEAGYVLTCVAYPTSDVTIETHKEEDLY-
No. 55 : Aphanizomenon flos-aquae        55 -AT-YKVTLI-DAEGTTTTIDCPDDTYILDAAEEAGLDLPYSCRAGACSTCAGKLVTGTI-DQSDQSFLDDDQVEAGYVLTCVAYPTSDVTIETHKEEDLY-
No. 56 : Chlorogloeopsis fritschii       56 -AT-YKVTLINDAEGLNQTIEVDDDTYILDAAEEAGLDLPYSCRAGACSTCAGKIKSGTV-DQSDQSFLDDDQIEAGYVLTCVAYPTSDCTIETHKEEELY-
No. 57 : Nostoc strain MAC I             57 -ATVYKVTLV-DQEGTETTIDVPDDEYILDIAEDQGLDLPYSCRAGACSTCAGKIVSGTV-DQSDQSFLDDDQIEKGYVLTCVAYPTSDLKIETHKEEDLY-
No. 58 : Nostoc strain MAC II            58 -AT-YKVRLFNAAEGLDETIEVPDDEYILDAAEEAGLDLPFSCRSGSCSSCNGILKKGTV-DQSDQNFLDDDQIAAGNVLTCVAYPTSNCEIETHREDAIA-
No. 59 : Nostoc muscorum                 59 -AT-FKVTLINEAEGTKHEIEVPDDEYILDAAEEEGYDLPFSCRAGACSTCAGKLVSGTV-DQSDQSFLDDDQIEAGYVLTCVAYPTSDVVIQTHKEEDLY-
No. 60 : Mastigocladus laminosus         60 -AT-YKVTLINEAEGLNKTIEVPDDQYILDAAEEAGIDLPYSCRAGACSTCAGKLISGTV-DQSDQSFLDDDQIEAGYVLTCVAYPTSDCVIETHKEEELY-
No. 61 : Spirulina maxima                61 -AT-YKVTLISEAEGINETIDCDDDTYILDAAEEAGLDLPYSCRAGACSTCAGKITSGSI-DQSDQSFLDDDQIEAGYVLTCVAYPTSDCTIQTHQDDGLY-
No. 62 : Spirulina platensis             62 -AT-YKVTLINEAEGINETIDCDDDTYILDAAEEAGLDLPYSCRAGACSTCAGTITSGTI-DQSDQSFLDDDQIEAGYVLTCVAYPTSDCTIKTHQEEGLY-
```

cluster and maintaining ferredoxin conformation, while many substitutions occurred in the peripheral portion of the molecule. The sequence of *Synechococcus* 6301* deduced from nucleotide sequence was not identical to that of ferredoxin isolated from the same cyanobacterium. The former sequence is a very unique one with extra sequence in the carboxyl terminus and less homologous to the others.

From the sequence alignment shown in Fig. 1, we made an amino acid difference matrix and constructed a phylogenetic tree according to the method by Fitch and Margoliash (1967). The relationship among the large groups of plants, such as higher plants, fern, horsetails, green algae, red algae and cyanobacteria, were mostly reasonable as shown in Fig. 2. However, we encountered some difficulties in constructing a phylogenetic tree of all ferredoxins; some of them appear on the

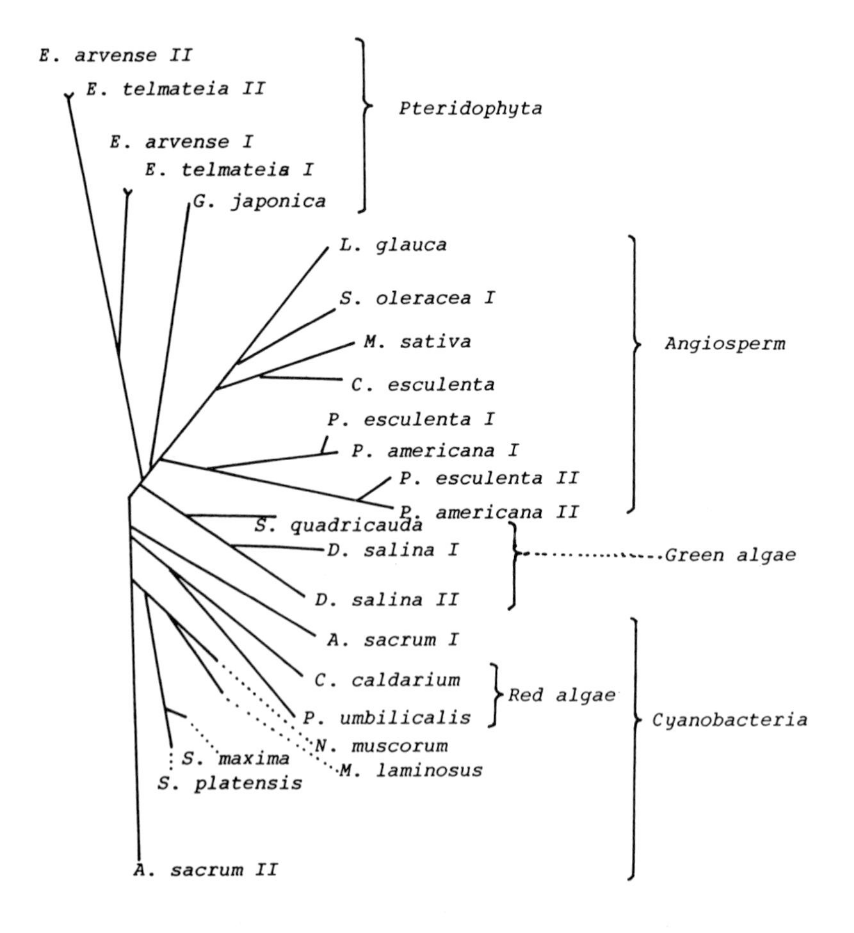

Fig. 2 A phylogenetic tree of 2Fe-2S ferredoxins from the representative plants which was constructed according to the method by Fitch and Margoliash.

phylogenetic tree. We have to keep in mind several important points when reading the phylogenetic tree of ferredoxin, as follows:
(1) Branching points in the phylogenetic tree of each plant groups are so close together that errors due to chance are inevitable.
(2) Branch lengths to each ferredoxin are rather longer compared with distances between branching points.
(3) Branch length in the phylogenetic tree may not be directly proportional to the real time elapsed.
(4) Isoferredoxins from a single species come out apart.

These imply that amino acid differences of ferredoxins do not always reflect the suggested taxonomic relationship among plants. We can list up 49 different species from which 62 ferredoxins were isolated and sequenced. We discuss here certain specific problems related to plant biosystematics.

III. Sequence Comparison and Amino Acid Deletions

As shown in Fig. 1, two deletions at positions 12 and 15 have to be placed in ferredoxin sequences from most higher plants and some green algae. Some exceptions are found in the lower order algae. Ferredoxins with only one deletion were also found. It is evident that when a ferredoxin has the deletion at position 12, it always has a proline residue at position 13, with only two exceptions of ferredoxin from green alga, *Bryopsis maxima*, belonging to *Codiaceae* and Synechococcus 6301*. A deletion at position 12 and Pro-13 appear to be coevolved, judging from the three-dimensional structure of ferredoxin. A polypeptide chain from the amino terminus runs to the right, turns and comes back to the left in the bottom of molecule. Ferredoxins without deletions have a rather blunt loop in this portion, but ferredoxins with deletions have a sharp turn, to which the proline residue contributes.

We cannot demonstrate at the present time which ferredoxin, with deletion or without deletion, is the primitive one. However, the ferredoxins without deletion may be an ancestral form, as these are generally found in lower order algae, such as cyanobacteria and primitive red algae. During the evolution of ferredoxin, occurrence of deletion at position 12 had to be compensated in some manner to retain the structure of the bottom of ferredoxin molecule. The most suitable response to this rare event would be the substitution of proline as the succeeding amino acid residue. This makes it possible to compensate for polypeptide-chain shortening and to turn it sharply. Therefore, the presence or absence of the deletion is considered to be an important character of the ferredoxin molecule.

The deletion of an amino acid in a certain position, which arises from the deletion of three adjacent bases, is a more improbable event than an amino acid substitution. Of the two isoferredoxins isolated from *Aphanothese sacrum*, one has the deletions at these positions and the other does not. It is very interesting in relation to the presence of ferredoxin isoproteins in many plants that these deletions have occurred in one ferredoxin gene line early in evolution after gene duplication. We found two unique sequences in the amino-terminal half of isoferredoxins from two horsetail species, *Equisetum arvense* and *E. telmateia*. One of them is -Thr-Val- at positions 7 and 8 of ferredoxin I and the other, -Leu-Ile- at positions 28 and 29 of ferredoxin II. These sites are usually -Val-Thr- and -Ile-Leu-. These substitutions are not considered to occur with a simple process, especially in the former case, where substitutions from -*GU*N*AC*N- to -*AC*N*GU*N- in nucleotide level are needed. The sequences found only in *Equisetum* ferredoxins indicate that *Equisetum* is independent and far from the other plants.

IV. Phylogenetic Relationships among Fern, Moss and Green Algae

The phylogenetic relationship between green algae and the lower land plants such as fern and moss is not clear. In order to consider the relationship among Bryophyta, Pteridophyta and Chlorophyta, we determined the sequence of ferredoxin from a liverwort, *Marchantia polymorpha* (Minami *et al.*, 1985). It had two deletions at positions 12 and 15 as observed in some green algae, fern and horsetail ferredoxins in addition to higher plant ferredoxins. The comparison of these ferredoxins with representative ferredoxins from some other plant groups provides a phylogenetic relationship as shown in Fig. 3. The branching point of the *M. polymorpha* ferredoxin was located between those of fern or horsetail ferredoxins and green algal ferredoxins. This is consistent with that established on the basis of morphology (Cronquist, 1971), which is generally accepted. Fig. 4 shows two proposed relationships of the three plant groups. The result from ferredoxin sequences is in accord with model A. However, Hori and Osawa (1979) have reported that model B fits 5S rRNA sequence data. More sequence analyses from fern or moss are required. We are now trying to isolate and to sequence ferredoxin from *Lycopodium clavatum* (*Lycopodiaceae*). A partial sequence of this ferredoxin is rather distinct from those of *Equisetum.*

V. Gene Duplication and Speciation

Equisetum arvense and *E. telmateia* are Japanese and American horsetails, respectively. Each two ferredoxin isoproteins I and II were isolated from the two horsetail species and their amino acid sequences were compared. Both of the isoproteins showed only one amino acid difference between species while, within the same species, ferredoxins I and II showed 29 (*E. arvense*) and 31 (*E. telmateia*) amino acid differences.

A similar relationship was obtained in the sequence comparisons of a pair of higher plants, the pokeweeds, *Phytolacca esculenta* and *P. americana.* The former is a species native to Japan; the latter is native to North America and introduced to

Japan by horticulturists about 100 years ago. In the case of *Phytolacca,* ferredoxins I and II of the two species showed two and six amino acid differences, respectively, but ferredoxins I and II within the same species showed 22 (*P. esculenta*) and 23 (*P. americana*) amino acid differences. As the average amino acid difference among all of ferredoxins sequenced so far from 15 species of angiosperms is 24, as shown in Table I, the difference values of isoferredoxins I and II within one species are large. These results lead us to conclude that speciation was preceded by gene duplication. Assuming that the amino acid substitution rate is, in principle, proportional to the time elapsed, speciation of *P. esculenta* and *P. americana* (or *E. arvense* and *E. telmateia*) occurred long after the gene duplication of ferredoxin I and II genes. Time of speciation of each species pair in the genera *Phytolacca* and *Equisetum* probably corresponds to disjunction of the American continent from the Eurasian one. Along the same line of study, we determined two ferredoxin sequences from Japanese *Trillium* (*T. kamtschaticum* and *T. smallii*). *Trillium* is a typical element of tertiary origin and there are three main species in Japan and several different species in North America. There are many investigations on the differentiation mechanisms of the Japanese species of *Trillium* (Haga, 1951, 1952, 1956). We found that there was no difference between two

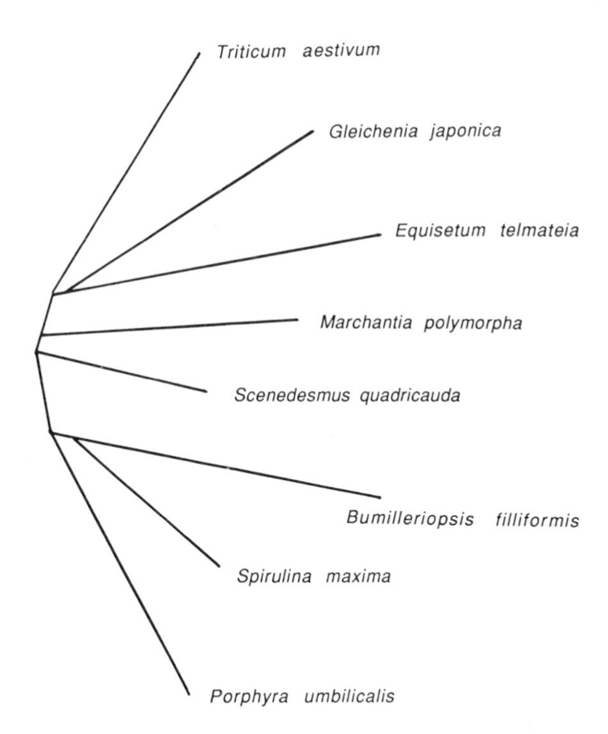

Fig. 3 A phylogenetic tree especially showing the relationship among ferredoxins from fern, horsetail, moss and green alga.

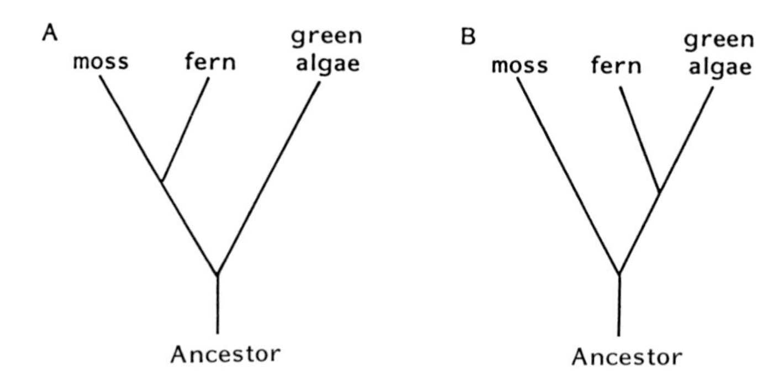

Fig. 4 Schematic phylogenetic trees among fern, moss and green algae.

ferredoxin sequences from the two *Trillium* species, though these two species were different in morphological characteristics, karyotype and distribution. This may indicate that they are so closely related to each other that they have not diverged enough to show even one amino acid difference or that pollen flow between species occurs to some extent, as it is reported that there is no barrier in the artificial cross-experiments (Ihara, 1981). This is being studied by examining the ferredoxin sequences from American *Trillium* species such as *T. ovatum, T. flexipes* and *T. grandiflorum,* which are not available yet.

VI. Ferredoxin Isoproteins and their Genes

The possession of two or more ferredoxin isoproteins by many plants is also available for study by biosystematists. The occurrence of isoferredoxins ranges from cyanobacteria to higher plants as listed in Table 1. As we reported in a previous paper (Wada *et al.*, 1985) the bandmorphs of ferredoxin isoproteins in polyacrylamide gel electrophoresis varied qualitatively, as well as quantitatively, by the subtle change of environmental factors such as light conditions, developmental stages and given stresses. This suggests that these isoferredoxins are encoded by different genes located in the nuclear genome. Moreover, we reported that the non-photosynthetic cells such as those of roots and fruits also have ferredoxins but which are distinct from the photosynthetic cells (Wada *et al.*, 1986, 1989). Fig. 5 is a sequence comparison between ferredoxins from radish, *Raphanus sativus,* leaf and root. Root ferredoxins I and II both have deletion at position 12 like those from *Bryopsis maxima* and *Synechococcus* sp. Position 15, at which most higher plant ferredoxins have a second deletion, was occupied with glutamine and glutamic acid, respectively. Root ferredoxins are very dissimilar to leaf ferredoxin and to the other angiosperm leaf ferredoxins, as shown in Table 2. The average number of differences between radish root ferredoxins and angiosperm leaf ferredoxins is 38, which conforms to the average difference between angiosperms and green algae.

TABLE 1
Ferredoxin Isoproteins

Plant Names	Number of Fd *	Remarks
Cyanobacteria		
Aphanothece sacrum	2 "	Separable on DEAE-cellulose column
Nostoc muscorum	2 '	Separable on DEAE-cellulose column
Nostoc strain MAC	2 "	Separable on DEAE-cellulose column
Nostoc verrucosum	2	Separable on DEAE-cellulose column
Aphanizomenon flos-aquae	3 '	Fd III binds to membrane tightly
Anabaena variabilis	2	One of them is specific in electron transfer of dark nitrogen fixation
Spirulina maxima	2 '	Different redox potentials
Green Algae		
Dunaliella salina	2 "	Cm-Fds are separable on DEAE-cellulose column
Equisetaceae		
Equisetum arvense	2 "	Japanese horsetail, separable on DEAE-cellulose column
Equisetum telmateia	2 "	American horsetail
Higher Plants		
Alocasia macrorhyza	2 "	Separable on butyl-Toyopearl column by $(NH_4)_2SO_4$ reverse gradient
Colocasia esculenta var. *Antiquorum*	3 "	Separable on DEAE-cellulose column and butyl-Toyopearl column
Glycine max	3	Separable on DEAE-cellulose column
petunia	2	
Phytolacca esculenta	2 "	Japanese pokeweed, separable on DEAE-cellulose colum
Phytolacca americana	2 "	American pokeweed, separable on DEAE-cellulose column
Phytolacca japonica	3	Japanese pokeweed, detected by electrophoresis
Pisum sativum	3	Fd III is found in etiolated plants
Phaseolus anguralis	2	found by electrophoresis
Raphanus sativus	4 or more '''	Some of them are root specific Fds. No distinctive functions are know yet.
Spinacia oleracea	3 "	Fd II is relatively rich in etiolated plants. One is root specific.
Triticum aestivum	2 '	Detected in seedlings but not in mature plants.
Vicia faba	3	Detected by electrophoresis
Zea mays	3	Detected in seedlings
tomato	3	One is fluit specific

* ''', " and ' marks mean the number of Fd sequenced.

```
                ''''"''''"''''"''''"''''"''''"''''"''''"''''"''''"'
1. Root Fd I    SAV-YKVKLIG-PDGQENEFDVPDDQYILDAAEEAGVDLPYSCRAGASCTC
2. Root Fd II   SAV-YKVKLIG-PDGQENEFDVPDDQYILDAAEEAGVDLPYSCRAGACSTC
3. Leaf Fd I    -AT-YKVKFIT-PEG-EQEVECDDDVYVLDAAEEAGIDLPYSCRAGSCSSC
                * *     * *    * * * **  * *        *         *  *
                '''"''''"''''"''''"''''"''''"''''"''''"''''"''''"'
1. Root Fd I    AGKIEKGQV-DQSDGSFLEDHHFEKGYVLTCVAYPQSDLVIHTHKEEELF
2. Root Fd II   AGQIVKGQV-DQSEGSFLEDDHFEKGFVLTCVAYPQSDCVIHTHKETELF
3. Leaf fd I    AGKVVSGSV-DQSDQSFLDDDQIAEGFVLTCAAYPTSDVTIETHREEDMV
                   * * *      *   *  ****      *   *  ** *  *  ***
```

Fig. 5 Sequence comparison of leaf and root ferredoxins from *Raphanus sativus*. The sites which show different amino acids in leaf ferredoxin from both root ferredoxins are marked with asterisks.

TABLE 2

Average Amino Acid Substitutions among Angiosperm Ferredoxins

	Angiosperm Ferredoxins	Rape and Radish Leaf Ferredoxins	Radish Root Ferredoxins
Angiosperm Ferredoxins	24 (S.D. : 5.8)	25 (S.D. : 3.7)	38 (S.D. : 3.5)
Rape and Radish Ferredoxins	25	5	35 (S.D. : 1.6)
Radish Root Ferredoxins	38	35	13

This large difference between leaf and root ferredoxins suggests strongly that the root ferredoxin(s) exist in all higher plants and that their genes have diverged from the leaf ferredoxin gene line a long time before speciation had occurred. It is not improbable that the origin of root ferredoxin gene traces back to the time that plant roots differentiated morphologically, as well as functionally as an organ of plants; namely when *Bryophyta* emerged from *Chlorophyta.*

In order to confirm the root ferredoxin, we isolated ferredoxin from spinach root and sequenced partially. Spinach root ferredoxin showed several amino acid substitutions even in the short region of amino terminus, compared with leaf ferredoxin, and was rather similar to those of radish root ferredoxins.

We also obtained interesting results with the maturation of tomato fruits. Tomato plants have at least three different isoferredoxins (I, II and III). The green and immature fruits as well as leaves have ferredoxins I and III. With maturation of fruits, the colour turns to pale green and then becomes red. Ferredoxin II increases with fruit maturation and, on the contrary, ferredoxins I and III, which are dominant in green fruits, decrease. It seems that change of ferredoxin pattern coincides with conversion of chloroplasts to chromoplasts and that the ferredoxin II gene begins to be expressed simultaneously with chromoplast formation. Ferredoxin II must be the ferredoxin of chromoplasts and functions in their redox systems.

Three results on ferredoxin isoproteins have been described here: these are (1) change of ferredoxin bandmorphs with environmental factors, (2) expression of unique ferredoxin molecules in specific tissues and (3) replacement of isoferredoxins with plastid conversion. These imply that the ferredoxin gene is not single but rather exists as a multigene family, although so far there is no case of two or more ferredoxin genes having been cloned from a plant. In the near future, studies on gene cloning and gene organization of ferredoxin may become useful and indispensable to solve questions on plant biosystematics.

Acknowledgement

I wish to acknowledge the encouragement and the helpful discussions of Prof. H. Matsubara, Osaka University. I am also indebted to Ms. Y. Sanada, Mr. S. Morigasaki, Ms. H. Sakai, and Ms. K. Kamide. This work was supported in part by Grant-in-Aid for Scientific Research on Priority Area of the Japanese Ministry of Education, Science and Culture (#62618001, #63618001, and #01618001 to KW).

References

Aketagawa, J. and Tamura, G. (1980). Ferredoxin-sulfide reductase from spinach. *Agric. Biol. Chem.* **44**, 2371-2378.

Alam, J., Whitaker, R. A., Krogman, D. W. and Curtis, S. E. (1987). Isolation and sequence of the gene for ferredoxin I from the cyanobacterium *Anabaena* PCC 7120. *In* "Progress in Photosynthesis Res." Vol. IV, (J. Biggins, ed.), pp. 793-795. Martinus Nijhoff, Dordrecht.

Arnon, D. I. (1988). The discoveries of ferredoxin: The photosynthetic path. *Trends Biochem. Sci.* **13**, 30-33.

Bowsher, C. G., Emes, M. J., Cammack, R. and Hucklesby, D. P. (1988). Purification and properties of nitrite reductase from roots of pea (*Pisum sativum* cv. Meteor). *Planta* **175**, 334-340.

Buchanan, B. B. (1980). Role of light in the regulation of chloroplast enzymes. *Annu. Rev. Plant Physiol.* **31**, 341-374.

Chen, T.-M., Hermodson, M. A., Ulrich, E. L. and Markley, J. L. (1983). Nuclear magnetic resonance studies of 2Fe-2S ferredoxins. 2. Determination of the sequence of *Anabaena variabilis* ferredoxin II, assignment of aromatic resonances in proton spectra and effects of chemical modification. *Biochemistry* **22**, 5988-5995.

Cronquist, A. (1971). Introductory botany 2nd ed. pp. 76-87, Harper Intl. Edition, Harper & Row, New York.

Fitch, W. M. and Margoliash, E. (1967). Construction of phylogenetic trees. *Science* **155**, 279-284.

Fukuyama, K., Hase, T., Matsumoto, S., Tsukihara, T., Katsube, Y., Tanaka. N., Kakudo, M., Wada, K. and Matsubara, H. (1980). Structure of *S. platensis* 2Fe-2S ferredoxin and evolution of chloroplast-type ferredoxin. *Nature* **286**, 522-524.

Haga, T. (1951) Genom and polyploidy in the genus *Trillium*. III.Origin of the polyploid species. *Cytologia* **16**, 243-258.

Haga, T. (1952). Genom and polyploidy in the genus *Trillium*. II. Morphology of a natural hybrid

and its parental plants in comparison. *Mem. Fac. Sci. Kyushu Univ.* Ser. E Biol. **1**, 1-12.
Hage, T. (1956). Genom and polyploidy in the genus *Trillium*. VI. Hybridization and speciation by chromosome doubling in nature. *Heredity* **10**, 85-98.
Hori, H. and Osawa, S. (1979). Evolutionary change in 5S rRNA secondary structure and a phylogenetic tree of 54 5S rRNA species. *Proc. Natl. Acad. Sci.* USA **76**, 381-385.
Ihara, M. (1981). Experimental analyses of the evolutionary processes in *Trillium*. I. Interspecific crossability and pollen flow in Japanese species. *Bot. Mag. Tokyo* **94**, 313-324.
Inoue, K., Hase, T., Matsubara, H., Fitzgerald, M. P. and Rogers, L. J. (1984). Amino acid sequence of a ferredoxin from *Rhodymenia palmata*, a red alga in the *floridophyceae*. *Phytochemistry* **23**, 773-776.
Masui, R., Wada, K., Matsubara, H., Williams, M. M. and Rogers, L. J. (1988a). Characterization, amino acid sequence and phylogenetic considerations regarding the ferredoxin from *Ochromonas danica*. *Phytochemistry* **27**, 2817-2820.
Masui, R., Wada, K., Matsubara, H. and Rogers, L. J. (1988b). Properties and amino acid sequence of the ferredoxin from the unicellular cyanobacterium *Synechococcus* 6307. *Phytochemistry* **27**, 2821-2826.
Matsubara, H. and Hase, T. (1983). Phylogenetic consideration of ferredoxin sequences in plants, particularly algae. *In* "Proteins and Nucleic Acids in Plant Systematics" (U. Jensen and D. E. Fairbrother, eds), pp. 168-181. springer-Verlag, Berlin and Heidelberg.
Minami, Y., Sugimura, Y., Wakabayashi, S., Wada, K., Takahashi, Y. and Matsubara, H. (1985a). Isolation, properties and amino acid sequence of ferredoxin from multinuclear unicellular green alga, *Bryopsis maxima*. *Physiol. Veg.* **23**, 669-678.
Minami, Y., Wakabayashi, S., Imoto, S., Ohta, Y. and Matsubara, H. (1985b). Ferredoxin from a liverwort, Marchantia polymorpha. Purification and amino acid sequence. *J. Biochem.* **98**, 649-655.
Schmitter, J.-M., Jacquot, J.-P., de Lamotte-Guery, F., Beauvallet, C., Dutka, S., Gadal, P. and Decottignies, P. (1988). Purification, proterties and complete amino acid sequence of the ferredoxin from a green alga, *Chlamydomonas reinhardtii*. *Eur. J. Biochem.* **172**, 405-412.
Smeekens, S., Binsbergen, J. V. and Weisbeek, P. (1985). The plant ferredoxin precursor; nucleotide sequence of a full length cDNA clone. *Nucleic Acid Res.* **13**, 3179-3194.
Takahashi, Y., Hase, T., Wada, K. and Matsubara, H. (1983). Ferredoxins in developing spinach cotyledons: The presence of two molecular species. *Plant Cell Physiol.* **24**, 189-198.
Uchida, A., Ebata, S., Wada, K., Matsubara, H. and Ishida, Y. (1988). Complete amino acid sequence of ferredoxin from Peridinium bipes (Dinophyceae). *J. Biochem.* **104**, 700-705.
Wada, K., Oh-oka, H. and Matsubara, H. (1985). Ferredoxin isoproteins and their variation during growth of higher plants. *Physiol. Veg.* **23**, 679-686.
Wada, K., Onda, M. and Matsubara, H. (1986). Ferredoxin isolated from plant non-photosynthetic tissues. Purification and characterization. *Plant Cell Physiol.* **27**, 407-415.
Wada, K., Masui, R., Matsubara, H. and Rogers, L. J. (1988). Properties and structure of the soluble ferredoxin from Synechococcus 6301 (*Anacystis nidulans*). Relationship to gene sequences. *Biochem. J.* **252**, 571-575.
Wada, K., Onda, M. and Matsubara, H. (1989). Amino acid sequences of ferredoxin isoproteins from radish roots. *J. Biochem.* **105**, 619-625.

11 A Protein Sequence Study of the Phylogeny and Origin of the Dicotyledons

PETER G. MARTIN AND JULIE M. DOWD

Department of Botany, University of Adelaide, Box 498, G.P.O. Adelaide, South Australia, 5001

I. Introduction

The problems of angiosperm phylogeny have been investigated using sequences of the N-terminal 40 amino acids of the small subunit of rubisco. In a sample of five gymnosperms, 20 monocotyledons and 310 dicotyledons we have found variation at 32 positions and, after translating into inferred nucleotide sequences using the genetic code in a parsimonious way, there is variation at 71 sites. This has been used to deduce phylogenetic trees. A preliminary account of this work was published (Martin and Dowd, 1989) when the survey of the dicotyledons was about two-thirds as large as now and, since it is unlikely that more sequences will be obtained in this laboratory, it is appropriate to give more details, although a complete account, including sequences, is in course of preparation.

The main emphasis in this paper is placed on our strategy for handling what is, by present standards in any field of molecular taxonomy, a sample which has not only a large number of species but also one in which there is only controversial guidance available from other taxonomic approaches about groupings higher than the level of families.

II. Methods

Biochemical methods were those given by Martin and Jennings (1983). Normally, 100g of fresh leaves were required.

BIOLOGICAL APPROACHES AND EVOLUTIONARY TRENDS IN PLANTS ISBN 0-12-402960-4

The computing methods used have, in general, been the same as those of Martin *et al.* (1983); in the earlier phases of our work, the main program used was MINTREE, i.e. the "branch and bound" program of Hendy and Penny (1982). With a Vax 785 computer, the usual limit for simultaneous analysis was 12 taxa, for which there are 654,729,075 possible trees. Recently the progam HENNIG86 (Farris, 1988) has been used in conjunction with a microcomputer (Microbyte 230) and this system usually allows the simultaneous analysis of 17 taxa, for which there are nearly ten million times as many trees as for 12 taxa. Although MINTREE has been superseded, its co-program ANALYZE is still used in conjunction with HENNIG86 because, unlike the latter, it possesses efficient routines for obtaining ancestral sequences and inter-nodal lengths.

Whenever possible we have used the *ie* (implicit enumeration) option of HENNIG86 which gives a reliable analysis and, in our experience, allows the simultaneous analysis of up to 17 taxa. Beyond this number, we have used the *bb* (branch swapping) routine but treat the result with caution that increases with the number of taxa. Another advantage of HENNIG86 is that it includes a program for successive weighting which often reduces the result to one or very few trees and this usually removes the necessity for deriving a consensus tree.

III. The Survey of the Dicotyledons

It being impractical for us to sample all families, the aim was to study 124, about half the total; in fact protein production failed for technical reasons in three so that the final number was 121. Of these families, 97 were chosen either because they are large families (e.g. at least 20 genera) or because, although small, they gave a more complete coverage of the total range of variation. Each of these families was sampled twice, as far as possible from as wide a range of variation as was available, e.g. by choosing from different sub-families or tribes. About 10% of protein preparations failed so that, when no substitute species was available, a few species were left unpaired. The other 24 families had been studied in different contexts during earlier phases of our research and were represented by various numbers of species; four monogeneric families were sampled only once; four families were sampled twice and 10 three times while Asteraceae, Fagaceae, Onagraceae, Papilionaceae, Proteaceae and Solanaceae had been the subjects of larger samplings.

An idea of the completeness of the coverage of the total range of variation can be gained from the fact that, of the 41 orders of dicotyledons in the phylogenetic system of Thorne (1983), all but three were sampled and two of these are parasitic and devoid of rubisco.

As explained above, the limitations of available computing programs make it desirable to confine the number of taxa considered simultaneously to about 17. While it was simple to reduce the number from 310 by deriving ancestral sequences of familial nodes, there was still a major problem in dealing with 121 families. As an initial step in tackling this problem, we have sought a consensus from four current phylogenies. Thorne (1983) and Dahlgren (1983) have super-orders as their major groups, the former nominating 19 and the latter 25. If these two authors agree that families are in the same super-order, then they have been grouped together in our scheme, with one proviso. Cronquist (1981) and Takhtajan (1983) have respectively 6 and 7 sub-classes as their major groups and these two authors have been allowed a veto; if either of them does not also agree that families are in the same sub-class, then they are left ungrouped. In this way we have divided 101 of the studied families into 25 Groups leaving 20 ungrouped because there is disagreement. Being reluctant to use a a formal term like super-order but needing to make clear that we use the word in a defined sense, we use Group with a capital G. The Groups are shown in Table 1. It will be seen that Groups range in size from two to nine families.

Before analysing to determine the arrangements within Groups, we first derived a base for the angiosperm tree that could be used as an out-group for other analyses. Inspection of sequences and a number of preliminary analyses reinforced other taxonomy in suggesting that the families of Groups 1, 2 and 3 were closest to the gymnosperms and, indeed, that Schisandraceae was closest so its junction with the gymnosperm tree was chosen as "Base" (Fig. 1) and its ancestral sequence used when analysing Groups. An example of the analysis of a small Group is given in Fig. 2 and of a larger one in Fig. 3. In Fig. 2, two families show dichotomous pairing while Moraceae was disjunct with one member, *Humulus*, at the base of the tree. In Fig. 3, *Hoya* was left a singleton by failure to purify protein in other members of Asclepiadaceae; the two members of Loganiaceae s.l. pair adjacently while the other four families show dichotomous pairing.

It will be seen that in Figs. 2 and 3, two or three nodes near the base of the tree have been numbered and the ancestral sequences of these have been used in the first of five stages of arranging the Groups into an overall tree. Although we have used the Groups in the initial stages above, we have not abandoned objectivity and this first stage is intended as a test of the validity of Groups. If a family is wrongly placed in a Group, it is likely to act like an out-group and take up a position near the base of a Group tree. Therefore, depending on the size and complexity of the Group, we have taken one, two or three nodes from near the base of each Group tree and have analysed them together; altogether there are 58 such nodes. If a family is mis-placed in a Group, then the nodes of that Group are likely to separate in the joint analysis while those from valid Groups should stay together. One of the weaknesses of this test is that it can only be done using the *bb* option of HENNIG86 which is unlikely to be completely reliable with 58 taxa. This analysis was done three times and inspection of the results indicated that some separation of nodes occurred in seven Groups, which were then further investigated using the test devised by Lake (1987). These tests gave no further grounds for doubting the integrity of two Groups but confirmed that the other five were really suspect.

TABLE 1

Families of dicotyledons grouped because they are placed in the same major taxon by all of Cronquist (1981), Dahlgren (1983), Takhtajan (1983) and Thorne (1983)

Group 1	Group 4	Group 9	Group 12	Group 17	Group 21
MAGNOLI	ULM	DIPTEROCARP	ERIC	CONNAR	LAMI
WINTER	MOR	ELAEOCARP	EPACRID	SAPIND	VERBEN
ANNON	URTIC	TILI	Group 13	ANACARDI	Group 22
MYRISTIC	Group 5	STERCULI	CUNONI	SIMAROUB	SOLAN
SCHISANDR	HAMAMELID	BOMBAC	ROS	MELI	CONVOLVUL
MONIMI	BETUL	MALV	SAXIFRAG	RUT	POLEMONI
LAUR	FAG	Group 10	Group 14	Group 18	Group 23
ARISTOLOCHI	CASUARIN	VIOL	CAESALPINI	HALORAG	SCROPHULAR
CALYCANTH	Group 6	FLACOURTI	MIMOS	RHIZOPHOR	GESNERI
Group 2	DILLENI	DATISC	PAPILIONI	Group 19	BIGNONI
BERBERID	THE	CUCURBIT	Group 15	ZYGOPHYLL	PEDALI
RANUNCUL	OCHN	SALIC	TRAP	GERANI	Group 24
LARDIZABAL	CLUSI	CAPPAR	LYTH	TROPAEOL	VALERIAN
MENISPERM	Group 7	BRASSIC	MYRT	MALPIGHI	CAPRIFOLI
PAPAVER	MYRIC	RESED	PUNIC	Group 20	Group 25
Group 3	JUGLAND	MORING	ONAGR	LOGANI	API
CABOMB	Group 8	Group 11	MELASTOMAT	GENTIAN	ARALI
NYMPHAE	CARYOPHYLL	SAPOT	COMBRET	APOCYN	
	NYCTAGIN	STYRAC	Group 16	ASCLEPIAD	
	AMARANTH	PRIMUL	OLAC	OLE	
	PHYTOLACC	MYRSIN	CELASTR	RUBI	
	CHENOPODI				

Families that do not fit into one of the Groups

ASTER	CORIARI	GOODENI	NELUMBON	POLYGON	THYMELAE
BUX	CROSSOSOMAT	HYDROPHYLL	PIPER	PROTE	VIT
CAMPANUL	ELAEAGN	LECYTHID	PLUMBAGIN	RHAMN	
CHRYSOBALAN	EUPHORBI	LOAS			

Note: "-aceae" omitted from all names

Lake's test is confined to four species, A, B, C, D, and uses chi-square to decide which is the most probable relationship of the three possibilities, i.e. A+B and C+D, A+C and B+D, A+D and B+C. For example, in Group 14, Caesalpiniaceae clustered with Group 13 (e.g. Rosaceae) and away from Mimosaceae and Papilionaceae which stayed together. In a Lake's test involving the representatives

Ceratonia, Pyrus, Leucaena and *Vicia* respectively, chi-square with one degree of freedom was 8 (the others were 0 and 1) and so strongly confirmed the closeness of Rosaceae and Caesalpiniaceae. While this was only preliminary evidence, Caesalpiniaceae was added to the list of unplaced families.

Similar action was taken with the other four suspect families. These were, first, in Group 4, as foreshadowed in Fig. 2, *Humulus* separated from the rest; second, in Group 22 Solanaceae separated from Polemoniaceae plus Convolvulaceae; third, the two members of Group 24, Caprifoliaceae and Valerianaceae, separated; fourth, and most tentative, in Group 5, Hamamelidaceae seemed to separate from the other three families. After relegation of some to the list of unplaced families, single ancestral nodes were chosen for the Groups.

At the second stage of deducing an overall tree, the remaining (single) Group nodes were analyzed simultaneously using the *bb* option several times and, by choosing nodes for pairs of Groups that clustered closely and constantly, the number of taxa was reduced to 16. At the third stage, each of the unplaced families was analysed (using *ie*) one at a time with the 16 nodes and note was taken in which third of the tree it took up its position, borderline taxa being placed in more than one third. Analyses of the individual Groups from each third, together with all relevant unplaced families, were then carried out and these led to the fourth stage, the redefinition of some Groups. In deciding whether a grouping was real, length of internodes within Groups was considered important because "long edges" are notoriously likely to join at inappropriate places (Penny *et al.*, 1987). In a few cases we could decide from the total tree length whether it was more appropriate for a

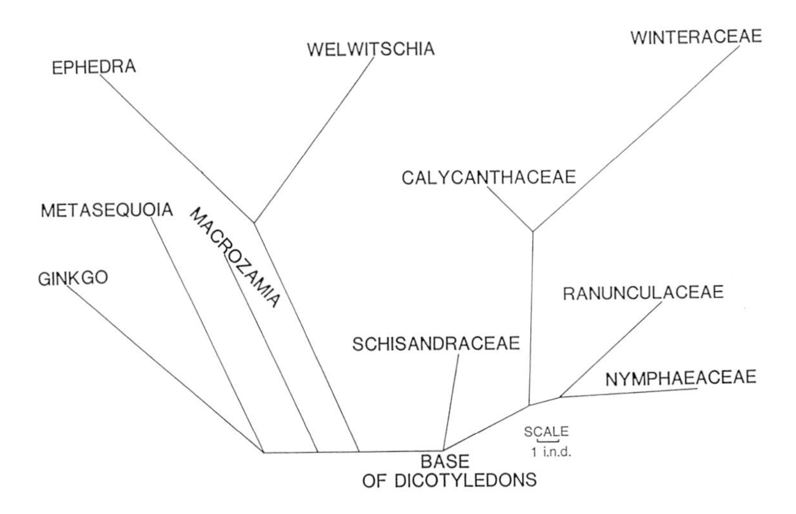

Fig. 1 Five gymnosperms analysed with familial nodes of five angiosperm families from Groups 1, 2, and 3. The ancestral sequence derived for the junction of Schisandraceae has been used as an out-group for analysing the Groups of dicotyledons.

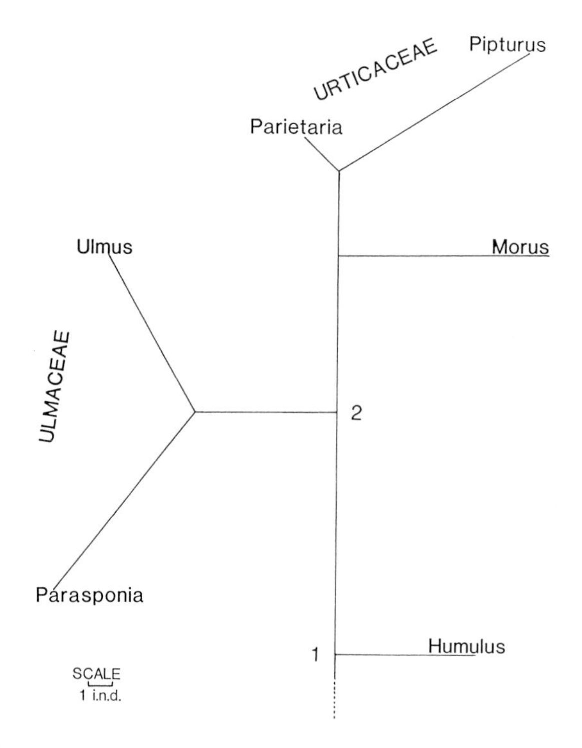

Fig. 2 Group 4, an example of a small Group. See text for the explanation of node numbers. Note that the members of Moraceae, *Morus* and *Humulus*, are disjunct with the latter at basal position.

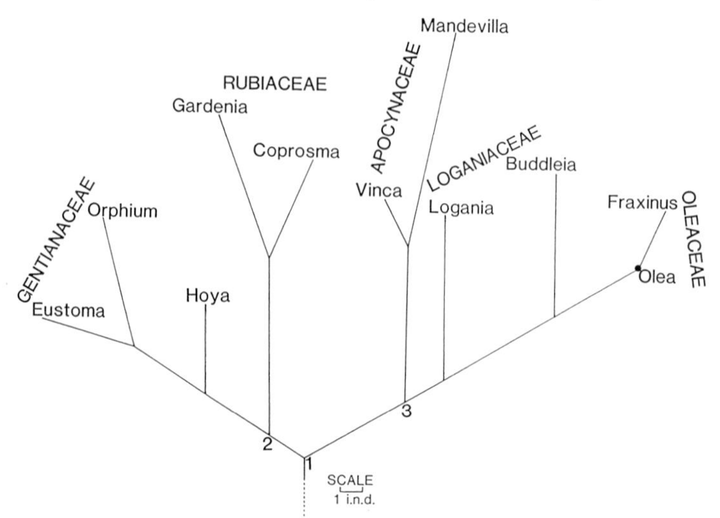

Fig. 3 Group 20, an example of a larger Group. *Hoya* is un-paired because protein production failed in other members of Asclepiadaceae.

family to join in one place or another. For example, the tree with Hamamelidaceae joining its original group 5 was three units longer than with it joining in its preferred site with Group 13, which was therefore selected.

As a consequence of these tests, only 15 of the original Groups remained unchanged, the other ten having increased, decreased or merged and, in addition four new Groups were formed. The positions of three families remained in doubt, viz. Plumbaginaceae, Buxaceae (from which *Simmondsia* had separated) and Loasaceae. The following summarises the changes:

Group 4A: *Humulus* removed.

Group 5A: Hamamelidaceae removed.

Group 8A: At the base of the original tree (Centrospermae) were added *Humulus* and Lecythidaceae.

Group 12A: To the original members (Ericaceae and Epacridaceae) were added Polemoniaceae and Convolvulaceae (from Group 22) and Polygonaceae.

Group 13: Dissolved.

Group 14A: The original members of Groups 13 and 14 plus Coriariaceae, Crossosomataceae, Hamamelidaceae and Proteaceae.

Group 18A: To the original members (Rhizophoraceae and Haloragaceae) were added Chrysobalanaceae, Vitaceae and, from Group 25, Apiaceae and Araliaceae.

Group 22A: Only Solanaceae remained.

Group 24: Dissolved.

Group 26: (new) Asteraceae, Campanulaceae, Caprifoliaceae and Goodeniaceae grouped.

Group 27: (new) Elaeagnaceae grouped with Rhamnaceae.

Group 28: (new) *Simmondsia* grouped with Euphorbiaceae.

Group 29: (new) Hydrophyllaceae, Thymelaeaceae and Valerianaceae formed a group.

As a result of these changes, 26 Groups remained and, when Group nodes had been derived, these were analyzed together using the *bb* option of HENNIG86. This tree was, as before, divided into three parts which, with generous overlaps, each contained 14 taxa. Each part was then analysed using *ie* and the three parts fitted together to give the final tree (Fig. 4).

IV. Discussion

A. General

In presenting the details of how we have arrived at our phylogenetic tree, our aim has been to show that our work is repeatable; any worker who follows our methods should arrive at the same end-point. We avoid the word "conclusion" because we

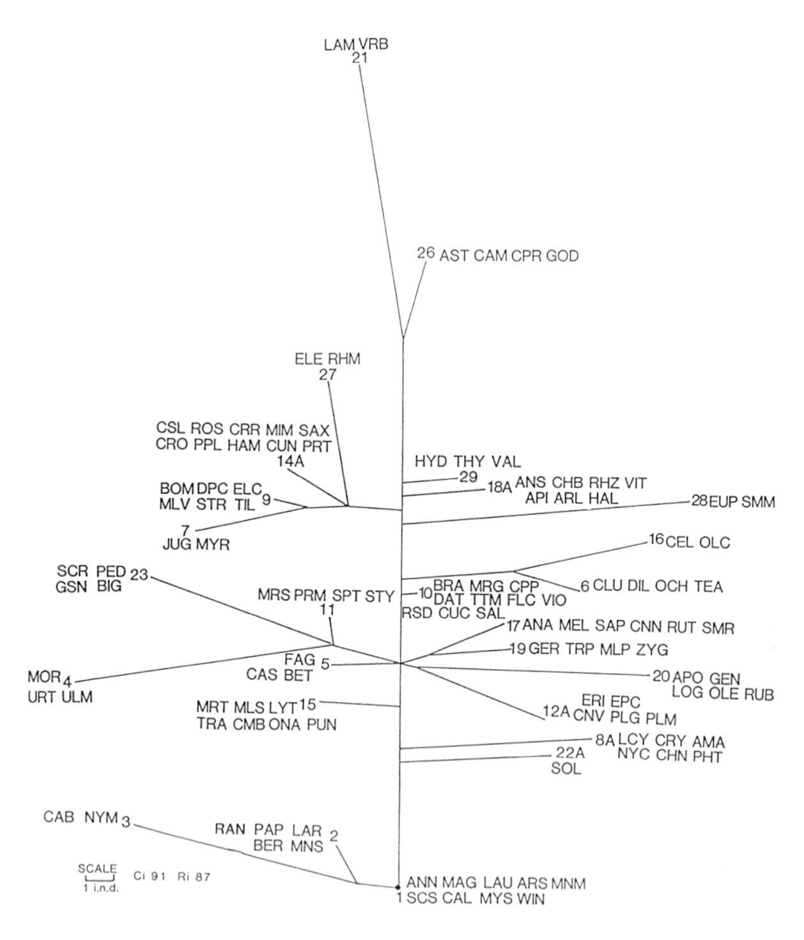

Fig. 4 The overall phylogenetic tree for the dicotyledons. Group numbers are the revised ones and members are indicated using the mnemonic code of Weber (1982).

do not claim that this work is conclusive. Rather it has led to new working hypotheses which, we hope, others will test with more extensive sampling and more data including the much longer sequences attainable with nucleic acid sequencing. To such investigators our analytical methods, whether perceived as successful or not, may prove a useful example.

The reader will validly ask how reliable our trees are. This can be best assessed by starting from the trees for Groups, i.e. the 15 unchanged ones and the 11 new or revised ones, and considering the degree of correct grouping within them. Of the 95 families with two or three representatives, only 11 showed disjunction. Of these, four were families *sensu lato* with some taxonomic opinions holding that they should be split. When it is further considered that one aberrant species can disrupt at least two families, we feel that we can fairly claim that our methods and other taxonomy give each other considerable support at the level of placing species into families.

If our methods, while not perfect, are good at the level of placing taxa into families, is there any reason why they should not be similarly reliable at higher levels? We have investigated this using the assumption that the probability of making errors will increase as internode lengths decrease. By measuring the average lengths of internodes within families, between families within Groups and between Groups in the final tree, we have found a ratio of 1.87:1.57:1. This suggests that there will indeed be a diminution of reliability as the taxonomic level becomes higher and we repeat that our results should have the status of working hypotheses, not firm conclusions.

B. The rate of evolution and the origin of angiosperms

Using a selection of the same data as has been discussed above, Martin and Dowd (1988) derived a molecular evolutionary clock with times drawn from the geophysical and biogeographical literature on the breakup of Gondwanaland. Pairs of species from the families Fagaceae, Proteaceae, Solanaceae and Winteraceae, putatively distributed to Australia, New Zealand, South Africa and South America by continental drift, were compared. Of two clocks that were considered, the preferred one suggested an evolutionary rate in a single lineage of one inferred nucleotide difference (i.n.d.) in 14 Ma.

In the overall tree for the dicotyledons (Fig. 4), there is a "trunk" from which branches depart at irregular intervals and in Fig. 5 we have arranged Groups in the

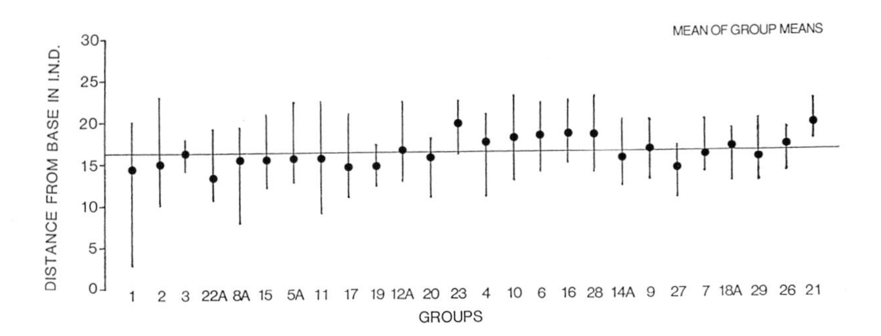

Fig. 5 The Groups are arranged in the order that they depart from the "trunk" of the overall tree for the dicotyledons (Fig. 4). The number of differences between individual species and the base of the dicotyledon tree has been measured and the mean and range (i.e. least and greatest) are shown for each Group as is the mean of all Group means. There is no obvious relationship between level of departure and number of differences.

order that they branch from the "trunk". For every species we have measured the number of differences (in i.n.d.) between it and the base of the angiosperm tree (Fig. 1). For each Group we show the mean of these distances and also the range from the smallest to greatest. The mean of all Groups is 16.2 i.n.d. We have also carried out an analysis of variance and shown that there is a significant ($P < .001$) variation between Groups. Thus, although the difference between a slowly evolving Group such as Group 3 (mean 14.1) and a rapidly evolving Group such as Group 21 (19.7) is not great, it is probably real.

The age of the dicotyledons can be derived from the product of the mean number of differences of species from base and the clock rate of 1 i.n.d. in 14 Ma but first an adjustment must be made. Because there is real variation in the rate of evolution, notice must be taken of the fact that the four families used to derive the clock all evolve more slowly than average; their mean number of differences from base is 14.7 i.n.d. The inferred age of the dicotyledons then becomes $14 \times 14.7 = 205$ Ma, i.e. near the beginning of the Jurassic. In this symposium, Knaak *et al.* quote two publications in press (Crane *et al.*, Wolfe *et al.*) which independently estimate the age of the angiosperms as 200 Ma. Thus if our deduction that the monocotyledons are derived from the dicotyledons is correct, there is good agreement.

Takhtajan (1969) suggested that the "cradle of the flowering plants" was " between Assam and Fiji". This concept was investigated by Schuster (1972) who produced maps showing the distributions of what he considered to be "by and large the most primitive and conservative extant Angiospermae". While these did indeed show concentrations between Assam and Fiji, there was a sharp discontinuity across Wallace's line, the families in south-east Asia being different from those in the south-west Pacific so that they paralleled the distributions of many groups of animals. Schuster concluded that Wallace's line represents the place where the Australian plate collided with Indonesia during its drift north in the tertiary, a conclusion now generally accepted. He also concluded that the two groups of primitive angiosperms had been located in their present ranges before collision and had migrated very little since collision. Schuster failed to draw the logical conclusion (Martin, 1977) that, if the angiosperms are monophyletic, these ancestral groups must have separated earlier. Recent developments in the earth sciences provide an explanation for this idea.

Audley-Charles (1987) and Audley-Charles *et al.* (1988) have produced reconstructions which show that at 160 Ma (Late Jurassic), rifting was just commencing separating Sumatra, the Thai-Malay Peninsula, Burma and South Tibet from the northern margin of Australia-New Guinea. Thus this tectonic event occurred after the time when the angiosperms probably originated and can account for the distribution pattern described by Schuster (1972). Although he does not refer to Schuster's maps, Takhtajan (1987) embraces this hypothesis.

Acknowledgements

This work has been supported by grants from the Australian Research Grants Scheme.

References

Audley-Charles, M.G. (1987). Dispersal of Gondwanaland: relevance to evolution of the angiosperms. *In* "Biogeographical evolution of the Malay Archipelago" (T.C. Whitmore, ed.), pp. 5-25. Clarendon Press, Oxford.

Audley-Charles, M.G., Ballantyne, P.D. and Hall, R. (1988). Mesozoic-Cenozoic rift-drift sequence of Asian fragments from Gondwanaland. *Tectonophysics* **155**, 317-330.

Cronquist, A. (1981). "An integrated System of Classification of Flowering Plants.". Columbia University Press, New York.

Dahlgren, R. (1983). General aspects of angiosperm evolution and macrosystematics. *Nord. J. Bot.* **3**, 119-149.

Farris, J.S. (1988). HENNIG86 reference: version 1.5. James S. Farris, Port Jefferson Station, New York.

Hendy, M.D. and Penny, D. (1982). Branch and bound algorithms to determine minimal evolutionary trees. *Math. Biosc.* **59**, 277-290.

Lake, J.A. (1987). A rate-independent technique for analysis of nucleic acid sequences: evolutionary parsimony. *Mol. Biol. Evol.* **42**, 167-191.

Martin, P.G. (1977). Marsupial biogeography and plate tectonics. *In* "The Biology of Marsupials" (B. Stenhouse and D. Gilmore, eds), pp. 98-115. Macmillan, London.

Martin, P.G. and Dowd, J.M. (1988). A molecular evolutionary clock for angiosperms. *Taxon* **37**, 364-367.

Martin, P.G. and Dowd, J.M. (1989). Phylogeny among the flowering plants as derived from amino acid sequence data. *In* "The Hierarchy of Life" (B. Fernholm, K. Bremer and H. Jornvall, eds), pp. 195-204. Elsevier.

Martin, P.G. and Jennings, A.C. (1983). The study of plant phylogeny using amino acid sequences of ribulose-1,5-bisphosphate carboxylase. 1. Biochemical methods and the patterns of variability. *Aust. J. Bot.* **31**, 395-409.

Martin, P.G., Dowd, J.M. and Stone, S.J.L. (1983). The study of plant phylogeny using amino acid sequences of ribulose-1,5-bisphosphate carboxylase. II. The analysis of small subunit data to form phylogenetic trees. *Aust. J. Bot.* **31**, 411-419.

Penny, D., Hendy, M.D. and Henderson, I.M. (1987). Reliability of evolutionary trees. Cold Springs Harbour Symposium on Quantitative Biology **52**, 857-862.

Raven, P.H. (1983). The migration and evolution of floras in the southern hemisphere. *Bothalia* **14**, 325-328.

Schuster, R.M. (1972). Continental movements, "Wallace's line" and Indomalayan-Australasian dispersal of land plants; some eclectic concepts. *Bot. Rev.* **38**, 3-86.

Takhtajan, A. (1969). "Flowering Plants. Origin and Dispersal." Oliver and Boyd, Edinburgh.

Takhtajan, A. (1983). The systematic arrangement of dicotyledonous families. *In* "Anatomy of the Dicotyledons" (C.R. Metcalfe and L. Chalk, eds), pp. 180-201. Clarendon Press, Oxford.

Takhtajan, A. (1987). Flowering plant origin and dispersal: the cradle of the angiosperms revisited. *In* "Biogeographical evolution of the Malay Archipelago" (T.C. Whitmore, ed.), pp. 26-31. Clarendon Press, Oxford.

Thorne, R.F. (1983). Proposed new realignments in the angiosperms. *Nord. J. Bot.* **3**, 85-117.

Weber, W.A. (1982). Mnemonic three-letter acronyms for the families of vascular plants: a device for more effective herbarium curation. *Taxon* **31**, 74-88.

12 Genome Organisation and Evolution in the Genus *Vicia*

S. N. RAINA

Department of Botany, University of Delhi, Delhi, India

I. Introduction

During the last twenty five years or so, rapid strides have been made towards better understanding of size, structure, organisation and evolution of plant genome. This has been made possible from the basic information gathered from chromosome research, linear differentiation of chromosomes, and the technical developments made in molecular biology. One of the most astonishing features now recognized about the, very complex, nuclear genome of higher plants is the variability in the amount and type of nuclear DNA between different species. The variation (> 2500 fold) in 1C DNA content in angiosperms ranges from about 0.05 in *Cardamine amara* to 127.4 picograms in *Fritillaria assyriaca* (Bennett, 1985). The data summarized by Bennett and Smith (1976) indicate thirty six fold range in the Leguminosae alone, and a ten fold variation within a single genus (Jones and Brown, 1976). Part of such variation is due to numerical changes in chromosomes but in many others there is wide-spread and substantial variation resulting from amplification and, perhaps deletion of DNA segments within the chromosomes. The theory of constancy of DNA amount within species was exploded when convincing examples of significant intraspecific variation was well recognized and it is now generally believed that such a phenomenon may no longer be regarded as exceptional (Bennett, 1985). It is a matter of time (results for 16 species obtained after 1980) before such variation will be known in many other species provided concerted efforts are made to measure DNA amounts of individual land races and/or genotypes of particular species.

BIOLOGICAL APPROACHES AND EVOLUTIONARY TRENDS IN PLANTS ISBN 0-12-402960-4

The information accumulated over the years about inter-specific DNA variation is that there is reasonably direct relationship between DNA content and changes in biophysical characters of the nucleus and the cell, and the rate and duration of DNA synthesis phase, mitotic cell cycle time, the duration of meiosis, pollen development, minimum generation time, seed weight, radiosensitivity, radiation induced mutation rates, ecological factors, and the optimum environment and geographical ranges of crop and noncrop species, and similarly, intraspecific variation in DNA content has many biological consequences (see Bennett, 1985; Grant, 1987).

The structural basis, functional significance and molecular composition of the DNA variation within chromosomes has been yet another field of considerable study ever since variety of molecular techniques such as thermal denaturation analysis, equilibrium density gradient analysis, analysis of the reassociation kinetics of dissociated DNA and *in situ* hybridization were developed. It was soon realized that DNA in plant chromosomes of diverse range of monocot and dicot species contains fast reassociating highly repetitive fraction, slow reassociating middle repetitive fraction and single copy sequences which follow a slow second order kinetics in reassociation. Repeated sequences are known to represent about 46-92 per cent of the genome in angiosperms (Flavell *et al.*, 1974). The inherent instability (movement, amplification) of repeated DNA sequences remains suppressed under normal conditions but in new genetic and/or physiological conditions the genomic instability increases (Flavell *et al.*, 1983; Flavell, 1985).

A very brief account, given above, about the abundant data now accumulated, has revealed that the plant genomes are undergoing a continuous process of variation and fixation. This is most apparent under *in vitro* system and/or genetic engineering process where, as a consequence of changed environment, the same plant genome generates extensive variability in shortest possible time. The understanding of genome organisation at molecular level, *in vivo* and in manipulated (plant and cell culture) systems, therefore, is essential not only for practical and beneficial ends in plant breeding and biotechnology and for procedures that could be developed to mitigate against unwanted source of genetic and phenotypic variability, but also in the discipline of biosystematics. Better agronomic performance, for example, was achieved in *Triticale* by manipulating significant reduction in DNA content in span of a few years (Bennett, 1985).

In this chapter, I have given a brief account of the evolution of genetic complexity, in, and divergence between, species of the genus *Vicia*, the changes in nuclear phenotype following extensive amplification or diminution of base sequences among the species and rigid constraints operating upon the evolutionary changes that would survive and in some way have adaptive significance. The genus comprises of about 120 species widely distributed throughout the temperate zones of both hemispheres. About 40 species are economically important (Harlan, 1956). The genus has been subdivided into four sections (*Ervum, Cracca, Vicia, Faba*) by Ball (1968) and Ehara (1950). Kupicha (1976) has subdivided the genus into two subgenera (*Vicilla, Vicia*). Subgenera *Vicilla* and *Vicia* have been further subdivided into 17 and 5 sections, respectively.

II. Chromosome Variation in Number, Size and Form

A. *Between species*

Chromosome numbers have been counted in some 75 species of *Vicia* (Kesavacharyulu *et al.*, 1982; Kesavacharyulu *et al.*, in prep.) and of these, the majority (36) are diploids with x=7. The second commonst basic number group is x=6 (19 species), and there are relatively few (3) with x=5. There are two tetraploids known with $2n=4x=24$. Another major kind of variation, which is particularly striking at interspecific level (Fig.1), is that concerning the size differences which are to be found between chromosome complements as a whole. The chromosome complements were also characterized by the differences in, relative size of chromosome within complements, centromeric indices, and number and nature of nucleolar chromosomes. While details about chromosome complements are not relevant here, mention may be made about some aspects concerning the present study. Arm length asymmetry and change in relative sizes of chromosomes within complements were most common in the species with $2n$=10,12. Evidently, there has been, during course of evolution, chromosomal repatterning resulting from pericentric inversion, inter-arm transposition and fusion, which has, apart from reduction in chromosome number, changed drastically the arm ratio and have created differences in relative lengths of non homologous chromosomes, the latter, particularly in complements with $2n$=10, 12. It is generally accepted that the primitive *Vicia* complement is $2n=2x=14$ and that the lower basic numbers (x=5,6) were derived as a result of Robertsonian fusion (Heitz, 1931; Coutinho, 1940; Hirayoshi and Matsumura, 1952; Raina and Rees, 1983a,b). Such translocation will convert two

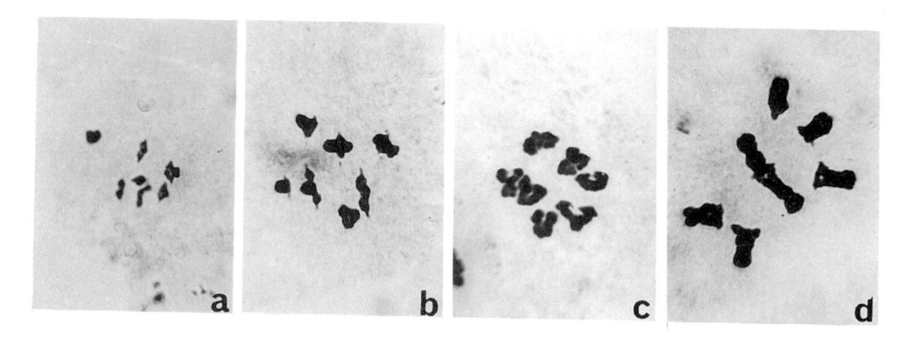

FIG.1 Metaphase I.a. *V. monantha* (3.85 pg), b.*V. ervilia* (8.41 pg), c. *V. narbonensis* (16.11 pg), d.*V. faba* (27.07 pg).

subtelocentric chromosomes to one large median/submedian chromosome resulting in aberrant karyotype as is the case in most of the species with $2n$=10,12 investigated by the present author (unpublished).

B. Within species

Interestingly, fifteen out of 75 species investigated so far are reported to have intraspecific variation in chromosome numbers. While the present author did not find any variation in the chromosome number within species, previous data on the morphology of chromosome complements and/or meiosis is so much limited and /or conflicting that no meaningful information regarding the trend in karyotype evolution could be made out in large number of species. The individual cases, however, where information regarding the relationship between different basic numbers within species is available, it is clear that they have been attributed to the derivation from a single basic number largely through centric fusion (Hanelt and Mettin, 1966; Plitmann, 1967; Hollings and Stace, 1974; Ladizinsky and Temkin, 1978; Rousi, 1961; Heitz, 1931; Coutinho, 1940).

In *V. narbonensis* (2n=14) four types (A,B,C,D) of karyotypes could be easily recognized, and on the basis of comparative karyomorphology and the meiotic data of F_1 hybrids (AxB, BxC, AxC) it was clearly evident that the alteration in the karyomorphology (Fig.2) between A, B and C is the result of segmental interchanges, where B is normal karyotype, and A and C are interchange homozygotes (Raina *et al.*, 1989). In karyotype D all the seven chromosomes in the haploid complement could be distinguished from A, B and C, and F_1 hybrids between AxD showed reduction in chiasma frequency by 53 per cent with an appreciable increase in the number of rod, as against ring, bivalents, implying thereby that genome D is well differentiated from A and possibly B and C, and deserve special status. Further, size difference in the chromosomes between the two

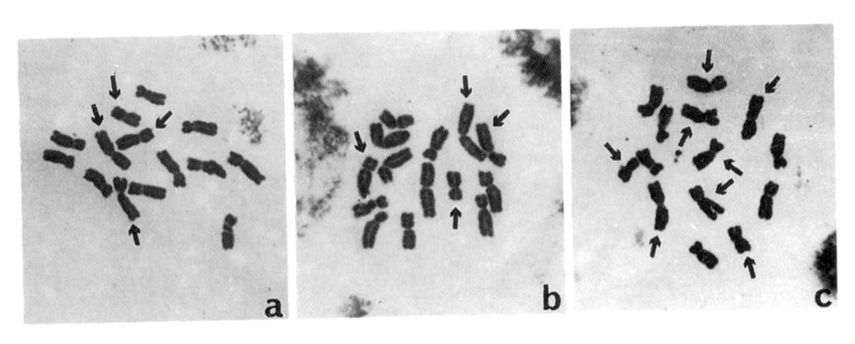

FIG.2 *V. narbonensis* a-c. Mitotic complements in F_1 AxB, BxC and AxC. Arrows indicate the chromosomes involved in the interchanges. (Reproduced from Raina *et al.*, 1989).

complements (A, D) is reflected in characteristic feature at metaphase I in the formation of highly heteromorphic bivalents. DNA has not yet been measured in D, but we might expect the difference between the two genomes in DNA content that would be distributed throughout the complement. In *V. macrocarpa* ($2n$=$2x$=12) also the intraspecific variation in the chromosome complements is the result of segmental interchanges (Liu *et al.*, 1988).

III. Variation in Genome Size

A. Between species

The variation in chromosome size between complements of different species strongly suggests a variation in nuclear DNA in association with the divergence and evolution of *Vicia* species. In the present work nuclear DNA measurements were made in 89 (75% of the total) species (Raina and Rees, 1983a; Raina *et al.*, in prep.). The lowest amount (3.85 pg) of DNA was observed in *V. monantha* ssp. *triflora* ($2n$=$2x$=14) and highest (27.07 pg) in *V. faba* ssp. *minor* ($2n$=$2x$=12). There is, therefore, seven fold DNA variation in the genus and differences among species are highly significant ($P<0.001$). Chooi (1971) had earlier reported six fold variation in DNA content in the *Vicia* species. The chromosome number in itself is no reliable guide to DNA quantity. The two polyploid species (*V. amoena* $2n$=$4x$=24, *V. tenuifolia* $2n$=$4x$=24) have smaller DNA amounts than many diploids of the same genus. It is also evident that there is no regular pattern either in relation to basic chromosome number groupings. Among diploids, for which chromosome numbers have been counted, within x=7, x=6, x=5 the range extends from a maximum of 20.68, 27.07, 20.02 pg to minimum of 3.85, 4.50 and 3.95 pg, respectively. That is to say, the largest chromosome complement, for example, among the diploid x=6 is six times greater, in terms of DNA amount, than that of the smallest.

The nuclear DNA variation in the diploid species within four sections, placed in order of morphological evolutionary advancement (Chooi, 1971), of the genus range from 5.10-13.43, 3.85-18.56, 3.95-20.68 and 9.98-27.07 pg in *Ervum, Cracca, Vicia and Faba*, respectively. In the most advanced section (*Faba*) the DNA amounts have been determined in all the seven species, and it will be seen that the two extremes of the DNA values are appreciably higher than those present in the other sections. It is, therefore, tempting to assume that massive increase in chromosomal DNA was achieved as species diverged and evolved.

B. Within species-complexes

Section *Faba*, considered as the genus *Faba* by Miller (1754) but later classified as a

TABLE 1
The 2C nuclear DNA amounts in *Vicia* species within *V. narbonensis* and *V. sativa* complexes

	Species	2n	DNA amount (picograms)
1.	*V. narbonensis* complex		
	V. johannis	14	14.14
	V. serratifolia	14	15.63
	V. narbonensis	14	16.11
	V. galilaea	14	16.51
	V. haeniscyamus	14	18.21
2.	*V. sativa* complex		
	V. cordata	10	3.95
	V. sativa	12	4.50
	V. nigra	12	4.75
	V. pilosa	14	4.78
	V. angustifolia	12	5.03
	V. macrocarpa	12	5.40

section of *Vicia* (Ledebour, 1843; Ascherson and Graebner, 1909) comprises the maximum of seven species, of which *V. bithynica* (9.98 pg) and *V. faba* (27.07 pg) are clearly distinct based on morphological features. The remaining closely related species, known as *V. narbonensis* complex (Kupicha, 1976), have been variously assigned to from two to five species (Plitmann, 1967, 1970; Ball, 1968; Ladizinsky, 1975; Schäfer, 1973). The DNA amounts in the five species are very different (Table 1) and the present findings support the view of Schäfer (1973) on distinguishing five species within the complex.

Similarly, *V. sativa* complex has been variously assigned to from one to fourteen separate species (Baker, 1970). According to Plitmann (1967) and Ball (1968) the complex was considered to represent one species but with five and six distinctive subspecies, respectively. Based on the crossability data Mettin and Hanelt (1964), Hanelt and Mettin (1966) and the present author (unpublished), on the other hand, consider these subspecies as distinct species. This is further supported by the fact that the DNA amounts among the taxa within the complex are very different (Table 1).

C. *Within species*

V. narbonensis (French vetch), considered to be wild progenitor of *V. faba* (Zohary and Hopf, 1973; Chapman, 1984) exhibits considerable morphological variation (Plitmann, 1967; Ladizinsky, 1975; Anonymous, 1987), and as mentioned before,

TABLE 2
Analysis of variance of DNA content in *V. narbonensis* accessions

Source	Sum squares	d.f.	Mean squares	F	P
Accessions	1.999	22	0.0908	11.49	0.01
Replicate	0.004	2	0.002	0.25	ns
Error	0.351	44	0.0079		

intraspecific karyotypic polymorphism as well. The hybridization between these two species for transfer of desired genes from *V. narbonensis* into highly developed commercial variaties of *V. faba* has met with little success (Roupakias, 1986; Haq, 1979; Roupakias and Tai, 1986; Ramsay and Pickersgill, 1984). Interestingly enough, homozygote A has enabled *V. narbonensis* to adapt to a wide range of secondary habitats in and far beyond the Meditterean region. Another major kind of variation in the genome of *V. narbonensis* was significant differences (Table 2) in DNA amounts between accessions collected from various geographic and ecological regions. The adaptive significance, if any, of the intraspecific variation (15.89-16.59 pg) of more than four per cent is being worked out.

The complete elimination of heterozygotes, establishment of homozygotes as a means of cytogenetic differentiation (Raina *et al.*, 1989), commonly found at interspecific level (Pal and Khoshoo, 1973), together with considerable morphological and nuclear DNA variation is a prelude to divergence from the ancestral type, and it is evisaged that further isolation and differential selection forces might lead to saltational speciation in *V. narbonensis*.

The two varieties of highly polymorphic species (*V. grandiflora*) differ much more markedly in DNA amounts than the accessions of *V. narbonensis*. 2C nuclei of *V. grandiflora* var. *grandiflora* (9.05 pg) contain about 25 per cent more DNA than *V. grandiflora* var. *kitaibeliana* (7.20 pg), morphologically a form intermediate between var. *grandiflora* and var. *biebersteiniana* (Plitmann, 1967).

IV. The Distribution of DNA Changes between the Complements

The original results of Raina and Bisht (1988) suggested rigid pattern of discontinuities in the spread of DNA values in 56 species. It is evident also now from a consideration of the DNA values from 89 species (75% of the total) represented in Fig. 3 that there is the same pattern of discontinuous and interrupted gradation of DNA amounts from the highest in diploid *V. faba*, down to the lowest in the diploid *V. monantha*. The 89 species clustered into ten separate groups. The DNA interval between successive nine groups, in increasing order of DNA amounts, was similar,

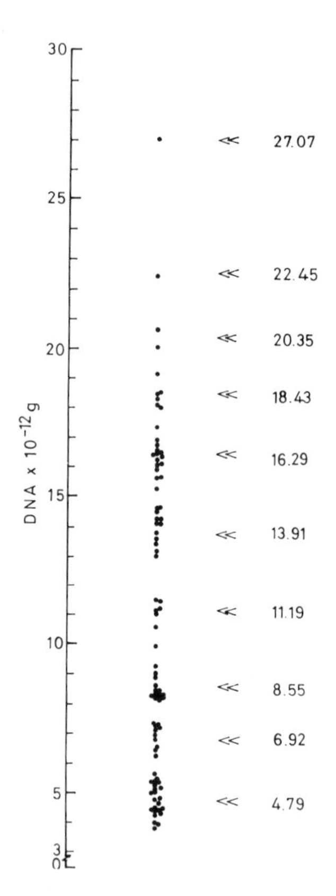

FIG.3 Mean DNA amounts in 89 species of *Vicia*.

the average (2.21) being the same as observed earlier in 56 species. The remaining 25 per cent species, for which DNA amounts are yet to be determined, are most unlikely to fill the first eight intervals, of the average of 2.21 pg, and the collective data presented here is indicative of the fact that the DNA amounts in these species are likely to fall within each cluster group and/or between groups 9 and 10 with average intervals between the groups remaining about the same.

The discontinuous mode of DNA distribution would suggest that there are basic constraints underlying the organisation of *Vicia* genome and only particular DNA packages were tolerable during speciation. Similar "quantum jumps" on consistent pattern have been reported in other plant genera (Rothfels *et al.*, 1966; Narayan, 1982, 1987; Raina *et al.*, 1986; Sparrow and Nauman, 1973).

V. The Distribution of DNA Changes within the Complements

The total variation in the 2C nuclear DNA amount in the diploid x=7, x=6 and x=5 species of *Vicia* is estimated to be about 540, 600 and 500 per cent, respectively, and the immediate question which arises from such amazing evolutionary change is how do the distribution of quantitative changes within chromosomes, that accompanied divergence and evolution within the genus, is achieved. The variation, being independent of the change in chromosome number, must arise from the loss or gain of DNA within chromosomes. There could be three possibilities, one that the extra DNA, between high and low DNA species, is distributed within the complement according to the existing size of chromosomes, two, the distribution of the extra DNA within the complement is random, and third, increase in nuclear DNA is achieved by equal DNA additions to each chromosome of the complement.

To investigate the possible mechanism about the distribution of the quantitative DNA changes within chromosomes in *Vicia* species, DNA amount was measured separately for each chromosome within complements of the species with x=5, x=6 and x=7. First, however, it was necessary to take account of and to allow for Robertsonian fusions which in a number of species with x=5 and x=6 confound the detection of changes in DNA amount. When allowance is made for such fusions, we find that the DNA differences between the species, be it with x=5, x=6, x=7, are equally distributed among all chromosomes within the complement (Raina and Rees, 1983a; Raina *et al.*, 1988). This would suggest that underlying the evolutionary changes there are rigid constraints, in some way of adaptive significance, upon changes in genome organisation during evolution. A comparable constraint has been detected in other plant genera (Seal and Rees, 1982; Narayan, 1982; Narayan and Durrant, 1983; Parida *et al.*, 1989; Narayan, 1988). Where an increase in total DNA is achieved by equal increments to all chromosomes, an inevitable consequence would be that the relative differences in the chromosome size within the complement would diminish, the complements would become more symmetrical (Fig.4).

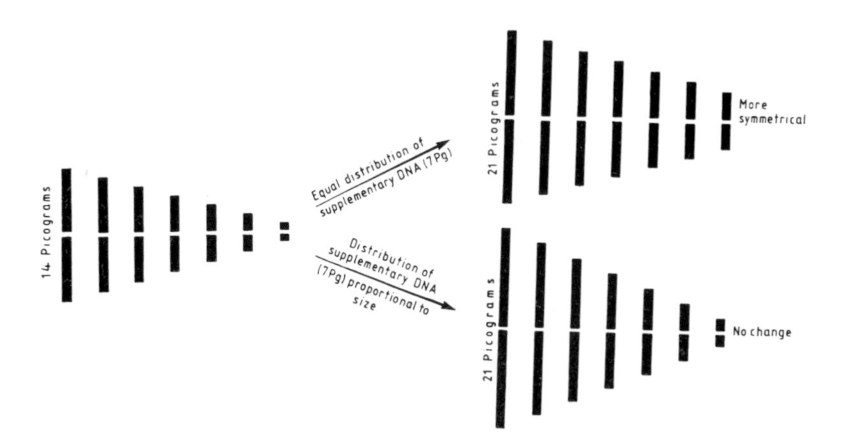

FIG.4 Consequences of the distribution of supplementary DNA amount in individual chromosomes within the complements.

VI. The Consequences of Variation in DNA Amount

A. Chromatin area and DNA density

As in nuclear DNA amounts, the *Vicia* species showed very highly significant differences with respect to chromatin area ($P< 0.001$) as well. The areas of chromatin are, as to be expected, positively correlated with DNA amounts ($P< 0.001$). The relationship is, however, not strictly linear one is borne out by the fact that DNA density per unit area of interphase nuclei increases with increasing DNA amounts (Raina and Bisht, 1988). The regression is positive and significant ($P< 0.001$). From the analysis of heterochromatin content in three *Vicia* species, showing four fold DNA variation, it was amply clear that disproportionate increase of heterochromatin content, 13.01 per cent from *V. eriocarpa* to *V. melanops* (Table 4), is not solely responsible for the increase in DNA concentration in the interphase nucleus as the DNA amount increases from one species to another. Evidently, there is condensation of at least part of the extra, supplementary DNA of euchromatin component as well (Nagl, 1982; Barlow, 1977) and then the DNA concentration in interphase nuclei should increase, as observed, with increasing nuclear DNA content. This would imply that the evolutionary changes in DNA amount and nuclear area occurs at different rates, and the condensed chromatin inactive with respect to RNA synthesis (Schmalenberger and Nagl, 1979; Hsu, 1962; Allfrey *et al.*, 1963; Derenzini *et al.*, 1978), could be interrupted as an adaptive constraint upon part of the supplementary DNA to remain in latent, but replicative (Setterfield *et al.*, 1978), form.

B. Chromosome size between complements

There is a large body of evidence showing that chromosome size (total length or volume of chromosome complement at metaphase) and amount of nuclear DNA, and relative chromosome length or volume and mean relative DNA content per metaphase chromosome are closely related (McLeish and Sunderland, 1961; Rothfels *et al.*, 1966; Martin and Shanks, 1966; Rees *et al.*, 1966; Rees and Jones, 1967; Jones and Rees, 1968; Barlow and Vosa, 1969; Haneen and Caspersson, 1973; Nishikawa, 1970; Bennett *et al.*, 1982). We also found that DNA amounts of the haploid chromosome sets plotted against their corresponding area give positive and a highly significant regression ($P< 0.001$) with no significant differences between their slopes (Raina and Bisht, 1988; Raina, 1983). They establish very clearly that DNA contents of chromosomes within complements are proportional to area. But when analysis of covariance was made, it was found that there were significant differences in mean chromatin area between species. After adjustments for differences in DNA there remained highly significant differences between species, eight species fall in two groups for adjusted mean areas (Table 3). The two

TABLE 3
Adjusted mean chromosome areas (Data from Raina and Bisht, 1988)

Species	Mean DNA value in picograms	Observed mean area in arbitrary units	Adjusted mean area	S.E. of adjusted mean area
V. dasycarpa	0.6312	5.9464	10.4855	0.2926
V. villosa	0.6675	6.0735	10.3762	0.2812
V. disperma	0.7405	6.1528	9.9803	0.2590
V. cordata	0.7896	6.5060	10.0138	0.2615
V. hirsuta	1.2787	9.020	9.3436	0.1478
V. bithynica	1.4255	10.8307	10.1985	0.1509
V. hyrcanica	2.5373	14.6058	6.7354	0.4666
V. lutea	2.5754	13.5928	5.4743	0.4759

species (*V. hyrcanica*, *V. lutea*) with higher DNA amounts have smaller mean areas than others, implying thereby that the chromosome area increases with increasing DNA amount but only upto a certain limit, beyond which the increase in DNA results in increased metaphase coiling. Such a remarkable change in the degree of metaphase coiling, in an orderly fashion, over a certain limit of DNA amount within a chromosome, might be an adaptation to facilitate large chromosomes to congregate at the equatorial plate, followed by orderly separation of sister chromatids.

VII. Molecular Composition of Nuclear DNA

A. Base composition

The three species showing four fold DNA variation had virtually identical melting profiles (Raina and Narayan, 1984; Raina, 1988). The ratio of the G+C content estimated from the melting profiles clustered closely around the mean value of 36.9 per cent (Table 4). The mean base compositional heterogeneity, which is a measure of the base pair distribution in nuclear DNA, was 18.7 per cent G+C. The alternative estimates of G+C ratios from buoyant density analysis in Caesium chloride gradient closely corresponded with those based on the T_m's, difference on an average by 2.26 per cent could result from the presence of unusual base pairs or due to the methylation of the deoxycytidine residues in the DNA. It is clear from the above that on the basis of base ratios and of base pair distribution, the fraction involved in the quantitative DNA changes associated with speciation is indistinguishable from that of the DNA complement as a whole.

B. Proportions of repetitive and non repetitive constituents

The amounts of repetitive and non repetitive constituents of the genomic DNA for three species, derived from cot reassociation experiments (Raina and Narayan, 1984; Raina, 1988), are listed in Table 4. It will be seen that in absolute amounts the three DNA components increase with increase in total nuclear DNA amounts. In proportional terms, however, the three DNA components differ in three species and, unlike closely related genus (*Lathyrus*) and many other plant genera (Narayan and Rees, 1976; Rees and Narayan, 1977; Hutchinson *et al.*, 1980), show no consistent correlation with their total DNA amounts. While *V. eriocarpa* and *V. johannis* have lower amounts of highly repetitive DNA, *V. melanops* has a significantly large amount, and among the middle repetitive and non repetitive DNA sequences, the rate of increase is greater for the middle repetitive fraction (Fig. 5). On plotting the degree of repetition in the middle repetitive DNA, estimated from the amount of middle repetitive DNA and their average kinetic complexity in base pairs (Table 4), of the three species against their absolute amounts, the regression was linear and highly significant ($P < 0.001$) which suggests that increase in middle repetitive DNA, reassociating over a broad range of *Cot* values, was achieved by extensive replication and base sequence divergence among families of repetitive sequences. The increase in the non repetitive fraction is due to steady accretion of highly diverged base sequences resulting from mutations, deletions, insertions and base se-

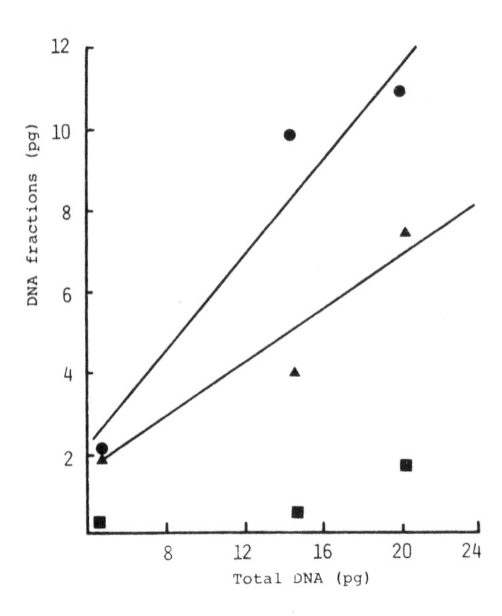

FIG. 5 The amounts of highly repetitive DNA (■) middle repetitive DNA (•) and non repetitive DNA (▲) are plotted against the total DNA of the three species (Reproduced from Raina and Narayan, 1984).

TABLE 4

Nuclear DNA (picograms) components of Vicia species (Modified from Raina and Narayan, 1984; Raina and Bisht, 1988)

Species	Total DNA	DNA in euchromatin		DNA in heterochromatin		% G+C Content		Base compositional heterogeneity
		Amount	%	Amount	%	From T_ms	From the buoyant densities	
V. eriocarpa	4.50	4.0	88.89	0.5	11.11	34.89	36.22	19.52
V. johannis	14.14	11.04	78.08	3.1	21.92	37.80	35.20	19.52
V. melanops	20.02	15.19	75.88	4.83	24.12	38.06	35.20	17.08

Highly rept DNA		Mid. rept. DNA			Non rept. DNA		
Amount	%	Amount	%	Average kinetic complexity	Amount	%	Average kinetic complexity
0.23	5	2.16	48	0.76×10^6	2.11	47	3.51×10^9
0.43	3	9.93	69	0.61×10^6	3.96	28	4.22×10^9
1.60	8	10.81	54	0.60×10^6	7.60	38	7.02×10^9

quence rearrangements among families of repetitive sequences as evidenced by Thompson and Murray (1980) in a few legume species.

C. Chromosomal location of DNA change

Repetitive DNA sequences, like satellite DNA sequences, with very low base sequence complexity are effective as molecular probes in the chromosomal location of DNA change. In neutral Caesium chloride density gradient analysis, *V. melanops* gave a light satellite DNA fraction which makes up 5.1 per cent of the genome. The average G+C content estimated for this fraction was 28 per cent which is 7 per cent lower than that of total genomic DNA. The satellite DNA constituted 62 per cent of highly repetitive fraction of *V. melanops* (Narayan *et al.*, 1985).

The chromosome complement of *V. melanops* has distinct C bands at telomeric regions in all the short arms and a long arm. When hybridized *in situ* to the chromosomal DNA the satellite DNA was predominantly located at the centromeric and intercalary regions in all chromosomes within the complement of *V. melanops*. The satellite DNA sequences in *V. melanops*, as also in several other plant species including *V. faba* (Bedrook *et al.*, 1980; Teoh *et al.*, 1983; Timmis 1975), appear to be present in very few copies in C bands. The same satellite DNA sequences were also located, to a lesser extent, in the chromosome complement of the species (*V. hybrida*) belonging to the same section (*Vicia*), but it did not hybridize with the chromosomal DNA of the species (*V. eriocarpa*, *V. johannis*), from two different sections (*Cracca, Faba*) of the genus. This shows that differential amplification of satellite DNA sequences in different species as well as base sequence divergence during the evolution of the genus were responsible for the observed variation in the amounts of satellite DNA sequences in *Vicia* species.

VIII. Conclusions

In the genus *Vicia* cytogenetic evolution has resulted in a numerous and diverse group of species. At the level of gross chromosome morphology, Robertsonian fusion and pericentric inversion have played major role in the chromosomal differentiation between species. The evidences are that the lower basic number (x=5, x=6) were derived, as a result of Robertsonian fusion, from the primitive *Vicia* complement with $2x$=14. At intraspecific level, four distinct types (A,B,C,D) of chromosome complements were recognized in wild progenitor species (*V. narbonensis*) of widely cultivated *V. faba*. The karyotypic differentiation observed between A, B and C is the result of unequal segmental interchange involving four out of seven chromosome pairs, and the interchange homozygote (A) is not only firmly established but also has enabled the species to spread further by adapting wide range of habitats. The genotype D is genetically very different from A, B and C and therefore, deserves special status.

The considerable variation in chromosome size among species is reflected in astonishing variation in nuclear DNA amount. A survey of 89 species (75% of the total) showed that the divergence and evolution of species was accompanied by a seven fold variation in the amount of chromosomal DNA, ranging from 3.85 in *V. monantha* to 27.07 pg in *V. faba*. In terms of DNA variation within a genus, *Vicia* ranks second after *Crepis*, where ten fold variation has been reported. The DNA amount in *Vicia* varies independently of chromosome number and of taxonomic grouping within the genus. The preliminary indications are change in DNA amount during evolution, especially in annual species, involved amplification of DNA sequences. There were significant differences in DNA amounts, between species within two (*V. narbonensis*, *V. sativa*) species complexes, and also between *V. narbonensis* accessions representing geographically, ecologically and morphologically diverse populations. The karyotypic polymorphism, elimination of heterozygotes by natural selection and intraspecific DNA variation in *V. narbonensis* have played significant role in the evolution of genetic architecture of the species, and to warrant the process of rapid speciation. The DNA distribution is under rigid constraint, it is discontinuous; 89 species cluster into nine separate DNA groups and the average nuclear DNA amount separating each successive pair is approximately the same. Allowing for Robertsonian fusion, the remarkable observation was that the differences in DNA amount between species are spread equally among all chromosomes of the complement. The genomic constraints underlying the DNA distribution among chromosomes have pronounced effect upon the evolution of karyotype, the species with high DNA amount would be more symmetrical than species with low DNA amount.

There were significant differences between species in chromatin area, and the disproportionate increase in the heterochromatin relative to euchromatin component is partly responsible for the steady increase in the compaction of DNA in interphase nuclei with increasing DNA amounts. Part of euchromatin component of supplementary DNA remains condensed and thus is inactive to RNA synthesis. The results show that changes in DNA content within chromosomes affect the degree of metaphase coiling in an orderly fashion, the chromosomes with higher DNA are, beyond a certain limit, more compact that their counterparts in low DNA species.

The quantitative changes in nuclear DNA amounts did not result in significant changes in base composition and base pair distribution. The changes were, however, achieved in the amounts of both repetitive and non repetitive DNA sequences. While the increase in the repetitive fraction is achieved by the proliferation of repetitive base sequences, the increase in non repetitive fraction is due to steady accretion of highly diverged base sequences as a consequence of mutations, deletions, insertions and base sequence rearrangements among families of repetitive sequences. Some of the highly repetitive sequences, resolved as satellite DNA in *V. melanops*, was located in all chromosomes within the complements of *V. melanops* and *V. hybrida*. It did not, however, hybridize *in situ* to the chromosomal DNA of species belonging to other sections of the genus. The evidence is that these sequences might have proliferated relatively recently during the evolution of the genus.

References

Allfrey, V.G., Littau, V.C. and Mirsky, A.E. (1963). On the role of histones in regulating ribonucleic acid synthesis in the cell nucleus. *Proc. Natl. Acad. Sci. US* **49**, 414-421.

Anonymous, (1987). Evaluation of germplasm collections. *In* "Genetic Resources Programme", pp. 23-34. ICARDA, Aleppo, Syria.

Ascherson, P.F.A. and Graebner, C.O. (1909). *In* "Synopsis der mitteleuropaischen Flora" 6. Engelman, Leipzig.

Baker, H.G. (1970). *In* "Taxonomy and the Biological Species Concept in Cultivated Plants" (O.H. Frankel and E. Bennett, eds), pp. 49-68. Blackwell Scientific Publications, Oxford.

Ball, P.W. (1968). *In* "Flora Europaea" (T.G. Tutin, V.H. Heywood, N.A. Burges, D.M. Moore, D.H. Valentine, S.M.Walters and D.A. Webb, eds), pp. 129-136. Cambridge University Press, London.

Barlow, P.W. (1977). Determinants of nuclear chromatin structure in angiosperms. *Ann. Sci. Nat. Bot. Biol. Veg. Ser.* **18**, 193-206.

Barlow, P.W. and Vosa, C.G. (1969). The chromosomes of *Puschkinia libanotica* during mitosis. *Chromosoma* (Berl.) **27**, 436-447.

Bedrook, J.R., Jones, J., O'Dell, M., Thompson, R.D. and Flavell, R.B. (1980). A molecular description of telomeric heterochromatin in *Secale* species. *Cell* **19**, 545-560.

Bennett, M.D. (1985). *In* "Plant Genetics" (M. Freeling, ed.), pp. 283-302. Alan R. Liss, New York.

Bennett, M.D. and Smith, J.B. (1976). Nuclear DNA amounts in angiosperms. *Phil. Trans. R. Soc. London Ser. B.* **274**, 227-274.

Bennett, M.D., Smith, J.B., Ward, J.P. and Finch, R.A. (1982). The relationship between chromosome volume and DNA content in unsquashed metaphase cells of Barley, *Hordeum vulgare* cv. Tuleen 346. *J.Cell Sci.* **56**, 101-111.

Chapman, G.P. (1984). *In* "Systems for Cytogenetic Analysis in *Vicia faba* L." (G.P. Chapman and S.A. Tarawali, eds), pp. 1-11. Martinus Nijhoff/Dr. W. Junk Publishers, Dordrecht.

Chooi, W.Y. (1971). Variation in nuclear DNA content in the genus *Vicia*. *Genetics* **68**, 195-211.

Coutinho, L.A. (1940). Racas cariologicas na *V. sativa*. *Agron. Lusit.* **2**, 379-403.

Derenzini, M., Novello, F. and Pession-Brizzi, A. (1978). Perichromatin fibrils and chromatin ultrastructural pattern. *Expl. Cell Res.* **112**, 443-454.

Ehara, K. (1950). "Forage Crops" I. Yokendo, Tokyo.

Flavell, R.B. (1985). *In* "Genetic Flux in Plants" (B. Hohn and E.S. Dennis, eds), pp. 139-156. Springer-Verlag, Wien and New York.

Flavell, R.B., Bennett, M.D., Smith, J.B. and Smith, D.B. (1974). Genome size and the proportion of repeated nucleotide sequence DNA in plants. *Biochem. Genet.* **12**, 257-269.

Flavell, R.B., Jones, J., Lonsdale, D., O'Dell, M. (1983). *In* "Advances in Gene Technology" (K. Downey, R.W. Voellmy, F. Ahmad and J. Schultz, eds), pp. 47-59. Academic Press, London and New York.

Grant, W.F. (1987). *In* "Differentiation Patterns in Higher Plants" (K.M. Urbanska, ed.), pp. 9-32. Academic Press, London and New York.

Hanelt, P. and Mettin, D. (1966). Cytosystematisiche Untersuchungen in der Artengruppe um *Vicia sativa* L. II. *Kulturpflanze* **14**, 137-161.

Haq, N. (1979). Interspecific hybridization in *Vicia* species. FABIS Newsletter 1, 19.

Harlan, J.R. (1956). "Theory and Dynamics of Grassland Agriculture". D. Van Nostrand Co., Inc., Princeton.

Heitz, E. (1931). Nukloeolen und chromosomen in der Gattung *Vicia*. *Planta* **15**, 495-505.

Haneen, W.K. and Caspersson, T. (1973). Identification of the chromosomes of rye by distribution patterns of DNA. *Hereditas* **74**, 259-272.

Hirayoshi, I. and Matsumura, M. (1952). Cytogenetical studies on forage plants I. Chromosome behaviour and fertility in F_1 hybrids, common vetch x karasunoendo (native wild vetch in Japan). *Jpn. J. Breed.* **1,** 219-222.

Hollings, E. and Stace, C.A. (1974). Karyotype variation and evolution in the *Vicia sativa* aggregate. *New Phytol* **73**, 195-208.

Hsu, T.C. (1962). Differential rate in RNA synthesis between euchromatin and heterochromatin. *Expl. Cell Res.* **27**, 332-334.

Hutchinson, J., Narayan, R.K.J. and Rees, H. (1980). Constraints upon the composition of supplementary DNA. *Chromosoma* (Berl.) **78**, 137- 145.

Jones, R.N. and Brown, L.M. (1976). Chromosome evolution and DNA variation in *Crepis. Heredity* **36**, 91-104.

Jones, R.N. and Rees, H. (1968). Nuclear DNA variation in *Allium. Heredity* **23**, 591-605.

Kesavacharyulu, K., Raina, S.N. and Verma, R.C. (1982). Cytogenetics of *Vicia* I. Male meiotic system in twelve species. *Cytologia* **47**, 511-523.

Kesavacharyulu, K., Raina, S.N. and Bisht, M.S. (In prep.). Male meiosis, evolutionary DNA variation and recombination within and between chromosome complements of *Vicia* species.

Kupicha, F.K. (1976). The infrageneric structure of *Vicia*. *Notes R. Bot. Gard. Edinburgh* **34**, 287-326.

Ladizinsky, G. (1975). On the origin of the broad bean, *Vicia faba* L. *Israel J. Bot.* **24,** 80-88.

Ladizinsky, G. and Temkin, R. (1978). The cytogenetic structure of *Vicia sativa* aggregate. *Theor. Appl. Genet.* **53**, 33-42.

Ledebour, C.F.V. (1843). "Flora Rossica" I. Schweizerbart, Stuttgart.

Liu, Y.G., Yamamoto, K. and Raina, S.N. (1988). Chromosome constitution of two karyotypes in *Vicia macrocarpa*. *Jpn. J. Breed.* **38**, 35-42.

Martin, P.G. and Shanks, R. (1966). Does *Vicia faba* have multistranded chromosomes ? *Nature* **211**, 650-651.

McLeish, J. and Sunderland, N. (1961). Measurements of deoxyribo- nucleic acid (DNA) in higher plants by Feulgen photometry and chemical methods. *Expl. Cell Res.* **24**, 527-540.

Mettin, D. and Hanelt, P. (1964). Cytosystematische Untersuchungen in der Artengruppe um *Vicia sativa*, I. *Kulturpflanze* **12**, 163-225.

Miller, P. (1754). "The Gardeners Dctionary". 4th ed. Rivington, London.

Nagl, W. (1982). *In* "Cell Growth" (C. Nicolini, ed.), pp. 171- 218. Plenum Press, New York.

Narayan, R.K.J. (1982). Discontinuous DNA variation in the evolution of plant species. The genus *Lathyrus*. *Evolution* **36**, 877-891.

Narayan, R.K.J. (1987). Nuclear DNA changes, genome differentiation and evolution in *Nicotiana* (Solanaceae). *Plant Syst. Evol.* **157**, 161-180.

Narayan, R.K.J. (1988). Evolutionary significance of DNA variation in plants. *Evol. Trends Plants* **2**, 121-130.

Narayan, R.K.J. and Durrant, A. (1983). DNA distribution in chromosomes of *Lathyrus* species. *Genetica* **61**, 47-53.

Narayan, R.K.J. and Rees, H. (1976). Nuclear DNA variation in *Lathyrus*. *Chromosoma* (Berl.) **54**, 141-154.

Narayan, R.K.J., Ramachandran, C. and Raina, S.N. (1985). The distribution of satellite DNA in the chromosome complements of *Vicia* species (Leguminosae). *Genetica* **66**, 115-121.

Nishikawa, K. (1970). DNA content of the individual chromosomes and genomes in wheat and its relatives. *Rep. Kihara Inst. Biol. Res.* **22**, 57-65.

Pal, M. and Khoshoo, T.N. (1973). Evolution and improvement of cultivated Amaranths. 7. Cytogenetic relationships in vegetable Amaranths. *Theor. Appl. Genet.* **43**, 343-350.

Parida, A., Raina, S.N. and Narayan, R.K.J. (1990). Quantitative DNA variation between and within chromosome complements of *Vigna* species. (communicated)

Plitmann, U. (1967). "Biosystematical Study of *Vicia* of the Middle East", pp. 1-128. Private Publications.

Plitmann, U. (1970). *In* "Flora of Turkey 3" (P.H. Davis, ed.), pp. 274-325. University Press, Edinburgh.

Raina, S.N. (1983). Are chromosomes with higher DNA content more compact? (contributory paper). Proc. XVth Int'l Cong. Genet., 363, New Delhi.

Raina, S.N. (1988). Nuclear DNA divergence among *Vicia* species. *Genome* **30**, 276.

Raina, S.N. and Bisht, M.S. (1988). DNA amounts and chromatin compactness in *Vicia*. *Genetica* **77**, 65-77.

Raina, S.N. and Narayan, R.K.J. (1984). Changes in DNA composition in the evolution of *Vicia* species. *Theor. Appl. Genet.* **68**, 187-192.

Raina, S.N. and Rees, H. (1983a). DNA variation between and within chromosome complements of *Vicia* species. *Heredity* **51**, 335- 346.

Raina, S.N. and Rees, H. (1983b). *In* "Kew Chromosome Conference II" (P.E. Brandham and M.D. Bennett, eds), pp. 360. Allen and Unwin, London.

Raina, S.N., Kesavacharyulu, K. and Parida, A. (1988). *In* "Proc. Int'l Conf. Research in Plant Sciences and its Relevance to Future". 199. New Delhi, India.

Raina, S.N., Parida, A. and Narayan, R.K.J. (In prep.). Evolutionary DNA variation in *Vicia*.

Raina, S.N., Srivastav, P.K. and Rama Rao, S. (1986). Nuclear DNA variation in *Tephrosia*. *Genetica* **69**, 27-33.

Raina, S.N., Yamamoto, K. and Murakami, M. (1989). Intraspecific hybridization and its bearing on chromosome evolution in *Vicia narbonensis* (Fabaceae). *Plant Syst. Evol.* **167**, 201-217.

Ramsay, G. and Pickersgill, B. (1984). *In* "Systems for Cytogenetic Analysis in *Vicia faba* L. (G.P. Chapman and S.A. Tarawali, eds), pp. 138-140. Martinus Nijhoff/ Dr. W. Junk Publishers, Dordrecht.

Rees, H. and Jones, G.H. (1967). Chromosome evolution in *Lolium*. *Heredity* **22**, 1-18.

Rees, H. and Narayan, R.K.J. (1977). Evolutionary DNA variation in *Lathyrus*. *Chromosomes Today* **6**, 131-139.

Rees, H., Cameron, R.F., Jones, G.H. and Hazarika, M.H. (1966). Nuclear variation between diploid Angiosperms. *Nature* **211**, 828- 830.

Rothfels, K., Sexsmith, E., Heimburger, M. and Krause, M.O. (1966). Chromosome size and DNA content of species of *Anemone* L. and related genera (Ranunculaceae). *Chromosoma* (Berl.) **20**, 54-74.

Roupakias, D.G. (1986). Interspecific hybridization between *Vicia faba* L. and *V. narbonensis* L.: Early pod growth and embryo sac development. *Euphytica* **35**, 175-183.

Roupakias, D.G. and Tai, W. (1986). Interspecific hybridization in the genus *Vicia* under controlled environment. *Z. Pflazenzuchtung* **96**, 177-180.

Rousi, A. (1961). Cytotaxonomical studies on *Vicia cracca* L. and *V. tenuifolia* Roth I. chromosome numbers and karyotype evolution. *Hereditas* **47**, 81-110.

Schäfer, H.I. (1973). Zur Taxonomie der *Vicia narbonensis* Gruppe. *Kulturpflanze* **21**, 211-273.

Schmalenberger, B. and Nagl, W. (1979). *In* "Genome and Chromatin: Organisation, Evolution, Function" (W. Nagl, V. Hembleben and F. Ehrendorfer, eds), pp. 119-125. Springer Verlag, Wien and New York.

Seal, A. and Rees, H. (1982). The distribution of quantitative DNA changes associated with the evolution of diploid Festuceae. *Heredity* **47**, 179-190.

Setterfield, G., Sheinin, R., Dardick, I., Kiss, G. and Dubsky, M. (1978). Structure of interphase nuclei in relation to the cell cycle. *J. Cell Biol.* **77**, 246-264.

Sparrow, A.H. and Nauman, A.F. (1973). Evolutionary changes in genome and chromosome sizes and in DNA content in the grasses. *Brookhaven Symp. Biol.* **25**, 367-389.

Teoh, S.B., Hutchinson, J. and Miller, T.E. (1983). A comparison of cloned repetitive DNA sequences in different *Aegilops* species. *Heredity* **51**, 635-641.

Thompson, W.F. and Murray, M.G. (1980). *In* "Plant Genome"(D.R. Davis and D.A. Hopwood, eds), pp. 3-45. The John Innes Charity, England.

Timmis, J.N., Deumling, B. and Ingle, J. (1975). Localisation of satellite DNA sequences in nuclei and chromosomes of two plants. *Nature* **257**, 152-155.

Zohary, D. and Hopf, M. (1973). Domestication of pulses in the old world. *Science* **182**, 887-894.

Part III

POPULATION BIOLOGY AND LIFE HISTORY EVOLUTION

(1) REPRODUCTIVE BIOLOGY OF PLANTS

13 Relationship between Plant Breeding Systems and Pollination

MARY T. KALIN ARROYO[1] AND FRANCISCO SQUEO[2]

[1]*Departamento de Biología, Facultad de Ciencias, Universidad de Chile, Casilla 653, Santiago, Chile*
[2]*Departamento de Biología, Universidad de La Serena, La Serena, Chile*

I. Introduction

The notion that plant breeding systems have been molded by pollinators has been a recurrent theme in plant evolutionary biology for over a century (c.f. Darwin 1877 and onwards). Obligately outcrossed species require pollination vectors; however the corollary that pollinator-limitation precipitates the evolution of self-compatibility is still unsettled. Studies conducted on short-lived self-compatible annuals (e.g., Lloyd, 1965; Arroyo, 1973; Wyatt, 1983) are not conclusive in that the alternate hypothesis that observed low pollinator activity is a result of the smaller and less attractive flowers on the autogamous derivatives than in the ancestral condition is difficult to test.

In the tropics, where longevity predominates and environmental conditions are amenable for biotic pollination, the hypothesis that wide spacing of forest trees might impede xenogamous breeding systems on account of insufficient inter-tree movement by pollinators has been a favorite (Bawa, 1974; Ruiz, and Arroyo, 1978; Sobrevila and Arroyo, 1982); however no strong empirical support has been forthcoming for this hypothesis. A glaring deficiency in much research where the relationship between breeding system and pollinator availability has been sought is the lack of attention given to life-history constraints versus life-history independent selective effects. It is now widely accepted that the primary role of self-incompatibility is to control levels of heterozygosity, the production of genetic variability being a result, rather than the cause of selection for outbreeding. In that plants have no germ line and all spore mother cells are derived from vegetative cell lines, the

BIOLOGICAL APPROACHES AND
EVOLUTIONARY TRENDS IN PLANTS ISBN 0-12-402960-4

quantity of mutations and "mistakes" accumulating in cells giving rise to gametes, all other things being equal, increases with plant size (Ledig, 1986). Such high genetic loads in large plants, should favor outbreeding in that heterozygosity masks the load by the dominance of normal over deleterious alleles. While the expected correlation between life-form (reflecting plant size) and self-incompatibility under this hypothesis has been evident for some time (e.g., Raven, 1979; Hamrick, *et al.*, 1979; Arroyo, 1981), the available data and circumstances under which they were taken are not conducive to separating true life-history constraints from pollinator selection and environmental-mediated selection for individual genotype fitness.

We suggest that the relative roles of these selective factors should be most easily detected in environments combining low pollinator availability selecting for self-compatibility, longevity selecting for outbreeding and strong physical habitat harshness selecting for high individual genotype fitness (where physical habitat harshness is equivalent to biotic competition in a productive environment). These circumstances are most conveniently given in high elevation ecosystems, where moreover there are ideal gradients in pollinator availability and habitat harshness, and the distribution and abundance of pollinators is strongly determined by abiotic environmental factors (Arroyo *et al.*, 1985, 1989b).

This chapter will be largely devoted to describing the breeding systems of the alpine flora of the Chilean Patagonia, latitude 50°S, southern South America, correlations between breeding systems and life-form and the distribution of breeding systems in relation to pollinator availability and habitat-harshness. We also compare our results with those for other temperate and tropical environments and point out some of the key problems for advancing in this area. We purposely chose to conduct our study in the Patagonian alpine for the extremely harsh abiotic and biotic conditions there.

II. The Study Area

The Chilean Patagonia lies in the stream of the westerlies, which ensure storm passages throughout the year. The alpine zone on the Baguales Range (450-1100m and 700-1600m at the western and eastern extremes), located to the east of the Southern Patagonian Icefield (Fig. 1), is characterized by a short growing season (2-2.5 months), low mean summer temperatures (6.6°C at 900m on Cerro Diente; 8.0°C at 1150m on Cerro Santa Lucía) and summer rainfall. Incessant powerful winds often reaching gale-force and storms occur throughout the summer (Arroyo and Squeo, 1987), evoking the highly appropriate descriptive term of "screaming fifties" for the region, the former producing a physiologically arid climate for plants (Pisano, 1974), unamenable conditions for pollinators and strong substrate erosion. Temperature, wind and soil conditions are most extreme on sites closest to the icefield, where species richness for the alpine belts is notably depressed. The principle

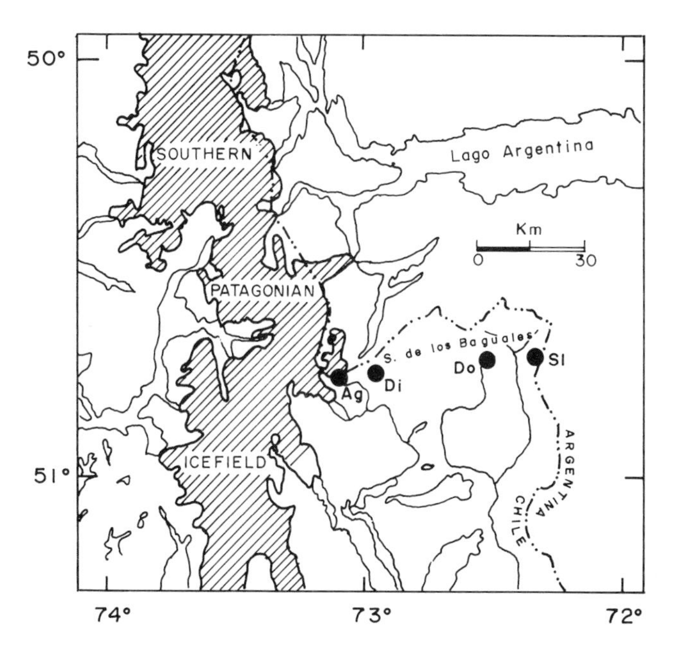

Fig. 1 Location of Sierra de los Baguales study area in the Chilean Patagonia, southern South America and specific sites referred to in text. Ag: Cerro Agudo; Di: Cerro Diente; Do: Cerro Donoso; Sl: Cerro Santa Lucía.

life-forms in the three vegetation belts (subalpine, low alpine and high alpine) and two habitat-types (nutrient-poor arid exposed slopes and nutrient-richer bog habitats) considered here are long-lived, slow growing low cushion herbs, extensively rhizomatous or stoloniferous perennial herbs, suffruticose perennial herbs, and low to prostrate shrubs. Important plant families and numbers of species present in the total native flora of 311 species are Asteraceae (55), Gramineae (51), Cyperaceae (18), Papilionaceae (16), Umbelliferae (16), Cruciferae (15) and Ranunculaceae (14) (Arroyo *et al.*, 1989a).

III. Pollination

Pollination mechanisms for the Baguales Range are shown in Fig. 2. A high proportion of the alpine flora is anemophilous. Dipterans, followed by lepidopterans and hymenopterans are the most important biotic pollinators, but their diversity and density is low. There are relative latitudinal increases in butterfly and wind pollination, and a strong decrease in the relative importance of bee-pollination. Flower visitation rates on the Santa Lucia (700-1300m) and Diente (not shown) sites in the Baguales Range are significantly lower than in the central (2200-3600m; 33°S) and

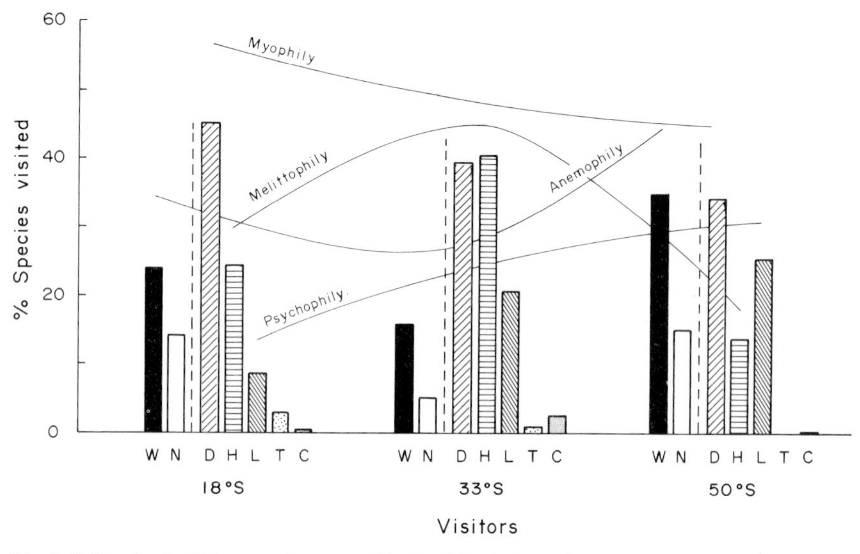

Fig. 2 Pollination in high mountain communities in Chile. Latitude 50°S is Cerro Santa Lucía, Sierra de los Baguales. W: wind-pollinated; N: non wind-pollinated - no visits recorded; D, H, L, T, C: visited extensively by dipterans, hymenopterans, lepidoterans, hummingbirds, coleopterans, respectively. From Arroyo *et al.* (1987). Original data for 18°S except for Asteraceae (Arroyo *et al.*, 1989b) unpublished; for 50°S unpublished.

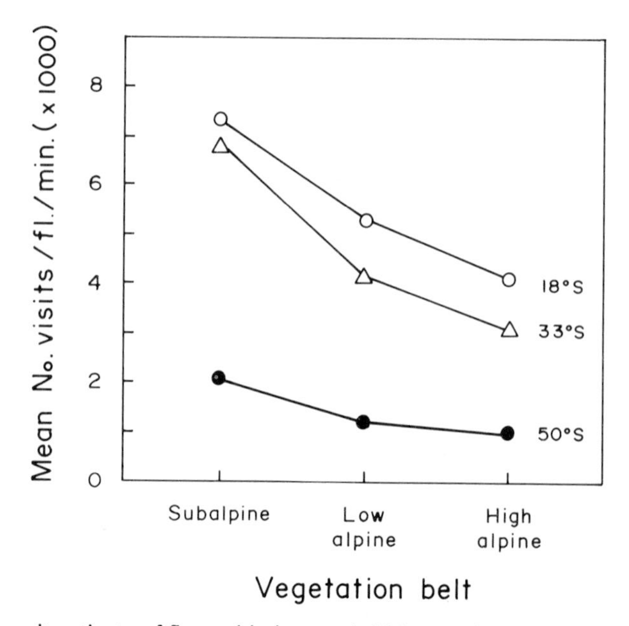

Fig. 3 Community estimates of flower visitation rates in high mountains at different latitudes in Chile. Latitude 50° is Cerro Santa Lucía. From Arroyo *et al.* (1985) and unpublished.

northern (3000-5000m; 18°S) Chilean Andes where temperature and humidity conditions are more favorable and wind is not an important factor (Fig. 3). The low visitation rates in the Patagonian alpine are not compensated for in general, nor along the altitudinal gradient, by greatly increased flower longevity (Squeo, unpublished) as occurs with elevation in the central Chilean mountains (Arroyo *et al.*, 1985).

IV. Sexual Dimorphism

Much debate has focused around whether sexual dimorphism evolves to promote outbreeding (e.g., Darwin, 1877; Bawa, 1980; Charlesworth and Charlesworth, 1987). Irrespective of why sexual dimorphism evolves, there is high dependence on external pollinating agents. Thus it was of much interest to determine the frequency of sexual dimorphism in the Patagonian alpine.

Sexual dimorphism characterizes 7.4% of the flora and dioecism fully 6.1% (Table 1) - all sexually dimorphic taxa are angiosperms (see Appendix 1 for listing). They belong to 13 genera and 12 families. Several are dominant elements in the above-treeline vegetation. The 5 species of *Hamadryas* (Ranunculaceae) belong to an exclusively dioecious endemic Patagonian genus, with only one other species in the Falkland Islands. Pseudo-viviparous *Poa alopecurus ssp. fuegiana* of the wetter western sites and agamospermous *Antennaria chilensis* with only female individuals, are derived from dioecious ancestors (Moore and Doggett, 1976; Moore, 1983). Inclusion of these would elevate the frequency of sexual dimorphism to 8.0%. There is significantly more sexual dimorphism and dioecy in woody than in herbaceous taxa (Table 1).

TABLE 1

Sexual dimorphism and dioecism in the total native Patagonian alpine flora on sites shown in Fig.1

	Total species	Sexually dimorphic[1]		Dioecious[2]	
		n	%	*n*	%
Total flora	311	23	7.4	19	6.1
Herbaceous species	282	14	5.0	10	3.5
Perennial herbs	273	14	5.1	10	3.7
Woody species	29	9	31.0	9	31.0
Shrubs	26	8	30.8	8	30.8

[1] $G = 15.53$; $d.f. = 1$; $P < 0.005$ for herbaceous versus woody species; [2] $G = 18.89$; $d.f. = 1$; $P < 0.005$ for herbaceous versus woody species.

Fig. 4 shows that there are trends for more sexual dimorphism at higher elevations on the arid exposed nutrient-poor slopes, and that the colder, windier and floristically-impoverished sites with poorer soils and less amenable conditions for pollinator activity to the west for this habitat type tend to have more sexual dimorphism. At some higher points, the local vegetation was comprised of up to 33% sexually dimorphic species and 50% sum frequency for sexual dimorphic species. Removing a small number of introduced species would increase the frequency of sexual dimorphism again at all altitudes (without altering the general trends shown in Fig. 4).

The data as presented above do not permit a distinction between sexual dimorphism being more abundant on the harsh habitats as a result of more long-lived species being present and true non-life history independent selection for sexual dimorphism.

Using the well studied Santa Lucía and Diente sites the frequency of sexual dimorphism generally increases with elevation within, as well as across individual life-forms at a given elevation (Table 2), as was shown for the total woody and herbaceous flora. These interesting and heretofore undescribed trends for alpine vegetation are striking in that highly significant negative correlations of % sexual dimorphism on total species-richness ($y = -0.17x + 31.13$; $F = 28.03$; $d.f. = 27$; P

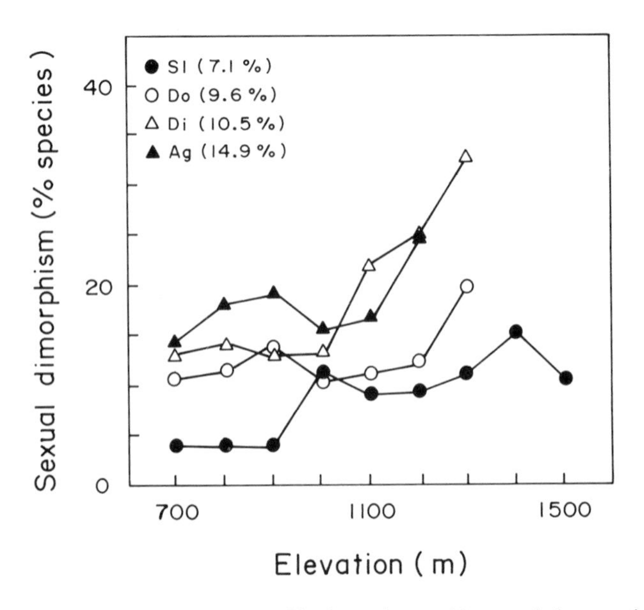

Fig. 4 Frequency of sexual dimorphism at 100m intervals on arid exposed slopes at 4 sites on the Baguales Range. Numbers in parentheses indicate frequency of sexual dimorphism in total flora of each altitudinal profile above 700m. East (Sl) and west (Ag) extremes significantly different (G = 3.91; d.f. = 1; P<0.05).

TABLE 2

Frequency (%) of sexual dimorphism according to life-form and vegetation belt (both habitat types included) at two sites in the Patagonian alpine. Numbers in parenthesis are total species for the life-form. SA: subalpine; LA: low alpine; HA: high alpine

	Santa Lucia			Diente		
Life-form	SA	LA	HA	SA	LA	HA
Caespit herbs	4.8(62)	3.5(57)	5.3(19)	0(33)	3.6(28)	7.7(13)
Rhiz./stol./cush. herbs	2.9(69)	7.6(79)	11.1(36)	5.3(57)	8.7(46)	9.1(33)
Suffrut. herbs	8.3(12)	10.0(10)	0(3)	12.5(8)	14.3(7)	33.3(3)
Shrubs	10.0(10)	45.5(11)	50.0(2)	38.0(21)	44.4(9)	100.0(2)

<0.001), ppm available nitrogen ($y = -1.29x + 32.48$; $F = 11.59$; $d.f. = 14$; $P < 0.005$) (two indicators of abiotic habitat conditions in an alpine system) and visitation intensity (percent of 10-minute observation periods on plants for which visiting was observed) ($y = -1.55x + 11.55$; $F = 108.44$; $d.f. = 4$; $P < 0.005$) (an indicator of pollinator availability) are obtained by pooling available data for all sites.

The traditional ecological correlates of sexual dimorphism (wind-pollination and fleshy fruits) (Bawa, 1980) are evident in the Patagonian alpine (Table 3). However

TABLE 3

Pollination and dispersal syndrome correlates of sexual dimorphism in the Patagonian alpine (four sites). Numbers in parentheses are percent of species in breeding system category with trait

	Species		Genera	
	Sexually dimorphic	Non sexually dimorphic	Sexually dimorphic	Non sexually dimorphic
Pollination				
Wind	13(56.5%)	105(36.5%)	9(69.2%)	40(31.0%)
Biotic	10(43.5%)	183(63.5%)	4(30.8%)	89(69.0%)
	$G = 3.44$ N.S.		G = 6.89**	
Dispersal				
Fleshy fruits	6(26.1%)	10(3.5%)	5(38.5%)	7(5.4%)
Dry fruits	17(73.9%)	278(96.5%)	8(61.5%)	122(94.6%)
	$G = 11.26$***		G = 9.11***	

** $P < 0.01$;*** $P < 0.005$.

wind-pollination is only correlated at the generic level. On account of the life-form turnover with elevation, and the tendency for fleshy fruits and wind-pollination to be more strongly concentrated among the woody sexually dimorphic species, none of these features adequately explains the elevational increases in sexual dimorphism in the Baguales Range.

V. Genetic Self-Incompatibility

A. Community trends and evidence for constraints

The breeding systems of non-sexually dimorphic species, belonging to 32 families and 71 genera from the Patagonian alpine are given in Table 4 (see Appendix I for taxa).

TABLE 4
Breeding systems in non-sexually dimorphic species in the Patagonia alpine listed in Appendix 1. SC: self-compatible; SI: self-incompatible; APO: apomictic

		Percent species (number)				
Life-form	N	APO	Highly SC	Partially SC	All[1] SC	SI
Annual herbs	7	0(0)	71.4(5)	28.6(2)	100.0(7)	0(0)
Perenn. herbs[2,3]	109	1.8(2)	43.2(47)	26.6(29)	69.7(76)	28.4(31)
Shrubs	8	0(0)	0(0)	25.0(2)	25.0(2)	75.0(6)
All species	124	1.6(2)	41.9(52)	26.6(33)	68.5(85)	29.8(37)

[1] Some SC species are probably partially agamospermous (see text and Appendix 1). [2] Percent SI in caespitose perennial herbs (n=34): 0%; in cushion herbs (n=13): 30.8%; in rhizomatous and stoloniferous perennial herbs (n=52): 38.5%; in suffruticose perennial herbs (n=10): 70.0%. [3] $G = 6.93$; $d.f. = 1$; $P < 0.01$ for frequency SI in herbaceous versus woody species.

Breeding systems were determined under field conditions using conventional bagging experiments. For some of the non-apomictic hermaphrodites it was a matter of corroborating records published earlier by Moore (1983) for Tierra del Fuegian populations. The record for *Antennaria chilensis* is from Moore (1983). Also assessed was natural (open pollination) fruit set. Representative samples of the species present in the three main vegetation belts were considered. However because of technical difficulties in working in a very windy climate and our centered interest on biotically-pollinated species, Gramineae (16-17% of species in each vegetation belt) and Cyperaceae (9% each in subalpine and low alpine, 4% in upper alpine) were not considered. A posteriori G-tests showed that our sample studied for self-incompatibility adequately reflected the life-form spectrum in each vegetation belt.

Many of the Patagonian alpine species (68%) are self-compatible, 42% being judged as strongly so (Table 4). Thirty-percent including the four heterostylous species in the sample (see Appendix 1) are self-incompatible and 2% reproduce apomictically. The frequency of self-incompatibility among families varied from 72% in Compositae, 60% in Iridaceae, 20% in Cruciferae, 15% in Papilionaceae and 0% in Caryophyllaceae, Ranunculaceae, Rosaceae and Scrophulariaceae (considering families only where 5 or more species were studied). The very small flowers of many species could not be reliably emasculated and furthermore as emasculation is only diagnostic for obligate agamospermy, the frequency of apomixis may have been underestimated. Some species of Rosaceae, Gentianaceae, Cruciferae, Saxifragaceae and Umbelliferae which have been assigned to one of the self-compatible categories on account of seed set per flower controlled in emasculated flowers not exceeding that by self-pollination, or the number of flowers tested being very small, indeed showed signs of agamospermy. Some fruit set on bagged female individuals of the dioecious *Hamadryas kingii* was obtained on the Diente site, however this result is preliminary. While it would be desirable to have a more accurate estimate of apomixis it should be pointed out that its underestimation is of little consequence in the present context, where the principle concern is whether outcrossing systems are maintained under physically harsh and pollinator-depauperate conditions.

In general the self-compatible species are strongly potentially autogamous (Appendix 1) and significant breeding system effects were seen on natural fruit set ($F = 42.66$; $d.f. = 2$; $P << 0.001$; 1-Way Anova), the highly self-compatible (Mean = 91.8%; $N = 22$) and partially self-compatible species (Mean = 79.0%; $N = 13$) producing significantly more fruit than the self-incompatible species (data is mostly for herbaceous species)(Mean = 40.1%; $N = 20$), but there being no difference for the two self-compatible categories.

Table 4 shows that self-incompatibility increases strongly across the life-forms from 0% in annuals to 28% in perennial herbs and 75% in shrubs as would be predicted from life-form constraints. A strong life-form correlation remains considering subdivisions of the major life-form categories (Table 4). Thus it appears that breeding systems are strongly constrained by life-history, even under these extreme conditions for biotic pollination. This result is highly significant considering that all life-forms in the Patagonian alpine are low in stature and have approximately equal access to pollinators.

Much of the inter-family variation in the frequency of self-incompatibility alluded to earlier persists when single life-forms are considered, suggesting the possibility of phylogenetic constraints on breeding system. The frequency of self-incompatibility among perennial herbs in Asteraceae (75%) is significantly higher than across all families ($G = 19.07$; $d.f. = 1$; $P < 0.001$), a trend that holds up even when the stricter set of non-suffruticose perennial herbs are compared ($G = 10.04$; $d.f. = 1$; P

< 0.001). Among perennial herbs in the Papilionaceae, in contrast, it is only 15% and does not differ from that expected from a random distribution in the species studied (G = 0.342; *d.f.*= 1; P = 0.558). Self-incompatibility is clearly under-represented in the perennial herb life-form in Ranunculaceae and Caryophyllaceae in comparison with the community at large. This kind of comparison nevertheless does not preclude possible unequal capacities of the plant families for attracting pollinators, which according to traditional dogma, could be reflected in more or less self-incompatibility. The frequency of self-incompatibility among the herbaceous legumes in the Patagonian alpine sample (15%) is intermediate between that in a comprehensive samples of 224 species (20%) for the temperate zone and 60 species for the tropics (8%) (Arroyo, 1981). Thus it cannot be legitimately concluded that the differences seen among families in the Patagonian alpine are due entirely to phylogenetic constraints. The higher frequency of self-incompatibility in the Patagonian and temperate legume species in general in comparison with the tropics, however, seems to suggest that there is less self-incompatibility in areas with high pollinator diversity. Of note, bees, the principal pollinators of the legume family (Arroyo, 1981), have been most affected by the adverse weather conditions in the Baguales Range (Arroyo *et al.*, 1987; Fig. 2).

B. *Altitudinal and habitat variation*

Whether self-incompatible and xenogamous breeding systems in general increase with habitat harshness and depauperate pollinator conditions, as was seen for sexual dimorphism, was evaluated by comparing the relative frequency of self-incompatibility and self-compatibility for the species we studied, according to their distribution in the three alpine vegetation belts (Santa Lucía and Diente sites) and combining this knowledge with data on sexual dimorphism. The majority of the Patagonian alpine species occur in 2 of the 3 vegetation belts - when a species occurred in more than one vegetation belt, it was assumed that its breeding system was invariant. To evaluate the relative effects of physical habitat harshness and reduced pollinator conditions on breeding systems, advantage was taken of the fact that hermaphrodite species, unlike sexually dimorphic taxa, are abundant in both of the two major habitat types on the Baguales Range. Most individual species in the Baguales Range occur in one only of these two habitat types. The plant species in these two habitat types (species occurring in both were considered twice for calculations in Table 6) do not vary in visitation intensity and both habitat types are well represented at all elevations, except in the third vegetation belt where bogs are poorly developed. Finding differences in the frequency of self-incompatibility on these two habitats types within the same life-form would provide strong evidence of non-pollinator, abiotic selective effects on breeding systems.

The uppermost vegetation belt, considering species from both habitat types, appears to have the highest frequency of self-incompatibility (Table 5) and this trend

TABLE 5

Altitudinal variation in the frequency of genetic self-incompatibility among non-sexually dimorphic species and the frequency of xenogamous breeding systems in general in the native alpine flora of the Baguales Range, Patagonia, 50°S. The estimated frequency of xenogamous breeding systems was calculated by adding the estimated number of self-incompatible (SI) species, based on the sample % of SI species to the known number of sexually dimorphic species in each vegetation belt and dividing that number by the total number of species present

			% (n) SI species in sample[1]			Estimated % xenogamy		
Vegetation belt	Total flora	Visitat.[2] intensity (%)	All life forms[3,4]	All perennial herbs[5]	Rhiz./Stolon. perennial herbs[6]	All life forms	All perennial herbs	Rhiz./stolon. perennial herbs
Subalpine	222	13.2	26.7(101)	23.9(88)	28.6(42)	32.0	27.2	30.5
Low alpine	185	9.1	33.7(92)	30.9(84)	43.9(41)	39.4	35.0	45.7
High alpine	77	8.3	44.2(43)	44.2(43)	72.2(18)	50.7	48.8	74.5

[1] Data from the Diente and Santa Lucía sites for zonal and azonal vegetation types; [2] Percent of 10-minute observation periods on pollinator activity from community pollination surveys (adjusted for rainy days) for which flower visiting was observed. [3] $G = 32.02$; d.f = 2; $P<0.001$ for the three vegetation belts. [4] Frequencies of life-forms in samples of hermaphrodite species with known breeding systems and in total hermaphrodite species in each vegetation belt not significantly different ($G = 1.190$; *d.f.*= 5; $P >>0.05$; $G = 3.078$; *d.f.* = 4; $P = >>0.05$; $G = 1.081$; *d.f.* =2; $P >>0.05$, for the subalpine, low alpine and high alpine belts, respectively). [5] $G = 5.470$; *d.f.* = 2; $P =<<0.10<0.05$, for three vegetation belts; $G = 5.38$, *d.f.* = 1; $P <0.05$ for altitudinal extremes; [6] $G = 9.995$; *d.f.* = 2; $P <0.01$ for three vegetation belts.

was generally paralleled at a taxonomic level. Among four families heterogeneous for breeding system (and for which more than 5 species were studied) there are trends for relatively more self-incompatibility at the higher elevations in Asteraceae, Iridaceae, and Cruciferae, with only the Papilionaceae showing less. The Asteraceae has much self-incompatibility, however the family was not over-sampled at the high elevations (G = 0.240, G = 0.984; G = 0.463, all N.S. for the subalpine, low alpine and high alpine belts, respectively comparing the sample and total flora). Other families contributing strongly to the self-incompatible species at high elevations are the Oxalidaceae (represented by the tristylous *Oxalis patagonica* and *O. loricata*), Rubiaceae (represented by the distylous *Oreopolus glacialis*) and Solanaceae (represented in particular by the endemic Patagonian genus *Benthamiella*). The wide taxonomic derivation of these self-incompatible species argues against a taxonomic artefact as an explanation for the altitudinal trend.

The frequency of self-incompatibility among species inhabiting the arid slopes is over three times higher than that for species inhabiting the more benign bog habitats (Table 6). Significantly, this trend remains when the single abundant life-form shared by both habitats is considered (Table 6).

TABLE 6

Comparison of the frequency of self-incompatibility, visitation intensity and ppm nitrogen among non-sexually dimorphic species inhabiting arid slopes versus bogs in the Patagonian alpine. Numbers in parentheses are sample sizes

	Visitation[1] intensity	ppm N	Percent self-incompatible spp.	
			All species[2]	Perennial herbs[3]
Arid slopes	15.20(53)	10-20	39.1(87)	36.4(77)
Bogs	11.86(19)	11-57	12.2(41)	10.8(37)

[1] From unpublished community pollination surveys (species observed for 10-minute intervals for 10 or more times on Santa Lucia site); t = 0.943; $d.f.$ = 70; P = 0.174 for arid slopes vs. bogs; [2] G = 10.37; $d.f.$ = 1; $P < 0.05$ for arid slopes versus bogs; [3] G = 8.87; $d.f.$= 1; $P < 0.05$ for arid slopes versus bogs.

The two co-occurring habitat types exhibit stronger breeding system differentiation than is seen over the entire elevational gradient (compare Tables 5 and 6). As indicated by nitrogen content, soil fertility is higher in the bog habitat as is plant cover and there is more organic matter present. That more self-incompatibility is being seen on the arid exposed slopes with similar pollinator conditions is consistent with outbreeding being favored with increasing habitat harshness.

A second indication that self-incompatibility is selected for under harsh conditions possibly beyond that expected from constraints imposed by life-history is the elevational increase in self-incompatibility within some life-forms (Table 5). This last situation, and indeed the general trend is slightly affected by a proportionately smaller bog flora (in which self-incompatible species tend to be less frequent) at the very extreme upper elevations. However even when the bog flora is removed there are steady increases in self-incompatibility from 34% to 49% at the altitudinal extremes for all life-forms combined and from 30% to 45% in perennial herbs, although these now fail to reach statistical significance ($G = 2.042$; *d.f.* = 1; for all life-forms; $G = 3.324$; *d.f.* = 1, for perennial herbs - extremes considered).

Clearly the trends seen in hermaphrodite species are remarkably similar to those detected for sexual dimorphism in the flora in general. The joint consideration of sexual dimorphism and self-incompatibility including species from both habitat types shows a very strong altitudinal increase in the frequency of xenogamous breeding systems in the entire flora and in some individual life-forms (Table 5). Most notably, 75% of the rhizomatous perennial herbs studied in the upper alpine belt are estimated to possess xenogamous breeding systems, a 100-fold increase over that in the subalpine. In summary, the distribution of xenogamous breeding systems in the Patagonian alpine provides no support for strong pollinator effects. On the other hand, strong life-history constraints are present, and these are apparently accentuated even further by selection for outcrossing systems within individual life-forms as physical habitat harshness increases.

VI. New Trends

A. Sexual dimorphism

Tropical latitudes generally provide ideal conditions for biotic pollination compared with the Patagonia and in the latter, conditions for biotic pollination are probably more extreme than in most other temperate latitudes. The trends found thus might be reflected on a regional scale. Tropical habitats are harsh for the organisms that inhabit them on account of strong biotic selection (Arroyo *et al.*, 1988) and longevity is commonplace. Thus, assuming that breeding systems are strongly determined by life-history constraints and non-pollinator selection, convergence in breeding system is predictable for the tropical and high temperate latitudes.

The frequency of dioecism in the Patagonian alpine is the highest for all temperate floras reported to date excluding the island floras of New Zealand and Hawaii and the Cape flora of South Africa (heavily influenced by the dioecious family Restionaceae), and it exceeds that in some tropical floras where woodiness is well developed (Table 7). More strikingly, the percentage of dioecism in shrubs in the Patagonian alpine is proportionately 2-3 higher than has been reported in three tropical floras where this life-form has been considered separately (Croat, 1979; Flores and Schemske, 1984; Bullock, 1985), around twice as high as in the shrubs of the temperate North Carolina flora (Bawa 1980), one and half to twice as high as

TABLE 7

Incidence of dioecism in various temperate and tropical floras ordered from the southern to the northern hemisphere

Flora	Dioecious (%)		Flora	Dioecious (%)	
Patagonian alpine	6.1		Camella, Mexico	12.3	(8)
New Zealand	12-13	(1)	India	6.7	(9)
South Australia	3.9	(2)	Hawaii	27.7	(10)
SW of W. Australia	4.4	(3)	Carolinas	3.7	(11)
South African Cape	6.7	(4)	California**	2.8	(12)
Ecuador	3.0	(5)	Northeastern USA*	5.4	(12)
Barro Colorado Is	9.0	(6)	British Isles	4.4	(13)
Puerto Rico/Virgins	6.1	(7)	Alaska**	3.9	(12)
			Alaskan Artic Slope	5.8	(12)

1. Godley (1979); 2. Parsons (1958); 3. McComb (1966); 4. Steiner (1988); 5. Bawa (1980); 6. Croat (1979); 7. Flores and Schemske (1984); 8. Bullock (1985); 9. Roy (1974); 10. Carlquist (1974); 11. Conn *et al.* (1980); 12. Fox (1985); 13. Kay and Stevens (1986). * without monocots. ** with introduced spp.

for all woody species in California and the northeastern USA (Fox, 1985) and as high or higher than has been reported for the tree life-form or tree and shrub life-form combined in some tropical communities (Table 8). Fox (1985) reported 31.3% and 40.9% dioecism for woody species including many trees, in the high latitude Alaskan and Alaskan Arctic Slope floras respectively. Our data for woody taxa is in line with the Alaskan data.

Paralleling the situation for woody species the frequency of dioecism among herbs (including annuals) found by us is around 3-4 higher than has been reported in the tropical studies mentioned earlier, and 2-3 times as high as in most other temperate floras, including the Alaskan Arctic Slope flora (Fox, 1985). Thus for equivalent life forms, sexual dimorphism is better developed in temperate latitudes and especially at high temperate latitudes, to the extent that the total amount of dioecism can exceed that in some tropical floras. Fox (1985) detected this general trend for temperate floras in the Northern Hemisphere, but it now appears to hold for the Southern Hemisphere and into tropical latitudes also.

The 30% self-incompatibility among hermaphrodites in general in the Patagonian alpine, increasing to 44% in the high alpine belt, is similar to the 25% among hermaphrodite species in a salt marsh, 29% in a sphagnum bog and 44% in a subalpine meadow in British Columbia obtained by Pojar (1974). Unfortunately no study conducted in the tropics has adequately considered the complete spectrum of life forms present (see however, Sobrevila and Arroyo, 1982). However the hermaphroditic shrubs in the Patagonia, exhibit as much or even more self-incompatibility than has been reported for the trees or trees and shrubs in some tropical forests (see Table 8). Likewise, the perennial herbs of this and other temperate communities (Pojar 1974) exhibit more self-incompatibility than the 11% that was found for 18 species in one tropical forest (J. Kress, pers. comm.). Clearly, when individual life-forms are considered, as was found for dioecism, there also seems to be more self-incompatibility in the temperate communities. This general trend, it will be recalled, was also suggested for a single family, the Leguminosae.

B. *Sexual dimorphism versus self-incompatibility*

The role of sexual dimorphism remain controversial in relation to outbreeding. Recent theoretical models developed by Charlesworth and Charlesworth (1987) however, show that sexual dimorphism is favored when the selfing rate and inbreeding depression are high and fruiting imposes a large energetic or resource burden. In the harsh Patagonian environment these conditions intersect. Scarce insects tend to concentrate on single large individuals, engendering a potential for high selfing rates. Secondly, because the plants are relatively large and long-lived in general, high selfing rates should result in larger than average inbreeding depressions for the reasons expressed above. Thirdly, seedlings are extremely scarce and thus their survival would be greatly facilitated by maximising seed reserves in response to the inherent cold temperature, low nutrients, etc, and a very short growing season. Obviously all of these features equally apply to hermaphrodites and sexually dimorphic species. However sexual dimorphism is likely to evolve when there are high demands on parental care (Charlesworth and Charlesworth, 1987). If the above arguments are correct there should be relatively more dioecism in relation to self-incompatibility among xenogamous species in the harshest environments.

TABLE 8

Comparison of ratio of dioecious to estimated number of self-incompatible species among woody taxa in some tropical forests and the Patagonian alpine. Percent SI is percentage of self-incompatible species in sample studied. Monoecious species and other floral sexuality types were grouped with the hermaphrodite species

Community	Total spp.	Percent dioecious	Percent SI	Ratio dioecious/ Estim. SI
Lowland rainforest (1)	333	23%	85.7	0.35
Semi-deciduous (2)	130	22%	79.4	0.36
Deciduous (3)	353	19%	75.7	0.31
Secondary deciduous (4)	21	19%	88.9	0.26
Montane cloud forest (5)	75	16%	40.7	0.47
Patagonian alpine (6)	27	33%	75.0	0.67

1. Bawa *et al.* (1985)-trees; 2. Bawa (1974)-trees; 3. Bullock (1985)-trees and shrubs; 4. Ruiz and Arroyo (1978)-trees; 5. Sobrevila and Arroyo (1982)-trees, shrubs, herbs; 6. Diente and Santa Lucia sites-shrubs (mostly) and trees.

Table 8 shows that this prediction holds up well for woody species comparing the Patagonian alpine with tropical communities where the appropriate data are available. Along the altitudinal gradient in the Patagonian alpine the corresponding figures for woody species are 0.59, 0.72 and a minimum of 3.61; however the ratio did not vary for perennial herbs (0.19-0.21) considering all sexually dimorphic species. Thus there is good support for the prediction regionally and some support locally.

VII. Problems in Evaluating the Relative Role of Life-History Constraints and Other Factors

Relatively more self-incompatibility and dioecism per life-form in the harsh Patagonian environment, which seems to be independent of pollinator selection, in comparison with the tropical latitudes suggests that life-history and pollinator independent selective effects are in operation on a regional scale as was suggested on local scale in the Patagonian alpine itself.

One problem that makes us inclined to be cautious about life-history independent/non-pollinator selective effects, particularly in comparing the temperate and tropical latitudes however, is that given life-forms are probably not equivalent from one region to another in relation to the factors that are critical in a breeding system analysis. Life-forms constitute a very crude and probably even misleading approximation of plant size/longevity. Some data available for longevity illustrate this problem very nicely. Callaghan and Emanuelsson (1985) reported maximum life-lengths of up to 400 years for trees, 120 years for shrubs and 100 years for perennial herbs in the Arctic. Although age determinations are unavailable for the Patagonian alpine we have no doubt that many of the species as in the Arctic, are extremely long-lived. Many species literally travel around via extensive interconnected underground rhizomes that are given to fragmentation. For the tropics primary rain-forest trees vary from maximum ages of around 250-300 years (Richards, 1952; Lieberman *et al.*, 1985). Pioneer tree species attain only 15-50 years (Richards, 1952; Brokaw, 1985). Clearly these values for primary forest trees are not very different from those reported for some trees in the Arctic, and those for pioneer trees are consistent with ranges given for many high latitude perennial herbs (Callaghan and Emanuelsson, 1985). This raises the possibility that the demographic equivalents of shrubs and perennial herbs in some temperate communities might often be closer to trees and shrubs in some tropical communities, in which case much of the regional variation in breeding system within life-forms might be spurious. Non-equivalence of life-forms may also explain why more self-incompatibility is found on the harsh arid slopes than in the more benign bog habitats in the Patagonian alpine - the bog species for any given life-form are generally smaller, and much less given to fragmentation, and accordingly should be under less intense life-history mediated selection for outcrossing.

It would be unwise even to reject variation in longevity within a given life-form as an explanation for the trend for increased xenogamy with habitat harshness within single habitats in the Patagonia. For example, members of any one life-form category could be more long-lived at the higher elevations. There is no clever way out of this dilemma at the present time. Clearly what is most needed are more sets of data on self-incompatibility and above all, better control of critical life-history features.

A second doubt that lingers is whether the self-incompatible species in the pollinator-depauperate Patagonian environment are really less serviced by pollinators than those in the pollinator-rich tropical communities. The fruit sets for self-incompatible species in the Patagonia were significantly lower than for self-compatible species, however they are higher than in two tropical communities (Bawa, 1974; Ruiz and Arroyo, 1978) and about the same in a third (Sobrevila and Arroyo, 1982). Nevertheless fruit set also tends to be lower in the self-compatible species in the tropics. Whether fruit set reflects pollinator availability has been hotly challenged considering such effects as resource limitation, life-form correlated post-fertilization abortion, and sexual selection on total flower number (Bawa and Webb, 1984). To some extent this doubt is logically redundant, because, unlike in an experimental situation, we are dealing with organisms that might have already evolved some kind of compensatory adaptations for the extreme conditions for pollination. The solution to this problem will depend on more accurate assessment of the probability of effective flower pollination, and on studies on these compensatory mechanisms.

VIII. Concluding Remarks

Life-history constraints acting through plant size, seems to be the main actor on the long-term breeding systems stage, with pollinator pressure being hierarchically less important. Strong abiotic and biotic selection characteristic of many natural habitats is suggested, but cannot be confirmed. Because harsh abiotic conditions favour longevity and longevity selects for xenogamy, greatest departures in breeding systems from that expected on the grounds of pollinator availability should be found in the types of environments we elected to study. In more benign alpine areas more self-compatibility and elevational increases in self-compatiblity are expected (c.f Moldenke, 1979). In the most extreme cases where wind-pollination, the ultimate resource for maintaining outcrossing is untenable, the long-lived species of abiotically- or biotically-harsh habitats are often likely to evolve towards apomixis, as has been described for hermaphrodite (Kauer *et al.*, 1978; Sobrevila and Arroyo, 1982) and dioecious (Ha *et al.*, 1988) woody species in wet or wet/cold tropical forests and as is seen in the presence of pseudo-vivipary in dioecious *Poa alopecurus* and in agamospermous fruit set in female individuals of *Hamadryas kingii* on the wetter sites in the Baguales Range. Unfortunately because apomixis is technically difficult to detect, its frequency has probably been grossly underestimated in many breeding systems studies, including ours.

Relaxed life-history constraints should also influence plant breeding systems. The self-compatibility of many annuals might have evolved more frequently in reponse to the latter than to pollinator-limitation (contra Arroyo, 1973).

Finally it should be borne in mind that ours is a community approach and what has been dealt with is an ecological snapshot of broad trends that have been produced over evolutionary time. Under the emulsion (percentages) are many present-day evolutionary experiments. These should provide the best clues to breeding system evolution.

Acknowledgements

Work supported by grants from FONDECYT, Chile, National Geographic Society, USA and DIB, Universidad de Chile, and partially conducted under the tenure of a John Simon Guggenheim Memorial Fellowship by MTKA. We thank many friends in the Patagonia who provided us with shelter, our technicians and many student helpers. The Japanese Society for the Promotion of Science is especially acknowledged for supporting travel to Japan for the first author.

References

Arroyo, Mary T. Kalin. (1973). A taximetric study of intraspecific variation in autogamous *Limnanthes floccosa* (Limnanthaceae). *Brittonia* **25**, 177-191.

Arroyo, Mary T. Kalin. (1981). Breeding systems and pollination biology in Leguminosae. *In* "Advances in Legume Systematics" (R.M. Polhill and P.H. Raven, eds), pp. 723-769. Royal Botanical Gardens, Kew.

Arroyo, Mary T. Kalin, Armesto, J. and Primack, R. (1985). Community studies in pollination ecology in the high temperate Andes of central Chile. II. Effect of temperature on visitation rates and pollination possibilities. *Plant Syst. Evol.* **149**, 187-203.

Arroyo, Mary T. Kalin, Squeo, F. and Lanfranco, D. (1987). Polinización biótica en los Andes de Chile: Avances hacia una síntesis. *In* "Ecología de la Reproducción e Interacciones Plant/Animal" (E. Forero, F. Sarmiento and C. La Rotta, eds). *Anales del IV Congreso Latinoamericano de Botánica* **2**, 55-76.

Arroyo, Mary T. Kalin and Squeo, F. (1987). Experimental detection of anemophily in *Pernettya mucronata* (Ericaceae) in western Patagonia. *Bot. Jahrbr.* **108**, 537-546.

Arroyo, Mary T. Kalin, Squeo, F. Armesto, J. and Villagrán, C. (1988). Effects of aridity on plant diversity in the northern Chile Andes. *Ann. Mo. Bot. Gard.* **75**, 55-78.

Arroyo, Mary T. Kalin and Peñaloza, A. (1989). Genetic self-compatibility in a South American species of *Ourisia* (Scrophulariaceae). *N.Z. J. Bot.* (submitted).

Arroyo, Mary T. Kalin and Squeo, F. (1990). Genetic self-incompatibility in the endemic Patagonia genus *Benthamiella* (Solanaceae). *Gayana (Bot.)* **47**, (in press).

Arroyo, Mary T. Kalin, Marticorena, C., Miranda, P., Matthei, O., Landero, A. and Squeo, F. (1989a). Contribution to the high elevation flora of the Chilean Patagonia: a checklist of species on mountains on an east-west transect at latitude 50°S. *Gayana (Bot.)* **46 (1-2)**, 121-151.

Arroyo, Mary T. Kalin, Ricardo Rozzi, Squeo, F. and Belmonte, E. (1989b). Pollination in tropical and temperate high elevation ecosystems: Hypotheses and the Asteraceae as a test case. *In* "Mount Kenya: Research, Conservation and Development" (M. Winiger, U. Wiesmann, J. Rheker, W. Lusigi and S.S. Ojany, eds), Geographic Bernensia, African Study Series (in press).
Bawa, K. (1974). Breeding systems of tree species of a lowland tropical community. *Evolution* **28**, 85-92.
Bawa, K. (1980). Evolution of dioecy in flowering plants. *Annu. Rev. Ecol. Syst.* **11**, 15-39.
Bawa, K., Perry, D.R. and Beach, J.H. (1985). Reproductive biology of tropical lowland rain forest trees. I. Sexual systems and incompatibility mechanisms. *Am. J. Bot.* **72**, 331-345.
Bawa, K. and Webb, C. (1984). Flower, fruit and seed abortion in tropical forest trees: implications for the evolution of paternal and maternal reproductive patterns. *Am. J. Bot.* **71(5)**, 736-751.
Brokaw, N. (1985). Treefalls, regrowth, and community structure in tropical forests. *In* "The Ecology of Natural Disturbance and Patch Dynamics" (S.T.A. Pickett and P.S. White, eds), pp. 53-69. Academic Press, London.
Bullock, S. H. (1985). Breeding systems in the flora of a tropical deciduous forest in Mexico. *Biotropica* **17**, 287-301.
Callaghan, T.V. and Emanuelsson, U. (1985). Population structure and processes of tundra plants and vegetation. *In* "The Population Structure of Vegetation" (J. White, ed.), pp. 399-439. W. Junk, Dordrecht.
Carlquist, S. (1974). Island Biology. Colombia University Press, New York Soc. **3**, 333-348.
Charlesworth, D. and Charlesworth, B. (1987). The effect of investment in attractive structures on allocation to male and female functions in plants. *Evolution* **41**, 948-986.
Conn, J.S., Wentworth, T.R. and Blum, U. (1980). Patterns of dioecism in the flora of the Carolinas. *Am. Midl. Nat.* **103**, 310-315.
Croat, T.B. (1979). The sexuality of the Barro Colorado Island flora. *Phytologia* **42**, 319-348.
Darwin, C. (1877). The different forms of flowers on plants of the same species. John Murray, London.
Flores, S. and Schemske, D.W. (1984). Dioecy and monoecy in the flora of Puerto Rico and the Virgin Islands: ecological correlates. *Biotropica* **16(2)**, 132-139.
Fox, J.F. (1985). Incidence of dioecy in relation to growth form, pollination and dispersal. *Oecologia* (Berl.) **67**, 244-249.
Godley, E.J. (1979). Flower biology in New Zealand. *N.Z. J. Bot.* **17**, 441-466.
Ha, C.O., Sands, V.E., Soepadmo, E. and Jong, K. (1988). Reproductive patterns in selected understorey trees in the Malaysian rain forest: the apomictic species. *Bot. J. Linn. Soc.* **97**, 317-331.
Hamrick, J.L., Linhart, Y.B. and Mitton, J.B. (1979). Relationships between life-history characteristics and electrophoretically detectable genetic variation. *Annu. Rev. Ecol. Syst.* **10**, 173-200.
Kauer, A., Ha., C., Jong, K., Sands, V., Chan, H., Soepadmo, E. and Ashton, P. (1978). Apomixis may be widespread among trees of climax rainforest. *Nature* **271**, 440-442.
Kay, Q.O.N and Stevens, P. (1986). The frequency, distribution and reproductive biology of dioecious species in the native flora of the Britain and Ireland. *J. Linn. Soc. Bot.* **92**, 39-64.
Ledig, F.T. (1986). Heterozygosity, heterosis, and fitness in outbreeding plants. *In* "Conservation Biology. The Science of Scarcity and Diversity" (M. Soulé, ed.), pp. 77-104. Sinauer, Sunderland.
Lieberman, D., Lieberman, M., Peralta, R. and Hartshorn, G. S. (1985). Mortality patterns and stand turnover rates in a wet tropical forest in Costa Rica. *J. Ecol.* **73**, 915-924.
Lloyd, D.G. (1965). Evolution of self-compatibility and racial differentiation in *Leavenworthia* (Cruciferae). *Contrib. Gray Herb. Harv. Univ.* **195**, 3-134.
McComb, J.A. (1966). The sex forms of species in the flora of the south-west of western Australia. *Aust. J. Bot.* **14**, 303-316.
Moldenke, A. (1979). Pollination ecology in the Sierra Nevada. *Phytologia* **42**, 223-281.
Moore, D. (1983). Flora of Tierra del Fuego. Anthony Nelson, England.
Moore, D. M. and Doggett. (1976). Pseudo-vivipary in Fuegian and Falkland Islands grasses. *Br. Antarct. Surv. Bull.* No. **43**, 103-110.

Parsons, P.A. (1958). Evolution of sex in the flowering plants of South Australia. *Nature* **181**, 1673-1674.

Pisano, E. (1974). Estudio ecológico de la región continental sur del área andino-patagónica. II. Contribución a la fitogeografía de la zona del Parque Nacional "Torres del Paine". *Ans. Inst. Pat. (Chile)* **5**, 59-104.

Pojar, J. (1974). Reproductive dynamics of four plant communities of southwestern British Colombia. *Can. J. Bot.* **52**, 1819-1834.

Raven, P.H. (1979). A survey of reproductive biology of Onagraceae. *N.Z. J. Bot.* **17**, 575-593.

Richards, P. W. (1952). The Tropical Rainforest. Cambridge University Press, Cambridge. 450 pp.

Roy, R.P. (1974). Sex mechanisms in higher plants. *J. Indian Bot. Soc.* **53**, 141-155.

Ruiz, T. and Arroyo, M. T. Kalin. (1978). Plant reproductive ecology of a secondary deciduous forest in Venezuela. *Biotropica* **10**, 221-230.

Sobrevila, C. and Arroyo, M. T. Kalin. (1982). Breeding systems in a tropical montane cloud forest in Venezuela. *Plant Syst. Evol.* **140**, 19-37.

Stebbins, G. L. (1958). Longevity, habitat and release of genetic variability in higher plants. *Cold Spring Habor Sympos. Quant. Biol.* **23**, 365-378.

Steiner, K. E. (1988). Dioecism and its correlates in the Cape flora of South Africa. *Am. J. Bot.* **75**, 1742-1754.

Tanner, E.V.J. (1982). Species diversity and reproductive mechanisms in Jamaican trees. *Biol. J. Linn. Soc.* **18**, 263-278.

Wyatt, R. (1983). Pollinator-plant interactions and the evolution of breeding systems. *In* "Pollination Biology" (L.A. Real, ed.), pp. 51-95. Academic Press, New York.

APPENDIX 1. Sexually dimorphic species (4 sites on Baguales Range) and tests on hermaphrodites in the Patagonian alpine (Santa Lucía and Diente sites). Each species is followed by its life-form and breeding system. Fl. number/number of individuals tested, % fr. set and seed number per fl. crossed or tested is given for each test. For species with one-seeded fruits, only % fruit set is given. A: ann. herb; P: caespitose perenn. herb; PC: cushion-forming perenn. herb; PR: perenn. herb with extensive rhizomes or stolons; PW: suffruticose perenn. herb; SH: shrub; T: tree. SC: highly self-compatible; SCP: partially self-compatible; SI: self-incompatible; AGAM: agamospermous; DIOEC: dioecious; DIOEC*: morphologically gynodioecious, functionally dioecious; GYNODIOEC: gynodioecious. S: hand self-pollinated fls.; SS: unmanipulated fls. tested for spontaneous self-pollination; C: hand cross-pollinated fls.; EM: fls. emasculated, unpollinated. Ruiz and Arroyo (1978) criteria for breeding systems were followed where self-pollination and cross-pollination data are available. Species with high fruit and seed sets in early SS tests were considered self-compatible without further testing. Fl. number for Asteraceae refers to total florets. Heads were pollinated 3-4 times. Cross-pollination for Asteraceae is minimum estimate because the pollination technique did not ensure the pollination of all florets present, whereas all florets were potentially open to automatic self-pollination.

AMARYLLIDACEAE: **Alstroemeria patagonica**(PR-SC): SC(26/4): 92.3%-13.54; EM(14/1)-0%. ASTERACEAE: **Abrotanella emarginata**(PC-PSC): S(560/3)-31.6%; SS(258/2)-36.4%.-**Adenocaulon chilense**(PR-SC): SS(93/3)-91.4%.-**Antennaria chilensis**(PR-females only, AGAM;) (Moore, 1983).-**Baccharis magellanica**(SH-DIOEC).-**Chiliotrichum diffusum**(SH-SI): S(222/3)-0.9%; SS(983/6)-0.5%; C(214/5)-6.5%.-**Erigeron leptopetalus**(PR-SI): S(148/2)-1.4%; SS(165/2)-0%; C(377/4)-63.4%.-**Gamochaeta spiciformis**(P-PSC): SS(695/5)-79.9%-**Hieracium antarcticum**(P-PSC): SS(185/2)-32.4%; EM(233/4)-0%.-**Hypochaeris arenaria**(PR-PSC): SS(309/6)-63.1%; EM(201/4)-0%.-**H. incana**(PR-SI): S(41/1)-0%; SS(488/6)-0.4%; C(167/4)-40.7%.-**Leucheria hahnii**(PR-SI): S(269/4)-1.5%; SS(312/14)-4.5%; C(179/4)-32.4%.-**L. leontopodiodes**(PR-SI): S(1310/6)-0.53%; SS(832/8)-1.6%; C(364/3)-32.7%.-**Nardophyllum bryoides**(SH-SI): S(149/4)-0%; SS(447/6)-0%; C(94/4)-32.7%.-**Nassauvia aculeata**(PW-SI): S(43/3)-0%; SS(274/6)-9.0%; C(58/3)-37.9%.-**N. lagascae**(PR-SI): S(794/1)-1.0%; SS(1017/3)-1.2%; C(48/1)-16.7%.-**N. magellanica**(PR-PSC): S(2330/6)-7.9%; SS(2430/6)-20.4%; C(2719/6)-47.4%.-**N. pygmaea**(PR-SI): S(1599/10)-1.5%; SS(2367/10)-1.0%; C(798/8)-8.4%.-**N. revoluta**(PR-SI): S(456/3)-2.9%; SS(291/3)-0.3%; C(643/3)-21.6%.-**Perezia megalantha**(PR-SI): S(341/7)-0.29%; SS(2033/40)-0%; C(356/4)-32.0%-**P. pilifera**(PR-SI, prob.): S(77/5)-2.6%; SS(115/7)-1.8%; C(119/8)-42.0% although some aborted.-**P. recurvata**(PW-SI): S(70/1)-0%; SS(178/3)-0.6%; C(132/4)-24.2%.-**Senecio alloeophyllus**(PW-SI): SS(426/3)-0%.; C(913/3)-46.1%.-**S. argyreus**(PW-SI): S(646/4)-0.5%; SS(511/5)-0.8%; C(284/3)-6.0%.-**S. kingii**(PR-SI): S(436/5)-2.3%; SS(698/5)-9.6%; C(527/3)-57.1%.-**S. lasegeui**(PR-SI): S(666/7)-5.3%; SS(1002/8)-2.5%; C(392/4)-38.3%.-**S. magellanicus**(PR-SI): S(816/5)-2.5%; SS(2338/8)-1.2%; C(1258/5)-28.9%.-**S. miser**(PW-SI): S(541/3)-1.5%; SS(3614/8)-2.0%; C(544/3)-25.9%.-**S. patagonicus**(PW-SI): S(205/4)-0%; SS(914/6)-1.1%; C(261/4)-54.0%.-**S. sericeonitens**(PW-SI): S(672/4)-6.4%; SS(864/2)-4.1%; C(684/5)-40.4%.-**S. tricuspidatus**(SH-PSC): S(809/6)-1.9%; SS(2402/6)-3.2%; C(209/3)-11.5%.BERBERIDACEAE: **Berberis buxifolia**(SH-SI): S(8/3)-0%; SS(7/2)-0%; C(13/5)-100%-10.60.-**B. empetrifolia**(SH-PSC): S(63/7)-12.7%-(0.44) SS(51/4)-15.7%-0.71; C(23/4)-39.13%-2.00. BORAGINACEAE:-**Myosotis stricta**(A-SC): SS(33/5)-93.93%-2.52.-**Plagiobothrys calandrinioides**(A-SC): SS(43/10)-100%-2.60.-**Plagiobothrys sp.**(A-PSC): SS(19/7)-78.9%-2.47. CALYCERACEAE: **Moschopsis rosulata**(PR-PSC): S(283/3): 1.1%; SS(66/1)-0%; C(504/2)-3.17%. CAMPANULACEAE: **Pratia longiflora**(PR-PSC): S(5/1)-60.0%-29.6; SS(19/3)-63.2%-13.05; EM(5/1)-14.3%-0.14. CARYOPHYLLACEAE: **Arenaria serpens**(P-SC): S(9/2)-88.9%-13.11.-**Cerastium arvense**(P-PSC): S(20/4)-55.0%-14.45; SS(52/10)-59.6%-6.56; EM(32/4)-3.1%-0.78.-**Colobanthus lycopodioides**(PC-SC): SS(218/8)-95.9%-12.92.-**C. quitensis**(P-SC): SS(65/7)-96.9%-25.04.-**C. subulatus**(PC-SC): SS(33/4)-93.9%-7.09.-**Sagina procumbens**(P-SC): SS(6/2)-83.3%-50.33.-**Silene**

chilensis(P-PSC): S(29/4)-44.1%-12.59; SS(42/6)-9.5%-3.21; EM(15/4)-0%.-**Stellaria debilis**(P-SC): SS(6/1)-83.3%-6.0. CELESTRACEAE: **Maytenus magallanica**(T-DIOEC). CRUCIFERAE: **Cardamine glacialis**(P-SC): S(14/1)-100%-16.07; SS(29/4)-75.9%-12.57; EM(19/3)-0%.- **Descurainia sophia**(A-SC): SS(10/1)-90%-NC.-**Draba funiculosa**(P-SC): SS(26/1)-84.6%-14.10; EM(2/1)-0%.-**D. magellanica**(P-SC-some AGAM): S(20/4)-95.0%-15.05; SS(59/4)-98.3%-16.09; EM(8/1)-12.5%-2.13.-**D. subglabrata**(P-SC): SS(117/4)-94.9%-15.98; EM(6/2)-0%.-**Eudema hauthalii**(PC-SI): S(46/5)-0%; SS(69/4)-1.45%-0.01; C(50/6)-22.0%-0.62; EM(5/1)-0%.-**Lepidium** sp.(P-PSC): SS(32/3)-78.1%-20.24.-**Menonvillea nordenskjoeldii**(PR-SI): S(25/4)-8.0%-0.08; SS(160/5)-3.1%-0.04; C(25/4)-40.0%-0.80; EM(5/1)-0%.-**Sisymbrium magellanicum**(P-PSC): S(38/5)-18.4%-2.66; SS(99/3)-9.1%-1.08; C(36/5)-33.3%-7.77.-**Thlaspi magellanicum**(P-SC): S(22/2)-90.9%-6.08; SS(31/2)-87.1%-4.81; EM(11/3)-0%. EMPETRACEAE: **Empetrum rubrum**(SH-DIOEC -sometimes polygamous). ERICACEAE: **Pernettya mucronata** (SH-DIOEC).-**P. pumila**(SH-DIOEC). GENTIANACEAE: **Gentiana prostrata**(A-Bi?-SC-some AGAM): SS(13/8)-100%-61.7; EM(10/7)-50.0%-43.5.-**Gentianella magellanica**(A-SC-some AGAM?): SS(48/13)-97.9%-43.38; EM(13/9)-23.1%-2.53. GERANIACEAE: **Geranium sessiliflorum**(P-PSC): SS(56/15)-67.9%-2.98; EM(17/7)-0%. GRAMININAE: **Poa alopecurus** spp. **alopecurus**(P-DIOEC).-**Poa alopecurus** ssp.**fuegiana**(P-PSEUDOVIVIPAROUS-see Moore & Doggett,1974).-**Poa ibari** (P-DIOEC).-**Poa pungionifolia** (PC-GYNODIOEC). GUNNERACEAE: **Gunnera magellanica**(PC-DIOEC). HYDROPHYLLACEAE: **Phacelia secunda**(P-PSC): S(11/1)-18.2%-0.36; SS(260/9)-31.5%-0.94. IRIDACEAE: **Phaiophleps biflora v. biflora**(PR-SI): S(19/7)-0%; SS(47/13)-10.6%-0.15; C(29/15)-65.5%-6.66.-**P. biflora v. lyckholmii**(PR-SI): S(3/2)-0%; SS(12/12)-8.3%-0.08; C(16/12)-68.8%-11.4. **Sisyrinchium junceum**(PR-SI): S(7/3)-14.3%-0.14; SS(19/9)-10.5%-1.37; C(8/3)62.5%-5.63.-**S.nanum**(PR-SC): S(6/1)-16.7-13.0; SS(58/13)-96.55%-14.31; EM(14/7)-0%.-**S. pearcei**(PR-SC): S(15/6)-86.7%-25.6; SS(41/17)-82.9%-25.6; C(2/2)-100%-26.5; EM(22/2)-0%. JUNCACEAE: **Luzula alopecurus**(P-SC): SS(536/4)-95.9%-NC.-**Luzula sp.**(PC-SC): SS(66/14)-84.9%-2.14. LOASACEAE: **Loasa bergii**(P-SC): S(4/1)-100%-131.8; SS(10/1)-100%-168.5; EM(4/1)-0%. MISODENDRACEAE: **Misodendrum punctulatum**(SH-DIOEC).-**M. quadrifolium**(SH-DIOEC). **Epilobium australe**(PR-SC): S(27/1)-100%; SS(117/7)-90.6%-47.89; EM(2/1)-0%.-**E. glaucum**(PR-PSC): SS(34/5)-67.7%-76.92; EM(4/1)-0%. ORCHIDACEAE: **Gavilea lutea**(P-SC): SS(26/2)-100%-NC. OXALIDACEAE: **Oxalis loricata**(PR-SI-tristylous): S(38/14)-2.6%-0.21; SS(9/6)-0%; C(intermorph)(27/14)-81.48%-6.74.-**O. patagonica**(PC-SI-tristylous): S(34)-2.9%-0.11; SS(98)-0%; C(intermorph)(23)-82.6%-7.98. PAPILIONACEAE: **Adesmia aurantiaca**(PR-SI): S(41/3)-0%.; SS(38/5)-2.6%-0.05; C(28/4)-28.6%-0.39; EM(5/1)-0%.-**A. burkartii**(PR-SC): S(6/1)-100%-3.50; SS(37/2)-94.6%-2.70.-**A. lotoides**(PR-PSC): S(12/4)-58.3%-1.75; SS(47/6)-8.5%-0.17; EM(26/6)-0%.-**A. parvifolia**(P-SC): S(23/4)-69.6%-2.57; SS(51/5)-86.3%-3.03; EM(10/3)-0%-**A. pumila**(PR-SC): S(21/4)-38.1%-2.00; SS(31/5)-32.3%-1.47; C(3/1)-66.7%-2.33;-EM(15/5)-0%.-**A. salicornoides**(PW-PSC): S(61/8)-27.9%-0.62; SS(563/18)-19.4%-0.48; C(48/7)-27.1%-0.79; EM(7/3)-0%.-**A. suffocata**(PR-SC): S(5/2)-100%-4.60.-**A. villosa**(PR-PSC): S(13/7)-15.4%-1.00; SS(20/7)-25.0%-1.16; C(10/5)-80.00%-4.70.-**Astragalus nivicola**(P-SC): S(34/4)-91.2%-8.26; SS(39/4)-87.7%-6.97; EM(5/1)-0%.-**Lathyrus magellanicus** v. **glaucescens**(PR-PSC): S(38/6)-28.9%-1.5; SS(96/2)-7.29%-0.30; C(61/8)-50.8%-3.04; EM(6/1)-0%.-**L. magellanicus** v. **magellanicus**(PR-SI): S(73/8)-4.1%-0.30; SS(44/3)-0%; C(50/6)-28.00%-2.56.-**Vicia bijuga**(PR-PSC): S(13/3)-15.4%-0.15; SS(25/5)-60.0%-2.24; EM(16/5)-0%.-**V. magellanica**(PR-PSC): S(33/4)-57.6%-3.93; SS(66/8)-81.8%-5.13; EM(35/3)-0%. PLUMBAGINACEAE: **Armeria maritima**(PC-SC): SS(376/6)-85.4%-0.85. PORTULACACEAE: **Calandrinia caespitosa**(P-PSC): SS(54/9): 74.1%-24.33; EM(14/4)-0%. PRIMULACEAE: **Primula magellanica**(PR-SC): S(5/5)-100%-57.80; SS(11/3)-81.2%-62.73. PROTEACEAE: **Embothrium coccineum**(SH-SI): S(8/1)-0%; SS(66/5)-9.1%-0.82; C(11/2)-63/4%-5.91.-RANUNCULACEAE: **Caltha appendiculata**(PC-SC-some AGAM?): SS(6)-100%-NC; EM(6/6)-16.7%-NC.- **Anemone multifida**(PR-SC): S(6/2)-100%-42.5; SS(27/8)-85.2%-70.03; EM(4/4)-0%. **Hamadryas delfinii**(PR-DIOEC).-**H. kingii**(PR-

DIOEC-some AGAM).-**H. magellanica**(PR-DIOEC).-**H. sempervivoides**(PC-DIOEC).-**Hamadryas sp. indet.**(PC-DIOEC).-**Ranunculus aquatilis**(P-SC): SS(9/1)-100%-13.22; **R. fuegianus**(PR-SC): SS(4/4)-100%-32.75; SS(8/5)-100%-38.0.-**R. peduncularis**(PR-SC-some AGAM): S(20/6)-95.0%-10.8; SS(11/5)-100%-9.27; EM(8/8)-50.0%-4.25.-**R. uniflorus**(PR-SC-some AGAM): SS(15/6)-92.3%-102.43; S(2/1)-100%-NC; EM(10/6)-40.0%-NC. ROSACEAE: **Acaena antarctica**(PW-SC): S(284/3)-96.1%; SS(631/5)-89.1%.-**A. lucida**(PW-SC): SS(58/2)-96.6%.-**A. magallanica**(PW-GYNODIOEC).-**A. ovalifolia**(P-SC): SS(118/1)-100%.-**A. pinnatifida**(P-SC): SS(214/3)-92.1%; EM(25/1)-0%.-**A. platyacantha**(P-SC): SS(606/7)-87.9%.-**Acaena sp.**(P-SC): SS(7/1)-100%.-**Geum magellanicum**(P-SC-some AGAM?): S(29/9)-89.7%-122.76; SS(24/6)-91.7%-123.33; EM(17/5)-15.8%-5.15. RUBIACEAE: **Galium antarcticum**(P-PSC): SS(31/4)-51.6%-0.65.-**Oreopolus glacialis**(PC-SI-distylous): S(86/7)-2.32%-0.02; SS(33/4)-0%; C(123/7)-52.8%-0.72.- SANTALACEAE: **Arjona patagonica**(PR-SI-distylous): S(50/14)-0%; SS(60/16)-0%; C(inter-morph)(8/4)-12.5%; C(intra-morph)(28/6)-0%.-**A. pusilla**(PR-SC): S(5/4)-80.0%; SS(2/2)-0%; EM(10/5)-0%. SAXIFRAGACEAE: **Escallonia alpina**(SH-SI): S(48/4)-0%; SS(48/3)-0%; C(20/4)-85.0%-101.36.-**Ribes culcullatum** (SH-DIOEC).-**Saxifraga magellanica**(P-SC-some AGAM?): SS(39/6)97.4%-62.05; EM(23/3)-17.4%-0.69.-**Saxifragella bicuspidata**(PC-PSC): SS(13/2)-61.5%-4.0. SCROPHULARIACEAE: **Calceolaria biflora**(PR-SC): S(9/3)-100%-623.0; SS(27/9)-92.6%-328.51; EM(9/4)-0%.-**C. tenella**(PR-SC): S(2/1)-100%-188.0; SS(60/3)-24.4%-51.61; EM(15/1)-0%.-**C. uniflora**(PC-PSC): S(30/8)-66.7%-28.5; SS(38/5)-52.6%-32.45; C(10/3)-80.0%-149.26;-EM(35/11)-0%.-**Euphrasia antarctica**(A-PSC): SS(18/9)-72.2%-7.11; EM(9/9)-0%.-**Limosella australis**(P-PSC): S(6/2)-66.7%-4.50.-**Ourisia poeppigii**(PR-SC): S(21/4)-100%-262.7; SS(17/4)-225.4; C(7/2)-100%-100.9; EM(PR-19/4)-0% (from Arroyo and Peñaloza 1989). SOLANACEAE: **Benthamiella nordenskjoldii**(PC-SI): S(162/6)-0%; SS(114/5)-0%; C(61/4)-39.3%-0.49 (from Arroyo and Squeo 1989). UMBELLIFERAE: **Azorella fuegiana**(PR-AGAM?): S(17/2)-70.6%-1.06; SS(45/3)-28.9%-0.42; C(25/3)-44.0%-0.64; EM(2/1)-100%-2.00.-**A. lycopodioides**(PC-PSC): S(31/5)-77.4%-0.87; SS(309/7)-23.9%-0.29; C(3/3)-66.7%-0.67.-**Bolax gummifera**(PC-DIOEC*).-**Mulinum valentinii** (S-DIOEC*).-**Osmorhiza depauperata**(P-SC): S(2/1)-100%-NC; SS(165/5)-92.7%-NC. VALERIANACEAE: **Valeriana carnosa**(P-GYNODIOEC).-**V. lapathifolia**(PR-GYNODIOEC).-**V. magellanica**(PC-DIOEC*). VERBENACEAE: **Junellia** sp.(SH-SI): S(59/5)-0%; SS(8/3)-0%; C(11/6)-36.36%-early fruit abortion seen-NC. VIOLACEAE: **Viola maculata**(PR-SC): SS(7/6)-100%-15.14; **V. reichei**(PR-PSC): S(chasmog. fls)(4/4)-0%; SS(chasmog. fls.)(12/12)-8.33%-0.50; SS(cleistog. fls)(32/18)-68.7%-14.66; EM(chasmog. fls.)(11/10)-0%.-**V. tridentata**(PR-PSC): SS(8/2)-62.5%-4.63.

14 Variation and Evolution of Mating Systems in Seed Plants

SPENCER C. H. BARRETT AND CHRISTOPHER G. ECKERT

Department of Botany, University of Toronto, Toronto, Ontario, Canada M5S 3B2

I. Introduction

The mating system involves those attributes of an organism that govern how gametes are united to form zygotes. The analysis of mating systems is usually concerned with the genetic relatedness of sexual partners and hence the degree of inbreeding practised by individuals within populations. Mating patterns in plant populations are influenced by numerous environmental, demographic and genetic factors. Of these, the type of breeding system (e.g. dioecism, heterostyly) that a population possesses is of prime importance (Richards, 1986). Quantitative studies of mating systems, until recently, were largely restricted to cultivated plants. However, the recent growth of population biology, and the advent of electrophoretic techniques has provided impetus for a growing number of investigations of mating patterns in natural plant populations (Schemske and Lande, 1985). Consequently, during the past two decades there has been a rapid increase in studies on the measurement of mating system parameters, particularly levels of self- and cross-fertilization.

The mating system is not a static property of an individual, population or species. Mating patterns are dynamic, and subject to modification by many forces, operating on both ecological and evolutionary time scales. Interest in the causes of shifts in mating systems arise because of the important consequences that such changes have on population genetic structure, selection response, and speciation. Models of mating system evolution have become increasingly complex as workers appreciate that selection on mating patterns does not occur in isolation from other features of the life history and demography of populations (e.g. Lloyd, 1980;

BIOLOGICAL APPROACHES AND
EVOLUTIONARY TRENDS IN PLANTS ISBN 0-12-402960-4

Charlesworth and Charlesworth, 1981; Holsinger, 1986; Campbell and Waser, 1987). A prerequisite for development of realistic models of mating system evolution is accurate information on the mating process and on the ecological and genetic context in which it occurs.

In this paper we examine some of the causes and consequences of mating system variation in flowering plants. We begin by discussing briefly how plant mating systems can be measured and what parameters are of most significance for understanding their evolution. We then present results from a survey of published outcrossing rates in natural populations and discuss their significance for models of mating system change. Finally, we focus on intraspecific studies, discuss their relevance for understanding the processes responsible for evolutionary changes in mating behavior, and conclude by outlining five areas of inquiry that may aid in future attempts to determine the causes of mating system change.

II. Measuring Mating Systems

Inferences on mating parameters, such as the selfing rate, can be made from observations of floral morphology. While this can be a useful guide in some species, such as those that are dioecious or possess cleistogamous flowers, it can be misleading in many self-compatible plants, particularly those with mixed mating systems. In these cases quantitative analysis, requiring large population samples and a statistical model of the mating process, is required to establish the true nature of mating patterns (Clegg, 1980; Ritland, 1983).

There are two types of information required to accurately describe the mating system. These involve measures of fertility and estimates of the kinds of mating events that occur (Brown, 1990). Measures of fertility attempt to estimate the relative contribution that individuals make through male and female gametes to the next generation. Mating events are usually classified as to whether zygotes result from selfing, outcrossing or apomixis. Among the outcrossed fraction it may also be important to determine the degree of biparental inbreeding (Waller and Knight, 1989).

Despite considerable progress in recent years in measuring plant mating systems we are still some way from being able to estimate all of the types of information required for an accurate depiction of the mating process, for even a single natural population. For example, because of the difficulties in measuring male fertility variation and the paternity of outcrossed offspring our knowledge of male mating success is rudimentary at best (Devlin *et al.*, 1989). This deficiency handicaps our ability to test models that invoke sexual selection on the male component of fitness in hermaphrodite plants (Charnov, 1982; Lloyd, 1984). In contrast, because of the relative ease with which mating events can be distinguished for maternal parents, considerable headway has been made in estimating the average frequencies of cross- and self-fertilization for populations of different plant species.

Many aspects of the ecological and genetic relationships of uniting gametes are embodied in the parameters t, the outcrossing rate and s, the selfing rate which are probably the most convenient measures of the mating system in plants (Lloyd,

1980). These parameters are central to testing models for the evolution of several breeding systems, including autogamy (Lloyd, 1979; Lande and Schemske, 1985), heterostyly (Charlesworth and Charlesworth, 1979), dioecy and gynodioecy (Lloyd, 1975; Charlesworth and Charlesworth, 1978a) as well as more generally the evolution of combined versus separate sexes (Lloyd, 1982). This is because in many species the frequency of self-fertilization has a direct bearing on the number and quality of offspring as a result of the phenomenon of inbreeding depression (Charlesworth and Charlesworth, 1987).

Quantitative estimates of outcrossing rate were first obtained from predominantly self-pollinating crops using morphological markers and a model which assumes that each zygote results from either self-fertilization, with fixed probability s, or fertilization by a pollen grain chosen at random from the population with probability $t = 1 - s$ (Jones, 1916; Fyfe and Bailey, 1951). Since then the mixed-mating model has been used to estimate outcrossing rates for a wide range of agricultural and natural populations of both inbreeding and outbreeding species. Brown *et al.* (1985) have reviewed the basic assumptions of the model and some of the difficulties that have emerged during its application.

Several refinements have been made to techniques for estimating outcrossing rates in recent years. Isozyme loci have largely replaced the use of morphological marker genes (Brown and Allard, 1970). This is because morphological markers are often difficult to find, exhibit dominant expression reducing statistical power, and there is also the possibility that the genes themselves (e.g. flower color loci) may directly influence the mating process. Allozyme variants have, in contrast, several distinct advantages. They are codominantly expressed, highly polymorphic, and unlikely to be under strong selection. Because of these advantages workers surveying allozyme variation in plant populations can now routinely estimate outcrossing rates, so long as samples are collected separately from individual maternal parents and not pooled.

The large number of marker loci available using allozyme polymorphisms has also led to the development of more sophisticated estimation procedures using multiple, rather than single loci (Green *et al.*, 1980). A single estimate of outcrossing based on the joint behavior of several marker loci, increases the probability of detecting an outcrossing event, in comparison with single locus estimates (Shaw *et al.*, 1981). In addition, multilocus estimates are less affected by selection and nonrandom outcrossing than single locus estimates, and under most circumstances, multilocus measures, using three to four loci, provide a reliable average estimate of the outcrossing rate at the population level (Ritland and Jain, 1981). More difficult problems are encountered, however, in estimating the outcrossing rates of individual plants within populations. As a result, less information on plant to plant variation in outcrossing rate is available for natural populations (Ritland and Ganders, 1985; Morgan and Barrett, 1990). Further discussion of the strengths and weaknesses of various estimation procedures used for measuring mating patterns in natural populations are given in Clegg (1980), Ritland (1983), and Brown (1990). A recent paper by Ritland (1990) provides details of computer algorithims used for estimating mating system parameters.

III. Models of Mating System Evolution

Theoretical interest in the evolution of self-fertilization began with Fisher's (1941) demonstration that a variant practising some degree of selfing will spread in an outbreeding population, as long as its ability to disseminate pollen to other individuals is not affected by its propensity to self-fertilize. The reason is simply that while the average individual in an outcrossing population transmits one haploid genome through each seed and one through each successful pollen grain, a self-fertilized individual passes on two haploid genomes through each seed while still transmitting one in each successful outcrossing pollen grain. An allele for complete self-fertilization is, therefore, one and a half times more likely to be transmitted to the next generation than an alternate allele for random mating. Given the observation that many flowering plants are predominantly outcrossing, theoretical work has sought to explain why selfing has not spread in more species through the 'automatic selection' of genes influencing the rate of self-fertilization (Jain, 1976) .

The spread of a selfing variant in an outcrossing population may be prevented if the proportion of pollen wasted or 'discounted' by self-pollination is equal to or higher than the variant's selfing rate (Lloyd, 1979; Holsinger *et al.*, 1984; Holsinger, 1988a). While this effect is of theoretical interest, there are, to date, no published estimates of pollen discounting in natural populations so it is difficult to evaluate its role in constraining the evolution of self-fertilization.

The explanation for the maintenance of outcrossing which has attracted most attention, involves the common observation that offspring resulting from close inbreeding are usually inferior to those from crosses between unrelated individuals (Darwin, 1876; Charlesworth and Charlesworth, 1987). If selfed progeny survive and reproduce only half as well as outcrossed progeny, selfing may no longer be advantageous. Although this relationship appears simple, the origin and maintenance of inbreeding depression in natural populations, and therefore its importance in determining levels of outcrossing has been the subject of much debate (Lande and Schemske, 1985; Campbell, 1986; Charlesworth and Charlesworth, 1987; Holsinger, 1988b).

Lande and Schemske (1985) proposed that predominant selfing and predominant outcrossing should be alternative states of the mating system in most plant populations. Their genetic models predict a bimodal distribution of outcrossing rates, owing to selection for the maintenance of outcrossing in historically large populations, with substantial inbreeding depression, and selection for selfing when increased inbreeding, through bottlenecks or pollinator failure, reduces inbreeding depression below one-half:

$$\text{i.e.} \quad \delta = (1 - \frac{\text{fitness of selfed offspring}}{\text{fitness of outcrossed offspring}}) < \frac{1}{2}$$

Schemske and Lande (1985) tested their models by surveying the distribution of outcrossing rates for 55 species of flowering plants. Although the distribution of t was somewhat bimodal, providing support for their models, a significant proportion of species (31%) exhibited mixed mating systems ($0.21 < t < 0.81$).

Several explanations have been advanced to account for the discrepancy between Lande and Schemske's theoretical predictions and the empirical evidence (Waller, 1986; Schemske and Lande, 1987). One possibility is that Lande and Schemske's models do not accurately depict the joint evolution of inbreeding depression and self-fertilization. More recent models indicate that the precise endpoint of mating system evolution may depend on 1) the architecture and strength of selection on genes controlling inbreeding depression and self-fertilization (Charlesworth and Charlesworth, 1987, 1990; Holsinger, 1988b; Charlesworth *et al.*, 1990); 2) the level of biparental inbreeding in populations (Uyenoyama, 1986); and 3) spatial variation in the strength of inbreeding depression (Holsinger, 1986). Unfortunately, at the present time the accumulation of theoretical ideas on inbreeding depression and mating system evolution has far outstripped empirical data from in natural plant populations.

What other factors besides inbreeding depression might be important in shaping the evolution of plant mating systems? Lloyd (1979) considered the role of reproductive assurance in favoring the evolution of self-fertilization. He argued that when pollen vectors are unreliable some selfing may be selected, despite levels of inbreeding depression exceeding one-half. Furthermore, he pointed out that it will always be advantageous for individuals to self unfertilized ovules after opportunities for outcrossing are exhausted. 'Delayed self-fertilization' should, therefore, evolve regardless of the level of inbreeding depression. Reproductive assurance is likely to be important under conditions of low density, and in short-lived species such as annuals that occur in environments with unreliable pollinator service (Baker, 1955; Stebbins, 1957; Jain, 1976). The main difficulty with empirical work on this problem is that most of the conditions favoring the evolution of selfing, such as small population size, low plant density, poor pollinator service and reduced inbreeding depression, often occur together. Teasing out their direct effects on the selfing rate is, therefore, a major challenge and probably requires experimental approaches.

IV. A Comparative Approach

A. *Mating system surveys*

Comparisons of the taxonomic distribution of a trait is a valuable tool in evolutionary biology. Finding similar traits in diverse taxa may indicate convergent responses to similar selection pressures (Pagel and Harvey, 1988). As discussed above, this approach was used by Schemske and Lande (1985) who surveyed outcrossing rates to assess their models for the evolution of self-fertilization. In the last five years the number of published estimates of t has more than doubled. Here we compile this larger data set to assess the importance of several genetic, ecological and life-history factors thought to be important in mating system evolution.

We collected outcrossing rates determined by analyzing the segregation of morphological (17% of 155 studies) or allozyme markers in five or more progeny arrays from either natural populations (94% of 155 studies) or from plants transplanted from nature into a common garden. Estimates derived from comparing the performance of selfed, outcrossed and open-pollinated progeny (Charlesworth, 1988), or calculated from F-statistics (Ritland, 1983), were excluded. We included multilocus estimates of t whenever available. In cases where more than one locus was used, and multilocus estimates were not given, we averaged t across loci. When estimates were available for more than one population of a given species, t was averaged across populations . To construct the distribution of t across taxa, estimates were classified into five equal classes following Schemske and Lande (1985). The data involved in this survey are available from the authors on request.

B. Interspecific variation in outcrossing rate

Outcrossing rates for 129 species of seed plants are presented in Figure 1A. The species are distributed among 67 genera from 33 families. The majority of our sample involved dicotyledonous angiosperms; the remainder were monocotyledons (12%) and gymnosperms (13%). Schemske and Lande's survey of outcrossing rates showed a highly non uniform distribution of t , with a deficiency of species in the intermediate categories. Our larger sample including more than twice as many species shows the same pattern ($\chi^2 = 52.90$, $d.f. = 4$, $P < .001$). When gymnosperms are removed from the analysis the distribution remains non uniform ($\chi^2 = 21.74$, $d.f. = 4$, $P < .001$). These results are consistent with the prediction that predominant selfing and predominant outcrossing are alternative stable endpoints of mating system evolution.

C. Outcrossing rate and longevity

Further support for the importance of inbreeding depression for mating system evolution is revealed by subdividing the distribution of t by life form (Fig. 1B). Since inbreeding depression is the expression of genetic load, we would expect life forms accumulating high genetic load to practice predominant outcrossing. If we assume that the per cell, per year mutation rate is constant across species, one would expect longer-lived organisms to accumulate higher genetic load. There is some evidence that genetic load is indeed higher in longer-lived species (Wiens, 1984; Ledig, 1986). Our data shows a significant association between outcrossing and longevity. The distribution of t for woody perennials ($N = 51$ species) is more strongly skewed towards outcrossing than that for herbaceous perennials and annuals (N = 76 species; Kolgomorov-Smirnov 2-sample test [Sokal and Rohlf, 1981, p. 443] $D = 0.498$, $P < .001$). The same effect is not evident in the comparison of herbaceous perennials ($N = 24$) and annuals ($N = 54$; $D = 0.292$, $P > .10$). This may be due to the smaller sample sizes involved, or may indicate factors in addition to longevity influence the accumulation of genetic load. For instance, the higher levels of gene flow in many trees should create larger effective population sizes which would, in turn, protect rare deleterious mutations from loss due to genetic drift (Loveless and Hamrick, 1984). Ledig (1986) also suggested that trees may accumulate more genetic load through somatic mutation by virtue of their large size (see also Klekowski, 1988). An additional explanation that may account for the correlation between longevity and outcrossing is also associated with differences in size among life forms. Because long-lived plants are often large, they would be more likely to self-fertilize (see below). As a result there may be stronger selection for mechanisms restricting self-fertilization (Maynard-Smith, 1978).

D. Outcrossing rate and pollination mode

Aide (1986) pointed out that when Schemske and Lande's original distribution of outcrossing rates was subdivided according to pollination mode, wind-pollinated species showed a clear bimodal distribution whereas animal-pollinated species did not. The distribution presented in Figure 1C shows the same effect, although there is also some trend towards bimodality among animal-pollinated species as well ($\chi^2 = 20.68$, $d.f. = 4$, $P < .05$). Aide (1986) explained the difference between the two pollination modes by suggesting that, while selection on outcrossing rates was strongly directional, as Lande and Schemske predicted, the intermediate t values recorded in animal-pollinated species were more influenced by environmental factors. Animal vectors are likely to show considerable spatial and temporal variation in abundance and behavior. Wind, on the other hand, would be expected to disseminate pollen more predictably.

Some support for this interpretation is provided by the much lower interpopulation variation in t values observed for wind-pollinated species compared to animal-pollinated species. Among the sample of wind-pollinated species ($N = 35$) only *Larix laricina* displayed much interpopulation variation in t ($t = .54$ -.91, Knowles *et al.*, 1985). In contrast, among animal-pollinated species ($N = 94$), 14 exhibited t values spanning an equivalent range, and of those, six spanned twice that range.

Not all animal-pollinated species with mixed mating systems exhibit conspicuous variation in outcrossing rates. For example, in *Eucalyptus* outcrossing rates measured in 10 species were similar both within and among ($\overline{X} = .76$, $SD = .08$) species (Moran and Bell, 1983). These data provide a challenge to the models of Lande and Schemske, particularly since strong inbreeding depression occurs in

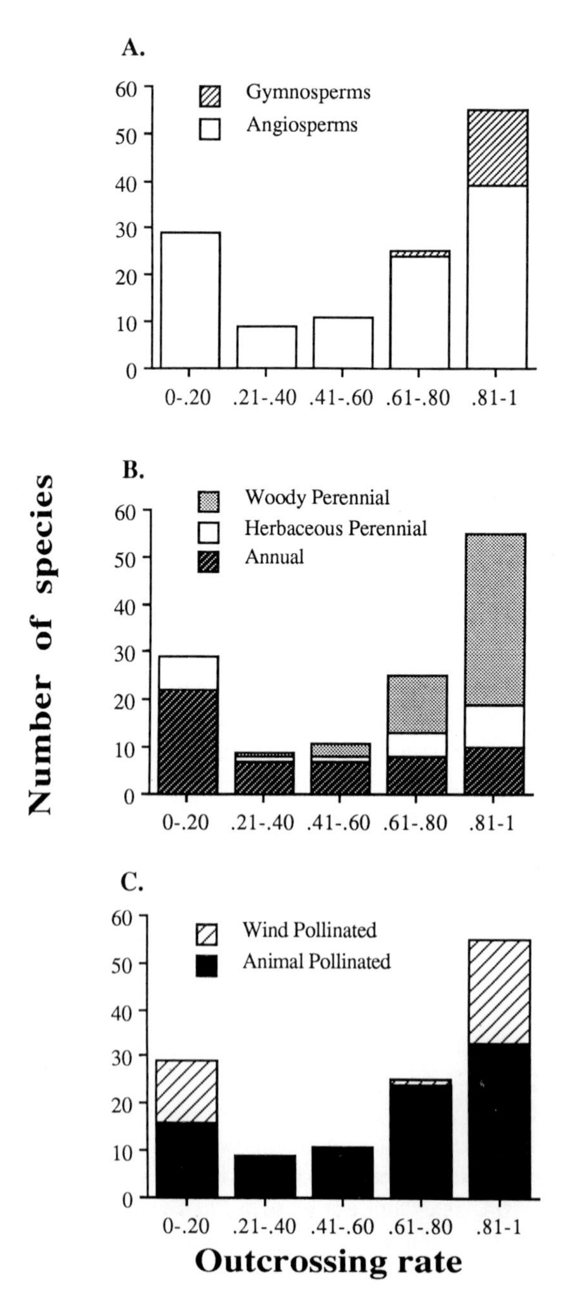

FIG. 1 The distribution of outcrossing rates in populations of 129 species of seed plants subdivided into A. angiosperms and gymnosperms; B. life forms; and C. animal- and wind-pollinated species.

several species of the genus (Hopper and Moran, 1981). This anomaly may be explained if there is little genetic variation within *Eucalyptus* for floral mechanisms that would increase outcrossing rates, a phylogenetic constraint that may have arisen during the evolution of this group's specialized floral morphology.

Unreliability of pollen vectors may also favor selfing rates higher than predicted by considering inbreeding depression alone, because selfing provides reproductive assurance. This should be most important in annuals where the evolutionary risks of pollinator failure are highest. This prediction is borne out by the distribution of t in Figure 1B. It is also apparent, however, that many animal-pollinated perennials practice moderate amounts of self-fertilization. This may be maladaptive and simply result from the vagaries of pollinator behavior. It may, on the other hand, result from selection for delayed selfing associated with reproductive assurance. Unfortunately, there has been little empirical work on the adaptive significance of delayed self-fertilization (but see Cruden and Lyon, 1989), and it remains an open question whether such mechanisms even occur in wind-pollinated plants.

Few of the species in our sample are known to have floral mechanisms which obviously promote delayed selfing. Genotypes of *Lupinus nanus* showed levels of autofertilility that were higher than their floral morphology or outcrossing rates would suggest (Horovitz and Harding, 1972). This kind of autofertility may result from selection for delayed self-fertilization. Mechanisms involving pollen-pistil interactions also play a role in regulating the degree of selfing and outcrossing (Weller and Ornduff, 1977; Glover and Barrett, 1986; Seavey and Bawa, 1986; Bowman, 1987). Whether or not these mechanisms are sensitive to the supply of outcrossed pollen remains largely unexplored. Studies testing for increases in t when the supply of outcross pollen is experimentally augmented would be valuable for assessing the prevalence and functional significance of delayed selfing.

The contrasting distributions of outcrossing rates between animal- and wind-pollinated species in our sample is intriguing. However, it is important to consider the possibility that this pattern is simply a sampling artifact. Outcrossing rates for wind-pollinated species come mostly from either habitually selfing, weedy grasses, or conifers which are predominantly outcrossing and possess high genetic loads (Ledig, 1986). Perhaps as a wider array of wind-pollinated families, particularly those with herbaceous taxa, are studied mixed mating systems will be revealed. Clearly, grasses producing both cleistogamous (selfing) and chasmogamous (presumably outcrossing?) flowers are potential candidates (Campbell *et al*., 1983; Schoen, 1984). However, the most telling evidence for the evolutionary stability of mixed mating systems in wind-pollinated species would come from hermaphroditic species where outcrossing and selfing entail similar energetic costs (Schoen and Lloyd, 1984).

E. Problems with a comparative approach

The potential bias in our sample of outcrossing rates from wind-pollinated species is common to applications of the comparative method. Indeed this problem is likely to be a confounding influence throughout our analysis. Ideally each point in the frequency distributions in Figure 1 should represent an independent evolution of a certain level of outcrossing. This is unlikely. While the comparative approach seeks robust generalities, we are faced with the problem of distinguishing similarity due to convergent evolution from that due to common ancestry (Pagel and Harvey, 1988). For example, if the ancestral conifer from which the Pinaceae differentiated was highly outcrossing, the mating systems of modern taxa may be bound by phylogenetic inertia due to a lack of genetic variance for t throughout the lineage (see also above). One might argue, however, that it is unlikely that no genetic variance exists for mating system modification in the family, since selfing is thought to have evolved repeatedly within many other taxonomic groups (Stebbins, 1974; Jain, 1976). Following this view it seems reasonable to assume that outcrossing is actively maintained by selection in modern conifers, presumably because of their high genetic loads.

Nevertheless, estimating the distribution of t and searching for correlations with life history and pollination mode should ideally involve a stratified random sample of outcrossing rates from as many families as possible. In contrast, among the relatively small number of families represented in our sample, some (e.g. Myrtaceae) are clearly overrepresented. To partly remedy this bias we also constructed distributions of t among genera and families choosing, at random, one species per genus and one genus per family as representatives. The distributions for all three taxonomic levels were not significantly different (3x5 contingency table: $G = 5.15$, $d.f. = 8$, $P = .74$). Still, attempting to minimize the confounding effect of phylogeny is usually a poor substitute for analyses that make use of information provided by cladistic approaches (Donoghue, 1989). Unfortunately, for most examples of mating system change phylogenetic data are unavailable or are fraught with difficulties of interpretation owing to the occurrence of homoplasy (Eckenwalder and Barrett, 1986; Wyatt, 1988).

V. Intraspecific Variation in Outcrossing Rates

Intraspecific variation in outcrossing rate may be more valuable for identifying the factors responsible for mating system change. Experimental approaches are often feasible, and work at the population level avoids some of the uncontrolled variables that hamper comparisons between species. Our survey of outcrossing rates added several species to those already discussed by Schemske and Lande (1985), in which variation in t spanned much of the range from predominant selfing to near complete outcrossing. In addition, a larger number of other species displayed moderate but still significant variation among populations in outcrossing levels. Most data on variation in outcrossing rates comes from comparisons among populations. We concentrate on this variation here with the knowledge that more meaningful data will come from studies of individual differences in mating patterns within populations.

Factors responsible for variation in outcrossing rates can be roughly classified as either genetic or ecological (Table 1). In practice, however, particular levels of outcrossing usually result from interaction between local environmental conditions, and the demographic, life history and genetic characteristics of populations. Historical factors and phylogenetic constraints may also play an indirect role by influencing the genetic structure of populations and the breeding system possessed by a particular species. Because of the complexity of these interactions it can be difficult to isolate the causes of mating system variation within species, particularly if experimental approaches cannot be employed (see Barrett and Husband, 1990).

TABLE 1

Genetic, demographic and environmental factors that contribute to intraspecific variation of outcrossing rate in natural plant populations

Factor	Species	Mean t	Range[1] of t	Relation[2] with t	Ref[3]
I. GENETIC FACTORS					
A. Quantitative floral traits					
Flower Size	*Lycopersicon pimpinellifolium*	.13	.00 -.40†	+	1
Herkogamy	*Turnera ulmifolia*	.34	.04 -.79†	$+^{NS}$	2
Dichogamy	*Clarkia tembloriensis*	.55	.08 -.83	+	3
	Gilia achilleifolia	.57	.15 -.95	+	4
B. Polymorphic floral traits					
Flower color					
pigmentation	*Ipomoea purpurea*	.65	.60 -.86†	+	5
blue reflectance	*Lupinus nanus*	.43	.27 -.34	+	6
Heterostyly					
frequency of homostyles	*Amsinckia spectabilis*	.29	.03 -.53	–	7
	Eichhornia paniculata	.55	.00 -.96	–	8
	Primula vulgaris	.85	.08 - 1.1†	–	9
Gynodioecy					
frequency of females	*Bidens menziesii*	.62	.58 -.65	+	10
	Plantago coronopus	.86	.62 -.98	+	11
Capitulum polymorphism					
frequency of radiate morph	*Senecio vulgaris*	.04	.03 -.04†	+	12

TABLE 1 (Cont'd)

Factor	Species	Mean t	Range of t	Relation with t	Ref
II. DEMOGRAPHIC FACTORS					
Plant size	*Malva moschata*	.65	.22 -1.0†	–	13
Plant density					
animal pollination	*Cavanillesia plantanifolia*	.46	.35 -.57†	+	14
	Echium plantagineum	.88	.73 -.97	+	15
	Helianthus annuus	.75	.65 -.86	–	16
wind pollination	*Pinus ponderosa*	.92	.85 -.96	+	17
	Plantago coronopus	.70	.62 -.92	+	11
Population size	*Eichhornia paniculata*	.55	.00 -.96	+	8
III. ENVIRONMENTAL FACTORS					
Moisture	*Abies lasiocarpa*	.89	.65 -.99†	+	18
	Bromus mollis	.09	.04 -.16†	+*NS*	19
	Hordeum spontaneum	.02	.00 -.02	+	20
	Picea englemanni	.86	.85 -.93†	+	18
Altitude	*Abies balsamea*	.89	.78 -.99	–	21

1: Ranges marked with '†' are from within-population studies; all others are among-population ranges.
2: The relationship of each factor with t is indicated as '+' = positive; and '–' = negative ('*NS* ' denotes a non-significant but consistent trend).
3: References: 1 = Rick *et al.*, 1978; 2 = Barrett and Shore, 1987; 3 = Vasek and Harding, 1976; 4 = Schoen 1982; 5 = Brown and Clegg, 1984; 6 = Horovitz and Harding, 1972; 7 = Ganders *et al.*, 1985; 8 = Barrett and Husband, 1990; 9 = Piper *et al.*, 1982; 10 = Sun and Ganders, 1988; 11 = Wolff, 1988; 12= Marshall and Abbott, 1982; 13 = Crawford, 1984; 14 = Murawski *et al.*, unpubl. MS.; 15 = Burdon and Brown, 1986; Burdon *et al.*, 1988; 16 = Ellstrand *et al.*, 1978; 17 = Farris and Mitton, 1984; 18 = Shea, 1987; 19 = Brown *et al.*, 1974; 20 = Brown *et al.*, 1978; 21=Neale and Adams, 1985.

A. Genetic control of mating system variation

There is considerable evidence from cultivated plants of both major gene control and quantitative inheritance of floral traits that directly influence levels of outcrossing. In particular, considerable work has been conducted on the genetics of self-incompatibility and self-compatibility and on sex expression (Frankel and Galun, 1977; Nettancourt, 1977). This variation has enabled crop breeders to modify mating patterns by artificial selection particularly towards increased levels of self-fertility (Jain, 1984). Less genetic data are available from natural plant populations, although both major gene and quantitative inheritance of mating system modification have been documented.

Population differentiation in floral traits such as flower size, style length, stamen length, and degree of herkogamy and dichogamy are among the commonest types of genetic variation influencing outcrossing rates. Unfortunately, few workers have conducted genetic studies to determine their inheritance in wild species. This approach has been used in *Lycopersicon pimpinellifolium* (Rick *et al.* 1978), *Mimulus* spp. (MacNair and Cumbes, 1989; C. B. Fenster and K. Ritland, unpubl.), and *Turnera ulmifolia* (Shore and Barrett, 1990). In each case quantitative control of variation was revealed. Few attempts, however, have been made to go beyond the simple demonstration of polygenic control. Preliminary work in *Mimulus* and *Turnera* involved estimating the minimum number of genes controlling floral traits using methods developed by Lande (1981) and Cockerham (1986). Since non-additive gene action was also detected in these studies, however, the validity of these approaches is questionable.

Major gene control of floral traits that influence outcrossing is easier to detect and is particularly evident in polymorphic sexual systems such as gynodioecism (e.g. Sun, 1987) and heterostyly (Piper *et al.*, 1984), the flower color polymorphism in *Ipomoea purpurea* (Schoen and Clegg, 1985), and the ray and disc floret polymorphism in *Senecio vulgaris* (Marshall and Abbott, 1982). The ease with which major gene systems can be modelled and studied experimentally has enabled microevolutionary studies of these polymorphisms, and provided some of the best evidence for the mechanisms responsible for mating system modification (see Clegg and Epperson, 1988).

This approach has been employed in studies of the self-compatible, tristylous, aquatic *Eichhornia paniculata* (Pontederiaceae). Outcrossing rates in this species are among the most wide-ranging reported from flowering plants (Fig. 2A). The variation largely results from the evolutionary breakdown of tristyly and the spread and fixation of self-pollinating homostylous variants. While mating system modification is under clear genetic control in this species, this does not preclude a role for both environmental and demographic factors. An analysis of the effects of population size and plant density on mating patterns demonstrated that both variables account for a significant proportion of the variation in t among the populations surveyed (Barrett and Husband, 1990). In addition, comparisons of the seed fertility of selfing and outcrossing morphs and the geographical distribution of homostylous populations provide strong support for the role of reproductive assurance as a selective mechanism for the evolution of self-fertilization (Barrett *et al.*, 1989).

Self-compatible heterostylous species may often provide opportunities for studying mating system variation in natural populations (Barrett, 1989). A recent geographical survey of style morph frequencies in the self-compatible, tristylous, clonal aquatic *Decodon verticillatus* (Lythraceae) revealed wide variation in both morph structure (Fig. 3) and the occurrence of floral variants with stamens and styles in close proximity (C. G. Eckert and S. C. H. Barrett, unpubl.). Evidence from other heterostylous species (e.g. Ganders *et al.*, 1985; Glover and Barrett, 1986) would suggest that this variation is likely to have a direct bearing on the levels of self-fertilization occurring within populations.

In *D. verticillatus* populations at the margins of the range, in Michigan and Ontario, are more likely to be dimorphic or monomorphic for style length than populations in more southerly areas such as Florida and Georgia. This difference may be associated with restricted opportunities for sexual recruitment and greater reliance on clonal reproduction in northern populations. Increased asexual reproduction has often been suggested in geographically or ecologically marginal portions of a species' range because of unfavorable conditions for seed germination and seedling establishment (Barrett, 1980). These effects may directly influence outcrossing rates because higher levels of self and geitonogamous pollination

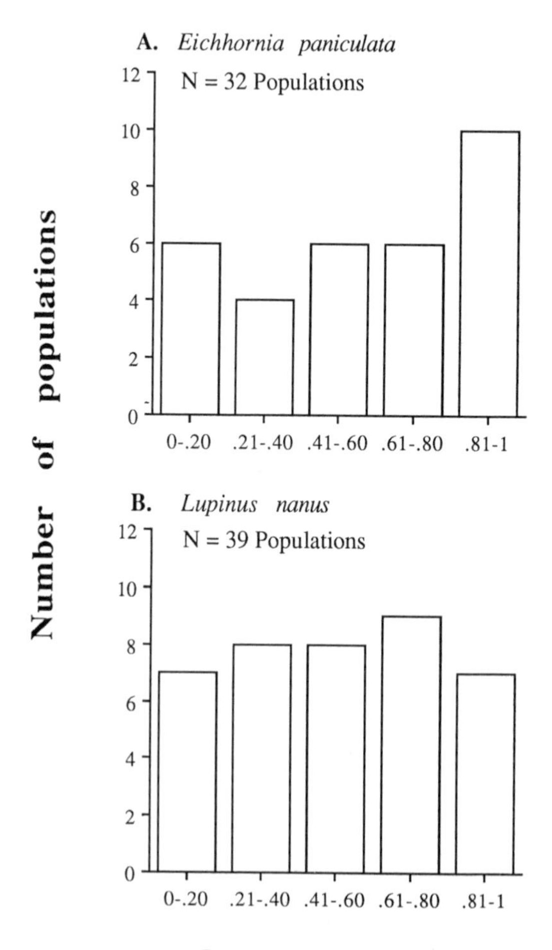

FIG. 2 The distribution of outcrossing rates in A. Thirty-two populations of *Eichhornia paniculata* (Pontederiaceae) from northeastern Brazil and Jamaica (after Barrett and Husband, 1990); and B. Thirty-nine populations of *Lupinus nanus* (Leguminosae) from California U.S.A. (after Harding, 1970; Harding *et al.*, 1974)

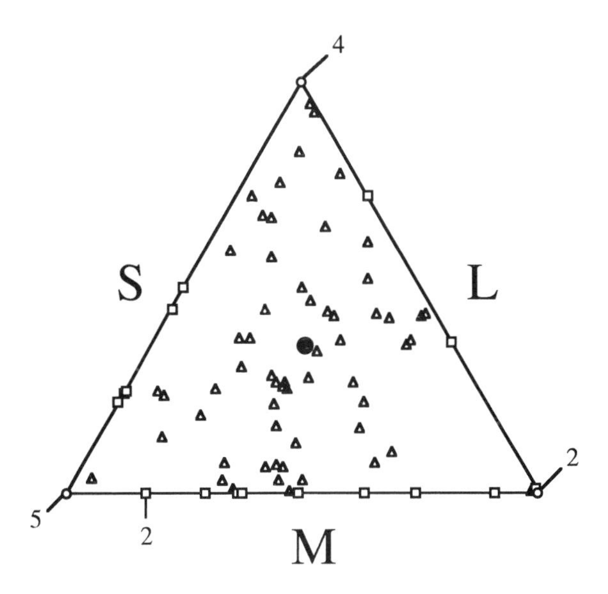

FIG. 3 Style morph frequencies in 93 populations of *Decodon verticillatus* (Lythraceae) sampled in eastern Ontario , Michigan, and southeastern USA. Each point represents morph frequencies of a single population. The distance of a population from an axis is proportional to the relative frequencies of morphs (long-, mid-, or short-styled morphs represented by L, M, S, respectively) in the population. Trimorphic, dimorphic and monomorphic populations are represented by triangles, squares and circles, respectively. Equal morph frequencies are indicated by the solid circle (C.G. Eckert and S. C. H. Barrett, unpubl.). The variation in morph structure in *D. verticillatus* is likely to be associated with considerable interpopulation variation in outcrossing rate.

would be likely in populations composed of large clones (Handel, 1985; Silander, 1985). In wide-ranging clonal species such as *D. verticillatus* this is likely to have important implications for geographical variation in mating patterns and opportunities for the evolution of self-fertilization.

Our discussion of genetic factors influencing intraspecific variation in mating systems has largely concerned models for the evolution of autogamy. Outcrossing and selfing rates are also of importance, however, for other models of mating system change, particularly those concerned with evolution of gender polymorphisms such as gynodioecism and dioecism. The selective force most commonly invoked to account for the spread of females in cosexual populations is inbreeding depression of progeny from hermaphrodites as a result of self-fertilization (Lewis, 1941; Lloyd, 1974, 1975; Charlesworth and Charlesworth, 1978). Resulting gynodioecious populations can remain evolutionarily stable, or form an intermediate stage in the evolution of dioecism (Bawa, 1980). Unfortunately, joint measures of selfing rates and inbreeding depression in gynodioecious species are unavailable (Charlesworth, 1989; Sun and Ganders, 1986; Kohn, 1988;), so it is premature to draw general conclusions on the validity of models for the evolution of sexual dimorphism based primarily on inbreeding depression .

A difficulty with testing models for the evolution of dioecism is that few species

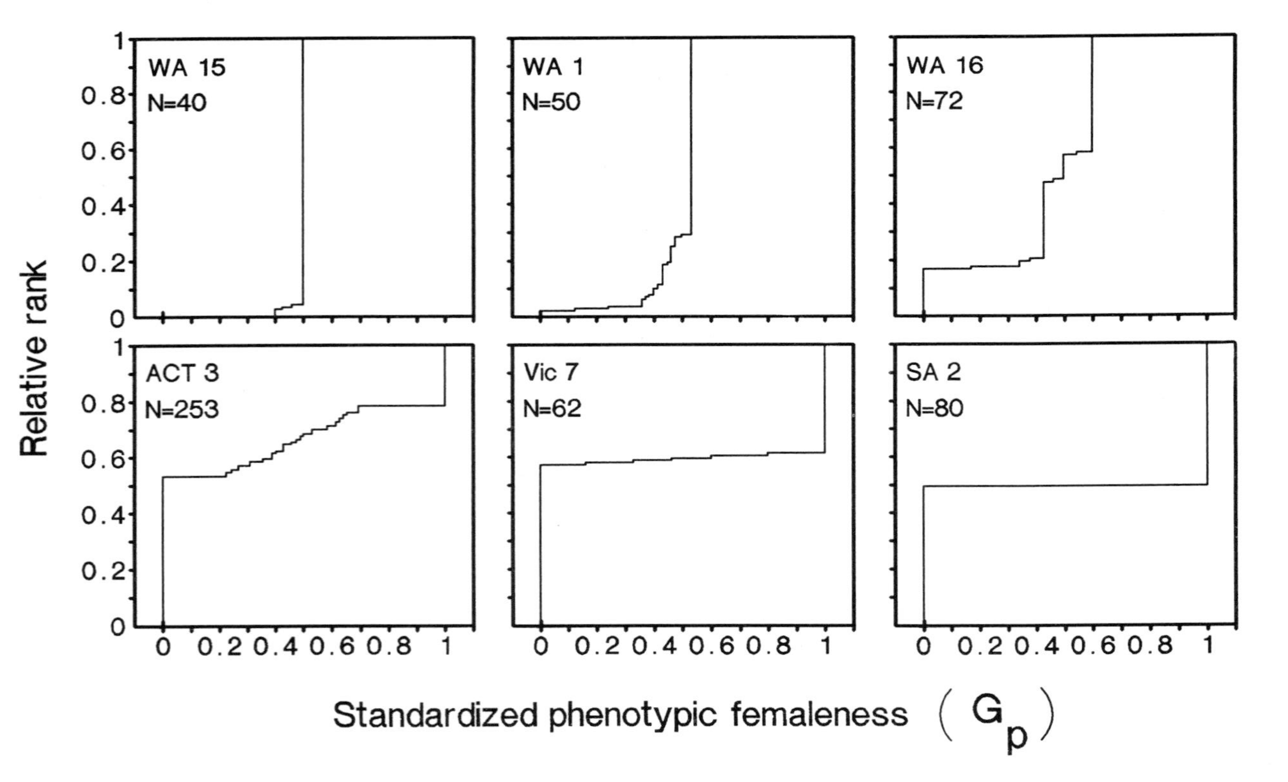

FIG. 4 Variation in phenotypic gender (G_p) among six populations of *Wurmbea dioica* (Liliaceae) in Australia. Gender measures were calculated following the method of Lloyd and Bawa (1984). Plants with a G_p value of 1 or 0 are purely female or male, respectively. Sample sizes refer to the number of plants sampled in each population (S.C.H. Barrett, unpubl. MS.).

are known that possess both cosexual and dioecious populations (e.g. *Sagittaria latifolia*, Wooten, 1971; *Cotula* spp., Lloyd, 1972; *Ecballium elaterium*, Costich, 1989). It appears that once dioecism arises it is often accompanied by speciation, perhaps because of strong character divergence associated with the acquisition of sexually dimorphic flowers. While quantitative data on selfing rates, inbreeding depression and allocation of resources to male and female function can often be obtained from related species, the value of these measures for mating system models are questionable because of the confounding influences of other taxonomic differences that result from adaptive differentiation.

The Australian geophytic lily *Wurmbea dioica* provides a useful system for testing hypotheses concerned with the evolution of dioecism. In this wide ranging, fly-pollinated species of rocky woodland slopes, gender expression varies widely both within and between populations (Fig. 4). The two extremes of gender represented by cosexual and dioecious populations are linked by gynodioecious and subdioecious populations, indicating the probable evolutionary pathway to sexual dimorphism involves a gynodioecious stage. In parts of western Australia cosexual and sub-dioecious populations grow in close proximity, enabling microevolutionary investigation of the factors maintaining gender polymorphism. The estimates of phenotypic gender illustrated in Figure 4, while useful in drawing attention to the quantitative nature of sex expression, give only a rough guide to mating patterns in each population. Rigorous testing of models of mating system evolution would require outcrossing estimates for cosexual and gynodioecious populations, as well as data on inbreeding depression, male and female fertility variation, and patterns of sex allocation. Nevertheless, geographical surveys of the type conducted in *Wurmbea* and *Decodon* provide an important first step in identifying promising species for intraspecific studies of mating system evolution. Systematists, because of their wide knowledge of geographical variation in floral traits, may be in the best position to provide population biologists with this kind of information.

B. Ecological influences on mating system variation

Where ecological factors play a major role in causing variation in outcrossing levels the effect of selection on the mating system is likely to be reduced. With low heritability of traits influencing mating patterns and unpredictable changes in those features of the environment that influence reproduction, the mating system may be unable to respond in an adaptive fashion. A wide array of ecological factors can influence levels of outcrossing in plant populations. Their effects may be direct such as the influence of environmental factors (e.g. temperature, humidity) on pollination and pollen-pistil interactions, or indirect such as the effect of local habitat on the size and spatial structure of populations.

In animal-pollinated plants a major source of variation appears to be the reliability of pollen vectors. Inclement weather during flowering may influence the number and types of pollinators that visit flowers. However, even where pollinator activity is reliable, changes in the species composition and foraging behavior of pollinators may have potentially subtle effects on mating patterns and reproductive success (Schemske and Horvitz, 1984; Wolfe and Barrett, 1988). Any consistent failure of pollinators to service outcrossing populations can potentially lead to the evolution of self-fertilization. It has been suggested that this is more likely to occur in small populations, at low density, or under ecologically or geographically marginal conditions (Lloyd, 1980). However, with few exceptions (e.g. Wyatt, 1986), there has been little effort to measure visitation levels in populations with contrasting mating systems, or to determine how often pollinator limitation occurs in outcrossing taxa in which selfing populations have evolved.

The large variation in outcrossing rates displayed by populations of several *Lupinus* species (e.g. *Lupinus nanus*, Fig. 2B) may largely result from differences in pollinator abundance within and among sites (Horovitz and Harding, 1972; Schemske and Lande, 1985). While some evidence for the genetic control of floral traits influencing mating patterns has been found in *Lupinus* spp., common garden studies strongly implicate differing pollinator levels as the major cause of variation in t (Horovitz and Harding, 1972). Moreover, studies of year to year variation in outcrossing rate detected several cases of dramatic changes in t (Harding, 1970; Harding *et al.*, 1974), a result unexpected if genetic factors were largely responsible for the observed levels of outcrossing. More work of the type conducted by Harding and colleagues on *Lupinus* is required to determine to what extent pollinator levels account for mating system variation in other animal-pollinated plants. In addition, the role that different pollinators play in influencing outcrossing rates would be particularly valuable.

Several other ecological factors have been implicated as the cause of mating system variation in plant populations (Table 1). However, the effects of these parameters may be highly idiosyncratic, depending on particular attributes of the species in question. Plant density and plant size are good examples of this problem. Studies in both animal- and wind-pollinated species have produced contrasting results with respect to the effect of density on outcrossing rate (Table 1). Similarly, the predicted relationship between plant size and outcrossing rate might appear straightforward, because of the propensity of large plants to experience high levels of geitonogamous pollination, however, the limited data available provide variable results. In *Malva moschata*, the selfing rate and size of floral display were positively correlated (Crawford, 1984) , whereas in *Sabatia angularis* this relationship was only obtained in one of four subpopulations that were investigated (Dudash and Barrett, 1989).

Most studies on the effects of ecological factors on outcrossing rates are based on correlations rather than experimental approaches. Because of the complex interactions of ecological variables it can be extremely difficult to isolate important factors affecting mating patterns using a correlative approach. The ease with which many plants can be experimentally manipulated should offer broad scope for controlled investigation in this regard.

VI. Future Directions

There has been impressive growth in studies of plant mating systems, particularly procedures for measuring mating parameters and the formulation of theoretical models for evolutionary change. Despite this progress, our discussion has highlighted a number of issues concerned with the evolution of mating systems where our existing body of knowledge is limited. In this concluding section we draw attention to several areas which we believe would repay detailed investigation in the future.

A. *Taxonomic and ecological surveys of outcrossing rates*

The 129 species in our survey of outcrossing rates is far from a random sample. Aside from a general appeal for data from taxonomic and ecological groups not studied to date, there are several gaps in the data of particular interest to models of mating system evolution. For example, what is the true distribution of t among wind-pollinated species, and will estimates from groups other than conifers and grasses support the strong bimodal distribution apparent in the current data? If pollen vector reliability is a key factor in explaining outcrossing rate distributions (Aide, 1986), estimates from hydrophilous plants may be particularly interesting because of parallels between wind and water in terms of their reliability as abiotic vectors of pollen. In addition, quantitative estimates of mating system parameters in Pteridophytes and Bryophytes would also be valuable and allow comparisons to be made with the patterns observed in seed plants.

B. *Intraspecific studies of spatial and temporal variation of outcrossing rates*

In 48% of the species in our sample t was estimated only once from a single population. More extensive sampling of populations is likely to detect variation, particularly in animal-pollinated, self-compatible species. For those species in which a mixture of self- and cross-fertilization occurs it would be particularly valuable to determine the extent of temporal variation in outcrossing rates both within (e.g. Burdon *et al.*, 1988) and between (e.g. Cheliak *et al.*, 1985) seasons. If outcrossing rates in populations with intermediate values are of similar magnitude over several years it would provide some evidence in support of the mixed mating system as an evolutionarily stable strategy. Perhaps the most critical gap in our knowledge concerns the pattern and extent of plant to plant variation in mating behavior within populations. To what extent do attributes such as size, spatial location and floral display influence outcrossing rates? Statistical procedures required to estimate individual outcrossing rates are limited by the need for large sample sizes. The search for a wider array of genetic markers, such as restriction fragment length polymorphisms, or hypervariable minisatellite sequences, may alleviate this problem to some extent.

C. Genetic analysis of mating system variation

More information is required on the genetic architecture of reproductive traits that influence mating patterns. At present we know little about the extent of heritable variation for reproductive traits within natural plant populations. Quantitative genetic approaches are required to determine the heritability of outcrossing rates and the extent to which patterns of genetic correlation among floral traits are likely to influence response to selection on the mating system. Of particular importance will be to determine how often selfing rates are under simple versus more complex modes of inheritance; this information has a direct bearing on the evolutionary dynamics of mating system modification (Lande and Schemske, 1985; Charlesworth *et al.*, 1990). Since inbreeding depression is a critical parameter in most models of the evolution of self-fertilization more empirical work is required on its occurrence and genetic basis, particularly in populations with different mating systems. Of particular interest will be to determine whether genetic variation for inbreeding depression commonly occurs within natural populations. Such variation has important consequences for the spread of genes influencing the selfing rate (Holsinger, 1989). Because the dynamics of genetic change for more than one locus in populations with partial selfing are so complex, predicting the course of mating system evolution is likely to be particularly difficult (Weir and Cockerham, 1973; Wright, 1987). Further theoretical work on the quantitative genetics of partial selfing may help clarify some of the issues concerned with the evolution of mixed mating systems.

D. Measurement of selection on mating systems

While the measurement of selection in natural populations has advanced greatly in the last decade (Endler, 1986), there have been few attempts to measure selection on traits likely to influence plant mating systems (see Campbell, 1989; Schemske and Horvitz, 1989). Estimating the strength and nature of selection, however, is central to testing models of mating system evolution. For instance, Lande and Schemske's (1985) models predict strong directional selection on the outcrossing rate within populations. Measurement of selection on the mating system is difficult because the outcrossing rate is a complex phenotypic trait reflecting interactions between reproductive characters, demography and environmental factors (Table 1). Although this appears complicated, quantitative methods such as path analysis (Schemske and Horvitz, 1988) and multivariate selection models (Lande and Arnold, 1983) may enable us to untangle some of this biological complexity.

E. *Phylogenetic analysis of mating systems*

Most work on mating systems has been conducted by geneticists and evolutionary biologists working at the population or species level. There is now a need to couple these microevolutionary studies with patterns of macroevolution. This is of particular importance since mating system traits are a special class of evolutionary characters. By influencing mating patterns, they not only affect their own transmission but also the transmission of all other genes in an organism. Studies on mating system variation among related species should be conducted in conjunction with estimation of phylogenetic relationships using cladistic methods (Olmstead, 1989). Molecular approaches that use various classes of DNA now offer a source of characters which can be used for reconstructing phylogenies and are independent of the mating system. Once phylogenies have been estimated, the distribution of mating systems can be mapped onto the phylogeny enabling various evolutionary questions to be addressed (Coddington, 1988). Of particular interest will be how often and in what sequence in relation to other reproductive traits breeding systems such as dioecism, heterostyly, self-incompatibility, and autogamy have evolved within lineages. Many controversies that relate to evolutionary origins will undoubtedly be resolved by the adoption of phylogenetic approaches. Systematists could play a leading role in these investigations.

Acknowledgements

We thank Brian Husband, Joshua Kohn and Kermit Ritland for helpful suggestions, James Hamrick for providing on unpublished manuscript, and the Natural Sciences and Engineering Research Council of Canada for financial support. This paper was prepared while the senior author was holding an E. W. R. Steacie Memorial Fellowship.

References

Aide, T. M. (1986). The influence of wind and animal pollination on variation in outcrossing rates. *Evolution* **40**, 434-435.

Baker, H. G. (1955). Self-compatibility and establishment after "long-distance" dispersal. *Evolution* **9**, 347-348.

Barrett, S. C. H. (1980). Sexual reproduction in *Eichhornia crassipes* (water hyacinth). II. Seed production in natural populations. *J. Appl. Ecol.* **17**, 113-124.

Barrett, S. C. H. (1989). Mating system evolution and speciation in heterostylous plants. *In* "Speciation and its Consequences" (D. Otte and J. A. Endler, eds), pp. 257-283. Sinauer, Sunderland, USA.

Barrett, S.C.H. (unpubl. MS.). Gender variation in *Wurmbea dioica* (Liliaceae) and its significance for the evolution of dioecy.

Barrett, S. C. H. and Husband, B. C. (1990). Variation in outcrossing rates in *Eichhornia paniculata*: The role of demographic and reproductive factors. *Plant Species Biol.* **5**, (in press).

Barrett, S. C. H., Morgan, M. T. and Husband, B. C. (1989). The dissolution of a complex genetic polymorphism: the evolution of self-fertilization in tristylous *Eichhornia paniculata. Evolution* **43**, 1398-1416.

Barrett, S. C. H. and Shore, J. S. (1987). Variation and evolution of breeding systems in the *Turnera ulmifolia* L. complex (Turneraceae). *Evolution* **41**, 340-354.

Bawa, K. S. (1980). Evolution of dioecy in flowering plants. *Annu. Rev. Ecol. Syst.* **11**, 15-39.

Bowman, R. N. (1987). Cryptic self-incompatibility and the breeding system of *Clarkia unguiculata* (Onagreaceae). *Am. J. Bot.* **74**, 471-476.

Brown, A. H. D. (1990). Genetic characterization of plant mating systems. *In* "Plant Population Genetics, Breeding, and Genetic Resources" (A. H. D. Brown, M. T. Clegg, A. L. Kahler and B.S. Weir, eds), pp. 145-162. Sinauer, Sunderland, USA.

Brown, A. H. D. and Allard, R. W. (1970). Estimation of the mating system in open pollinated maize populations using isozyme polymorphisms. *Genetics* **66**, 133-145.

Brown, A. H. D., Barrett, S. C. H. and Moran, G. F. (1985). Mating system estimation in forest trees: Models, methods and meanings. *In* "Population Genetics in Forestry" (H. -R. Gregorius, ed.), pp. 32-49, Proceedings of the meeting of the IUFRO Working Party, August 21-24. Springer-Verlag, Gottingen, W. Germany.

Brown, A. H. D., Marshall, D. R. and Albrecht, L. (1974). The maintenance of alcohol dehydrogenase polymorphism in *Bromus mollis L. Aust. J. Biol. Sci.* **27**, 545-559.

Brown, A. H. D., Zohary, D. and Nevo, E. (1978). Outcrossing rates and heterozygosity in natural populations of *Hordeum spontaneum* Koch in Isreal. *Heredity* **41**, 49-62.

Brown, B. A. and Clegg, M. T. (1984). Influence of flower color polymorphism on genetic transmission in a natural population of the common morning glory, *Ipomoea purpurea. Evolution* **38**, 796-803.

Burdon, J. J. and Brown, A. H. D. (1986). Population genetics of *Echium plantagineum* L. — target weed for biological control. *Aust. J. Biol. Sci.* **39**, 369-378.

Burdon, J. J., Jarosz, A. M. and Brown, A. H. D. (1989). Temporal patterns of reproduction and outcrossing in weedy populations of *Echium plantagineum. Biol. J. Linn. Soc.* **34**, 81-92.

Campbell, C. S., Quinn, J. A., Cheplick, G. P. and Bell, T. J. (1983). Cleistogamy in grasses. *Annu. Rev. Ecol. Syst.* **14**, 411-441.

Campbell, D. R. (1989). Measurements of selection in a hermaphroditic plant: Variation in male and female pollination success. *Evolution* **43**, 318-334.

Campbell, D. R. and Waser, N. M. (1987). The evolution of plant mating systems: Multilocus simulations of pollen dispersal. *Am. Nat.* **129**, 593-609.

Campbell, R. B. (1986). The interdependence of mating structure and inbreeding depression. *J. Theor. Biol.* **30**, 232-244.

Charlesworth, B. and Charlesworth, D. (1978a). A model for the evolution of dioecy and gynodioecy. *Am. Nat.* **112**, 975-997.

Charlesworth, D. (1988). A method for estimating outcrossing rates in natural populations of plants. *Heredity* **61**, 469-471.

Charlesworth, D. (1989). The population biology of gynodioecy in *Silene vulgaris. Am. J. Bot.* **76** (Suppl.), abstract no. 195.

Charlesworth, D. and Charlesworth, B. (1979). A model for the evolution of distyly. *Am. Nat.* **114**, 467-498.

Charlesworth, D. and Charlesworth, B. (1981). Allocation of resources to male and female functions in hermaphrodites. *Biol. J. Linn. Soc.* **15**, 57-74.

Charlesworth, D. and Charlesworth, B. (1987). Inbreeding depression and its evolutionary consequences. *Annu. Rev. Ecol. Syst.* **18**, 237-268.

Charlesworth, D. and Charlesworth, B. (1990). Inbreeding depression with heterozygote advantage and its effect on selection for modifiers changing the outcrossing rate. *Evolution* (in press) .

Charlesworth, D., Morgan, M. T. and Charlesworth, B. (1990). Inbreeding depression, genetic load and the evolution of outcrossing rates in a multi-locus system with no linkage. *Evolution* (in press).

Charnov, E. L. (1982). "The Theory of Sex Allocation." Princeton University Press, Princeton, USA.

Cheliak, W. M., Danick, B. P., Morgan, K., Yeh, F. C. H. and Strobeck, C. (1985). Temporal variation of the mating system in a natural population of jack pine. *Genetics* **109**, 569-584.
Clegg, M. T. (1980). Measuring plant mating systems. *Bioscience* **30**, 814-818.
Clegg, M. T. and Epperson, B. K. (1988). Natural selection of flower color polymorphisms in morning glory populations. *In* "Plant Evolutionary Biology" (L. D. Gottlieb and S. K. Jain, eds), pp. 255-273. Chapman and Hall, London, UK.
Cockerham, C. C. (1986). Modifications in estimating the number of genes for a quantitative character. *Genetics* **114**, 659-664.
Coddington, J. A. (1988). Cladistic tests of adaptational hypotheses. *Cladistics* **4**, 3-22.
Costich, D. E. (1989). Differences in seed size and seed number in the dioecious and monoecious subspecies of *Ecballium elaterium* across a climatic gradient in Spain. *Am. J. Bot.* **76** (Suppl.), abstract no. 252.
Crawford, T. J. (1984). What is a population? *In* "Evolutionary Ecology" (B. Shorrocks, ed.), pp. 135-173. British Ecol. Soc. Symp. No. 23, Blackwell Scientific, Oxford, UK.
Cruden, R. W. and Lyon, D. L. (1989). Facultative xenogamy: Examination of a mixed mating system. *In* "The Evolutionary Ecology of Plants" (J. H. Bock and Y. B. Linhart, ed.), pp. 171-208. Westview Press, Boulder, USA.
Darwin, C. R. (1876). "The Effects of Cross and Self-Fertilization in the Vegetable Kingdom". John Murray, London, UK.
Devlin, B., Roeder, K. and Ellstrand, N. C. (1989). Fractional paternity assignment: theoretical development and comparison to other methods. *Theor. Appl. Genet.* **76**, 369-380.
Donoghue, M. J. (1989). Phylogenies and the analysis of evolutionary sequences, with examples from seed plants. *Evolution* **43**, 1137-1156.
Dudash, M. R. and Barrett, S. C. H. (1989). The influence of environmental and demographic factors on the mixed mating system of *Sabatia angularis*. *Am. J. Bot.* **76** (Suppl.), abstract no. 261.
Eckenwalder, J. E. and Barrett, S. C. H. (1986). Phylogenetic systematics of the Pontederiaceae. *Syst. Bot.* **11**, 373-391.
Endler, J. A. (1986). "Natural Selection in the Wild." Princeton University Press, Princeton, USA.
Farris, M. A. and Mitton, J. B. (1984). Population density, outcrossing rate, and heterozygote superiority in ponderosa pine. *Evolution* **38**, 1151-1154.
Fisher, R. A. (1941). Average excess and average effect of a gene substitution. *Ann. Eugen.* **11**, 53-63.
Frankel, R. and Galun, E. (1977). "Pollination Mechanisms, Reproduction and Plant Breeding". Springer-Verlag, New York, USA.
Fyfe, J. L. and Bailey, N. T. J. (1951). Plant breeding studies in leguminous forage crops. I. Natural cross breeding in winter beans. *J. Agric. Sci.* **41**, 371-378.
Ganders, F. R., Denny, S. K. and Tsai, D. (1985). Breeding system variation in *Amsinckia spectabilis* (Boraginaceae). *Can. J. Bot.* **63**, 533-538.
Glover, D. E. and Barrett, S. C. H. (1986). Variation in the mating system of *Eichhornia paniculata* (Spreng.) Solms. (Pontederiaceae). *Evolution* **40**, 1122-1131.
Green, A. G., Brown, A. H. D. and Oram, R. N. (1980). Determination of outcrossing in a breeding population of *Lupinus albus* L. *Z. Pflanzenzüchtung* **84**, 181-191.
Handel, S. N. (1985). The intrusion of clonal growth patterns on plant breeding systems. *Am. Nat.* **125**, 367-384.
Harding, J. (1970). Genetics of *Lupinus*. II. The selective disadvantage of the pink flower color mutant in *Lupinus nanus*. *Evolution* **24**, 120-127.
Harding, J., Mankinen, C. B. and Elliot, M. (1974). Genetics of *Lupinus*. VII. Outcrossing, autofertility, and variability in natural populations of the Nanus group. *Taxon* **23**, 729-738.
Holsinger, K. E. (1986). Dispersal and plant mating systems: The evolution of self-fertilization in subdivided populations. *Evolution* **40**, 405-413.
Holsinger, K. E. (1988a). The evolution of self-fertilization in plants: Lessons from population genetics. *Oecol. Plant.* **9**, 95-102.

Holsinger, K. E. (1988b). Inbreeding depression doesn't matter: The genetic basis of mating-system evolution. *Evolution* **42**, 1235-1244.

Holsinger, K. E., Feldman, M. W. and Christiansen, F. B. (1984). The evolution of self-fertilization in plants: A population genetic model. *Am. Nat.* **124**, 446-453.

Hopper, S. D. and Moran, G. F. (1981). Bird pollination and the mating system of *Eucalyptus stoatei*. *Aust. J. Biol.* **29**, 625-638.

Horovitz, A. and Harding, J. (1972). Genetics of *Lupinus*. V. Intraspecific variability for reproductive traits in *Lupinus nanus*. *Bot. Gaz.* **133**, 155-165.

Jain, S. K. (1976). The evolution of inbreeding in plants. *Annu. Rev. Ecol. Syst.* **7**, 69-95.

Jain, S. K. (1984). Breeding systems and the dynamics of plant populations. *In* "Genetics: New Frontiers. IV. Applied Genetics." (V. L. Chopra, B. C. Joshi, R. P. Sharma and H. C. Bansal, eds), pp. 291-316. Oxford & IBH, New Delhi, India.

Jones, D. F. (1916). Natural cross-pollination in the tomatoe. *Science* **43**, 509-510.

Klekowski, E. J. J. (1988). "Mutation, developmental selection and plant evolution". Columbia University Press, New York, USA.

Knowles, P., Furnier, G. R., Aleksiuk, M. A. and Perry, D. J. (1985). Significant levels of self-fertilization in natural populations of tamarack. *Can. J. Bot.* **65**, 1087-1091.

Kohn, J. R. (1988). Why be female? *Nature* **335**, 431-433.

Lande, R. (1981). The minimum number of genes contributing to quantitative variation between and within populations. *Genetics* **99**, 541-553.

Lande, R. and Arnold, S. J. (1983). The measurement of selection on correlated characters. *Evolution* **37**, 1210-1226.

Lande, R. and Schemske, D. W. (1985). The evolution of self-fertilization and inbreeding depression in plants. I. Genetic models. *Evolution* **39**, 24-40.

Ledig, F. T. (1986). Heterozygosity, heterosis, and fitness in outbreeding plants. *In* "Conservation biology: The Science of Scarcity and Diversity" (M. E. Soulé, ed.), pp. 77-104. Sinauer, Sunderland, USA.

Lewis, D. (1941). Male sterility in natural populations of hermaphrodite plants. *New Phytol.* **40**, 56-63.

Lloyd, D. G. (1972). Breeding systems in *Cotula* L. (Compositae, Anthemideae). I. The array of monoclinous and diclinous systems. *New Phytol.* **71**, 1181-1194.

Lloyd, D. G. (1974). Theoretical sex ratios of dioecious and gynodioecious angiosperms. *Heredity* **32**, 11-34.

Lloyd, D. G. (1975). The maintenance of gynodioecy and androdioecy in angiosperms. *Genetica* **45**, 325-339.

Lloyd, D. G. (1979). Some reproductive factors affecting the selection of self-fertilization in plants. *Am. Nat.* **113**, 67-79.

Lloyd, D. G. (1980). Demographic factors and mating patterns in Angiosperms. *In* "Demography and Evolution in Plant Populations" (O. T. Solbrig, ed.), pp. 67-88. Blackwell, Oxford, UK.

Lloyd, D. G. (1982). Selection of combined versus separate sexes in seed plants. *Am. Nat.* **120**, 571-585.

Lloyd, D. G. (1984). Gender allocations in outcrossing cosexual plants. *In* "Perspectives on Plant Population Ecology" (R. Dirzo and J. Sarukhán, eds), pp. 277-303. Sinauer, Sunderland, USA.

Lloyd, D. G. and Bawa, K. S. (1984). Modification of the gender of seed plants in varying conditions. *Evol. Biol.* **17**, 255-338.

Loveless, M. D. and Hamrick, J. L. (1984). Ecological determinants of genetic structure in plant populations. *Annu. Rev. Ecol. Syst.* **15**, 65-95.

MacNair, M. R. and Cumbes, Q. J. (1989). The genetic architecture of interspecific variation in *Mimulus*. *Genetics* **122**, 211-222.

Marshall, D. F. and Abbott, R. J. (1982). Polymorphism for outcrossing frequency at the ray floret locus in *Senecio vulgaris* L. I. Evidence. *Heredity* **48**, 227-235.

Maynard-Smith, J. (1978). "The Evolution of Sex". Cambridge University Press, Cambridge, UK.

Moran, G. F. and Bell, J. C. (1983). Eucalyptus. *In* "Isozymes in Plant Genetics and Breeding, Part B" (S. D. Tanksley and T. J. Orton, eds), pp. 423-441. Elsevier, Amsterdam, Netherlands.

Morgan, M. T. and Barrett, S. C. H. (1990). Outcrossing rates and correlated mating within a population of *Eichhornia paniculata* (Pontederiaceae). *Heredity* (in press).

Murawski, D. A., Hamrick, J. L., Hubbell, S. P. and Foster, R. B. (unpubl. MS.). Mating systems of two bombacaceous trees of a neotropical moist forest.

Neale, D. B. and Adams, W. T. (1985). Allozyme and mating-system variation in balsam fir (*Abies balsamea*) across a continuous elevational transect. *Can. J. Bot.* **63**, 2448-2453.

Nettancourt, D. de (1977). "Incompatibility in Angiosperms". Springer-Verlag, Berlin, W. Germany.

Olmstead, R. (1989). Phylogeny, phenotypic evolution, and biogeography of the *Scutellaria angustifolia* complex (Lamiaceae): Inference from morphological and molecular data. *Syst. Bot.* **14**, 320-338.

Pagel, M. D. and Harvey, P. H. (1988). Recent developments in the analysis of comparative data. *Quart. Rev. Biol.* **63**, 413-440.

Piper, J. G., Charlesworth, B. and Charlesworth, D. (1984). A high rate of self-fertilization and increased seed fertility of homostyle primroses. *Nature* **310**, 50-51.

Richards, A. J. (1986). "Plant Breeding Systems". George Allen & Unwin, London, UK.

Ritland, K. (1983). Estimation of mating systems. *In* "Isozymes in Plant Genetics and Breeding, Part A" (S. D. Tanksley and T. J. Orton, eds), pp. 289-302. Elsevier, Amsterdam, Netherlands.

Ritland, K. (1990). A series of FORTRAN computer programs for estimating plant mating systems. *J. Hered.* (in press).

Ritland, K. and Ganders, F. R. (1985). Variation in the mating system of *Bidens menziesii* (Asteraceae) in relation to population substructure. *Heredity* **55**, 235-244.

Ritland, K. and Jain, S. K. (1981). A model for the estimation of outcrossing rate and gene frequencies using *n* independent loci. *Heredity* **47**, 35-52.

Schemske, D. W. and Horvitz, C. C. (1984). Variation among floral visitors in pollination ability: a precondition for mutalism specialization. *Science* **225**, 519-521.

Schemske, D. W. and Horvitz, C. C. (1988). Plant-animal interactions and fruit production in a neotropical herb: A path analysis. *Ecology* **69**, 1128-1137.

Schemske, D. W. and Horvitz, C. C. (1989). Temporal variation in selection on a floral character. *Evolution* **43**, 461-465.

Schemske, D. W. and Lande, R. (1985). The evolution of self-fertilization and inbreeding depression in plants. II. Empirical observations. *Evolution* **39**, 41-52.

Schemske, D. W. and Lande, R. (1987). On the evolution of plant mating systems: A reply to Waller. *Am. Nat.* **130**, 804-806.

Schoen, D. J. (1982). The breeding system of *Gilia achilleifolia:* Variation in floral characteristics and outcrossing rate. *Evolution* **36**, 352-360.

Schoen, D. J. (1984). Cleistogamy in *Microlaena polynoda* (Gramineae): An examination of some model predictions. *Am. J. Bot.* **71**, 711-719.

Schoen, D. J. and Clegg, M. T. (1985). The influence of flower color on outcrossing rate and male reproductive success in *Ipomoea purpurea*. *Evolution* **39**, 1242-1249.

Schoen, D. J. and Lloyd, D. G. (1984). The selection of cleistogamy and heteromorphic diaspores. *Biol. J. Linn. Soc.* **23**, 303-322.

Seavey, S. R. and Bawa, K. S. (1986). Late-acting self-incompatibility in angiosperms. *Bot. Rev.* **52**, 195-219.

Shaw, D. V., Kahler, A. L. and Allard, R. W. (1981). A multilocus estimator of mating system parameters in plant populations. *Proc. Natl. Acad. Sci. USA* **78**, 1298-1302.

Shea, K. L. (1987). Effects of population structure and cone production on outcrossing rates in Engelmann spruce and subalpine fir. *Evolution* **41**, 124-136.

Shore, J. S. and Barrett, S. C. H. (1990). Quantitative genetics of floral characters in homostylous *Turnera ulmifolia* var. *angustifolia* Willd. (Turneraceae). *Heredity* (in press).

Silander, J. A. J. (1985). Microevolution in clonal plants. *In* "Population Biology and Evolution of Clonal Organisms" (J. B. C. Jackson, L. W. Buss and R. E. Cook, eds), pp. 107-152. Yale University Press, London, UK.

Sokal, R. R. and Rohlf, F. J. (1981). "Biometry". 2nd ed. W. H. Freeman and Company, New York, USA.

Stebbins, G. L. (1957). Self-fertilization and population variability in the higher plants. *Am. Nat.* **91**, 337-354.
Stebbins, G. L. (1974). "Flowering Plants: Evolution Above the Species Level". Belknap Press, Cambridge, MA, USA.
Sun, M. (1987). Genetics of gynodioecy in Hawaiian *Bidens. Heredity* **59**, 327-336.
Sun, M. and Ganders, F. R. (1986) Female frequencies in gynodioecious populations correlated with selfing rates in hermaphrodites. *Am. J. Bot.* **73**, 1645-1648.
Sun, M. and Ganders, F. R. (1988) Mixed mating systems in Hawaiian *Bidens* (Asteraceae). *Evolution* **42**, 516-527.
Uyenoyama, M. K. (1986). Inbreeding and the cost of meiosis: The evolution of selfing in populations practicing biparental inbreeding. *Evolution* **40**, 388-404.
Vasek, F. C. and Harding, J. (1976). Outcrossing in natural populations. V. Analysis of outcrossing, inbreeding, and selection in *Clarkia exilis* and *Clarkia tembloriensis. Evolution* **30**, 403-411.
Waller, D. M. (1986). Is there disruptive selection for self-fertilization? *Am. Nat.* **128**, 421-426.
Waller, D. M. and Knight, S. E. (1989). Genetic consequences of outcrossing in the cleistogamous annual, *Impatiens capensis.* II. Outcrossing rates and genotypic correlations. *Evolution* **43**, 860-869.
Weir, B. S. and Cockerham, C. C. (1973). Mixed self and random mating at two loci. *Genet. Res.* **21**, 247-262.
Weller, S. G. and Ornduff, R. (1977). Cryptic self-incompatibility in *Amsinckia grandiflora. Evolution* **31**, 47-51.
Wiens, D. (1984). Ovule survivorship, brood size, life history, breeding systems and reproductive success in plants. *Oecologia* (Berl.) **64**, 47-53.
Wolfe, L. M. and Barrett, S. C. H. (1988). Temporal changes in the pollinator fauna of tristylous *Pontederia cordata*, an aquatic plant. *Can. J. Zool.* **66**, 1421-1424.
Wolff, K., Friso, B. and van Damme, J. M. M. (1988). Outcrossing rates and male sterility in natural populations of *Plantago coronopus. Theor. Appl. Genet.* **76**, 190-196.
Wooten, J. W. (1971). The monoecious and dioecious conditions in *Sagittaria latifolia* L. (Alismataceae). *Evolution* **25**, 549-553.
Wright, A. J. (1988). Some applications of the covariances of relatives with inbreeding. *In* "Second International Conference on Quantitative Genetics" (B. S. Weir, E. J. Eisen, M. M. Goodman and G. Namkoong, eds), pp. 10-20. Sinauer, Raleigh, USA.
Wyatt, R. (1986). Ecology and evolution of self-pollination in *Arenaria uniflora* (Caryophyllaceae). *J. Ecol.* **74**, 403-418.
Wyatt, R. (1988). Phylogenetic aspects of th e evolution of self-pollination. *In* "Plant Evolutionary Biology" (L. D. Gottlieb and S. K. Jain, eds), pp. 109-131. Chapman and Hall, London, UK.

15 Reproductive Biology of Milkweeds (*Asclepias*): Recent Advances

ROBERT WYATT AND STEVEN B. BROYLES

Department of Botany, University of Georgia, Athens, GA 30602, USA

I. Introduction

"Pod of the Milkweed"
"Calling all butterflies of every race
From source unknown but from no special place
They ever will return to all their lives,
Because unlike the bees they have no hives,
The milkweed brings up to my very door
The theme of wanton waste in peace and war
As it has never been to me before ...

But waste was of the essence of the scheme.
And all the good they did for man or god
To all those flowers they passionately trod
Was leave as their posterity one pod
With an inheritance of restless dream ...

Where have those flowers and butterflies all gone
That science may have staked the future on?
He seems to say the reason why so much
Should come to nothing must be fairly faced"
Robert Frost

An astute observer of nature, New England poet Robert Frost (1962) was among the first to note the extremely low fruit-set that characterizes milkweeds. Hypotheses to explain this phenomenon have centered on extrinsic factors related to effective pollination and on intrinsic factors related to successful fertilization and

BIOLOGICAL APPROACHES AND EVOLUTIONARY TRENDS IN PLANTS ISBN 0-12-402960-4

allocation of resources to maturing seeds. Our understanding of these processes and attempts to test various hypotheses have grown over the past 20-25 years. This paper describes and discusses some of the recent progress on this topic.

II. Structure and Function of Milkweed Flowers

The extremely specialized morphology of milkweed flowers is rivalled only by that of the orchids. These groups are the only dicot and monocot family, respectively, whose pollen grains are cohered together and transported as a unit, termed a "pollinium." Flowers of *Asclepias* consist of five showy, reflexed petals covering five smaller, greenish sepals (Fig. 1A). Two free, superior ovaries are joined by their styles to form a gynostegium with five lateral stigmatic surfaces (Fig. 1B,C). These surfaces are enclosed by the tightly abutting wings of adjacent anthers to produce five stigmatic chambers. Five coronal extensions from the base of the stamen (hoods), each of which usually surrounds an arching structure (horn), serve as reservoirs for nectar secreted by nectaries located within the stigmatic chambers (Galil and Zeroni, 1965; Fig. 1A, B). There are five pollinaria (Lynch, 1977), each of which consists of paired pollen sacs (pollinia) from adjacent anthers joined by translator arms to a corpusculum situated just above the alar fissure, a narrow opening into the stigmatic chamber (Fig. 1A, B, D). Pollination is a two-stage process: (1) removal of a pollinarium occurs when a groove in the corpusculum catches on a bristle or other appendage of an insect and is forcibly pulled from the flower, and (2) insertion is effected when a pollinium lodges in a stigmatic chamber (Wyatt, 1978). Following insertion, pollen tubes emerge only where the convex surface of the pollinium contacts the stigmatic surface (Galil and Zeroni, 1969; Sreedevi and Namboodiri, 1982). Pollen tubes subsequently grow down the stylar canal and finally enter the ovary (Corry, 1883; Frye, 1902).

Until recently, the relationship between the five stigmatic chambers and two subtending ovaries with respect to pollen tube growth has been a mystery. By examining cleared and sectioned gynostegia from cross-pollinated flowers of *A. amplexicaulis*, Sage *et al.* (1990) concluded that three adjacent stigmatic chambers transmit pollen tubes to one of the two separate ovaries and the remaining two chambers transmit to the second ovary. These observations therefore confirm Woodson's (1954) predictions. Despite the potential for pollen tubes to cross over at the point of fusion of the two styles, Sage *et al.* (1990) never observed this phenomenon, which was reported as a rare anomaly observed in one case out of hundreds examined by Sparrow and Pearson (1948). Sage *et al.* (1990) also discovered a furrow on the adaxial surface of the gynostegium that enables one to identify which stigmatic chambers transmit to particular styles and ovaries: three stigmatic chambers transmitting to one ovary are on one side of this furrow, and the two chambers transmitting to the second ovary are on the opposite side.

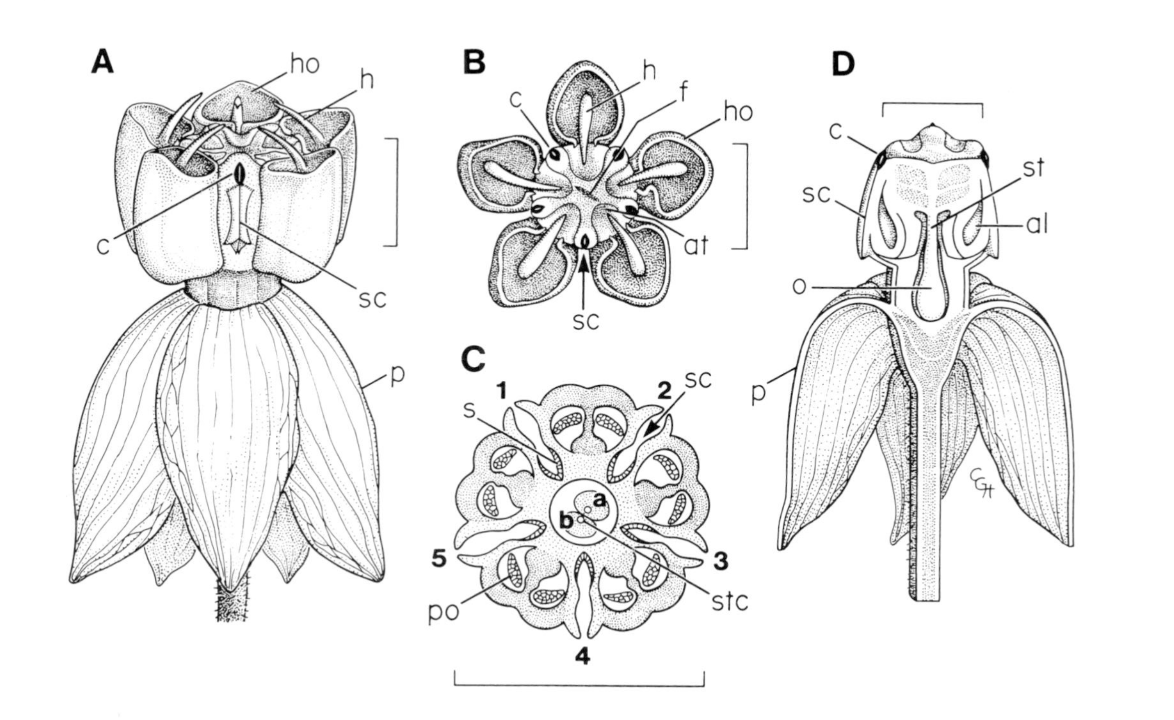

Fig. 1. Flower morphology of *Asclepias amplexicaulis*. A. Whole Flower, showing reflexed petals, corona of hoods and horns, and one surface of the gynostegium. B. Top view of flower, showing location of the hoods and horns and entrances to the stigmatic chambers relative to the furrow in the gynostegium. C. Transverse section of the flower, showing the location of the stigmatic chambers relative to the styles. Chambers 1, 2, and 3 will transmit pollen tubes to style a, while chambers 4 and 5 will transmit pollen tubes to style b. D. Longitudinal section of the flower along the axis of the furrow, showing the location of the gynostegium relative to the style and ovary. In all figures, the scale delimits 3 mm. Abbreviations are: al, anther locule; at, anther tip; c, corpusculum; f, furrow; h, horn; ho, hood; o, ovary; p, petal; po, pollinium; s, stigmatic surface; sc, stigmatic chamber; st, style; and stc, stylar canal.

TABLE 1
Comparison of probabilities of twin fruits from multiple insertions with observed frequencies of fruit initiation (data for *Asclepias syriaca* from Sparrow and Pearson, 1948)

Type of Insertion	Expected Probability of x fruits		Observed Frequency of x fruits		
	1	2	0	1	2
1	1.0	0.0	0.466	0.554	0.000
2 adjacent	0.6	0.4	0.307	0.508	0.185
2 opposite	0.2	0.8	0.318	0.284	0.398
3 adjacent	0.2	0.8	0.313	0.289	0.398
3 oppostie	0.0	1.0	—[1]	—	—
4	0.0	1.0	0.167	0.111	0.722
5	0.0	1.0	0.000	0.091	0.909

[1]Sparrow and Pearson (1948) did not distinguish between adjacent triple and opposite triple insertions.

Milkweeds rarely mature follicles from both ovaries of a single flower. In *A. syriaca*, for example, Sparrow and Pearson (1948) estimated the frequency of "twin" fruits at 5.4%. Moore (1947) reported higher estimates, ranging from 9.5% to 24.4%. In our experience with *A. tuberosa*, *A. exaltata*, and other milkweeds in the eastern United States, percentages are typically < 5%. Failure of both fruits to mature may be attributed in part to spatial aspects of pollinium insertion. Given that three stigmatic chambers transmit pollen tubes to one ovary and the remaining two, to the second, the probability of obtaining twin fruits is twice as great for opposite double insertions as for adjacent ones (Table 1). This expectation is borne out by observed frequencies of fruit-set in experimental cross pollinations performed by Sparrow and Pearson (1948). Their opposite double insertions yielded twice as many twin fruits as adjacent double insertions, the value actually being nearly identical to that for triple insertions (Table 1).

Interestingly, opposite double insertions occur more frequently than adjacent ones in most milkweeds. In *A. syriaca*, 88 of 153 double insertions were opposite (Sparrow and Pearson, 1948) and in *A. tuberosa*, 41 of 68 (Wyatt, 1976). In *A. purpurascens*, however, 30 of 57 double insertions were adjacent, and adjacent triple insertions outnumbered opposite ones 28 to 9 (Wyatt, unpublished).

III. Explanations for Low Fruit-set in Milkweeds

In attempting to explain low fruit-set in milkweeds, investigators have focused on two hypotheses: (1) that effective levels of pollination are limiting and (2) that resources for fruit and seed maturation are limiting. Proponents of resource limitation have argued that pollination levels are generally high in milkweed populations and that abortion of developing fertilized fruits is common (Willson and

Rathcke, 1974; Willson and Price, 1977; Queller, 1983, 1985). Furthermore, addition of inorganic fertilizer increased fruit set in *A. syriaca* and *A. verticillata*, while shading and leaf removal decreased it (Willson and Price, 1980). Queller (1985) also observed higher survival of fruits after application of fertilizer to *A. exaltata*. It is unclear, however, what side-effects each of these experimental treatments might have had on nectar production and pollinator activity. For example, it is possible that nutrient supplementation could have increased nectar production, which in turn could have increased pollination rates. Similarly, while Chaplin and Walker (1982) showed that flower and fruit production in *A. quadrifolia* are influenced primarily by resources stored in the taproot, it is not known if these resources also affect attraction of pollinators to the plant.

Many studies have reported that milkweed flowers receive high levels of pollination (e.g., Lynch, 1977; Willson and Bertin, 1979; Shannon and Wyatt, 1986a). Nevertheless, the proportion of these that represent self-pollinations or that fail for other reasons to result in pollen germination and fertilization of ovules is unknown. Controlled hand-pollinations increased fruit-set in *A. tuberosa* from levels of 0.33% in natural populations to 14.8% (Wyatt, 1976, 1981). Similarly, hand-pollinations of *A. exaltata* increased fruit-set to 19.7% from 2.5% in natural populations (Queller, 1985). In *A. syriaca*, supplemental hand-pollination of open-pollinated flowers doubled fruit production as compared to control flowers, leading Morse and Fritz (1983) to conclude that the plants are pollen-limited. As in the case of tests of resource limitation, there are potential problems with the interpretation of these experiments (see Zimmerman and Pyke, 1988). Wyatt (1976, 1980, 1981, 1982) has argued that both pollination and resources can limit fruit-set in milkweeds and has presented a model that incorporates both extrinsic and intrinsic factors to predict fruit-set levels close to those observed in nature.

IV. Nectar Production in Milkweeds

To some extent the high rates of pollination of milkweed flowers may be due to their long life spans and copious production of nectar. The reproductive phase of individual flowers of milkweeds, typically 4-8 days is long compared with most other flowering plants (Primack, 1985). An average flower of *A. tuberosa* lasts 7.4 $\pm$ 0.34 (mean $\pm$ standard deviation) days (Wyatt, 1981); of *A. exaltata;* 6.2 $\pm$ 0.85 days (Wyatt and Shannon, 1986); of *A. incarnata*, 4.87 $\pm$ 1.66 or 3.87 $\pm$ 1.04 days (early versus late in the season: Kephart, 1987); of *A. syriaca*, 5.18 $\pm$ 1.24 or 5.30 $\pm$ 0.82 days (Kephart, 1987); and of *A. verticillata*, 6.30 $\pm$ 2.02 or 5.14 $\pm$ 1.18 days (Kephart, 1987).

Comparisons of nectar production between different milkweed species are complicated because workers use different sampling methods and ways of expressing components of production. Wyatt and Shannon (1986) found that an average flower of *A. exaltata* produced 63.5 µl of nectar over its six-day life span, a value much higher than that for *A. verticillata* (Willson *et al.*, 1979), *A. quadrifolia* (Pleasants and Chaplin, 1983), *A. curassavica* (Wyatt, 1980; Opler, 1983) and *A. syriaca* (Willson and Bertin, 1979; Morse, 1982; Southwick, 1983). In Southwick's (1983) study, for example, total nectar production ranged from 3.8 to 17.8 µl of nectar per flower. Similarly, the 23.6 mg of sucrose produced by an average flower of *A. exaltata* (Wyatt and Shannon, 1986) greatly exceeds sugar production by *A. syriaca* (Willson and Bertin, 1979; Southwick *et al.*, 1981; Southwick, 1983), *A. quadrifolia* (Pleasants and Chaplin, 1983), and *A. verticillata* (Willson *et al.*, 1979).

In *A. exaltata* most nectar is produced overnight and increases in concentration during the day (Wyatt and Shannon, 1986; Wyatt, Broyles, and Derda, unpublished). This is also the pattern in *A. verticillata* (Willson *et al.*, 1979; Bertin and Willson, 1980), in which maximum nectar secretion occurred between 1800-2200 hr, and in *A. syriaca,* in which maximum production occurred between 1400-2200 hr (Willson and Bertin, 1979) or between 2000-0800 hr (Southwick, 1983). In contrast, Pleasants and Chaplin (1983) found that nectar production in *A. quadrifolia* peaked in the morning and that virtually no nectar was secreted overnight. Nectar concentrations in *A. exaltata* are low in the morning, averaging less than 30%, but increase steadily to 40-60% late in the afternoon (Wyatt and Shannon, 1986). This agrees in general with the results of Southwick (1983) for *A. syriaca,* whose nectar concentrations increased nearly twofold from morning values of 18.6% to evening values of 36.2%. In Illinois populations of *A. syriaca,* Willson and Bertin (1979) reported an increase from minimal concentrations of 10-17% to maximal concentrations of 29-42% in repeatedly sampled flowers. For flowers sampled only once, maximum concentrations reached 57%. Similarly, nectar concentrations in *A. verticillata* increased from about 5% to 20% in repeatedly sampled flowers and from 20% to 60% in those sampled once. Diurnal changes in milkweed nectar are strongly correlated with temperature and humidity and apparently are caused by passive evaporation of the nectar (Wyatt, Broyles, and Derda, unpublished).

In *A. exaltata,* plants that produced more concentrated nectar showed higher levels of reproductive success (Wyatt and Shannon, 1986). Nectar concentration was positively correlated with both number of pollinia inserted per flower and number of pollinaria removed per flower, the latter being highly statistically significant. Furthermore, plants that produced concentrated nectar matured more fruits and had higher levels of fruit-set per flower than plants that produced dilute nectar. This observation fits previous studies that found that bees prefer concentrated nectar (Heinrich, 1975; Corbet, 1978). Waller (1972) discovered that honeybees preferred sucrose concentrations between 30-50% when offered solutions ranging from 10-65% and that maximum foraging rate increased as concentrations increased to 50%.

V. Pollen Germination in Milkweeds

Contrary to the apparent beneficial effect of concentrated nectar on pollinator behavior is its influence on pollen germination. Among others, Shannon and Wyatt (1986b) reported that a 30% sucrose solution gave the best germination of pollinia of *A. exaltata*. In *A. syriaca*, Eisikowitch *et al.* (1987) found that germination was inhibited by sucrose concentrations above 30%. They further showed in laboratory experiments that pollinia can be stored in a 60% glucose solution for at least 48 hr before beginning to lose viability. This suggests a possible mechanism to maximize the beneficial effects of gametophytic competition. Regardless of when they were inserted into stigmatic chambers, pollinia would be inhibited from germinating until new nectar was secreted or atmospheric humidity increased so that the concentration of nectar in the chambers dropped to 30%.

Sage *et al.* (1990) observed that nearly half of all insertions performed by hand on *A. amplexicaulis* failed to result in pollen germination after 24 hr. It is possible that some of these failures were due to concentrated nectar within the stigmatic chamber. On the other hand, unless nectar concentrations vary dramatically from chamber to chamber within a single flower, performance of pollinia in double insertions should have been nearly identical, especially because all pollinia were taken from a single donor plant. Another factor that may result in failure of pollinia to germinate is contamination of the nectar by growth of microorganisms. Eisikowitch *et al.* (1990) found that yeast-infected nectar completely inhibited pollen germination in laboratory experiments. Therefore, failure of inserted pollinia to germinate may constitute a major contributing factor to low fruit-set in *Asclepias*.

Morse (1982) suggested that the durable covering of milkweed pollinia might allow a long residence time on pollinators, which would promote long-distance dispersal of pollen. This assumes that pollinia are highly resistant to desiccation and are slow to lose viability following their removal from flowers. To test this assumption, Shannon and Wyatt (1986b) performed in vitro pollen germination experiments using *A. exaltata*. Germinability varied greatly among plants; some plants produced pollinia that germinated with twice the average success of other pollen donors. Flower age had little effect on pollen germinability, although germination success of 3- and 4-day-old flowers was slightly less than for 0-, 1-, and 2-day-old flowers. Pollinia that were removed from flowers and allowed to dry lost viability rapidly. After 24 hr, their germinability was reduced by more than half. In *A. syriaca*, Eisikowitch *et al.* (1987) found that pollinia retain maximum germinability for at least four days under field conditions. It appears, therefore, that pollinia of milkweed species, such as *A. syriaca*, that inhabit open sites are more resistant to desiccation than those of species, such as *A. exaltata*, that grow in moist forests and meadows.

In addition to considering whether pollen is viable when it reaches a stigma, it is relevant to ask whether the stigma is receptive when pollen arrives. In agreement with Shannon and Wyatt's (1986b) findings for *A. exaltata,* Morse (1987) found that pollen viability of *A. syriaca* did not change significantly over the 5-day life span of flowers. Stigma receptivity, however, decreased more than threefold over this period. Coupled with Morse's (1987) observation that pollinators visit older flowers more often early in the season than in mid-season, this suggests the possibility of decreased reproductive success because of pollinium insertion into less receptive flowers.

VI. Levels of Self-pollination in Milkweeds

Another possible cause of low fruit-set in milkweeds is self-pollination (i.e., insertion of a pollinium into a stigmatic chamber of the same genetic individual). Self-insertion can lead to fruit-set only if that plant is self-compatible. Delpino (1865), Hildebrand (1866), and Corry (1883) believed that all species of *Asclepias* are self-incompatible, and experimental crosses of *A. syriaca* by Moore (1946) and Sparrow and Pearson (1948) support this view. Woodson (1954) also maintained this view and arbitrarily discounted reports of successful self-pollination in *A. syriaca* (Plotnikova, 1938; Stevens, 1945) and *A. incarnata* (Fischer, 1941). More recently, Wyatt (1976) reported that 2% of self-pollinations of *A. tuberosa* resulted in fruit-set, and Kephart (1981) reported 29% and 4% success for self-pollinations of *A. incarnata* and *A. syriaca*, respectively. She found *A. verticillata* to be completely self-incompatible. Thus, in most cases, we should expect self-insertions of milkweed pollinia to be ineffective in fruit-set.

The technical difficulties associated with marking and following pollinia have prevented researchers from obtaining accurate quantitative estimates of levels of self-insertion in natural populations of milkweeds. Recently, Pleasants *et al.* (1990) have proposed using ^{14}C as a label to track dispersal of milkweed pollinia and to estimate levels of self-pollination. Unfortunately, this approach needs to be refined before it can be applied in a field situation, and, as yet, no estimates of self-insertion rates are available. Pleasants and Ng (1987) attempted to estimate levels of self-pollination in *A. syriaca* by comparing numbers of insertions in emasculated umbels to those in umbels with intact corpuscula. Over a range of umbel sizes, they calculated that 36% of inserted pollinia represented self-insertions. There is a problem, however, with the assumption that emasculation has only the effect of removing a source of self-pollen, as Wyatt (1978) has shown in similar emasculation experiments on *A. tuberosa* that the presence of an intact corpusculum increases the likelihood of successful insertion. Thus, an alternative interpretation of Pleasants and Ng's (1987) result is simply that removal of pollinaria decreased the overall level of successful insertions.

VII. Pollen-mediated Gene Flow in Milkweeds

Morse (1982, 1987) has argued that the potentially long residence time of pollinia on pollinators suggests that pollen is routinely carried long distances. To investigate pollen-mediated gene flow in milkweeds, Broyles and Wyatt (1990b) used genetic markers to identify pollen parents of seeds produced in a natural population of *A. exaltata.* The mean for realized pollen dispersal (4.33 $\pm$ 2.57 m) determined from the paternity exclusion analysis was more than three times greater than the mean for potential pollen dispersal (1.30 $\pm$ 1.40 m) predicted from pollinator observations (Fig. 2). The distribution of interplant flight distances for flower visitors was leptokurtic and skewed right ($g_1 = 2.60$, $t = 10.1$, $P < 0.005$; $g_2 = 10.4$, $t = 19.3$, $P < 0.005$): 60% of all flights occurred between plants separated by distances < 1 m. Only 5% of the flights were longer than 5 m. On the other hand, 35% of the realized pollen dispersal distances exceeded 5 m, and the distribution was platykurtic and skewed right ($g_1 = 0.30$, $t = 8.83$, $P < 0.005$; $g_2 = -0.42$, $t = -6.08$, $P < 0.005$). The distribution of realized pollen dispersal was strikingly similar to a random distribution of all interplant distances. Insect flight distances were much shorter than expected if the bees and butterflies were visiting plants at random within the population. Broyles and Wyatt (1990b) attributed the disparity between realized pollen dispersal and pollinator flight distances to pollen carryover.

Broyles and Wyatt (1990b) also found that mating between plants was random with respect to genetic similarity between individuals of *A. exaltata*. Furthermore, there was no spatial pattern to genetic distances between plants, and mating occurred at random with respect to interplant distances. Despite apparent panmixia in the population, matings between genetically similar individuals resulted in fruits with fewer seeds, a higher proportion of which were inviable (Table 2). This

TABLE 2

Means, standard deviations, and sample sizes for number of seeds per fruit and percent viability of seeds of *Asclepias exaltata*. Plants are grouped into classes on the basis of genetic distances (Nei, 1972). Along columns, means connected by lines are not significantly different ($P > 0.05$)

Genetic Distances Between Mating Pairs	Number of Pairs	Seed Number	Percent Viability
0.000 - 0.040	14	47.79 ± 24.87	64.36 ± 24.25
0.041 - 0.080	39	55.95 ± 17.15	81.33 ± 18.17
0.081 - 0.120	26	60.50 ± 21.54	81.46 ± 20.69
0.121 - 0.160	7	69.71 ± 23.68	87.14 ± 18.39
> 0.161	2	69.00 ± 12.73	94.50 ± 6.36

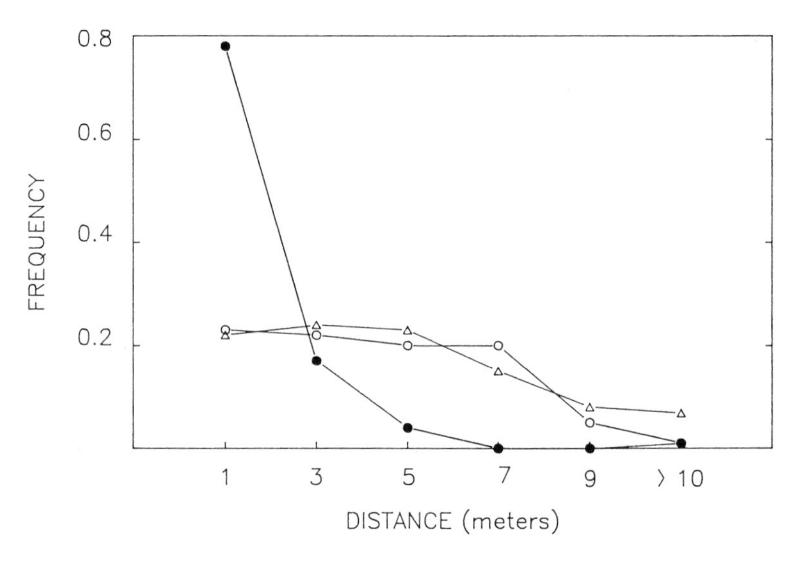

Fig. 2 Distributions of interplant pollinator flight distances (closed circles), realized pollen dispersal (open circles), and interplant distances (triangles) in the Mt. Lake population of *Asclepias exaltata* in 1986.

suggests that inbreeding depression may play a role in determining seed-set in milkweed fruits. There was no evidence of "outbreeding depression" (*sensu* Price and Waser, 1979) in *A. exaltata* .

VIII. Hybridization in Milkweeds

Another aspect of pollination that could contribute to low fruit-set in milkweeds is interspecific gene flow. It is not uncommon for several species of *Asclepias* to occur sympatrically, to overlap phenologically, and to share the same large, generalist, Hymenopteran and Lepidopteran pollinators (Kephart and Heiser, 1980; Kephart, 1981, 1983, 1987; Kephart *et al.*, 1988). Despite Kephart and Heiser's (1980) strong rejection of the view that mechanical ("lock and key") isolation is highly effective in milkweeds, it is nonetheless true that interspecific hybridization is rare (but see Kephart *et al.*, 1988) due to well-developed physiological barriers to crossing (Kephart, 1981). Therefore, interspecific insertions, which accounted for up to 60% of all pollinations in some mixed populations of milkweeds (Kephart and Heiser, 1980), may represent a major loss of potential fecundity. Such crosses represent a wastage of pollen as well as a decrease in potential female success due to preemption of stigmatic chambers and/or loss of ovules (Kephart, 1981; Kephart and Heiser, 1981).

IX. Genetic Structure of Milkweed Populations

There is little information available regarding the genetic structure of milkweed populations. Broyles and Wyatt (1990b) found no significant spatial differentiation within a population of 55 genetically distinct flowering plants of *A. exaltata* in southwestern Virginia. The plants occupied an area of approximately 150 m^2, and average density was 0.5 plants/m^2. Based on pollen dispersal, Broyles and Wyatt (1990b) estimated neighbourhood size at 81.1 plants and neighbourhood area at 159.6 m^2. Thus, neighbourhood size was larger than the number of genetically distinct individuals in the population. Note, however, that this estimate probably represents a lower limit because seed dispersal will contribute to an expanded neighborhood. Milkweeds have wind-dispersed, comose seeds that are capable of long distance dispersal (Morse and Schmitt, 1985; Sacchi, 1987).

Our detailed electrophoretic analysis of *A. exaltata* in Virginia has also revealed the capacity of this species to propagate clonally to a limited extent (Broyles and Wyatt, unpublished). In addition, our analyses of *A. syriaca* have confirmed its ability to form extensive clones, and Kephart (1981) noted this propensity in *A. verticillata.* Given that populations of at least some milkweeds include genetically identical individual plants, it is important that we keep this fact in mind as we interpret field experiments designed to estimate rates of self-insertion and fruit-set. For example, in interpreting any experiments that involve following the transfer of marked pollinia between plants, we need to know which transfers actually involve crosses between stems of the same genet, which are still effectively self-insertions. Similarly, the very erratic fruit-set observed in Willson and Rathcke's (1974) experiment in which stems of *A. syriaca* were joined together to form artificially large inflorescences may have resulted from their sometimes joining together stems of different clones, so that a source of outcross pollen was present within the single large umbel. In other cases, low fruit-set resulted because stems were in fact members of the same clone, as intended by the experimenters.

X. Evolution of Inflorescence Size in Milkweeds

Some recent authors have regarded low fruit-set in milkweeds as a consequence of sexual selection. They argue that large inflorescences are favored because they attract more pollinators and thereby increase rates of pollen removal (Willson and Rathcke, 1974; Willson and Price, 1977; Queller, 1983; Bell, 1985; Wolfe, 1987). Support for this "pollen donation" hypothesis comes from studies that have demonstrated that large inflorescences mature fewer fruits per unit investment in flowers while pollen removal rates increase. There have been no unambiguous tests of the pollen donation hypothesis, however, because male reproductive success is difficult to measure directly. Broyles and Wyatt (1990a) tested the pollen donation hypothesis by quantifying the number of seeds sired by individual genotypes in a natural population of *A. exaltata.* There was a linear relationship between number of flowers per plant and number of seeds produced per plant (female success: Fig. 3). Similarly, number of seeds sired by a plant was linearly related to flower number per plant (male success: Fig. 4). The slope of the regression line for females was significantly greater than that for males ($F = 5.51$, $P < 0.05$, d.f. = 1,106), contrary to the prediction of the pollen donation hypothesis that male success should rise more steeply as a function of increasing inflorescence size.

Broyles and Wyatt (1990a) also calculated relative frequencies of inflorescence size (flower number per plant) and proportions of male and female reproduction associated with each size class. Large plants with > 100 flowers comprised only 25% of the population, yet they sired 63% and produced 55% of all seeds. Also contrary to the pollen donation hypothesis, calculations of functional gender showed that individuals with large inflorescences did not contribute relatively more genes to the next generation through pollen than through ovules.

XI. Multiple Paternity and Seed Number in Milkweeds

Contrary to findings for other flowering plants (e.g., Marshall and Ellstrand, 1985; Ellstrand and Marshall, 1986; Schnabel, 1988), it appears that multiple paternity within fruits of milkweeds is rare. Of a total of 103 fruits of *A. exaltata* screened by Broyles and Wyatt (1990a), only two (1.9%) showed directly observable multiple paternity. In each instance, < 20% of the seeds were sired by pollen from the second father. Several authors have argued that multiple paternity within fruits increases the vigor of progeny by enhancing gametophytic selection (Mulcahy and Mulcahy, 1975; Mulcahy, 1979; Schemske and Paulter, 1984; Ellstrand, 1984; Ellstrand and Marshall, 1986; Winsor *et al.*, 1987). The intensity of competition between pollen from different plants should be less in plants like milkweeds that produce singly-sired fruits. While inferior pollen donors are likely to sire some seeds in species with high levels of multiple paternity within fruits, inferior pollen donors in milkweeds are likely to experience complete failure as males, especially in years when pollination rates are high. Also by maturing a few singly-sired fruits with many seeds, milkweeds may produce offspring with a decreased range of genetic variation. High genetic similarity of these full-siblings may lead to cooperative interactions that result in reduced intersib competition and uniform seed size (Kress, 1981).

Interestingly, both seed size and seed number are relatively uniform within species of *Asclepias* (Wilbur, 1977), and in many cases seed number is actually less variable than seed weight, which is often cited as one of the least phenotypically plastic of all plant characters (Harper, 1977). For example, in two populations of *A. asperula* from Texas, coefficients of variation for seed weight were 36.2% and 45.4%, while corresponding figures for seed number were 22.2% and 27.5% (Wyatt, unpublished). As in the case of multiple paternity, the low level of

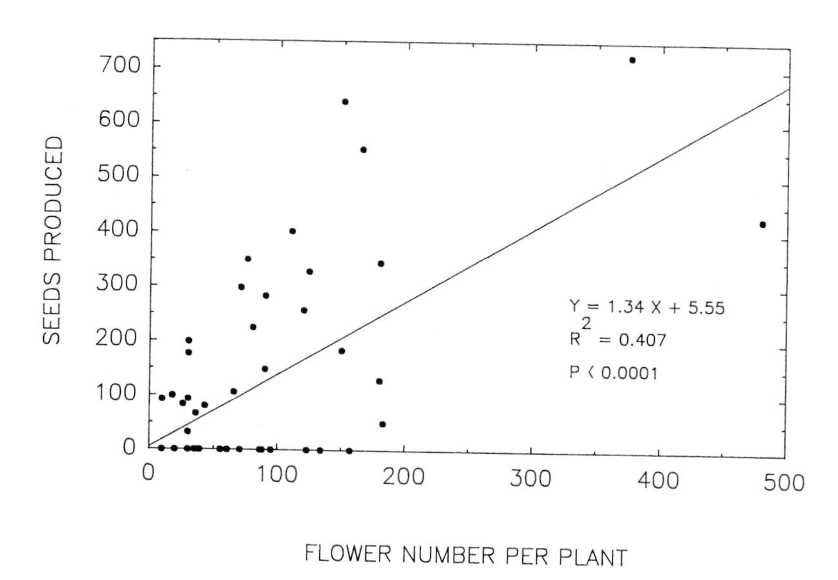

Fig. 3 Number of seeds produced as a function of the number of flowers per plant ($n = 55$) in the Mt. Lake population of *Asclepias exaltata*.

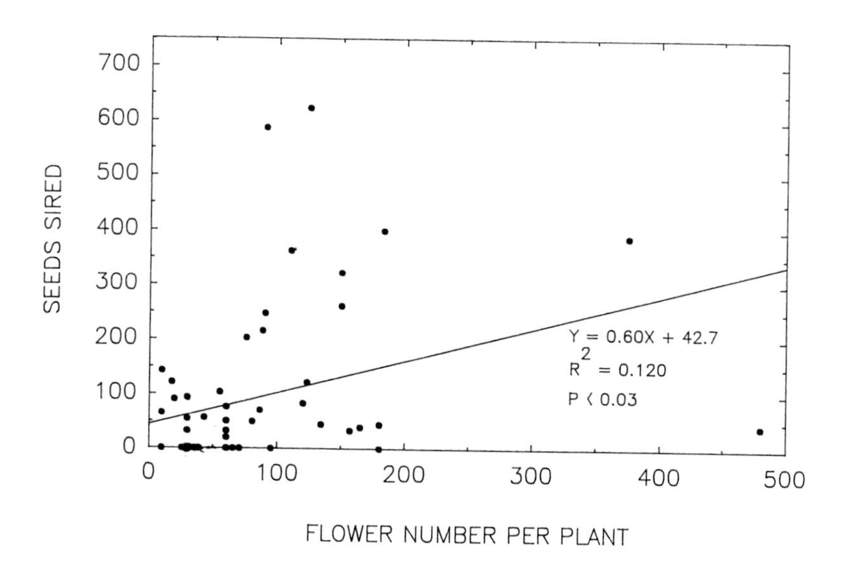

Fig. 4 Number of seeds sired per plant as a function of the number of flowers per plant ($n = 55$) in the Mt. Lake population of *Asclepias exaltata*.

variation in seed number is probably due to the unique pollination system of milkweeds, in which a single pollination event ensures delivery of more than enough pollen grains to fertilize all of the ovules within an ovary. Also contributing to reduced variability in number of seeds per fruit is selective maturation of fruits with high seed numbers. Bookman (1983) showed that *A. speciosa* selectively aborts fruits that contain few seeds. Putting this observation together with the assumption that this species is not pollen-limited, Bookman (1984) argued that the production of "surplus" flowers in milkweeds is a mechanism to allow selective maturation of fruits based on offspring quantity and quality.

XII. Summary

Milkweeds are rivalled in floral complexity only by orchids, which also transmit their pollen grains in large groups within pollinia. Unlike orchids, however, milkweeds are characterized by low fruit-set, typically averaging 1-5%. Both extrinsic factors related to pollination and intrinsic factors related to fertilization and fruit maturation appear to be involved in limiting fruit production in *Asclepias*.

Twin fruits, representing successful fertilization and maturation of the two separate ovaries within a single flower, are extremely rare in milkweeds. One reason for this has been revealed by recent anatomical work which has demonstrated that three of the five stigmatic chambers of a milkweed flower transmit pollen tubes to one ovary, while the other two chambers transmit pollen tubes only to the second. Electrophoretic analyses of seeds from twin fruits are consistent with this model, as they demonstrate that multiple pollinations by genetically distinct fathers are involved.

Milkweed flowers are long-lived and produce copious amounts of nectar, which flows from nectaries within the stigmatic chambers to fill the hoods, which act as nectar reservoirs. Rates of pollination and fruit production increase with increased nectar concentration in *A. exaltata.* Nectar concentrations above 30% appear to inhibit pollen germination, as does the growth of yeasts in nectar. Pollen germinability varies up to two-fold among plants and is sharply reduced by drying for 24-48 hours. Milkweeds are mostly self-incompatible, and it appears that levels of self-insertion of pollinia are high.

Realized gene flow through insect-mediated pollen dispersal is high within populations of *A. exaltata*, as shown by paternity analysis. It also appears that dispersal of the comose seeds of *A. syriaca* is very effective both in colonizing new sites and in fostering gene flow between isolated populations. Within populations, despite the potential for clonal propagation in some species, large patches of homogeneous genotypes do not appear to develop.

Contrary to prevailing dogma, it appears that interspecific hybridization does occur in milkweeds. Morphological and biochemical data support the view that there is limited, localized hybridization between sympatric species. Thus, it appears that the concept of well-developed reproductive isolation due to mechanical, "lock and key" barriers has been overemphasized.

Early attempts to explain the evolution of inflorescence size were hampered by failure to consider the genetic basis of the variation observed and failure to determine the unit on which selection should act. Direct tests of the "pollen donation" hypothesis have cast doubt on the validity of the view that flower number and other floral traits evolved primarily to enhance male reproductive success.

In milkweeds, seed number is often less variable than seed weight. This may be due to the unique pollen delivery system of milkweeds, which ensures that each pollination event delivers more than enough pollen grains to fertilize all of the ovules within an ovary. In addition, there is some evidence that milkweeds may selectively mature fruits that contain a full complement of filled seeds.

Acknowledgements

We thank N. C. Ellstrand, W. E. Friedman, J. L. Hamrick, R. F. Sage, S.L. Sherman-Broyles, A. A. Snow, and C. J. Williams for comments on earlier drafts of portions of this manuscript. We are indebted to C. R. Werth and C. W. dePamphilis for sharing their expertise in electrophoresis and to our coworkers, G. S. Derda, S. R. Kephart, T. L. Sage, and T. R. Shannon, for assistance in many phases of this research. We thank the staff at Mountain Lake Biological Station for encouragement, financial support, and the use of laboratory equipment. We thank the National Park Service for allowing us to study milkweeds in Shenandoah National Park. We also thank W. E. Friedman for use of his laboratory facilities. This research was supported financially by grants from the Whitehall Foundation and National Science Foundation (to RW) and fellowships from the DuPont Corporation, University of Virginia, and the University of Georgia (to SB). A fellowship from the John Simon Guggenheim Foundation (to RW) also helped make this research possible.

References

Bertin, R. I. and Willson, M. F. (1980). Effectiveness of diurnal and nocturnal pollination of two milkweeds. *Can. J. Bot.* **58**, 1744-1746.

Bookman, S. S. (1983). Costs and benefits of flower abscission and fruit abortion in *Asclepias speciosa*. *Ecology* **64**, 264-273.

Bookman, S. S. (1984). Evidence for selective fruit production in *Asclepias*. *Evolution* **38**, 72-86.

Broyles, S. B. and Wyatt, R. (1990a). Paternity analysis in a natural population of *Asclepias exaltata*: Multiple paternity, functional gender, and the "pollen donation hypothesis." *Evolution* (In press).

Broyles, S. B. and Wyatt, R. (1990b). Pollen-mediated gene flow in a natural population of *Asclepias exaltata*: The influence of pollinator behavior, spatial distribution, and genetic similarity. *Am. Nat.* (in press).

Chaplin, S. J. and Walker, J. L. (1982). Energetic constraints and adaptive significance of the floral display of a forest milkweed. *Ecology* **63**, 1857-1870.

Corbet, S. A. (1978). Bee visits and the nectar of *Echium vulgare* L. and *Sinapis alba* L. *Ecol. Entomol.* **3**, 25-37.

Corry, T. H. (1883). On the mode of development of the pollinium in *Asclepias cornuti* Decaisne. *Trans. Linn. Soc. London* Ser. **2**, Bot. 75-84.

Delpino, F. (1865). Relazione sull 'apparecchio della fecondazione nelle Asclepiadee. Torino.

Eisikowitch, D., Kevan, P. G. and Lachance, M.A. (1990). The nectar-inhabiting yeasts and their effect on pollen germination in common milkweed, *Asclepias syriaca* L. *Israel J. Bot.* **39**, 217-226.

Eisikowitch, D., Kevan, P. G., Fowle, S. and Thomas, K. (1987). The significance of pollen longevity in *Asclepias syriaca* L. under natural conditions. *Pollen Spores* **29**, 121-128.

Ellstrand, N. C. (1984). Multiple paternity within the fruits of the wild radish, *Raphanus sativus*. *Am. Nat.* **123**, 819-828.

Ellstrand, N. C. and Marshall, D. L. (1986). Patterns of multiple paternity in populations of *Raphanus sativus*. *Evolution* **40**, 837-842.

Fischer, E. (1941). Der Anbau einer Faser- und Bienen-futterpflanze. *Pflanzenbau* **17**, 212-218.

Frost, R. (1962). Pod of the Milkweed. *In* R. Frost. In the Clearing. Holt, Rinehart, and Winston, New York.

Frye, T. C. (1902). A morphological study of certain Asclepiadaceae. *Bot. Gaz.* **34**, 389-413.

Galil, J. and Zeroni, M. (1965). Nectar system of *Asclepias curassavica*. *Bot.Gaz.* **126**, 144-148.

Galil, J. and Zeroni, M. (1969). On the organization of the pollinium in *Asclepias curassavica*. *Bot. Gaz.* **130**, 1-4.

Harper, J. L. (1977). Population Biology of Plants. Academic Press, New York.

Heinrich, B. (1975). Energetics of pollination. *Annu. Rev. Ecol. Syst.* **6**, 139-170.

Hildebrand, R. (1866). Über die Befruchtung von *Asclepias cornuti*. *Bot. Zeit.* **24**, 376-378.

Kephart, S. R. (1981). Breeding systems in *Asclepias incarnata* L., *A. syriaca* L., and *A. verticillata* L. *Am. J. Bot.* **68**, 226-232.

Kephart, S. R. (1983). The partitioning of pollinators among three species of *Asclepias*. *Ecology* **64**, 120-133.

Kephart, S. R. (1987). Phenological variation in flowering and fruiting of *Asclepias*. *Am. Midl. Nat.* **118**, 64-76.

Kephart, S. R. and Heiser, C. B. (1980). Reproductive isolation in *Asclepias*: Lock and key hypothesis reconsidered. *Evolution* **34**, 738-746.

Kephart, S. R., Wyatt, R. and Parrella, D. (1988). Hybridization in North American *Asclepias*. I. Morphological evidence. *Syst. Bot.* **13**, 456-473.

Kress, W. J. (1981). Sibling competition and evolution of pollen unit, ovule number and pollen vector in angiosperms. *Syst. Bot.* **6**, 101-112.

Lynch, S. P. (1977). The floral ecology of *Asclepias solanoana* Woods. *Madroño* **24**, 159-177.

Marshall, D. L. and Ellstrand, N. C. (1985). Proximal causes of multiple paternity in wild radish, *Raphanus sativus*. *Am. Nat.* **126**, 596-605.

Moore, R. J. (1947). Investigations on rubber-bearing plants. V. Notes on the flower biology and pod yield of *Asclepias syriaca* L. *Can. Field Nat.* **61**, 40-46.

Morse, G. H. (1982). The turnover of milkweed pollinia on bumblebees and implications for outcrossing. *Oecologia* (Berl.) **60**, 190-197.

Morse, D. H. (1987). Roles of pollen and ovary age in follicle production of the common milkweed *Asclepias syriaca*. *Am. J. Bot.* **74**, 851-856.

Morse, D. H. and Fritz, R. S. (1983). Contributions of diurnal and nocturnal insects to the pollination of common milkweed (*Asclepias syriaca* L.) in a pollen-limited system. *Oecologia* (Berl.) **60**, 190-197.

Morse, D. H. and Schmitt, J. (1985). Propagule size, dispersal ability, and seedling performance in *Asclepias syriaca*. *Oecologia* (Berl.) **67**, 372-379.

Mulcahy, D. L. (1979). The rise of the angiosperms: A genecological factor. *Science* **206**, 20-23.

Mulcahy, D. L. and Mulcahy, G. B. (1975). The influence of gametophytic competition on sporophytic quality in *Dianthus chinensis*. *Theor. Appl. Genet.* **46**, 277-280.

Nei, M. (1972). Genetic distance between populations. *Am. Nat.* **106**, 283-292.

Opler, P. A. (1983). Nectar production in a tropical ecosystem. *In* "The Biology of Nectaries" (B. Bentley and T. Elias, eds), pp. 30-79. Columbia Univ. Press, New York.

Pleasants, J. M. and Chaplin, S. J. (1983). Nectar production rates of *Asclepias quadrifolia*: Causes and consequences of individual variation. *Oecologia* (Berl.) **59**, 232-238.

Pleasants, J. M. and Ng, G. (1987). The relationship between inflorescence size and self-pollination in the milkweed, *Asclepias syriaca*. *Iowa Acad. Sci.* (Abstract).

Pleasants, J. M., Horner, H. T. and Ng, G. M. Y. (1990). Tracking pollen dispersal: Use of ^{14}C as a label for milkweed pollinia. *Am. J. Bot.* (in press).

Plotnikova, T. (1938). An experiment in self-pollination of *Asclepias cornuti*. *Ukraine Acad. Sci., Inst. Bot. J.* No. **26-27**. (English summary).

Price, M. V. and Waser, N. M. (1979). Pollen dispersal and optimal outcrossing in *Delphinium nelsonii* . *Nature* **277**, 294-296.

Primack, R. B. (1985). Longevity of individual flowers. *Annu. Rev. Ecol. Syst.* **16**, 15-38.

Queller, D. C. (1983). Sexual selection in a hermaphroditic plant. *Nature* **305**, 706-707.

Queller, D. C. (1985). Proximate and ultimate causes of low fruit production in *Asclepias exaltata*. *Oikos* **44**, 373-381.

Sacchi, C. F. (1987). Variability in dispersal ability of common milkweed, *Asclepias syriaca*, seeds. *Oikos* **49**, 191-198.

Sage, T. L., Broyles, S. B. and Wyatt, R. (1990). The relationship between the five stigmatic chambers and two ovaries of milkweed (*Asclepias amplexicaulis* Sm.) flowers: A three-dimensional assessment. *Israel J. Bot.* **39**, 187-196.

Schnabel, A. (1988). Genetic Structure and Gene Flow in *Gleditsia triacanthos*. Ph.D. Dissertation, Univ. of Kansas, Lawrence.

Schemske, D. W. and Paulter, L. P. (1984). The effects of pollen composition on fitness components in a neotropical herb. *Oecologia* (Berl.) **62**, 31-36.

Shannon, T. R. and Wyatt, R. (1986a). Reproductive biology of *Asclepias exaltata* . *Am. J. Bot.* **73**, 11-20.

Shannon, T. R. and Wyatt, R. (1986b). Pollen germinability of *Asclepias exaltata:* Effects of flower age, drying time, and pollen source. *Syst. Bot.* **11**, 322-325.

Southwick, E. E. (1983). Nectar biology and nectar feeders of common milkweed, *Asclepias syriaca* L. *Bull. Torrey Bot. Club* **110**, 324-334.

Southwick, E. E., Lopez, G. M. and Sadwick, S. E. (1981). Nectar production, composition, energetics, and pollinator attractiveness in spring flowers of western New York. *Am. J. Bot.* **67**, 994-1002.

Sparrow, F. K. and Pearson, N. L. (1948). Pollen compatibility in *Asclepias syriaca*. *J. Agr. Res.* **77**, 187-199.

Sreedevi, P. and Namboodiri, A. N. (1982). The germination of pollinium and the organization of germ furrow in some members of Asclepiadaceae. *Can. J. Bot.* **60**, 166-172.

Stevens, O. A. (1945). Cultivation of milkweeds. *North Dakota Agr. Exp. Sta. Bull.* **333**, 1-19.

Waller, G. D. (1972). Evaluating responses of honey bees to sugar solutions using an artificial flower feeder. *Annu. Entomol. Soc. Amer.* **65**, 857-862.

Wilbur, H. M. (1977). Propagule size, number, and dispersion pattern in *Ambystoma* and *Asclepias*. *Am. Nat.* **111**, 43-68.

Willson, M. F. and Bertin, R. I. (1979). Flower visitors, nectar production and inflorescence size of *Asclepias syriaca*. *Can. J. Bot.* **57**, 1380-1388.

Willson, M. F. and Price, P. W. (1977). The evolution of inflorescence size in *Asclepias* (Asclepiadaceae). *Evolution* **31**, 495-511.

Willson, M. F. and Price, P. W. (1980). Resource limitation of fruit and seed production in some *Asclepias* species. *Can. J. Bot.* **58**, 2229-2233.

Willson, M. F. and Rathcke, B. J. (1974). Adaptive design of the floral display in *Asclepias syriaca* L. *Am. Midl. Nat.* **92**, 47-57.

Willson, M. F., Bertin, R. I. and Price, P. W. (1979). Nectar production and flower visitors of *Asclepias verticillata*. *Am. Midl. Nat.* **102**, 23-35.

Winsor, J. A., Davis, L. E. and Stephenson, A. G. (1987). The relationship between pollen load and fruit maturation and the effect of pollen load on offspring vigor in *Cucurbita pepo*. *Am. Nat.* **129**, 643-656.

Wolfe, L. M. (1987). Inflorescence size and pollinaria removal in *Asclepias curassavica* and *Epidendrum radicans*. *Biotropica* **19**, 86-89.

Woodson, R. E. (1954). The North American species of *Asclepias* L. *Ann. Mo. Bot. Gard.* **41**, 1-211.

Wyatt, R. (1976). Pollination and fruit-set in *Asclepias:* A reappraisal. *Am. J. Bot.* **63**, 845-851.

Wyatt, R. (1978). Experimental evidence concerning the role of the corpusculum in *Asclepias* pollination. *Syst. Bot.* **3**, 313-321.

Wyatt, R. (1980). The reproductive biology of *Asclepias tuberosa*: I. Flower number, arrangement, and fruit-set. *New Phytol.* **85**, 119-131.

Wyatt, R. (1981). The reproductive biology of *Asclepias tuberosa*. II. Factors determining fruit-set. *New Phytol.* **88**, 375-385.

Wyatt, R. (1982). Inflorescence architecture: How flower number, arrangement, and phenology affect pollination and fruit-set. *Am. J. Bot.* **69**, 587-596.

Wyatt, R. and Shannon, T. R. (1986). Nectar production and pollination of *Asclepias exaltata*. *Syst. Bot.* **11**, 326-334.

Zimmerman, M. and Pyke, G. H. (1988). Reproduction in *Polemonium*: Assessing the factors limiting seed set. *Am. Nat.* **131**, 723-738.

16 Biology of Asexually Reproducing Plants

KRYSTYNA M. URBANSKA

Geobotanical Department, Swiss Federal Institute of Technology
Zürich, Switzerland

I. Introduction

Reproduction in plants is exceedingly complex, species with only one type of reproductive behaviour being exception rather than a rule. It is therefore not surprising that there is a lack of agreement in defining both the phenomenon itself as well as the units involved.

Some biologists would insist that without knowledge of the **genetic** foundation, all other study is fruitless. The recognition of genetic causes and effects which influence the reproductive behaviour of plants is indeed essential as far as the evolutionary time scale is concerned. However, for a better understanding of population dynamics on an ecological time scale which includes several generations, reproductive **biology** is of primary importance, even in the absence of a detailed genetic assessment. In this respect I subscribe to the opinion expressed not so long ago by Willson (1983).

The present paper deals with asexually reproducing angiosperm plants; however, some basic notions have to be considered before the actual subject is tackled. These considerations are indispensable because the literature abounds with diverse terms and interpretations, and it is often not quite clear what a given author had exactly in mind.

Another point to be stressed is the general character of the present paper; detailed case studies will not be discussed here on account of the limited space allotted. I propose first to consider some definitions and assessment formulae, and then to compare briefly some important biological aspects of sexual and asexual reproduction as well as their bearing on the population dynamics of the species concerned.

BIOLOGICAL APPROACHES AND
EVOLUTIONARY TRENDS IN PLANTS ISBN 0-12-402960-4

II. Reproduction in the Angiosperms - Different Ways and Means, the Same Effect

Reproduction has a very special place amongst the basic functions of a live organism: it secures the formation of the offspring. Reproductive behaviour of flowering plants is frequently characterized both by sexual and asexual processes, and influenced not only by genetic make-up but also phenotypic plasticity. All these differences notwithstanding, a successful reproduction has but one outcome viz. the appearance of active descendants contributing to the population turnover.

The essential rôle of reproduction has been recognized in the numerous definitions which identify the reproductive process with the formation of new individuals, but some authors still emphasize the production of propagules rather than the appearance of the new plants (Table 1). Last but not least, some scientists (e.g. Grime, 1979) renounce the term "reproduction" and propose to replace it by "regeneration", whereas others (e.g. Harper, 1977) limit the reproduction to some specified conditions (Table 1).

TABLE 1
Some definitions of reproduction

Definition	Author(s)
Ability of organism to produce new individuals of the same kind; this process is essential to presërvation of species.	Hartmann, 1959
Reproduction involves the formation of a new individual from a single cell; this is usually (though not always, e.g. apomicts) a zygote. In this process, a new individual is "reproduced" by the information that is coded in all that cell.	Harper, 1977
Production of sexual or asexual propagule which starts a new developmental cycle and a new organism.	Dawkins, 1982; Tuomi and Vuorisalo, 1989
Propagation of parental genes in time and space resulting in formation of physiologically independent descendants which represent a new generation in the demographic sense.	Urbanska, 1985, 1989

I should like to emphasize the need of considering reproduction as the process resulting in formation of active offspring and not completed by the production of propagules alone. In most cases there is no continuous developmental pattern linking seed/vegetative propagule with seedling/plantlet, the two stages of plant life history being separated by more or less distinct cryptobiotic phase (Amen, 1966). A complete development of an offspring individual may be strongly influenced by an enforced dormancy even in truly viviparous plants. This functional discontinuity means that a successful production of dispersal units may not necessarily bring about a reproductive success.

Another doubtful aspect is the origin of the offspring individual from a single preprogrammed cell suggested by Harper (1977) as the sole criterion of reproduction (Table 1). The hypothesis of Harper is certainly correct, but the limitations suggested (zygote) do not stand well a closer scrutiny: the production of a new plantlet or of a specialized vegetative propagule from a meristematic cell seems to be related, too, to a specific information coded in that cell, otherwise the meristeme would have given rise to a new plant organ (e.g. leaf) and not to a characteristic reproductive unit. It is not only the zygote that carries the information required.

The above mentioned aspects considered, I propose to define reproduction as follows:

Reproduction is the biological process resulting in formation of physiologically autonomous offspring individuals.

This definition represents a departing point for further considerations offered in the present paper.

III. What is an Individual?

The traditional concept of selection assumes that the individual is a coherent structural, physiological and genetic unit. This concept holds for many organisms but it does not apply to numerous plants (Eriksson, in press). In unitary organisms an individual represents a distinct level of biological organization, but e.g. in modular organisms (thus majority of higher plants) the conceptual distinction between individual and group selection is not self-evident (Tuomi and Vuorisalo, 1989). Further problems arise when clonal plants are studied because many clones can and do become fragmented and their parts follow then independent fates.

In plant population biology, the **genetic individual** or the genet was usually considered as the basic evolutionary unit (Harper, 1977, 1980, 1981; Janzen, 1977; Cook, 1979; White, 1979; Pitelka and Ashmun, 1985; Jackson and Coates, 1986). The genet is considered by Harper as the product of a single zygote; the sexual origin of the genetic individual is included, too, in the definition proposed by Jackson *et al.* (1985). On the other hand, Silvertown (1987) defined the genet as "a plant of whatever size and structure originating from a seed" and it is not clear whether he associated the seed production uniquely with the sexual process or regarded the propagule type as the determining element in the concept of individual.

According to the opinion of Harper and his followers, the genetic individual can be anything from a single growing plant to a large clone in which the clone members i.e. **ramets** are physically disconnected. The latter situation was termed by Jerling (1985) a "split-up clone".

Harper argued that "the ability of some plant species to form fragmented phenotypes of a single genotype is just one of the variety of successful ways of playing the game of being a plant". This argument, albeit interesting, is not very helpful in studies dealing with plant populations biology or ecology. In particular, demography which accounts for individuals within a population requires a revised approach. The genetic concept of individual is hardly applicable to e.g. monozygotic twins (Randall and Rick, 1945). It is not satisfactory, either, when ramets are separated from each other and their performance has to be evaluated individually (Hartnett and Bazzaz, 1983, 1985a, b; Urbanska *et al.*, 1987, 1988; Tschurr, 1988; Hasler, unpubl.). Perhaps the most extreme cases which demonstrate the insufficiency of the genetic individual concept in the field application are agamospermous plants which produce individually "packed" but uniformly maternal progeny, or the facultatively apomictic plants which may give birth to both biparental and purely maternal offspring individuals in the same generation (see also further parts of the present paper).

The sexual origin was not always considered an obligatory criterion in the concept of individual. For example Hartmann (1957, 1959) regarded both sexual and asexual reproduction as equally important ways of formation of new individuals; his elegant data on vegetative reproduction in high-alpine plants support well this opinion.

Some more recent concepts of individual take at least partly account of the asexual reproduction and specifically of clone fragmentation; for example Cook (1985) distinguished between clonal growth through formation of daughter ramets and reproduction through formation of physiologically independent plants. These concepts give the **physiological autonomy of an individual a priority over its sexual or asexual origin**; the same rang of priorities was emphasized by myself (Urbanska, 1985, 1989) and accentuated, too, in the definition of a "**structural individual**" (Vuorisalo and Tuomi, 1986; Tuomi and Vuorisalo, 1989).

The animated discussion concerning the concept of individual shall undoubtedly go on. Whatever its final outcome, the recent publications and meetings demonstrate that a functional definition of individual is urgently needed in studies on plant population. I propose therefore to **distinguish between a genetic individual** (the genet *sensu* Harper) **and a biological individual.** The former concept is theoretically useful but clearly not applicable in numerous plant groups; the latter one admittedly has some limitations (esp. for clonal populations with well-defined ramets/ramet groups interconnected by subterranean structures) but might be rather widely used in field and experimental studies. I suggest the following definition:

Biological individual is a structurally coherent, physiologically autonomous organism of sexual or asexual origin.

The biological individual can thus be

. single genet or a single ramet

. group or ramets corresponding either to a whole clone or to a clone part.

The biological individual may originate from

. sexually produced seed
. asexually produced seed
. specialized vegetative propagule
. clone fragmentation.

IV. Patterns and Processes of Asexual Reproduction in the Angiosperms

Asexual reproduction in flowering plants includes diverse mechanisms. Both the classification and the terminology are rather arbitrarily used to date. For example, the term "apomixis" originally used by Winkler (1908) in a broad sense is now often limited to design only one aspect of asexual reproduction viz. the agamospermy.

One way to classify the asexual reproduction in the Angiosperms is based on different units of dispersal and survival which also act as the population founders. This biological approach permits to distinguish between agamospermy and vegetative reproduction, the latter group including both the reproduction by specialized vegetative propagules as well as the clone fragmentation (Table 2).

TABLE 2
Asexual reproduction in the Angiosperms

Type of reproduction			Reproductive unit
1.	Agamospermy		seed / fruit
2.	Vegetative reproduction:		
	(a)	reproduction by specialized vegetative propagules	vegetative propagule
	(b)	clone fragmentation	single ramet / ramet group

A. *Agamospermy*

The asexual reproduction by seed results normally in appearance of maternal or metromorphous offspring constisting of genetically identical copies of the mother plant. This characteristic contrast between physiological/structural autonomy of individual plants and the overall genetic uniformity of the whole offspring generation is well demonstrated e.g. in *Hieracium villosum* ($2n$=27) where a single ramet may produce up to several hundred seedlings, all of the same maternal type (Urbanska, unpubl.).

Agamospermy includes some essentially different patterns (Table 3). In **gametophytic apomixis** (Nogler, 1984) the alternation of generations occurs, but there is no alternation of nuclear phases in female function: both the sporophyte and the gametophyte carry the same chromosome number. On the other hand, male function - if present - is mostly characterized by the alternation both on the generation as well as the nuclear phase level, although the pollen production may be sometimes quite irregular and deficient. This pattern is observable both in plant reproducing by **autonomous agamospermy** as well as in **pseudogamous** taxa which require fertilization of the endosperm nucleus for a successful seed development. In **adventitious embryony** the formation of embryo is purely sporophytic, but the endosperm is produced in embryo sacs, autonomously or after fertilization.

TABLE 3
Patterns of asexual reproduction by seed

Type	Alternation of generations	Male contribution	Origin of embryo	Origin of endosperm
1. Gametophytic apomixis:				
(a) autonomous agamospermy	yes	none	maternal	maternal
(b) pseudogamy	yes	endosperm	maternal	biparental
2. Adventitous embryony	embryo: none endosperm: yes	none or endosperm	maternal	maternal or biparental

Autonomous agamospermy consists of two complementary phases viz. 1/ formation of unreduced female gametophyte by displospory or apospory, and 2/parthenogenetic development of the egg cell accompanied by spontaneous development of endosperm (for more details see Gustafsson, 1946-47; Urbanska, 1974; Nogler, 1984). The seed formation is thus entirely dependent on female function and male contribution is superfluous. In spite of this well-defined pattern, numerous autonomously agamospermous plants do produce pollen, sometimes at a considerable expense. The other end of the diversity spectrum is represented by dioecious agamospermous species often known only as female plants and forming characteristic female-biased populations (e.g. Bayer and Stebbins, 1983, 1987; Urbanska, 1984).

Pseudogamy represents a quite different proposition. Formation of unreduced female gametophyte and parthenogenetic development of the egg cell closely resemble the corresponding phases in autonomously agamospermous plants, but male function is indispensable because the endosperm does not develop without fertilization (e.g. Fagerlind, 1944; Gustafsson, 1946-47; Christen, 1950; Rutishauser, 1954, 1965; Skalinska, 1959; Rychlewski, 1961, 1967; Izmailow, 1970; Nogler, 1972, 1984). This important aspect of reproductive biology of pseudogamous plants deserves a special attention because the **pollen donors do not sire the actual progeny**. In this respect, some reports on **male-sterile pseudogamous**

plants which depend on cross-pollination for the production of maternal offspring are particulary intriguing (e.g. Nygren, 1950; Smith, 1963; Rychlewski, 1961, 1967). An additional interesting aspect represents a possible "multiple endosperm paternity" as the endosperm nuclei in pseudogamous plants may occasionally be fertilized by several male gametes and more than one donor may thus be involved.

Adventitious embryony is exceedingly complex from biological point of view. The embryo is formed from a sporophytic cell but grows to full term in the embryo sac and is nourished by the endosperm which may or may not require fertilization (e.g. Osawa, 1912; Afzelius, 1932; Pijl, 1934; Gustafsson, 1946-47). In this group female function obviously plays the main rôle but male function may be important, too, as suggested by the data on e.g. *Nothoscordum fragrans*, diploid *Citrus* forms or *Nigritella nigra.*

B. *Vegetative reproduction*

While the asexual reproduction by seed is sometimes neglected but frequently recognized, the vegetative reproduction remains to date a rather controversial subject. Numerous authors agree that plants are able to reproduce both by sexual and vegetative means, but there is some confusion as to the character and meaning of vegetative reproduction. In some papers vegetative reproduction is referred to but the authors concerned probably have in mind the clonal growth (e.g. Billings and Mooney, 1968); an "other way round" interpretation was suggested by Harper (1977) who discarded completely the notion of vegetative reproduction and argued that it should be regarded as a form of growth.

All these problems may be related, on the other hand, to the great variety of units involved in vegetative reproduction; on the other hand, data on behaviour of vegetatively reproducing plants are largely insufficient and only few groups have been studied in detail to date, the excellent monography of the duckweed family (Landolt, 1986; Landolt and Kandeler, 1987) being one of the rare exceptions. Reproductive biology of vegetatively reproducing plants was handled so far mostly in a descriptive way, quantitative data being frequently only fragmentary. The brilliant study of Kawano and Nagai (1975) in which descriptions match the detailed numerical evaluations represents in this respect a model work for future students.

The tentative classification of vegetative reproduction proposed in Table 4 takes into account some important biological aspects.

Vegetative reproduction in the Angiosperms typically follows precise recurrent patterns, the reproductive cycle being defined by 1/ formation of reproductive units, 2/ their dispersal, and 3/ the subsequent germination/establishment of new plants. Γhe only form of vegetative reproduction which can be considered opportunistic is he enforced clone fragmentation influenced e.g. by water impact or forces of frost and gravity; the latter phenomenon occurs rather frequently in the Arctic and strongly influences the vegetation pattern (Murray, 1987).

Two further elements which are very important in biology of vegetatively reproducing plants are 1/ timing of the release of propagules or daughter ramets and 2/ dormancy or lack of formancy of reproductive units at the time of their dispersal. In this respect vegetative reproduction is quite comparable to the reproduction by seed, sexual and asexual alike.

TABLE 4
Vegetative reproduction

Type	Pattern	Propagule/ramet independence	Dormancy at dispersal
1. Reproduction by specialized propagules	recurrent	early	present
	recurrent	early or retarded	absent
2. Reproduction by clone fragmentation:			
(a) natural	recurrent	early	present or absent*
	recurrent	late	absent
(b) enforced	opportunistic	immediate	absent

* dormancy usually associated with formation of turions

Specialized vegetative propagules formed above or/and underground usually carry nutrients in form of complex carbohydrates (Urbanska, 1981) and most frequently are dormant at the time of their separation from the mother plant. In general, dormant vegetative propagules esp.those formed above ground are released soon after having reached maturity, but non-dormant propagules developing into plantlets may remain attached to flowering shoots of the mother plant for a longer time (e.g. *Poa alpina*, Urbanska, 1985).

Reproduction by natural clone fragmentation which is not enforced by environmental factors but forms an integral part of behaviour of a given species is characterized by formation of single daughter ramet or ramet group and its subsequent separation from the mother plant. In most cases daughter ramets become established before the actual link between them and the mother individual disappears as a result of die-back of the interconnecting structure; however, a spontaneous clone fragmentation accompanied by the biosynthesis of abscisic acid(ABA) is well-known, too, e.g. in the Lemnaceae family. Reproduction by natural clone fragmentation is, to the best of my knowledge, always related to non-dormant offspring except for the formation of turions where the dormancy seems to be induced by particular ecological conditions (see e.g. Urbanska, 1984, 1985).

In some plants which reproduce by vegetative means, a prolonged period of attachment between the mother plant and the daughter ramet(s) is apparently influenced by some developmental insufficiency of the growing offspring individual, more maternal care being thus required. For example, observations on natural populations of *Geum reptans* in the Swiss Alps show that single daughter ramets produced at the end of leafy runners remain linked to the mother plant during at least two years before they become independent; prematurely detached young ramets do not survive (Urbanska, unpubl.). One cannot but think of the mammals offspring that have to be carried to term and are nourished via the umbilical cord until they reach a sufficient developmental stage, premature birth resulting in decease.

It seems therefore that not only agamospermy but also various types of vegetative reproduction are precisely preprogrammed and remain under control of finely-balanced factors. There apparently is nothing left to chance in the formation of offspring individuals by vegetative means.

V. Assessment of Reproductive Strategies in Sexual and Asexual Plants

Reproductive strategies of flowering plants have until recently been considered mostly in terms of resource allocation to reproduction called **reproductive effort.** Reproductive effort may be estimated in three possible ways as 1/ total energy as propagules relative to total energy as starting capital plus gross assimilation, 2/ total energy as propagules relative to total energy as starting capital plus net production and 3/ total weight of propagules relative to total biomass weight at the propagule-bearing stage. The first two evaluations are respectively termed gross and net reproductive effort, whereas the third one equals net reproductive effort with crude reproductive efficiency (Harper and Ogden, 1970).

In a recent paper I proposed another approach, based on performance of reproductive units (Urbanska, 1989). Its components are 1/ reproductive offer RO, 2/ reproductive efficiency EFF and 3/ germination percentage γ. The concept presented and the general notion of reproductive effort *sensu* Harper and Ogden are by no means mutually exclusive but complement one another. While the measurements of reproductive effort indicate a global resource allocation to reproduction, data on e.g. reproductive offer demonstrate how was invested some part of this allocation. In fact, both approaches may be indispensable to a complete assessment of life-history strategies in plants.

The approach proposed was originally worked out for plants reproducing by seed. In the present paper I propose to outline the original concept and to consider its application to various forms of asexual reproduction in the Angiosperms.

A. Definition and evaluation of reproductive offer RO, reproductive efficiency EFF and reproductive success RS

Reproductive offer may be defined as amount of reproductive units at early premating stages. It corresponds to the number of primary pollen grains and the number of ovules, respectively.

Reproductive efficiency is the ability to carry to term viable reproductive units. The terminating point is defined in male function by the onset of pollen dispersal, whereas female reproductive efficiency should be measured at the seed dispersal.

Reproductive success is determined by appearance of active offspring individuals. In male function, reproductive success related to the germination percentage of the seeds fathered by a given individual/population members. Female reproductive success relates to the germination percentage of the seeds produced by a given mother individual/population.

Evaluation formulae are proposed separately for male and female function.

Female reproductive offer ♀RO is determined by number of ovules and number of flowers. It may be calculated for an individual as follows:

$$♀RO_I = \sum_{i=1}^{n_I} ov_i \approx \overline{ov} \cdot \overline{n}_I = \overline{ov}_I \qquad \text{(a)}$$

where

n_I = number of flowers in an individual

$\overline{n}_I$ = average number of flowers per individual

ov_i = number of ovules within the ith flower of the individual

$\overline{ov}$ = average number of ovules per flower

$\overline{ov}_I$ = average number of ovules per individual.

For a population, female reproductive offer $♀RO_P$ can be calculated according to the formula

$$♀RO_P = \sum_{j=1}^{P} ♀RO_{Ij} \approx P \cdot \overline{ov} \cdot \overline{n}_I \qquad \text{(b)}$$

where

P = number of individuals within a given population

$♀RO_{Ij}$ = female reproductive offer of the jth population member.

Male reproductive offer ♂RO can be calculated for an individual with the equation:

$$♂RO_I = \sum_{i=1}^{n_I} pg_i \approx \overline{pg} \cdot \overline{n}_I = \overline{pg}_I \qquad \text{(c)}$$

where

n_I = number of flowers in an individual
$\bar{n}_I$ = average number of flowers per individual
pg_i = number of primary pollen grains within the ith flower of the individual
$\overline{pg}$ = average number of primary pollen grains per flower
$\overline{pg}_I$ = average number of primary pollen grains per individual.

For a population, male reproductive offer ♂ROP can be calculated as follows:

$$♂RO_P = \sum_{j=1}^{P} ♂RO_{Ij} \approx P \cdot \overline{pg} \cdot \bar{n}_I \quad (d)$$

where
P = number of individuals within a given population
$♂RO_{Ij}$ = male reproductive offer of the jth population member.

Female reproductive efficiency ♀EFF i.e. the ability to carry viable seed to term may be expressed by the number of fertile seeds at dispersal relative to the female reproductive offer ♀RO. For an individual, it can be thus calculated by the formula:

$$♀EFF_I = \frac{\sum_{i=1}^{n_I} s_i}{♀RO_I} \approx \frac{\bar{n}_I \cdot \bar{s}}{♀RO_I} = \frac{\bar{s}_I}{\overline{ov}_I} \quad (e)$$

where
n_I = number of flowers in an individual
$\bar{n}_I$ = average number of flowers per individual
s_i = number of seeds within the ith flower of an individual
$\bar{s}$ = average number of seeds per flower
$\bar{s}_I$ = average number of seeds per individual
$\overline{ov}_I$ = average number of ovules per individual.

For a population, female reproductive efficiency $♀EFF_P$ may be considered as an average individual efficiency, thus

$$♀EFF_P = \frac{\sum_{j=1}^{P} ♀EFF_{Ij}}{P} \quad (f)$$

where
P = number of individuals forming a given population
$♀EFF_{Ij}$ = female reproductive efficiency of the jth population member.

Male reproductive efficiency ♂EFF, i.e. the ability to carry viable pollen to term, may be expressed by the number of viable pollen grains at anthesis relative to the male reproductive offer ♂RO; it can thus be calculated for an individual as

$$♂EFF_I = \frac{\sum_{i=1}^{n_I} vp_i}{♂RO_I} \approx \frac{\bar{n}_I . \overline{vp}}{♂RO_I} = \frac{\overline{vp_I}}{\overline{pg_I}} \qquad (g)$$

where

n_I = number of flowers in an individual

$\bar{n}_I$ = average number of flowers per individual

vp_i = number of viable pollen grains within the ith flower of the individual

$\overline{vp}$ = average number of viable pollen grains per flower

$\overline{vp}_I$ = average number of viable pollen grains per individual

$\overline{pg}_I$ = average number of primary pollen grains per individual.

For a population, male reproductive efficiency $♂EFF_P$ may be regarded as an average individual efficiency, thus

$$♂EFF_P = \frac{\sum_{j=1}^{P} ♂EFF_{Ij}}{P} \qquad (h)$$

where

P = number of individuals forming a given population

$♂EFF_{Ij}$ = male reproductive efficiency of the jth population member.

Female reproductive success ♀RS can be calculated for an individual by the formula

$$♀RS_I = ♀EFF_I \cdot \gamma_{MI} \approx \frac{\overline{s_I}}{\overline{ov_I}} \cdot \gamma_{MI} \qquad (i)$$

where

$♀EFF_I$ = female reproductive efficiency of an individual

γ_{MI} = germination percentage of the seeds produced by a given mother individual.

For a population, female reproductive success $♀RS_P$ can be expressed as follows:

$$♀RS_P = ♀EFF_P \cdot \gamma_{MP} \qquad (j)$$

where

$♀EFF_P$ = female reproductive efficiency of a given population

γ_{MP} = germination percentage of the seeds produced within this population.

Male reproductive success ♂RS can be calculated for an individual according to the formula

$$\sigma RS_I = \frac{\sigma EFF_I \cdot s_{FI} \cdot \gamma_{FI}}{\sum_{i=1}^{n_I} vp_i} \approx \frac{\bar{s}_{FI} \cdot \gamma_{FI}}{\overline{pg}_I} \quad (k)$$

where

σEFF_I = male reproductive efficiency of an individual
s_{FI} = number of seeds fathered by a given individual
$\bar{s}_{FI}$ = average number of seeds fathered by a given individual
γ_{FI} = germination percentage of the seeds fathered by a given individual.

For a population, male reproductive success can be expressed as follows:

$$\sigma RS_P = \frac{\sigma EFF_P \cdot s_{FP} \cdot \gamma_{FP}}{\sum_{j=1}^{P} (\sum_{i=1}^{n_I} vp_{ij})} \approx \frac{s_{FP} \cdot \gamma_{FP}}{P \cdot \overline{pg}_I} \quad (l)$$

where

σEFF_P = male reproductive efficiency of a given population
s_{FP} = number of seeds fathered by members of a given population
γ_{FP} = germination percentage of these seeds
vp_{ij} = number of viable pollen grains in the ith flower of the jth population member.

B. *Assessment of reproductive strategies in asexually reproducing Angiosperms*

Reproductive strategies in asexually reproducing flowering plants may be assessed according to the approach proposed above, but some evaluation formulae have to be adapted to particular form of reproduction.

As far as **autonomously agamospermous** plants are concerned, taxa which produce some pollen should be assessed separately as to female and male function, even if there are *no returns on male investment*. Reproductive success is to be evaluated in female function only. The general sequence is therefore as follows:

♀RO → ♀EFF → ♀RS
♂RO → ♂EFF.

In autonomously agamospermous populations consisting exclusively of female (pistillate) individuals, female reproductive parameters equal the global values. The general sequence is therefore

♀RO → ♀EFF → ♀RS.

Reproduction in **pseudogamous** plants is characterized by *hidden returns on male investment* (the offspring is maternal, but male contribution is indispensable to the seed development). The corresponding evaluation formulae are:

♀RO → ♀EFF → ♀RS

♂RO → ♂EFF. ♂RS is not measurable with the current methods.

Plants which form seeds by **adventitious embryony** but produce pollen should be evaluated separately for female and male function. The embryo being purely sporophytic, no female reproductive offer will be evaluated. The corresponding evaluation formula of female reproductive efficiency for an individual is adapted as follows:

$$♀EFF_I = \sum_{i=1}^{n_I} s_i \approx \bar{n}_I \cdot \bar{s} = \bar{s}_I$$

where

n_I = number of flowers in an individual

$\bar{n}_I$ = average number of flowers in an individual

s_i = number of seeds in the ith flower of the individual

$\bar{s}$ = average number of seeds per flower

$\bar{s}_I$ = average number of seeds per individual.

For a population, female reproductive efficiency $♀EFF_P$ may be considered as an average individual efficiency, thus

$$♀EFF_P = \frac{\sum_{j=1}^{P} ♀EFF_{Ij}}{P}$$

where

P = number of individuals forming a given population

$♀EFF_{Ij}$ = female reproductive efficiency of the jth population member.

The general evaluation sequence is:

♀EFF → ♀RS

♂RO → ♂EFF. ♂RS is not measurable with the current methods.

As far as the **vegetative reproduction** is concerned, no specific gender is involved but the offspring production is traditionally viewed as female function. No female reproductive offer will be considered. Both the specialized vegetative propagules as well as daughter ramets are sporophytic; the evaluation formulae should be therefore of the same type as that proposed for assessment of adventitious embryony, only the symbols being adapted as follows for an individual:

$$♀EFF_I = \sum_{i=1}^{n_I} d_i \approx \bar{n}_I \cdot \bar{d} = \bar{d}_I$$

where

n_I = number of reproducing ramets in an individual

$\bar{n}_I$ = average number of reproducing ramets in an individual

d_i = number of daughter ramets/propagules produced by the ith reproducing ramet of the individual

$\bar{d}$ = average number of daughter ramets/propagules per reproducing ramet

$\bar{d}_I$ = average number of daughter ramets/propagules per individual.

For a population, female reproductive efficiency $♀EFF_P$ may be considered as an average individual efficiency, thus

$$♀EFF_P = \frac{\sum_{j=1}^{P} ♀EFF_{Ij}}{P}$$

where

P = number of individuals forming a given population

$♀EFF_{Ij}$ = female reproductive of the jth population member.

Reproductive success can be calculated for an individual according to the formula

$$♀RS_I = ♀EFF_I \cdot \beta_{MI} \quad \text{or} \quad ♀EFF_I \cdot \alpha_{MI}$$

where

$♀EFF_I$ = female reproductive efficiency of an individual

β_{MI} = establishment percentage of daughter ramets produced by a given mother individual

α_{MI} = germination percentage of vegetative propagules produced by a given mother individual.

For a population, female reproductive success $♀RS_P$ can be expressed as follows

$$♀RS_P = ♀EFF_P \cdot \beta_{MP} \quad \text{or} \quad ♀EFF_P \cdot \alpha_{MP}$$

where

$♀EFF_P$ = female reproductive efficiency of a given population

β_{MP} = establishment percentage of daughter ramets produced within this population

α_{MP} = germination percentage of vegetative propagules produced within this population.

The general evaluation sequence is

$$♀EFF \rightarrow ♀RS.$$

VI. Some Biological Features of Sexual and Asexual Reproduction in the Angiosperms

As far as population biology and/or ecology is concerned, sexually and asexually reproducing Angiosperms closely resemble each other in numerous aspects. I propose to compare briefly some important features.

A. Reproductive output

While both the general morphology of reproductive units and the potential reproductive output are primarily determined by genetic factors, the actual number of seeds, specialized vegetative propagules, or daughter ramets apparently remain at least partially under environmental influence. Ecological factors operate both directly via the condition of the mother plant related to the availability of nutrients and other resources, and indirectly via meteorological conditions occurring during the propagule maturation period (for reproduction by seed, see Urbanska and Schütz, 1986). Agamospermous plants behave in this respect like the sexual taxa, seeds developed in less extreme conditions being fuller and germinating much better than those originating from extreme and unpredictable environment (Schütz, 1988, 1989; Urbanska and Schütz, 1986).

The variation in number of specialized vegetative propagules per individual in vegetatively reproducing plants follows the same pattern. For instance, number and size of non-dormant propagules formed in flowering shoots by *Poa alpina* is influenced both by health of the mother plant as well as the air humidity (Urbanska, 1985, and unpubl.). On the other hand, the number of reproductive structures and units in plants reproducing by recurrent clone fragmentation may be mostly influenced by the condition of the mother plant rather than a direct environmental situation (e.g. *Geum reptans*, Hartmann, 1957, 1959; Urbanska, unpubl.).

B. Dispersal

Dispersal patterns in asexually reproducing plants do not offer any specific features that would be different from those observed in sexual populations. Seeds formed by agamospermy disperse via identical mechanisms (compare e.g. anemochory in *Taraxacum alpinum* or *Hieracium villosum* with seed dispersal of sexual taxa belonging to these genera). Specialized vegetative propagules may be monochorous but often show polychory, and do not differ perceptibly in this respect from seeds; for example bulbils in *Polygonum viviparum* may just fall of the mother plant onto the ground below, but they are also dispersed by wind, water, or some alpine animals (Müller-Schneider, 1986; Urbanska, 1985, and unpubl.).

The limited dispersal radius observed e.g. in plants reproducing by some daughter ramets or underground formed bulbs is not exclusive of asexual reproduction since atelechorous sexual plants are, too, characterized by establishment rather than dispersal (Pijl, 1982); for example *Amphicarpum purshii* almost abolished dispersal by relying on its subterranean diaspores originating from cleistogamous flowers (e.g. Cheplick and Quinn, 1982). Incidentally, even non-specialized ramets may sometimes be dispersed over longer distances (note e.g. dispersal of *Spirodela polyrrhizza* by water fowl, Landolt, 1986).

The differences in timing of release of reproductive units from mother plant occurring among asexually reproducing species are comparable, too, to patterns observed in sexual plants where in tachysporous plants ("tachy" = fast) diaspores are set free immediately after maturation, whereas bradysporous plants are characterized by dispersal taking place only after a long delay ("brady" = slow).

C. Dormancy and life-span of reproductive units

The seed dormancy was studied by numerous authors and the classical categories of innate, induced, and enforced dormancy proposed by Harper (see e.g. Harper, 1977) apply both to sexual and agamospermous plants. On the other hand, dormancy occurring in specialized vegetative propagules is so far largely unexplored. It is worth mentioning that vegetative propagules in numerous plants are dispersed in dormant state and possibly have various dormancy types; for example the turion dormancy in *Spirodela polyrrhiza* is exceedingly complex and at least partly induced (see Urbanska, 1984). Dormancy in the bulbils of *Polygonum viviparum* seems to be partly enforced and partly innate as suggested by germination trials carried out in laboratory conditions (Urbanska, 1985, and unpubl.).

The data on life-span of asexually produced reproductive units are very incomplete and the reports deal mostly with seeds of agamospermous plants (Schütz, 1988, 1989). As far as the vegetative propagules are concerned, further data are very desirable. Bulbils of *Polygonum viviparum* survive two years when stored in refrigerator but their germinability decreases gradually, and the tetrazolium test is uniformly negative in the propagules which are about three years old (Urbanska, unpubl.). It should be noted, however, that seeds of sexually reproducing plants may, too, have a rather short life span and the changes due to aging are reflected both in the germination pattern as well as the response to some seed pretreatments (Schütz, 1988, 1989; Urbanska *et al.*, 1988).

D. Colonization and establishment

Asexually produced reproductive units function well as population founders and many of them may have better chances of further development than sexually produced seeds. Vegetative propagules usually have a "good start in life" on account of the resource supply they are carrying, and non-dormant vegetative propagules as well as daughter ramets are photosynthetically active even before the separation from the mother plant. On the other hand, both the sexually and asexually reproducing plants require safe-sites for a successful establishment: safe-sites are indispensable not only to seed development and seedling establishment but also to the success of specialized vegetative propagules and daughter ramets. This important biological aspect is clearly documented in the elegant studies of Schütz (1988, and unpubl.) dealing with a spontaneous immigration of diaspores into experimental plots above timberline: the non-dormant propagules of *Poa alpina* immigrated rapidly into ski run plots where safe-site conditions were simulated, but not a single individual succeed in colonization of non-protected sites. The vegetative reproduction of the young immigrant plants set in very rapidly.

The data discussed above demonstrate that sexual and asexual reproduction do not correspond to two biologically different processes, some patterns of population dynamics being strikingly similar.

VII. Conclusions

The question of asexual reproduction in the Angiosperms should be re-examined. Formation of the offspring by asexual means should be recognized as biologically equivalent to the sexual process. I suggest that

- genetic aspects of a given reproduction type have no direct bearing on spatial and temporal patterns of population dynamics,
- biological function of dispersal and survival does not depend on sexual or asexual origin of a given reproductive unit.

Asexual reproduction is successful and therefore important in numerous cases. It should not be confounded with the clonal growth i.e. **the formation of an interconnected ramet system.** Clonal growth offers potential for reproduction by clone fragmentation but by no means should it be identified with the reproductive process.

References

Afzelius, K. (1932). Zur Kenntnis der Fortpflanzungsverhältnisse und Chromosomenzahlen bei *Nigritella nigra*. *Sv. Bot. Tidskr.* **26**, 365-369.

Amen, R. (1966). The extent and role of dormancy in alpine plants. *Quart. Rev. Biol.* **41**, 271-281.

Bayer, R. J. and Stebbins, G. L. (1983). Distribution of sexual and apomictic populations of *Antennaria parlinii*. *Evolution* **37**, 555-561.

Bayer, R. J. and Stebbins, G. L. (1987). Chromosome numbers, patterns of distribution and apomixis in *Antennaria* (Asteraceae: Inulae). *Syst. Bot.* **12**, 305-319.

Billings, D.W. and Mooney, H. A. (1968). The ecology of arctic and alpine plants. *Biol. Rev.* **43**, 481-529.

Cheplick, G. P. and Quinn, J. A. (1982). *Amphicarpum purshii* and the "pessimistic strategy" in amphicarpic annuals with subterranean fruit. *Oecologia* (Berl.) **52**, 327-332.

Christen, H. R. (1950). Untersuchungen über die Embryologie pseudogamer und sexueller *Rubus*-Arten. *Ber. Schweiz. Bot. Ges.* **60**, 153-198.

Cook, R. E. (1979). Asexual reproduction: A further consideration. *Am. Nat.* **113**, 769-772.

Cook, R.E. (1985). Growth and development in clonal plant populations. *In* "Population Biology and Evolution of Clonal Organisms" (J. B. C. Jackson, L. W. Buss and R. E. Cook, eds), pp. 259-296. Yale Univ. Press, New Haven CT.

Dawkins, R. (1982). "The Extended Genotype. The Gene as the Unit of Selection". Freeman, Oxford.

Eriksson, O. Hierarchical selection in clonal plants. *Sommerfeltia* (in press).

Fagerlind, F. (1944). Die Samenbildung und die Zytologie bei agamospermischen und sexuellen Arten von *Elatostema* und einigen nahestehenden Gattungen nebst Beleuchtungen einiger damit zusammenhändender Probleme. *K. Sven. Vetensk. Handl.* **21**(4), 1-130.
Grime, J. P. (1979). "Plant Strategies and Vegetation Processes". Wiley & Sons, New York.
Gustafsson, A. (1946-47). Apomixis in higher plants. I-III. *Lund Univ. Arskr.* N.F. **42**(3), **43**(2), **43**(12).
Harper, J. L. (1977). "Population Biology of Plants". Academic Press, London.
Harper, J. L. (1980). Plant demography and ecological theory. *Oikos* **35**, 244-253.
Harper, J. L. (1981). The concept of population in modular organisms. *In* "Theoretical Ecology" (R. M. May, ed.), pp. 53-77. Sinauer, Sunderland MA.
Harper, J. L. and Ogden, J. (1970). The reproductive strategy of higher plants. I. The concept of strategy with special reference to *Senecio vulgaris* L. *J. Ecol.* **58**, 681-698.
Hartmann, H. (1957). Studien über die vegetative Fortpflanzung in den Hochalpen. *Jahr. Naturforsch. Ges. Graubündens* **48**.
Hartmann, H. (1959). Vegetative Fortpflanzungsmöglichkeiten und deren Bedeutung bei hochalpinen Blütenpflanzen. *Die Alpen* **55**, 173-184.
Hartnett, D.C. and Bazzaz, F. A. (1983). Physiological integration among intraclonal ramets of *Solidago canadensis*. *Ecology* **64**, 779-788.
Hartnett, D. C. and Bazzaz, F. A. (1985a). The genet and ramet population dynamics of *Solidago canadensis* in an abandoned field. *J. Ecol.* **73**, 407-413.
Hartnett, D. C. and Bazzaz, F. A. (1985b). The integration of neighborhoud effect by clonal genets of *Solidago canadensis*. *J. Ecol.* **73**, 415-427.
Izmailow, R. (1970). Cytogenetic studies in the apomictic species *Ranunculus cassubicus* L. *Acta Biol. Crac. Ser. Bot.* **13**, 37-50.
Jackson, J. B. C. and Coates, A. G. (1986). Life cycles and evolution of clonal (modular) animals. *Phil. Trans. R. Soc. Lond.* B **313**, 7-22.
Jackson, J. B. C., Buss, L. W. and Cook, R. E., eds (1985). "Population Biology and Evolution of Clonal Organisms". Yale Univ. Press, New Haven CT.
Janzen, D. H. (1977). What are dandelions and aphids? *Am. Nat.* **111**, 586-589.
Jerling, L. (1985). Are plants and animals alike? A note on evolutionary plant population ecology. *Oikos* **45**, 150-153.
Kawano, S. and Nagai, Y. (1975). The productive and reproductive biology of flowering plants. I. Life history strategies of three *Allium* species in Japan. *Bot. Mag. Tokyo* **88**, 281-318.
Landolt, E. (1986). The family of Lemnaceae - a monographic study. I. *Veröff. Geobot. Inst. ETH, Stiftung Rübel, Zürich* **71**.
Landolt, E. and Kandeler, R. (1987). The family of Lemnaceae - a monographic study. II. *Veröff. Geobot. Inst. ETH, Stiftung Rübel, Zürich* **95**.
Müller-Schneider, P. (1986). Verbreitungsbiologie der Blütenpflanzen Graubündens. *Veröff. Geobot. Inst. ETH, Stiftung Rübel, Zürich* **85**.
Murray, D. F. (1987). Breeding systems in Arctic flora. *In* "Differentiation Patterns in Higher Plants" (K. M. Urbanska, ed.), pp. 239-262. Academic Press, London.
Nogler, G. A. (1972). Genetic der Aposporie bei Ranunculus auricomus. II. Endospermzytologie. *Ber. Schweiz. Bot. Ges.* **82**, 54-63.
Nogler, G. A. (1984). Gametophytic apomixis. *In* "Embryology of Angiosperms" (B. M. Johri, ed.), pp. 475-518. Springer-Verlag, Berlin.
Nygren, A. (1950). Cytological and embryological studies in arctic *Poae*. *Symb. Bot. Ups.* **17**, 1-105.
Osawa, J. (1912). Cytological and experimental studies in *Citrus*. *Journ. Coll. Agric. Imp. Univ. Tokyo* **4**, 83-116.
Pijl, van der, L. (1934). Über die Polyembryonie bei *Eugenia*. *Rec. Trav. Bot. Neerl.* **31**, 113-187.
Pijl, van der, L. (1982). "Principles of Dispersal in Higher Plants". Springer-Verlag, Berlin.

Pitelka, L. F. and Ashmun, J. W. (1985). Physiology and integration of ramets in clonal plants. *In* "Population Biology and Evolution of Clonal Organisms" (J. B. C. Jackson, L. W. Buss and R. E. Cook, eds), pp. 399-435. Yale Univ. Press, New Haven CT.
Randall, T. E. and Rick, C. M. (1945). A cytogenetic study of polyembryony in *Asparagus officinalis* L. *Am. J. Bot.* **32**, 560-569.
Rychlewski, J. (1961). Cyto-embryological studies in the apomictic species *Nardus stricta* L. *Acta Biol. Crac. Ser. Bot.* **4**, 1-23.
Rychlewski, J. (1967). Karyological studies on *Nardus stricta* L. *Acta Biol. Crac. Ser. Bot.* **10**, 55-72.
Rutishauser, A. (1954). Die Entwicklungserregung des Endosperms bei pseudogamen *Ranunculusarten. Mitt. Naturforsch. Ges. Schaffhausen* **25**, 1-45.
Rutishauser, A. (1965). Genetik der Pseudogamie bei *Ranunculus auricomus* s.l. W. Koch. *Ber. Schweiz. Bot. Ges.* **75**, 157-182.
Schütz, M. (1988). Genetisch-ökologische Untersuchungen an alpinen Pflanzenarten auf verschiedenen Gesteinsunterlagen: Keimungs- und Aussaatversuche. *Veröff. Geobot. Inst. ETH, Stiftung Rübel, Zürich* **99**.
Schütz, M. (1989). Keimverhalten alpiner Compositae und ihre Eignung zur Wiederbegrünung von Skipistenplanierungen oberhalb der Waldgrenze. *Ber. Geobot. Inst. ETH, Stiftung Rübel, Zürich* **55**, 131-150.
Silvertown, J. W. (1987). "Introduction to Plant Population Ecology". Longmann Sci. Techn. Burnt Mill, Harlow.
Skalinska, M. (1959). Embryological studies in *Poa granitica* Br. Bl., an apomictic species of the Carpathian range. *Acta Biol. Crac. Ser. Bot.* **1**, 91-112.
Smith, G. L. (1963). Studies in *Potentilla* L. I. Embryological investigations into the mechanism of agamospermy in British *P. tabernaemontani* Schers. *New Phytol.* **62**, 264-282.
Tschurr, F. R. ((1988). Zur Regeneration bei einigen alpinen Pflanzen. *Ber. Geobot. Inst. ETH, Stiftung Rübel, Zürich* **54**, 111-140.
Tuomi, J. and Vuorisalo, T. (1989). What are the units of selection in modular organisms? *Oikos* **54**, 227-233.
Urbanska, K. M. (1974) L'agamospermie, système de reproduction important dans la spéciation des Angiospermes. *Bull. Soc. Bot. fr.* **121**, 329-346.
Urbanska, K. M. (1981). Reproductive strategies in some perennial angiosperms. *Viert. Jahr. Naturforsch. Ges. Zürich* **126**, 269-284.
Urbanska, K. M. (1984). Plant reproductive strategies. *In* "Plant Biosystematics" (Grant, W.F., ed.), pp. 211-228. Academic Press, Canada.
Urbanska, K. M. (1985). Some life history strategies and population structure in asexually reproducing plants. *Bot. Helv.* **95**, 81-97.
Urbanska, K. M. (1989). Reproductive effort or reproductive offer? - A revised approach to reproductive strategies of flowering plants. *Bot. Helv.* **99**, 49-63.
Urbanska, K. M. and Schütz, M. (1986). Reproduction by seed in alpine plants and revegetation research above timberline. *Bot. Helv.* **96**, 43-60.
Urbanska, K. M., Hefti-Holenstein, B. and Elmer, G. (1987). Performance of some alpine grasses in single-tiller cloning experiment and in the subsequent revegetation trials above the timberline. *Ber. Geobot. Inst. ETH, Stiftung Rübel, Zürich* **53**, 64-90.
Urbanska, K. M., Schütz, M. and Gasser, M. (1988). Revegetation trials above the timberline - an exercise in experimental population ecology. *Ber. Geobot. Inst. ETH, Stiftung Rübel, Zürich* **54**, 85-110.
Vuorisalo, T. and Tuomi, J. (1986). Unitary and modular organisms: criteria for ecological division. *Oikos* **47**, 382-385.
White, J. (1979). Plant as a metapopulation. *Annu. Rev. Ecol. Syst.* **10**, 109-145.
Willson, M. J. (1983). "Plant Reproductive Ecology". Wiley & Sons, New York.
Winkler, H. (1908). Über Parthenogenesis und Apogamie im Pflanzenreiche. *Progr. Rei. Bot.* **2**, 293-454.

17 The Demographic Consequences of Sexuality and Apomixis in *Antennaria*

PAULETTE BIERZYCHUDEK

Department of Biology, Pomona College, Claremont, CA 91711, USA and Rocky Mountain Biological Laboratory, Gothic, CO 81224 , USA

I. Introduction

Since its discovery in 1840 (Gustafsson, 1946a), apomixis has been the focus of considerable attention from botanists. Despite its status as only "an occasional aberration" in higher plants (Stebbins, 1950), the study of apomictic complexes has provided many keys to our understanding of evolution in plants in general. Babcock and Stebbins (1938) conducted a landmark study of variation and evolution in *Crepis*, which was followed by Gustafsson's (1946a,b, 1947) thorough, and still authoritative, review. Agamic complexes continue to be the focus of many biosystematic studies, probably because they pose extremely challenging problems for the biosystematist.

But about 10 years ago, a different group of biologists began to show an interest in apomictic taxa: population biologists and evolutionary biologists. Their concern with these groups was generated by a revival of interest in an old evolutionary question: why is sexual reproduction so prevalent? The accepted answer had been that asexual groups, because they lacked the ability to generate genetic variation except through mutation, would be more likely than sexual groups to go extinct when faced with environmental change (Fisher, 1930; Muller, 1932; Crow and Kimura, 1965). Beginning in the 1960's, this explanation began to be challenged, because it depended on natural selection working at the level of groups rather than on individuals (Williams, 1966, 1975; Stearns, 1987). This challenge was provoked by the recognition that, because asexual individuals produce no sons and expend little or no energy on male function (an expense often termed "the cost of sex"), they ought to be able to produce up to twice as many daughters as sexual individuals can, and

BIOLOGICAL APPROACHES AND EVOLUTIONARY TRENDS IN PLANTS ISBN 0-12-402960-4

thus increase rapidly at the expense of sexual individuals (Williams, 1975; Maynard Smith, 1978; Lloyd, 1980; Uyenoyama, 1984). In the face of this twofold advantage for asexual individuals, the group selection explanation was judged insufficient (Stearns, 1987). Attention turned from "why are so many species sexual?" to "why doesn't a mutation permitting asexual reproduction rapidly spread through a population of sexual individuals?"

Theoreticians met this challenge by offering a variety of possible advantages that might be gained by individuals producing genetically variable offspring, relative to individuals producing genetically uniform offspring. These explanations can be summarized as follows (see Bierzychudek, 1987b and references therein for details):

1. One possible advantage of sexual reproduction is that it generates rare genotypes (the minority advantage hypothesis). Levin (1975) first pointed out that predators, pathogens, and parasites would be expected to decrease the fitness of common genotypes to a greater extent than that of rare ones. This idea has been expanded upon by many others (e.g. Jaenike, 1978; Glesener, 1979; Hamilton, 1980, 1982; Hamilton *et al.*, 1981; Hutson and Law, 1981; Bell, 1982; Tooby, 1982; Rice, 1983; Bremermann, 1987), who have described the advantage of rarity as either reduced apparency to visual predators or as escape from exploitation by parasites and pathogens through the possession of novel defenses.

2. If the intensity of competition between two individuals is correlated with the degree of genetic similarity between them, then, when siblings compete with each other, a group of genetically dissimilar (sexual) siblings should have reduced competition, greater probabilities of survival, and better performance than a group of genetically identical (asexual) siblings (the sibling competition hypothesis [Williams, 1975; Maynard Smith, 1978; Bulmer, 1980; Young, 1981; Price and Waser, 1982; Barton and Post, 1986]).

3. Many different hypotheses for the advantage of sex contend that the production of genetically variable offspring will be the most successful strategy in spatially and/or temporally variable environments. Most of these hypotheses, termed "lottery models", depend on some biologically rather unlikely assumptions about the nature of that variability (see Bierzychudek, 1987b and references therein).

In light of these hypotheses, the comparative population biologies of asexual/sexual taxa suddenly became important. In 1978, Glesener and Tilman reviewed geographical patterns of parthenogenesis, with the hope that those patterns would help to differentiate among these hypotheses. One of the observations they made was that parthenogenetic taxa nearly always had more extensive ranges, ranging to higher latitudes and higher elevations than their sexual counterparts, a pattern that was termed "geographical parthenogenesis" by Vandel (1928, 1940). Because the twofold advantage of parthenogenetic individuals only manifested itself in marginal areas where, they assumed, biotic interactions were minimal, Glesener and Tilman (1978) concluded that this pattern provided support for the notion that such biotic interactions (i.e. the minority advantage hypothesis) were the most important selective factors promoting sexual reproduction. In 1982, Bell published an extensive review of geographic and ecological patterns of parthenogenesis, and interpreted the pattern similarly.

Both reviews ignored plants entirely, an omission remedied by Bierzychudek (1987a), who concluded that the same trends identified by Glesener and Tilman (1978) and Bell (1982) could be found among plants. I pointed out, however, that the observed patterns of geographic parthenogenesis are consistent with at least two other explanations (Bierzychudek, 1987a). First, in both animals and plants, breeding system differences are confounded with ploidy level differences. Asexual taxa are nearly always polyploid, while their sexual progenitors are typically diploid. It may be polyploidy rather than asexuality that permits occupancy of extreme environments. Indeed, it has long been recognized that polyploids increase in frequency with latitude (de Wet, 1980). Second, asexual forms may more readily colonize new environments, because of their ability to reproduce uniparentally. This ability may also allow them to expand their ranges more rapidly than sexual forms can.

In order to use data from sexual and asexual plants or animals to test hypotheses about the advantages of sex, and in order to understand patterns of geographic parthenogenesis, we need to know much more about the comparative population biologies of closely related sexual and asexual taxa. In particular, we need to know: 1) what are the relative rates of seed production of the two forms? Is there really a twofold "cost of sex"? 2) what are the geographical and ecological patterns of distribution of the two forms? and 3) what differences -- ecological and physiological -- might exist between the two forms that might help to explain their distributional patterns? This paper attempts to provide such a comparison for an agamic complex I have been studying for 6 years: *Antennaria parvifolia* (Asteraceae).

II. Study Organism

Antennaria Gaertner is a genus of perennial herbs distributed primarily throughout the temperate and arctic regions of the northern hemisphere (Bayer and Stebbins, 1987). Many of the taxa form agamic complexes; the sexual species are dioecious. The biosystematic relationships among the North American taxa have been studied by Bayer (1984, 1985a,b), Bayer and Crawford (1986) and Bayer and Stebbins (1982, 1987).

I have chosen to focus on *A. parvifolia*, an agamic complex that includes both apomictic and sexual forms. The two forms are morphologically identical (except that apomictic individuals are never staminate) and both are highly polyploid, ranging from octoploid to decaploid (Bayer and Stebbins, 1987). The sexual individuals are obligately outcrossed, because they are dioecious, and in this group apomixis appears to be obligate as well (Bierzychudek, pers. obs.). Apomixis in *Antennaria* is ameiotic (Stebbins, 1932), and so the progeny of apomictic individuals are normally genetically identical to one another and to their mothers.

A. parvifolia ranges over much of western North America, from the northern borders of Alberta and Saskatchewan to southern Arizona and New Mexico, and from central Nevada east nearly to the Missouri River (Bayer and Stebbins, 1987). Bayer and Stebbins (1987) have collected this group over much of its range and have demonstrated that populations north of the Canadian border are exclusively apomictic. My own studies have concentrated more intensely on populations in central Colorado, the center of the range and the probable place of origin of the genus (Bayer and Stebbins, 1987). Here *A. parvifolia* grows in alpine meadows and montane meadows and forests. Populations at and below 2,750 m are either mixtures of sexual and apomictic individuals or else consist solely of sexual males and females. Sexual individuals are uncommon above 2,750 m, however, and are quite rare above 3,100 m. By contrast, apomictic populations are common even as high as 3,500 m (Bierzychudek, pers. obs.). Much of the data reported here come from a population of mixed sexual and apomictic individuals at Cement Creek, Colorado, in Gunnison Co., USA, a sagebrush meadow at an elevation of approximately 2,740 m.

III. Demographic Differences Between the Two Forms

Bayer and Stebbins (1987) used the frequency of staminate individuals as a rough indicator of whether a particular population comprised sexuals, apomicts, or both. In our own studies, it has been crucial to distinguish between sexual and apomictic individuals, which we have done by determining whether seed set occurs in capitula protected from pollen transfer by dialysis tubing (for details see Bierzychudek, 1987c).

A. *Seed production*

For *Antennaria parvifolia*, there is a clear and significant "cost of sex". Apomictic females produce significantly more viable seeds per capitulum (and thus more seeds per individual) than sexual females do (Table 1). Indeed, because half of the sexual females' seed will produce sons, apomictic females typically produce more than 2 times as many daughters per year as sexual females do. There are several reasons for this difference in seed production. First, apomictic females produce significantly more ovules per capitulum than sexual females do (Table 1). Thus, even if every sexual ovule were fertilized, apomicts would have the potential to make more seed. Secondly, on average, less than half of a sexual female's ovules are fertilized (Table 1), which depresses the seed production of sexual females even further relative to apomicts. Not all of an apomictic female's ovules develop into mature seed

TABLE 1

Mean numbers of ovules and mean numbers of viable seed per capitulum, +/– the standard deviation of the mean. Means represent counts from 2 to 6 flowering stems (depending on availability) from each individual; numbers of individuals in parentheses. Significance determined by a Student's t-test, * = *P* < .05, ** = *P* < .01

	total ovules			viable seeds		
yr	sexual	apomictic	sig.	sexual	apomictic	sig.
1986	78.3±29.7(52)	88.8±18.5(19)	**	30.5±22.7(52)	40.6±18.2(19)	**
1987	63.5±21.2(55)	75.7±24.5(29)	**	28.5±19.9(55)	40.9±22.6(29)	**
1988	61.1±19.6(54)	66.4±17.4(28)	*	33.0±20.3(54)	37.9±18.3(28)	*

either; water stress at the time of ovule and seed production limits the reproduction of apomicts (Bierzychudek, pers. obs.). This resource limitation, however, takes effect at levels of fecundity that sexual females rarely achieve.

Why are so few of the sexual females' ovules fertilized? At Cement Creek, where these comparisons were made, plants grow quite densely; the average distance from a sexual female to the nearest male is 1.32 m (*s.d.* = 0.94, N = 45). Despite their proximity to male individuals, the seed production of sexual females appears to be pollen-limited. Indirect evidence of pollen limitation comes from observations of pollinator behavior on staminate and pistillate capitula. At Cement Creek, *A. parvifolia* is one of the first plants to bloom, in late May. It attracts a wide variety of flower visitors, including flies (which comprise over half the visitors), wasps and solitary bees, butterflies and moths, and beetles. These flower visitors avoid pistillate capitula, which lack nectar as well as pollen (Bierzychudek, 1987c). When visitors do land on pistillate heads, they remain there for a much briefer time than they do on staminate heads (Bierzychudek, 1987c).

To gather more direct evidence that seed production by sexual females might be pollen-limited, we conducted experimental hand-pollinations. Seventeen sexual pistillate clones were chosen that had been divided naturally into at least two physically-independent portions through decay of the rhizome connections between them (since plants can translocate resources from unpollinated to pollinated stems, this physiological separation is important; see Stephenson, 1981). Beginning midway through the blooming period, inflorescences of one portion of each clone (6-7 inflorescences) received supplemental pollen from 3 to 4 staminate plants located within 3 m (the staminate capitula themselves were used as "brushes"). Pollinations were repeated every day until the end of the flowering period, which lasts approximately 2 weeks. Control inflorescences were left unmanipulated, except that they were brushed with a camels'-hair brush at the same time as a control for the physical effect of pollen delivery. When seeds were mature, experimental and control inflorescences were collected and the numbers of viable seeds in one capitulum from each of 5 randomly-chosen stems from each treatment were counted.

Data were analyzed by pairing each plant's experimental and control data and performing a Wilcoxon signed-ranks test. Hand-pollinated females produced over 50% more seeds than their controls, a significant difference ($N = 17$, $P = .006$). This result confirms quite dramatically the hypothesis that the low seed production of sexual females is limited by the amount of pollen they receive.

The significance of this difference for discussions about the adaptive significance of sexual reproduction is considerable. Theory predicts that asexual and sexual females should have equal amounts of energy to devote to reproduction, but that, because sexual females must produce sons as well as daughters (or expend energy on male as well as female function), sexual females will produce only half as many daughters as asexual females, and therefore will have only half their intrinsic rate of increase. Because of pollen limitation, the "cost of sex" in *A. parvifolia* is considerably higher than theory predicts. If half of a sexual female's progeny are male, then sexual females produce only 35 - 43% as many daughters as apomictic females (see Table 1). Explaining how sexual and apomictic females can coexist thus requires a very large fitness advantage for those sexually-produced seeds.

Because many apomicts are found in extreme or marginal environments, pollen limitation is a very real possibility for many of them; this would increase the "cost of sex" above theoretical expectations, as is the case for *A. parvifolia* and for the only two other species that have been studied. Michaels and Bazzaz (1986) found that apomictic populations of *Antennaria parlinii* produced significantly more seeds/individual than sexual populations did. And in *Boehmeria*, which has both sexual and apomictic forms, the apomicts produce about 3 times as many seeds as the sexual individuals (Yahara, 1990). On the other hand, many apomicts retain pollen production, either because they are pseudogamous and require pollen for fertilization of the endosperm, or else because mutations for the loss of male function have not occurred. For these taxa, the "cost of sex" may be lower than theory predicts, because apomicts are still investing energy in male function. More information on the actual "cost of sex" would clearly be very valuable to population biologists interested in the maintenance of sexual reproduction.

B. *Causes of differences in geographical ranges*

What forces are responsible for the differential geographical and ecological distributions of sexual and apomictic *Antennaria*? I have tested two non-mutually-exclusive hypotheses. First, because apomictic individuals can produce seed without fertilization, even a single individual is capable of establishing a population. Thus, apomictic individuals are expected to be better colonizers (Bayer and Stebbins, 1983). The significant association between possession of apomictic breeding systems and occupancy of formerly-glaciated areas (Bierzychudek, 1987a) may be explained by apomicts' being the first to colonize the newly-opened areas made available after glacial retreat (Bayer and Stebbins, 1983). The second hypothesis proposes that apomictic individuals might be more likely to possess what Baker (1965) called "general-purpose genotypes" (see also Lynch, 1984). There are two reasons to expect apomicts to possess more robust physiologies. First, apomictic individuals are often created by hybridization (Stebbins, 1950), which could endow them with some of the physiological traits of both parental lines (Clausen *et al.*, 1946, 1947). Second, because obligately apomictic individuals do not undergo recombination, selective forces select among clones; apomictic clones are reproduced intact over many generations, whereas a sexual genotype never outlives an individual. Though sexual individuals tolerant of environmental conditions cannot pass that (presumably polygenic) trait to their offspring, tolerant apomictic individuals will certainly do so. Narrowly-adapted asexual clones will arise, but will go extinct relatively rapidly (Templeton, 1982), while apomictic clones with physiologies permitting them to tolerate environmental extremes will survive over many generations. Thus, in the long run, apomictic individuals ought to be more likely to possess genotypes that are "general-purpose".

1. Colonization ability

To compare the ability of sexual and apomictic individuals to found new populations, I selected 4 sites near the Rocky Mountain Biological Laboratory in Gothic Co. that were above the elevational limit of sexual *A. parvifolia*. At each site I transplanted apomictic individuals, sexual females, and sexual males. The plants (vegetative divisions of adult plants) were planted in blocks of 2, 4, or 8 individuals. Each block consisted either entirely of apomictic females or else of equal numbers of sexual males and females, planted 5 cm from one another; different blocks were separated from one another by 5 m. Blocks of different sizes and sexual compositions were alternated with one another; each site contained 1 block of 8 individuals, 2 blocks of 4, and 4 blocks of 2 individuals of each breeding system. Each block in this design represented a new colony, so that I could ask: "does the success of a colony in founding a new population depend on a) the number of individuals it contains, and b) the breeding system of those individuals?". I used numbers of viable seeds produced as a measure of success.

Because not all plants survived or flowered (there were no differences between sexual and apomictic females in their probabilities of survival or flowering), block size as originally planted was not a very accurate indication of the number of individuals in a colony. Therefore, I asked whether the number of seeds a plant produced depended on the number of flowering stems (male and female) within its colony, and whether the answer to this question was different for sexual and apomictic females. The colonization hypothesis predicts that the seed production of apomictic plants will be independent of the numbers of its neighbors, while that of sexual plants will be strongly dependent on its neighbors (especially since these small colonies are likely to be very strongly pollen-limited).

The results supported the colonization hypothesis (see Fig. 1). For sexual females, the regression of seed numbers on number of flowering stems in the colony was positive and significant (the regression equation was: viable seeds = 2.06(number of flower stems in block) - 7.30; $F_{1,17} = 9.13$, $P = .008$). For apomictic females, there was no relationship between number of seeds produced and number of flowering stems in the block ($F_{1,18} = 1.39$, $P = .25$). Morever, sexual females produced only 15.6 seeds per individual, while apomictic females produced on average 71.2 viable seeds per individual. To compare these figures with the seed production rates of the two kinds of females in well-established populations (e.g. Table 1), I have divided them by the average number of stems per individual and the average number of capitula per stem. The sexual females in this experiment produced a mean of 1.37 seeds/capitulum, the apomicts 7.73 seeds/capitulum (compare with Table 1). While both kinds of females show sharply reduced fecundity in comparison with Cement Creek, perhaps because these transplants are not as vigorous as long-established plants, or because these sites are not as favorable for growth as their home sites, the fecundity of sexual females is much more reduced than that of apomicts. The seed production ratio at Cement Creek ranged from 1.15 to 1.44, depending on the year; the ratio among these transplants is 5.64. Clearly, the extent of pollen limitation is much more pronounced in these "pioneer" populations. In the face of these extremely low levels of seed production, the probability of establishment of a new sexual population is very much lower than that for a new apomictic population.

2. General-purpose genotypes

I tested the hypothesis that apomictic individuals were more likely to possess general-purpose genotypes with both laboratory and field experiments. For the laboratory experiment, apomictic and sexual pistillate clones of *A. parvifolia* were collected from the Sawatch and West Elk Mountains in central Colorado and taken to the Phytotron at Duke University, where they were divided into replicates (see Bierzychudek, 1989, for methodological details) and distributed among 6 environmental chambers representing all pairwise combinations of 3 temperature regimes and 2 moisture regimes. The photoperiod was chosen to represent summer growing conditions in Colorado and the temperature regimes to represent limits somewhat higher and lower than those encountered by plants at relatively low and high elevations, respectively. Each genotype was represented by 2 to 6 replicates in each environment. Plants were allowed to grow under these conditions for 6 months, censused weekly for mortality and flowering, then harvested, dried, and weighed.

To compare the response of the sexual and the apomictic genotypes to their environmental conditions, I analyzed 3 measures of performance: 1) survival (the probability of living until the experiment was terminated), 2) flowering (mean number of capitula produced in each environment by survivors), and 3) biomass (the mean dry weight of both above- and below-ground parts attained by each genotype). For survival, a particular genotype's "sensitivity" to its environmental conditions (Falconer, 1981) was calculated as the geometric mean of its performance across the six environments. If each environment is assumed to represent a climatologically different year, the geometric mean is equal to the probability that a genotype will survive that series of 6 years. For flowering and biomass, sensitivity of each genotype was measured as the coefficient of variation of its performance across the 6 environments (using the coefficient of variation assures that the measure of variation will be independent of the sample mean, to avoid confusing high sensitivity with high average performance). The general-purpose genotype hypothesis predicts that the sensitivity (i.e. coefficient of variation) of apomictic genotypes will be significantly lower than that of sexual genotypes.

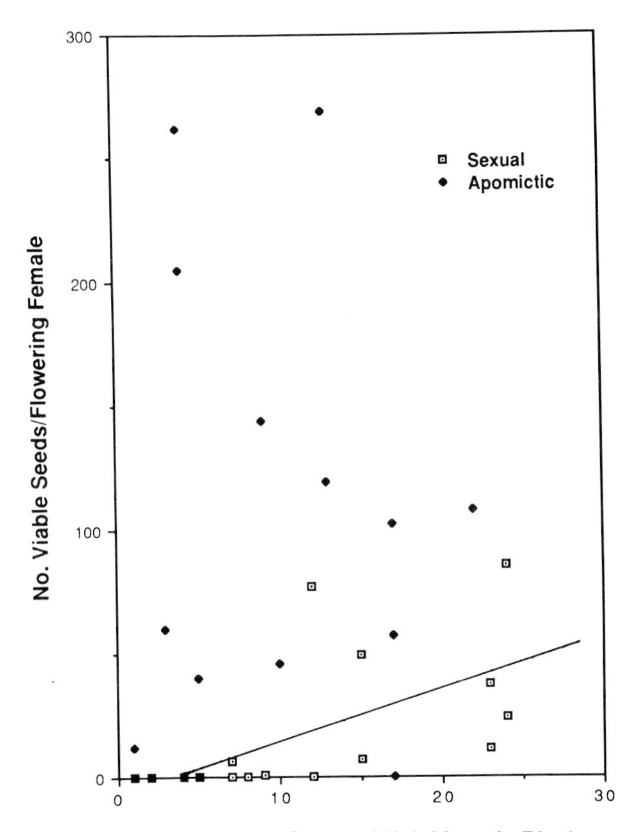

Fig. 1 Relationship between the number of flowering stems in a block (see text) and the number of viable seeds produced by flowering females in that block. Sexual females are represented by open squares, apomictic females by closed diamonds; closed squares are points occupied by both a sexual and an apomictic female. The diagonal line represents the regression of seed number on flower stems of neighbors for sexual plants (viable seeds = 2.06[number of flower stems in block] - 7.30; $F_{1,17} = 9.13$, $P = .008$). The regression for apomictic plants was not significant.

This experiment's results supported the hypothesis that apomictic genotypes were significantly more likely to be "general-purpose" than sexual genotypes were. For survival, the geometric mean was significantly higher for the apomictic genotypes (see Table 2). This result was due in part to the failure of many of the sexual genotypes to survive the hot-dry environment. The coefficient of variation of biomass across the 6 environments was also significantly lower for apomictic genotypes than for sexual genotypes (Table 2), i.e. the ability of the average apomictic genotype to achieve a high biomass was less dependent on its environmental conditions than was the case for the average sexual genotype. There were no significant differences (Table 2) between the two groups with respect to the coefficient of variation for flowering across the different environments, but this was largely because half of the sexual genotypes failed to flower in any of the environments, and thus could not be included in the test (see Bierzychudek, 1989, for details). Finally, not only were the apomictic genotypes less sensitive to the qualities of their environment, but they also achieved higher average performances: mean survival, flowering, and biomass of apomicts all significantly exceeded that of sexuals (see Table 2 and Bierzychudek, 1989, for details).

For the field experiment, sexual and apomictic females of *A. parvifolia* (different genotypes from those in the previous experiment) were cloned and planted into the field in Colorado at 4 different sites that differed in elevation, and therefore temperature and aridity, with 5 replicates of each genotype at each site: Cement

TABLE 2

Summary of results of laboratory experiment to test the "general-purpose genotype" hypothesis. Means are the average means for each breeding system, calculated from the mean performance of each genotype (n in parentheses) across all 6 environments. Coefficients of variation are the average coefficients of variation for each breeding system, calculated from the variation in performance of each genotype across all environments. Test of significance Mann-Whitney non-parametric test

	sexuals	apomicts	diff
mean survival	0.76 (11)	0.88 (14)	A>S $P = .002$
geom. mean survival	0.07 (11)	0.73 (14)	A>S $P = .0002$
mean reproduction (capitula no.)	0.18 (11)	1.04 (14)	A>S $P = .004$
c.v. reproduction	1.80 (5)	1.79 (14)	A=S
mean plant size (biomass in g)	1.74 (11)	2.55 (14)	A>S $P = .011$
c.v. plant size	0.82 (11)	0.58 (14)	A>S $P = .0003$

Creek (2740 m, relatively warm and arid); Kettle Ponds (2877 m, relatively cool and moist); 401 trail (3000 m, cool and arid); and Schofield Park (3125 m, cold and moderately arid). Except for the Cement Creek site, the transplant sites were above the elevational limit for sexual plants. Plants were censused yearly for 1) survival, 2) number of flower stems produced, and 3) area of plant cover. This experiment has been ongoing for 3 years, and none of the plants have yet been harvested. As was the case in the previous study, apomictic and sexual genotypes were compared by calculating their geometric means of survival and their coefficients of variation across the 4 transplant sites for number of inflorescences produced and area of plant cover in 1986, 1987, and 1988.

The results for the field experiment were quite different from the results obtained in the growth chamber experiment. While trends were typically in the same direction as the laboratory experiment, there were no significant differences between geometric means of survival or between the coefficients of variation (across the 4 sites) for sexual and apomictic genotypes for any year or performance measure. Furthermore, there were no significant differences between the mean performances of sexual and apomictic plants for any of these measures. Table 3 presents the results for 1988, when the maximum differences should be apparent.

Why did plants grown under field conditions not display the large, significant differences between apomictic and sexual genotypes so striking in the laboratory experiment? One possible explanation is that the 4 different field environments did not differ from one another as much as did the 6 growth chamber environments;

TABLE 3

Summary of results of field experiment to test the "general-purpose genotype" hypothesis. Means are the average means for each breeding system, calculated from the mean performance of each genotype (n in parentheses) across all 4 environments. Coefficients of variation are the average coefficients of variation for each breeding system, calculated from the variation in performance of each genotype across all environments. Test of significance Mann-Whitney non-parametric test

	sexuals	apomicts	diff
mean survival	0.53 (9)	0.54 (11)	A=S
geom. mean survival	0.27 (9)	0.31 (11)	A=S
mean reproduction (inflorescence no.)	1.30 (9)	1.00 (11)	A=S
c.v. reproduction	99.4 (9)	87.7 (11)	A=S
mean plant size (area of cover in cm^2)	20.10 (9)	18.85(11)	A=S
c.v. plant size	53.1 (9)	41.5 (11)	A=S

extremes of temperature and moisture were not quite as great, and the field environments resembled one another in a variety of biotic selective factors, such as competition from other plants. Differences in environmental sensitivity may only be apparent when plants are subjected to greater environmental extremes, such as might occur in a climatologically aberrant year, once in every decade or more.

But the two experiments are different not only with respect to the *variance* in performance of sexuals and apomicts; mean performance differed as well. Though mean performance of apomicts significantly exceeded that of sexuals in the growth chamber, this was not the case in the field. For survival, this can be explained as follows: in the growth chamber, plants died because they were physiologically incapable of coping with their particular environmental conditions; drought stress was a frequent cause of death. In the field, most deaths occurred when plants were buried by the burrowing of pocket gophers, which were abundant at all 4 sites. Such deaths are not selective; they occur randomly with respect to plant genotype. When such a high proportion of plant mortality is non-selective, very large sample sizes are necessary to detect any mortality differences between apomicts and sexuals that are genetically-based. Though this experiment's sample sizes were too small to detect any such differences, they nevertheless might be very important in nature.

IV. Conclusion

For *Antennaria parvifolia,* a species in which the apomictic form has a much more extensive range than the sexual, apomictic plants produce more than twice as many daughters as sexual females do, because sexual seed production is pollen-limited. Thus, because of their lack of dependence on pollinators to achieve fertilization, apomictic individuals of *A. parvifolia* are significantly more likely than sexual individuals to found new colonies. Finally, apomictic individuals are, under at least some circumstances, significantly less sensitive to environmental extremes. All these differences likely contribute to their more extensive ranges, especially to their exclusive occupancy of high latitude and high elevation sites. It is critical to know how general these results might be. If other apomictic taxa display similar abilities, the phenomenon of "geographical parthenogenesis" becomes much easier to explain, but explaining the persistence of sexual reproduction becomes an even more difficult problem. Future tests of hypotheses for the adaptive significance of sexual reproduction will have to take account of these differences in the biologies of sexual and apomictic taxa.

Acknowledgements

Field and laboratory work on *Antennaria* has been supported by several grants from the National Science Foundation (RII-8310325, BSR-8407468, and BSR-8706318) and the Seaver Science Research Fund. This work has benefitted from the field and laboratory assistance of J. Casey, G. Hammond, P. Kim, L. McLennan, D. Norman, W. Reiswig, B. Roy, T. Wade, and J. Winterer. Comments from B. Roy clarified the manuscript. I am grateful to the Japanese Organizing Committee of the 4th International Symposium of Plant Biosystematics and to Pomona College for making the presentation of this paper possible.

References

Babcock, E. B. and Stebbins, G. L. (1938). The American species of *Crepis*. Their interrelationships and distribution as affected by polyploidy and apomixis. Carnegie Inst. Wash. Publ. No. 504.

Baker, H. G. (1965). Characteristics and mode of origin of weeds. *In* "Genetics of Colonizing Species" (H. G. Baker and G. L. Stebbins, eds), pp. 137-172. Academic Press, N.Y.

Barton, N. H. and Post, R. J. (1986). Sibling competition and the advantage of mixed families. *J. theor. Biol.* **120**, 381-7.

Bayer, R. J. (1984). Chromosome numbers and taxonomic notes for North American species of *Antennaria* (Asteraceae: Inuleae). *Syst. Bot.* **9**, 74-83.

Bayer, R. J. (1985a). Investigations into the evolutionary history of the polyploid complexes in *Antennaria* (Asteraceae: Inuleae). I. The *A. neodioica* complex. *Plant Syst. Evol.* **150**, 143-163.

Bayer, R. J. (1985b). Investigations into the evolutionary history of the polyploid complexes in *Antennaria* (Asteraceae: Inuleae). II. The *A. parlinii* complex. *Rhodora* **87**, 321-339.

Bayer, R. J. and Stebbins, G. L. (1982). A revised classification of *Antennaria* (Asteraceae: Inuleae) of the eastern United States. *Syst. Bot.* **7**, 300-313.

Bayer, R. J. and Stebbins, G. L. (1983). Distribution of sexual and apomictic populations of *Antennaria parlinii*. *Evolution* **37**, 555-561.

Bayer, R. J. and Stebbins, G. L. (1987). Chromosome numbers, patterns of distribution, and apomixis in *Antennaria* (Asteraceae: Inuleae). *Syst. Bot.* **12**, 305-319.

Bell, G. (1982). The Masterpiece of Nature. The Evolution and Genetics of Sexuality. Univ. California Press, Berkeley, Ca.

Bierzychudek, P. (1989). Environmental sensitivity of sexual and apomictic *Antennaria*: do apomicts have general-purpose genotypes? *Evolution* **43**, 1456-1466.

Bierzychudek, P. (1987a). Patterns in plant parthenogenesis. *In* "The Evolution of Sex and its Consequences" (S. C. Stearns, ed.), pp. 197-217. Birkhauser-Verlag, Basel.

Bierzychudek, P. (1987b). Resolving the paradox of sexual reproduction: a review of experimental tests. *In* "The Evolution of Sex and its Consequences".(S. C. Stearns, ed.), pp. 163-174. Birkhauser-Verlag, Basel.

Bierzychudek, P. (1987c). Pollinators increase the cost of sex by avoiding female flowers. *Ecology* **68**, 444-447.

Bremermann, H. J. (1987). The adaptive significance of sexuality. *In* "The Evolution of Sex and its Consequences" (S. C. Stearns, ed.), pp. 135-161. Birkhauser-Verlag, Basel.

Bulmer, M. G. (1980). The sib-competition model for the maintenance of sex and recombination. *J. theor. Biol.* **82**, 335-345.

Clausen, J., Keck, D. D. and Heisey, W. M. (1945). Experimental taxonomy. *Carnegie Inst. Wash. Yearbook* **44**, 71-83.

Clausen, J., Keck, D. D. and Hiesey, W. M. (1947). Heredity of geographically and ecologically isolated races. *Am. Nat.* **81**, 114-133.

Crow, J. F. and Kimura, M. (1965). Evolution in sexual and asexual populations. *Am. Nat.* **99**, 439-450.

Falconer, D. S. (1981). Introduction to Quantitative Genetics, 2nd ed. Longman, London, U. K.

Fisher, R. A. (1930). The Genetical Theory of Natural Selection. Dover, N. Y.

Glesener, R. R. (1979). Recombination in a simulated predator-prey environment. *Am. Zool.* **19**, 763-771.
Glesener, R. R. and Tilman, D. (1978). Sexuality and the components of environmental uncertainty: clues from geographical parthenogenesis in terrestrial animals. *Am. Nat.* **112**, 659-673.
Gustafsson, A. (1946a). Apomixis in higher plants. I. The mechanism of apomixis. *Acta Univ. Lund* **42**, 1-66.
Gustafsson, A. (1946b). Apomixis in higher plants. II. The causal aspect of apomixis. *Acta Univ. Lund* **43**, 71-178.
Gustaffson, A. (1947). Apomixis in higher plants. III. Biotype and species formation. *Acta Univ. Lund* **43**, 183-370.
Hamilton, W. D. (1980). Sex versus non-sex versus parasite. *Oikos* **35**, 282-290.
Hamilton, W. D. (1982). Pathogens as causes of genetic diversity in their host populations. *In* "Population Biology of Infectious Diseases" (R. M. Anderson and R. M. May, eds), pp. 269-296. Springer-Verlag, Berlin.
Hamilton, W. D., Henderson, P. A. and Moran, N. (1981). Fluctuation of environment and coevolved antagonist polymorphism as factors in the maintenance of sex. *In* "Natural Selection and Social Behavior" (R. D. Alexander and D. W. Tinkle, eds), pp. 363-381. Chiron Press, N. Y.
Hutson, V. and Law, R. (1981). Evolution of recombination in a population experiencing frequency-dependent selection with time delay. *Proc. R. Soc. London. Ser. B* **213**, 345-359.
Jaenike, J. (1978). An hypothesis to account for the maintenance of sex within populations. *Evol. Theory* **3**, 191-194.
Levin, D. D. (1975). Pest pressure and recombination systems in plants. *Am. Nat.* **109**, 437-451.
Lloyd, D. G. (1980). Benefits and handicaps of sexual reproduction. *Evol. Biol.* **13**, 69-111.
Lynch, M. (1984). Destabilizing hybridization, general-purpose genotypes and geographic parthenogenesis. *Quart. Rev. Biol.* **59**, 257-290.
Maynard Smith, J. (1978). The Evolution of Sex. Cambridge Univ. Press, Cambridge, U.K.
Michaels, H. J. and Bazzaz, F.A. (1986). Resource allocation and demography of sexual and apomictic *Antennaria parlinii*. *Ecology* **67**, 27-36.
Muller, H. J. (1932). Some genetic aspects of sex. *Am. Nat.* **66**, 118-138.
Price, M. V. and Waser, N. M. (1982). Population structure, frequency-dependent selection, and the maintenance of sexual reproduction. *Evolution* **36**, 83-92.
Rice, W. R. (1983). Parent-offspring pathogen transmission, a selective agent promoting sexual reproduction. *Am. Nat.* **121**, 187-203.
Stearns, S. C. (1987). Why sex evolved and the differences it makes. *In* "The Evolution of Sex and its Consequences" (S. C. Stearns, ed.), pp. 15-31. Birkhauser-Verlag, Basel.
Stebbins, G. L. (1932). Cytology of *Antennaria*. II. Parthenogenetic species. *Bot. Gaz.* **94**, 322-345.
Stebbins, G. L. (1950). Variation and Evolution in Plants. Columbia Univ. Press, N. Y.
Stephenson, A. G. (1981). Flower and fruit abortion: proximate causes and ultimate functions. *Annu. Review Ecol. Syst.* **12**, 253-279.
Templeton, A. R. (1982). The prophecies of parthenogenesis, *In* "Evolution and the Genetics of Life Histories" (H. Dingle and J. P. Hegmann, eds), pp. 75-101. Springer-Verlag, N. Y.
Tooby, J. (1982). Pathogens, polymorphism, and the evolution of sex. *J. theor. Biol.* **97**, 557-576.
Uyenoyama, M. K. (1984). On the evolution of parthenogenesis: a genetic representation of the "cost of meiosis". *Evolution* **38**, 87-102.
Vandel, A. (1928). La parthenogenese geographique. Contribution a l'etude biologique et cytologique de la parthenogenese naturelle. *Bull. Biol. France Belg.* **62**, 64-281.
Vandel, A. (1940). La parthenogenese geographique. IV. Polyploidie et distribution geographique. *Bull. Biol. France Belg.* **74**, 94-100.
de Wet, J. M. J. (1980). Origin of polyploids. *In* "Polyploidy: Biological Relevance" (W. H. Lewis, ed.), pp 3-15. Plenum Press, N. Y.
Williams, G. C. (1966). Adaptation and Natural Selection. Princeton Univ. Press, Princeton, N. J.
Williams, G. C. (1975). Sex and Evolution. Princeton Univ. Press, Princeton, N. J.

Yahara, T. (1990). Evolution of agamospermous races in *Boehmeria* and *Eupatorium*. *Plant Species Biol.* **5** (in press).
Young, Y. P. W. (1981). Sib competition can favor sex in two ways. *J. theor. Biol.* **88**, 755-756.

18 Mating Systems and Speciation in Haplontic Unicellular Algae, Desmids

TERUNOBU ICHIMURA[1] AND FUMIE KASAI[2]

[1]*Institute of Applied Microbiology, University of Tokyo, Tokyo 113, Japan*
[2]*Microbial Culture Collection, National Institute for Environmental Studies, Tsukuba, Ibaraki 305, Japan*

I. Introduction

In haplontic unicellular algae, haploid cells or vegetative individuals may multiply indefinitely by asexual reproduction and thick-walled resistant diploid cells or zygospores are formed in a relatively limited number by sexual reproduction. Mating systems in these organisms are primarily concerned with the problem of how and when, under what kinds of intrinsic and extrinsic factors, they should start and accomplish their sexual reproduction to form such dormant zygospores as may survive harsh environmental conditions or be transported to new habitats, instead of continuing the multiplication of their vegetative individuals.

Wiese (1984) argues, "Because the continuity of their existence has been protected by asexual reproduction, these algae have been able to explore their full evolutionary potential with respect to sexual differentiation and reproduction, leading to a great variety in their sex phenomena." His review on mating systems in unicellular algae contains excellent descriptions of variation in sex phenomena with pertinent information on their physiological, molecular and ultrastructural aspects.

Like those in higher plants and animals, mating systems in haplontic algae are also concerned with the problem of sexual combination of genetic lineages. From this standpoint, we may well consider algal mating systems simply as either *homothallism* (zygospore formation within a single clone), *heterothallism* (zygospore formation between two complementary clones) or *parthenosporism* (parthenospore formation without syngamy but with some process of sexual reproduction, mostly within a single clone). It may be too naive to consider homothallic algae as true inbreeders, but in the present state of knowledge on these genetic systems, we may

BIOLOGICAL APPROACHES AND EVOLUTIONARY TRENDS IN PLANTS ISBN 0-12-402960-4

well call heterothallic algae outbreeders, homothallic algae inbreeders and parthenosporic algae apomicts.

In the haplontic life cycle, any genic or chromosomal mutation is selected as soon as its appearance, inasmuch as it affects the vegetative phase, whereas mutations in genes controlling the sexual phase (including both haploid gametic and diploid zygotic phase) are not subjected to natural selection as long as the asexual reproduction continues. Wiese and Wiese (1977) proposed a hypothesis on speciation by evolution of gametic incompatibility, which might occur through mutations in sexual recognition systems. One may consider that sexual isolation is the most common and important isolating mechanism among algal species based on the fact that there are many reports on sexual isolation in various algae but fewer on post-zygotic isolation (cf. Ichimura and Kasai, 1987). However, hybrid inviability seems to be common between closely related biological species between which no or only partial sexual isolation can be recognized (Ichimura, 1983, 1985; Ichimura and Kasai, 1984a).

In this chapter, we first illustrate some historical problems of desmid biology. Second, we review available information on sexual differentiation in desmids and on reproductive isolation within one taxonomic species of desmids, *Closterium ehrenbergii*. Then we show character divergences and ecological niches of several biological species of the *C. ehrenbergii* complex. Finally, we discuss some genetic and cytological problems in speciation of desmids.

II. Desmids and Their Ecological and Cytological Characteristics

Desmids are freshwater green algae and classified into five families (Mesotaeniaceae, Gonatozygaceae, Peniaceae, Closteriaceae and Desmidiaceae) of the Zygnematales (Růžička, 1977). Taxonomy depends largely on vegetative cells, which reproduce asexually by the unique mode of bipartition, and also on zygospores, to a limited extent, which are formed sexually by conjugation of two complementary sex cells, both of which resemble vegetative cells in morphology (Fig. 1). Desmids abound in species and individuals in shallow stagnant waters such as ponds, bogs and paddy fields rather than in large and deep lakes or in running waters. A single body of soft water rich in humic substances, especially in tropical to subtropical regions, usually contains an enormous number of desmid species (e. g. Therezien, 1985; Thomasson, 1986).

Among unicellular green algae, desmids are famous for the great diversity in shape and size of vegetative cells and chloroplasts. Zygospores of desmids are characteristic spherical cells with several thick walls, of which the outermost or the middle may have spines or warts in some genera. Such zygospores are remarkable objects for microscopists, but they are only sporadically found among vegetative cells collected from natural fields. The number of zygospore-known species are rather small, compared to the large number of described species.

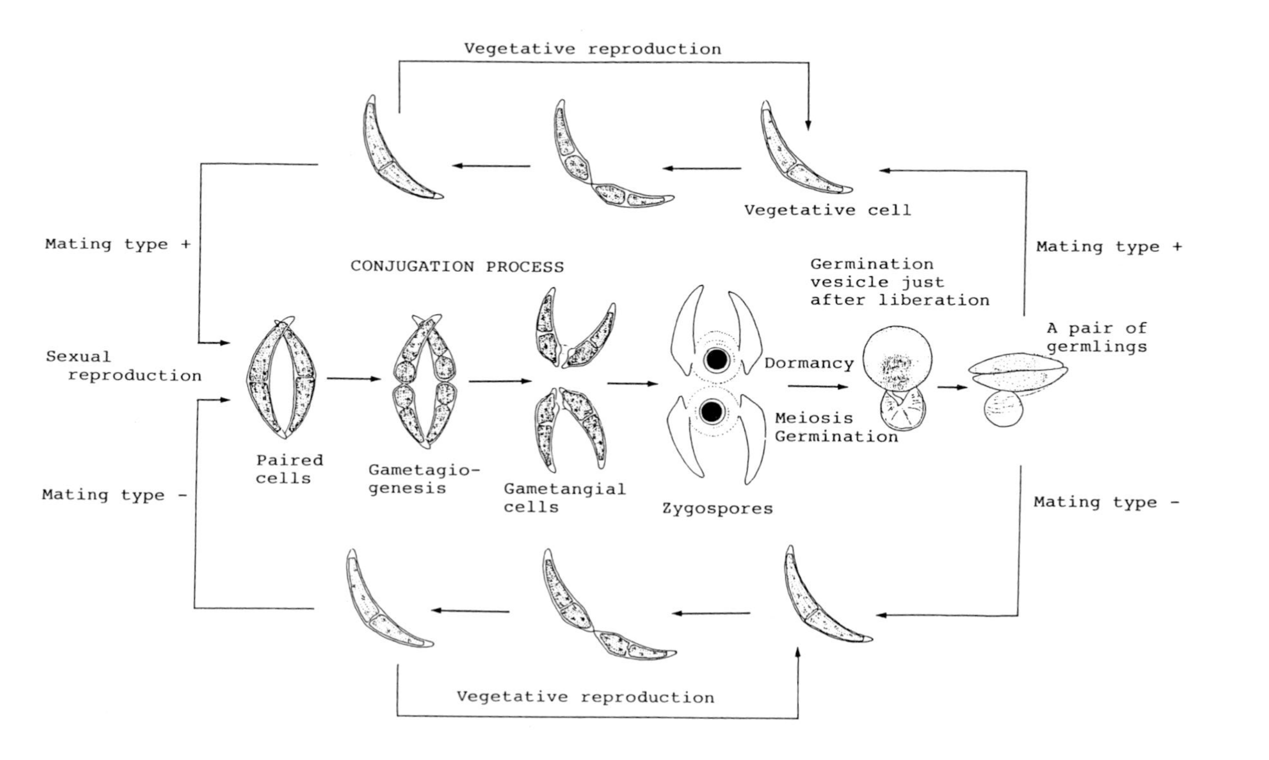

FIG. 1 Life cycle of a haplontic unicellular alga, *Closterium ehrenbergii*, of heterothallic mating system.

Because of the rarity of finding zygospores in nature, Fritsch (1930) suggested that there is a general tendency towards the elimination of sexual reproduction in the desmids that are morphologically most differentiated, like some species of *Micrasterias* and *Euastrum*. Recent workers also suggest that sexual reproduction occurs only sporadically in nature in the large majority of desmids and that only asexually-reproducing clonal populations may be prevalent in some habitats and some geographical areas (e. g. Coesel, 1974, 1988; Francke and Coesel, 1985).

Most desmids have a large number of small chromosomes (n=20-200, excepting a few species with n=8-10 or n=ca. 592), and many chromosome races are known within single taxonomic species (Sarma, 1983). Godward (1966) proposed that chromosomes of desmids, like those of the Zygnemataceae, are polycentric and their meiosis is postreductional (Brandham and Godward, 1965b), and she argued that chromosome races easily arise by agmatoploidy, which may cause some desmids to lose sexuality or to change mating types.

Nevertheless, recent culture studies on sexual reproduction show clear evidence for monocentric chromosomes and the conventional type of meiosis in *Pleurotaenium* (Ling and Tyler, 1976), *Micrasterias* (Blackburn and Tyler, 1980, 1981), *Closterium* (Kasai and Ichimura, 1983) and *Triploceras* (Ichimura and Kasai, 1989a).

III. Sexual Differentiation in Desmids

In the sexual reproduction of desmids, vegetative cells differentiate into sexually active cells, after (single zygospore in the majority of species) or before (twin zygospores in the minority) so-called gametangiogenetic division, which can approach, pair and conjugate with respective complementary sex cells, although they are still similar in shape to vegetative cells (cf. Ichimura, 1971; Ichimura and Kasai, 1984b). In such gametangiogamy, two amoeboid gametes of equal size fuse to form a zygospore midway between the two gametangia in the majority of desmids, but in a few limited species and genera a male amoeboid gamete migrates into a female gametangium to fuse with the stationary female gamete (Ichimura and Kasai, 1985). Thus, mating systems viewed from the morphological differentiation of sex cells are rather uniform in desmids, compared to the great variety that has been fully explored by other algal groups (Wiese, 1984).

Although they have no flagella, desmid cells are gliding on the substratum, slowly but ceaselessly, and the movement appears to be especially active when they gain sexual competence (Kies, 1964; Ichimura and Kasai, 1984b, 1987). Under experimental conditions, it has been observed that sexually active cells come close together to form a clump of cells before the formation of conjugating pairs. Directional approach between the two complementary sex cells has been interpreted as chemotactic attraction (Brandham, 1967; Coesel and Jong, 1986). Mechanisms of sexual clumping and subsequent processes leading to cell fusion have been studied biochemically and genetically in non-desmid algae, *Chlamydomonas* (cf. Wiese, 1984; Goodenough, 1985). Although we have no such information for desmids, it is known that differences in genetic systems controlling these prezygotic sexual processes cause sexual isolation between populations from different habitats and geographical areas within the same taxonomic species of desmids (Ichimura and Kasai, 1987). The genetic systems should involve components that respond to environmental factors such as nutrients, irradiance and temperature, as well as the ones that actually control the sexual processes by intrinsic factors, such as pheromonal substances, because degrees of sexual isolation observed between a given pair of populations or strains change considerably under different environmental conditions.

In order to adapt to the habitat, every desmid species has to develop its own genetic system that can switch from the machinery for asexual reproduction to the one for zygospore or parthenospore formation before the environmental conditions become too deleterious. The environmental cues for the switching differ from species to species. The so-called precocious homothallic strains that tend to form zygospores after only a few vegetative divisions even in growth media are often found in small temporary pools, whereas zygospore formation in most other strains is possible only in media of no or low nitrogen source. We show, for example, growth and mating media suitable for several species of *Closterium* of various mating systems in Table 1.

IV. Reproductive Isolation and Character Divergence in Speciation of Desmids

A. *Mating group as biological species*

The taxonomic species *Closterium ehrenbergii* is of particular interest with respect to speciation studies (Ichimura, 1981, 1985). There are many records with descriptions of its beautiful lunate vegetative cells. This species appears to not only flourish in warmer climatic regions but also expand into cool climatic regions.Růžička (1977) recognizes only five varieties for this species, which shows great morphological variation.

Clonal cultures of *C. ehrenbergii* have been studied based on many natural populations sampled from various habitats in Japan, Taiwan, Malaysia, Indonesia, Nepal, Australia, Denmark, Holland, France, England, the U. S. and Mexico (Lippert, 1967; Ichimura, 1981; Ichimura and Kasai, 1984a; Coesel, 1988, 1989). Most natural samples contained exclusively heterothallic clones, but a few from limited habitats contained homothallic or parthenosporic clones. Two sampling methods were applicable to study natural populations. One depended on liquid samples containing vegetative cells and the other on soil samples containing dor

TABLE 1
Mating systems and growth and mating media for several species of *Closterium*

Species	Mating system	Growth medium[1)]	Mating medium[1)]
C. acerosum	Homo.	C	MIH
C. acutum	Homo.	CA	VT
C. calosporum	Homo.	CA	MIH
	Hetero.	CA	MIH
C. cynthia	Homo.	CA	MIH
	Hetero.	CA	MIH
C. dianae			
var. *minus*	Homo.	CA	VT
C. ehrenbergii	Hetero.	C, CA	MIH
C. gracile	Hetero.	CA	MIH
C. incurvum	Homo.	CA	VT
C. intermedium	Hetero.	CA	MIH
C. libellula			
var. *intermedium*	Homo	CA	MIH
C. lineatum	Homo.	CA	VT
C. moniliferum	Partheno.	C	MIH
	Homo.	C	MIH
	Hetero.	C	MIH
C. pleurodermatum	Homo.	CA	MIH
	Hetero.	CA	MIH
C. praelongum			
var. *brevius*	Homo	CA	VT
C. pseudolunula	Hetero.	CA	MIH
C. pusillum			
var. *maius*	Hetero	C	MIH
C. rostratum			
var. *subrostratum*	Homo	CA	MIH
C. ralfsii	Partheno.	CA	CA
C. strigosum	Hetero.	C	MI
	Homo.	C	MI
C. submoniliferum	Hetero.	C	MIH
C. tumidum	Homo.	C	MIH
C. navicula	Homo.	CA	MIH
C. venus	Homo.	CA	VT
C. wallichii	Homo.	C	MIH

[1)] C, CA and VT media contain nitrogen source, whereas MI and MIH contain no nitrogen source. See Ichimura (1971, 1983); Ichimura and Watanabe (1974).

mant zygospores. Soil samples always yielded plus and minus mating type individuals in a 1:1 ratio, whereas some liquid samples contained only one mating type individual, either plus or minus.

Mating experiments with heterothallic clones of *C. ehrenbergii* showed various degrees of sexual isolation under the conventional culture conditions for inducing sexuality in microalgae (no- choice mating method). Therefore, we defined a mating group as a group of natural populations (a group of clones derived from them) within which no sexual isolation (free zygospore formation) has been observed (Ichimura, 1981). Thus, the mating groups of *C. ehrenbergii* were recognized primarily based on sexual isolation, and so far more than a dozen mating groups are known (Ichimura, 1985). Complete sexual isolation (no zygospore formation) was shown between mating groups with different zygospore characters: smooth-walled zygospores in Groups A, B, C, H, K, L, M and P *vs.* scrobiculated zygospores in Groups D, E, I and J. On the other hand, incomplete sexual isolation (sporadic zygospore formation) is known between mating groups of the same zygospore character, and these mating groups are considered to be closely related. These are the results that have been obtained by the no-choice mating method. However, the multiple-choice mating method combined with time-lapse microphotographic techniques has revealed almost complete sexual isolation even between the two closely related mating groups, A and B (Ichimura and Kasai, 1987), between which many combinations of clones yield a considerable number of zygospores by the no-choice mating method (Ichimura, 1981, 1983). We can estimate only a minimum degree of sexual isolation by the no-choice method. Therefore, other closely related mating groups, between which a few combinations of clones formed a small number of zygospores, are likely to be isolated completely by sexual isolation under more natural conditions than those used in the previous studies mentioned above. On the other hand, field data suggest that closely related mating groups are isolated from each other by geographical barriers or by ecological factors and that only some of the remotely related mating groups are frequently found intermingling in the same and one habitat (Ichimura, 1981, 1985). In addition, genetic studies show that the integrity of each of the mating groups is maintained also by such a postzygotic isolating mechanism that sporadically-formed intergroup hybrid zygospores can yield only extremely rare viable F_1 progeny (Ichimura, 1983; Ichimura and Kasai, 1984a). It is known that meiotic events in intergroup hybrid zygospores are abnormal (Kasai and Ichimura, 1984). Thus, every mating group of *C. ehrenbergii* is concluded to be an independent biological species.

B. Character divergence and ecological niche

It is well known that sibling species are found much more often in microbes than in higher organisms (Mayr, 1963). This may hold true for unicellular algae since reproductive isolation is frequently found among different strains of the same morphological species (e. g. Goldstein, 1964; Coleman, 1977; Wiese and Wiese, 1977). In some desmids, reproductive isolation seems to be correlated with morphological divergence (Ichimura and Watanabe, 1974; Watanabe, 1978; Watanabe and Ichimura, 1978; Kasai and Ichimura, 1986; Coesel, 1989), but not in other desmids (Starr, 1959; Brandham and Godward, 1965a; Blackburn and Tyler, 1987).

In order to learn under what circumstances reproductive isolation is correlated with morphological divergence, we have examined as many clones as possible for seven closely related mating groups of *C. ehrenbergii* (Kasai and Ichimura, 1986, unpubl.). Four main taxonomic characters of vegetative cells (length, width, length to width ratio and curvature) were measured for 50 (or 20) cells in each clone which had been cultured under a set of uniform conditions (Group P clones, however, were cultured at 15 °C and all other clones at 25 °C; cf. Kasai and Ichimura, 1986). For each of these four characters measured, we use mean value of each clone for statistical analysis of variation among clones of the same and different mating groups (Table 2). To illustrate cell size variations of the seven mating groups, we plot the range in length of each group versus the range in width (Fig. 2), and to illustrate cell shape variations, the range in curvature versus the range in length to width ratio (Fig. 3).

These seven mating groups are closely related, having similar smooth-walled zygospores. In general, their cell shapes are similar (Fig. 3), and though mean values are significantly different, their cell sizes intergrade with each other and completely overlap between Groups C and H (Fig. 2). No wonder the current taxonomic treatment includes all these forms in one morphological species. The range shown in the diagnostic description of *C. ehrenbergii* (Růžička, 1977) is larger than the total range of all these seven mating groups together. (Each of the five infraspecific taxa is described to show more or less similarly large dimensions.) However, we doubt whether one single species might show such a wide range of morphological variability and have such a world-wide geographical distribution (from tropical to arctic regions) as described in the taxonomic monograph. Though we have very limited information as yet, we can suggest that each has its own, more limited geographical distribution and ecological niche for the seven mating groups.

For example, natural populations of Groups A and B are principally distributed in paddy fields and small ponds from Honshu to Okinawa, Japan. Outside Japan, Group A are known from Queensland, Victoria and Tasmania, Australia, and Group B from Taiwan. In contrast, Group P populations are distributed more widely, though limited in cool waters such as streams and spring-fed ponds, from Hokkaido to northern Honshu, Japan, and boreal countries in Europe and the northern part of the United States (Ichimura and Kasai, unpubl.). Experimental data on temperature optima and ranges for vegetative growth of several representative clones indicate that both Groups A and B grow well from 20 to 30 °C but not below

TABLE 2

Statistical values of morphological characters of seven closely related mating groups of the *Closterium ehrenbergii* species complex (Ichimura and Kasai, 1989b)

Mating	Width (μm)			Length (μm)			L/W			Curvature (°)		
groups	Range	Mean	SD	Range	Mean	SD	Range	Mean	SD	Range	Mean	SD
A (47)[1]	38-60	50	4.2	216-310	250	21.6	4.5-6.5	5.0	0.4	112-156	137	8.4
B (44)	56-84	67	5.2	311-567	404	52.5	4.9-6.8	6.0	0.4	110-133	122	4.8
C (9)	52-55	54	1.0	303-350	323	13.6	5.8-6.4	6.0	0.2	111-137	121	8.0
H (38)	49-64	57	3.2	276-390	333	27.2	5.1-6.5	5.8	0.4	107-131	121	5.1
K (20)	60-80	73	5.4	317-420	363	26.6	4.4-5.5	5.0	0.3	129-157	145	7.6
M (17)	28-46	36	4.1	130-238	183	23.6	4.5-5.8	5.1	0.4	104-138	124	9.7
P (30)	75-109	93	7.7	381-650	513	62.4	4.4-6.4	5.5	0.5	104-148	126	10.1

[1] Number of clones measured.

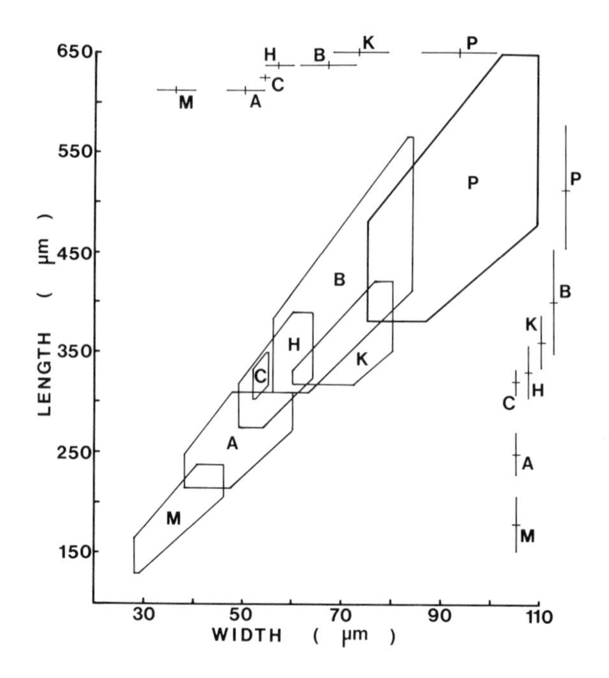

FIG. 2 Variation ranges in cell size for seven closely related mating groups of the *Closterium ehrenbergii* complex; attached bars indicate means and standard deviations (plotted from the data shown in Table 2).

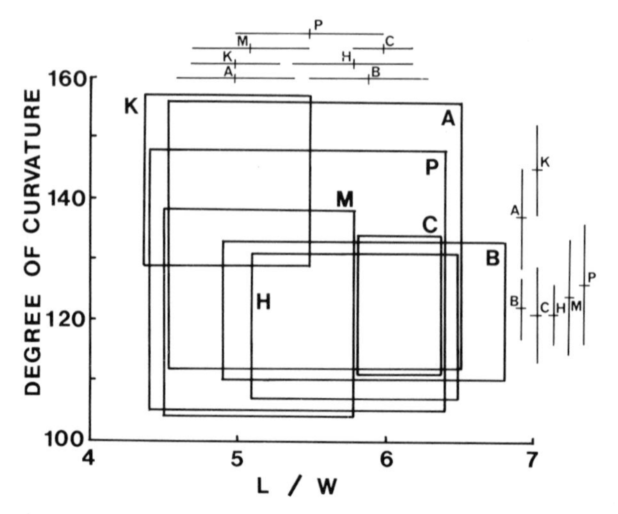

FIG. 3 Variation ranges in cell shape for seven closely related mating groups of the *Closterium ehrenbergii* complex; attached bars indicate means and standard deviations (plotted from the data shown in Table 2).

10 °C, whereas Group P grows well from 5 to 15 °C but very poorly above 25 °C (Kasai and Ichimura, in press). The life history patterns in nature of Groups A and B appear to be more or less similar in that they grow asexually when their habitats contain water from late spring to early autumn and they form zygospores before paddy fields become dry in late autumn. Although the two mating groups grow in the same warmer seasons in superficially similar habitats, they are seldom or rarely found intermingling with each other in single population samples collected at the same spot. They are actually isolated by habitat preference: Group A prefers slightly acidic waters rich in humus, whereas Group B prefers more alkaline waters rich in calcium (Ichimura, 1982). In contrast, Group P populations seem to continue growing asexually the whole year in more or less constant cool waters and are assumed to reproduce sexually only very occasionally in most localities (Ichimura and Kasai, unpubl.). Thus, it is clear that Group A, B and P each has its own distinct ecological niche.

As to Groups C, H, K and M, we have little information except as follows. It is supposed to be isolated either by habitat preference or by geographical barriers, because they are also seldom or rarely found coexisting in the same habitat. Groups C and H seem to be endemic to Japan and Nepal, respectively. Group K is known from Japan and Australia, and Group M from Japan and Nepal (Ichimura, 1985). However, we are still attempting to get more samples for the purpose of understanding their ecological niches and geographical distribution patterns.

Now we must consider, at all events, ecological and evolutionary implications of the significant morphological differences among the six mating groups, from small to large, M, A, C, K, B and P, all of which are distributed in Japan. Their cell sizes are particularly well diverged, compared to their cell shape. But Group H, which is allopatric to them, shows no difference from Group C and only slight differences from Group B (Fig. 2). If we plot cell sizes of exclusively Japanese populations (Fig. 4), then differences become clearer with gaps among Groups M, A, C, B and P, though only Group K remains overlapping, substantially with Group B and slightly with Group P. Group K, however, differs in cell shape from Group B (Fig. 5). Thus sympatric populations of all these six closely related mating groups turned out to be identifiable by the morphological characters without testing sexual isolation. However, their allopatric populations, especially between Groups M and A (Fig. 2) show some difficulty in making such morphological distinction.

We interpret the morphological divergence (with character displacement?) among the closely related sympatric groups as an evolutionary consequence that has been brought about with the differentiation of ecological niches. The closely related allopatric mating groups could have acquired more or less similar morphological characteristics as those seen between Groups C and H (or some populations of Groups B and H), since they have occupied similar (but not necessarily identical) types of habitats in different geographical areas. The nature of the habitats for these organisms should be studied not only from the viewpoint of physico-chemical

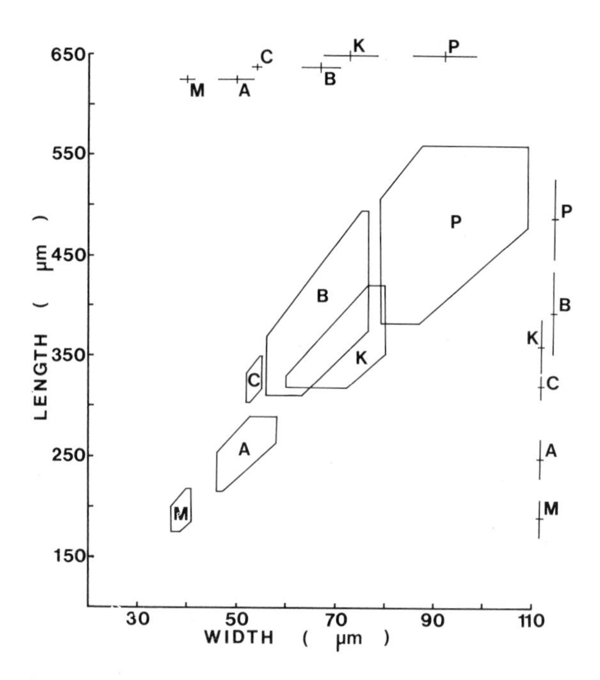

FIG. 4 Variation ranges in cell size for the closely related mating groups of the *Closterium ehrenbergii* complex shown in Fig. 2, when populations distributed outside Japan are excluded.

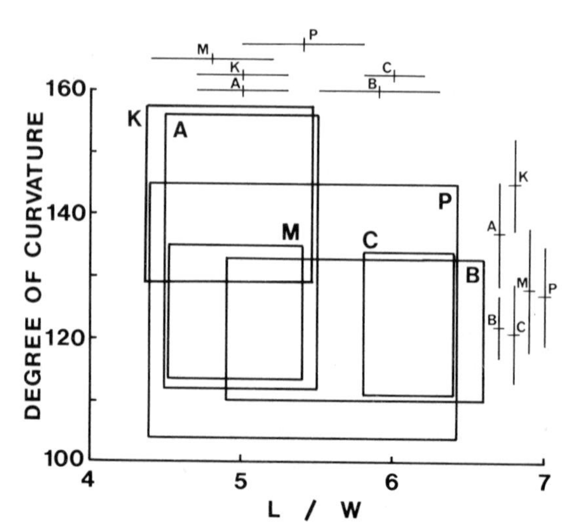

FIG. 5 Variation ranges in cell shape for the closely related mating groups of the *Closterium ehrenbergii* complex shown in Fig. 3, when populations distributed outside Japan are excluded.

(nutrients, pH, temperature, irradiance, etc.) and biotic (herbivores, parasites, pathogens, competitors, etc.) factors during their growing season but also from the viewpoint of temporal patterns of water levels that limit the duration of their growth and dormancy. Therefore, we do not mean that two different mating groups with the same morphology have the same genetic systems, but we assume that different genetic systems may be able to develop the same or similar morphology in somewhat different ways. It is clear, however, that we need more information on morphological variations not only for allopatric populations of closely related mating groups but also for sympatric populations of remotely related mating groups that are often found to coexist in the same habitat.

According to Hutchinson (1968), all characters (with the apparent exception of secondary sexual characters) either are adaptive or represent other adaptive characters. He considered that "the appendage length of rotifers (*Filina*) may have something to do with the degree of planktonic habitats, and so of water depths." At present, however, we cannot give any appropriate explanation for the adaptive nature of the cell shape in *Closterium*, especially for the differences in cell shape between Groups B and K, though we wish to do as did Hutchinson. But why are there so many different and fancy cell forms in dismids? West (1899) considered that desmids have evolved from filamentous conjugates by loss of the filamentous condition, accompanied by the development of specialized morphological characters in order to resist the attacks of many forms of aquatic animals. In addition to the viewpoint of predation pressure, the cell size differences between mating groups may be studied from the viewpoint of the evolutionary consequences of adopting different ecological strategies such as *r*- and *K*-selection (cf. Cavalier-Smith, 1985).

V. Genetic and Cytological Problems in Speciation

A. *Polyploidy*

It is premature to speculate about the scenario of speciation for the mating groups of *C. ehrenbergii*. However, it is possible to suggest that polyploidy has played a great role in their evolution (and also in the evolution of most other desmid species, as suggested by their high chromosome numbers). We have attempted to estimate haploid chromosome numbers of the mating groups from meiotic observations in germinating zygospores. However, it was not easy to count definitely a large number of small dot-like chromosomes, because of the difficulty of spreading the chromosomes and of distinguishing between bivalents and univalents. We have so far been able to estimate the chromosome numbers n=100-106 for Group A, n=97-104 for Group B, n=105-127 for Group H and n=181-246 for Group P (Kasai and Ichimura, 1984, unpubl.). Clones belonging to the same mating group may have different chromosome numbers, even if we discount some possible counting errors. In particular, Group P appears to be composed of clones of two distinct cyto-types.

In various crosses within Group P, however, we have confirmed that such differences in chromosome numbers affect mating efficiency only slightly and F_1 viability (Ichimura and Kasai, unpubl.). From this, one may wonder whether the high chromosome numbers known in many species of dismids could have evolved through complete genome duplication or polyploidy. But it seems to us to be doubtful that agmatoploidy has played the main role in evolution of desmids, as is suggested by classical cytologists (Godward, 1966; Sarma, 1983).

In several genera of desmids, extremely large so-called giant cells have been found among many normal cells in vegetative cultures and at germination of zygospores. From chromosome counts, these giant cells are known to be diploid, if we call normal cells haploid. Studies on giant cells in homo- and heterothallic species indicate that both mating systems can be maintained without any disturbance in sexual expression through genome duplication; neither homo- nor heterothallism changes at all (Starr, 1958; Brandham, 1965; Ling and Tyler, 1976; Ichimura and Kasai, 1987).

In a study in which we showed rather high viability and almost regular, diploid-like meiosis in "triploid" zygospores in Group A of *Closterium ehrenbergii*, we concluded that the normal vegetative cells of natural populations are not truly haploid but polyploid by their cytological nature and in their phylogenetic origin (Kasai and Ichimura, 1987). This conclusion is in accord with the fact that many multi- as well as uni- and bivalents were observed in meiosis of the intergroup crosses between closely related mating groups (Kasai and Ichimura, 1984). Hence we assume the presence of a genetic system that promotes exclusive bivalent formation by preventing homoeologous pairing in intragroup crosses but does not work well in intergroup crosses.

Recently we found that certain plus clones of Group P carry a genetic factor(s) that cause(s) considerable meiotic irregularity and hence F_1 inviability (F_1 viability: 40 % *vs*. 85 % in normal) even in intragroup crosses. The factor(s) seem(s) to be transmitted only to F_1 progeny of mating type plus (Ichimura and Kasai, unpubl.). We can consider two alternative interpretations (not necessarily mutually exclusive) for this finding. Group P might be a newly formed allopolyploid, as suggested on the basis of the eco-geographical distribution pattern, which may recall some allopolyploids of higher plants invading newly opened niches in the vast boreal areas after the Pleistocene (cf. Stebbins, 1971), and also on the basis of the large cell size and the large chromosome number (ca. twice the numbers of Groups A, B and H). The other interpretation is that the genetic factor in question might be a synaptic mutation(s) which could be maintained only in such populations as those in Group P, in which asexual reproduction continues for many years, and sexual reproduction may not occur except very occasionally in that habitat. We need further information to decide which is more plausible.

In studies of intergroup crosses between Groups A and B, for example, we have noticed that cell sizes of rarely surviving F_1 progeny are larger than the maximum of Group B (larger parental mating group), and that mating type minus F_1 progeny are predominant among such survivors (Ichimura, 1983). Preliminary cytological observations showed more or less irregular meiosis still in B_1 and B_2 when hybrids were backcrossed to either of the parental mating groups. Therefore we could not count chromosome numbers of the F_1 hybrids. However, we assume from their cell sizes that they are amphiploids or something like amphiploids that have almost all chromosomes of both Groups A and B. We have not yet encountered such intergroup hybrids in nature during our many years of surveying. This is no wonder, from the biological natures of Groups A and B mentioned above. However, if hybridization might have occurred, almost no hybrids between such well-differentiated mating groups as Groups A and B could survive in nature, because of their lower growth rates. Furthermore, their crossability with both their parental mating groups will not so easily allow the establishment of a new and distinct hybrid population.

In *Pleurotaenium*, several natural populations from lakes in Tasmania and Victoria, Australia, were studied under the names *P. ehrenbergii*, *P. mamillatum* and *P. coronatum*, and determined to belong to the same single biological species from the result of free zygospore formation among them in cultures (Ling and Tyler, 1974). Since then it has turned out that Australian and European populations of *P. ehrenbergii* and also of *P. coronatum* are different in certain morphological characters, and they will not cross at all. Nevertheless, it is of particular interest that the haploid chromosome number estimated from meiotic observations was n=53 for the *P. ehrenbergii* and the *P. mamillatum* and n=145 or 147 for the *P. coronatum* population (Ling and Tyler, 1976). Vegetative cells of the *P. coronatum* population were morphologically very similar to those of diploid clones (n=106) of the *P. ehrenbergii* population, which were spontaneously originated in cultures. Thus Ling and Tyler (1976) considered that the *P. coronatum* population might be a natural diploid of the *P. ehrenbergii*. However, it might be a derivative from an amphiploid formed by hybridization between closely related but cytologically differentiated *P. ehrenbergii* populations, as suggested from the chromosome number and some other cytological characteristics they have described.

Such giant cell formation or genome duplication as is found occasionally in cultures is likely to occur also in nature not so infrequently, as suggested by Brandham (1965). Especially in a newly colonized habitat, the frequency might be higher than the original habitat since slightly different environmental factors, e. g. chemical compositions of water, would act as stress that evokes such cytological abnormalities (cf. Kasai and Ichimura, 1987, 1990). As populations of a single mating type, either plus or minus, are frequently found in nature (Ling and Tyler, 1974; Ichimura, 1981), it is possible for such cytological variants or polyploids to maintain vegetative populations in some habitats for many years, even if they may not be able to encounter their potential mates.

In summary, because of the haplontic life cycle and the mating systems mentioned above, both auto- and allopolyploidy should have contributed greatly to evolution of the high chromosome numbers of desmids, although the genome duplication itself could not be a direct or instantaneous factor in speciation.

B. Chromosomal rearrangements

According to Cavalier-Smith (1985), cell size and nuclear DNA content are positively correlated. Therefore, we assume from the cell size mentioned above that nuclear DNA contents vary among the closely related mating groups of *C. ehrenberghii*. Groups A, B and H, however, have approximately the same levels of chromosome numbers in spite of the differences in their cell sizes, hence their nuclear DNA contents. Therefore, in general, chromosomes should be small in Group A, intermediate in Group H and large in Group B. However, we are discouraged from studying such differences with quantitative data for those organisms having large numbers of small dot-like chromosomes. At present we cannot show direct cytological evidence for chromosomal rearrangements among these three mating groups.

Nevertheless, we can get a clue for solving the problem from the analysis of F_1 viabilities in systematic crosses among clones of the same mating group, in which we expect regular meiosis. Table 3 shows the result so far obtained in several intragroup crosses of Group A (Ichimura and Kasai, 1989b). The first group of eleven crosses showed high F_1 viabilities (more than 85 %), suggesting homosequential homologous chromosomes among these clones. The second of five crosses showed lower F_1 viabilities (55-75 %), and the third of four crosses still lower (50-70 %). The lower values in these nine crosses are ascribed to particular clones, namely J6-73-3 and J6-73-4 in the second and J5-48-2 in the third, because their partner clones all yielded high F_1 viabilities when crossed with any clone other than these particular three, as shown in the first group. Since all parental clones grew well, each of the particular clones should have had a completely sound genotype for itself, in other words no deleterious genes for vegetative growth. Therefore, unless the factor is a synaptic mutation, there should be chromosomal rearrangements in the particular clones that result in so-called cryptic structural hybridity in the zygospores (cf. Stebbins, 1971). This argument, however, is valid only when their genomes have been demonstrated to be functionally haploid.

One line of evidence for functional haploidy has come from a study of sex determination and mating-type gene (Kasai and Ichimura, 1990). Another line of evidence is as follows. Starr (1954) reported a lethal factor in *Cosmarium botrytis* var. *subtumidum*. Although undetectable in the vegetative cell cycle or in the heterozygous condition, the zygotes formed by conjugation of two cells both of which carry the lethal factor always collapse shortly after their formation. This factor is known to be inherited as a single recessive gene, segregating independently from the mating-type gene. Among many clonal isolates from natural populations of Group H of *Closterium ehrenbergii* collected from lakes in the Pokkala Valey of Nepal, we

TABLE 3
Viability and mating type of F1 progeny in crosses within Japanese populations of Group A (Ichimura and Kasai, 1989b)

Cross Plus x minus	Individuals isolated	Percent survival	Mating type(+:–)	No. of pairs (+ –)	(+ +)	(– –)
M-16-4a x M-16-4b	418	94.0	191:202	183	1	2
x R-11-16	250	96.4	121:120	116	1	0
x R-13-20	128	93.0	57:62	54	0	1
80-1-2 x M-16-4b	212	84.0	89:89	82	0	0
GN-4-29 x	44	84.1	17:20	16	0	0
J5-48-19 x	320	94.7	153:150	144	0	0
x R-13-20	146	94.5	67:71	65	0	0
R-11-4 x M-16 4b	444	91.0	208:196	183	3	2
R-13-131 x	422	97.4	202:209	195	2	4
x 80-1-1	102	94.1	49:47	45	0	0
x R-13-20	76	88.2	35:32	30	0	0
Subtotal	2562	93.2	1189:1198	1113	7	9
M-16-4a x J6-73-4	102	70.6	34:38	23	1	0
R-11-4 x	72	75.0	25:29	20	0	1
J6-73-3 x M-16-4b	76	60.5	21:25	12	1	1
x 80-1-1	24	54.2	5:8	2	0	0
x R-11-16	164	64.6	53:53	33	0	1
Subtotal	438	66.4	138:153	90	2	3
M-16-4a x J5-48-2	238	67.7	77:83[1)]	51	1	1
80-1-2 x	32	68.8	14:8 [1)]	7	1	0
J5-48-19 x	108	50.9	19:36[1)]	13	0	0
R-13-131 x	100	56.0	30:26[1)]	12	0	0
Subtotal	478	61.3	140:153	83	2	1
Total	3478	85.4	1467:1504	1285	11	13

[1)] Aberrant mating type ratio and/or selfers in minus progeny.

found one plus clone from a lake and one minus from another lake carrying the same kind of genetic factor as in *Cosmarium botrytis* var. *subtumidum* (Ichimura and Kasai, unpubl.). From the simple disomic inheritance of these recessive lethal factors we conclude that genetically the vegetative cell is haploid and the zygote (zygospore) is diploid in these organisms.

C. Sex determination and mating-type gene

During the dormancy period in the desmid life cycle, the zygospore retains the two unfused haploid gamete nuclei until the time of germination, when it releases a single spherical vesicle. Cytological studies show that the two gamete nuclei fuse to form one diploid nucleus just before the release of the germination vesicle, in which meiosis takes place. After the second meiotic division, two or three nuclei degenerate in some species whereas all four nuclei survive in some others; consequently, one, two or four haploid juvenile vegetative cells (germlings) are formed from each germination vesicle in the respective species (Starr, 1959; Biebel, 1964; Brandham and Godward, 1965b; Lippert, 1967; Ling and Tyler, 1976; Blackburn and Tyler, 1980; Kasai and Ichimura, 1983; Ichimura and Kasai, 1985, 1989a). Genetic studies, however, are confined to a few heterothallic species in which a single zygospore produces a pair of germlings, as in *C. ehrenbergii* (Fig. 1). It is known that the majority of F_1 pairs, which are of opposite mating type, result from first-division segregation of mating-type alleles, whereas the minority of F_1 pairs, which are of either plus or minus mating type, are the result of second-division segregation caused by crossing-over between the mating-type locus and the centromere (Starr, 1954; right-hand three columns in Table 3). Thus mating types in these organisms are controlled by a pair of alleles (mt^+ and mt^-) and normally segregate in a 1:1 ratio through the sexual cycle.

We found, however, that diploid (instead of haploid) F_1 pairs occasionally obtained, from single germination zygospores, are all, without exception, of minus mating type (Kasai and Ichimura, 1987). Unless a zygospore that produces a pair of diploid minus germlings is formed from an exceptional conjugation between two minus cells, each member of such a diploid pair is expected to have both mt^+ and mt^- alleles. Although conjugation between like mating-type cells is known in ciliates (cf. Miyake, 1981), it is unlikely in *Closterium* (Ichimura and Kasai, 1984b, 1987). Such putative mt^+/mt^- heterozygous diploid clones have proved to cross with any haploid plus clone of the same mating group, resulting in many triploid zygospores which yield plus and minus F_1 progeny in a 1:1 ratio (Kasai and Ichimura, 1987). On the other hand, the triploid zygospores formed between a haploid mt^+ and a homozygous mt^-/mt^- diploid (obtained from a minus vegetative cell by hypertonic treatment) were shown to yield plus and minus F_1 progeny in a 1:4.8 ratio. Hence it has proved to be clear that mt^- is dominant to mt^+ and that mating-type inheritance in these zygospores follows the trisomic pattern (Kasai and Ichimura, 1990). Thus, the genetic studies on triploid zygospores involving the heterozygous and the homozygous diploid suggest that a sex-determining mechanism similar to the so-called dominant Y type (cf. White, 1978) is working effectively in these organisms which have evolved through polyploidization.

D. Genetic basis of mating systems

We note here that the four crosses involving the J5-48-2 clone yielded an aberrant F_1 mating-type ratio and some selfers in minus F_1 progeny (Table 3). In our terms, a "selfing minus" clone forms not only many zygospores with a normal plus clone but also some zygospores with a normal minus clone or within its clonal population. Preliminary experiments on some selfing minus F_1 clones of M-16-4a x J5-48-2 showed the following results (Ichimura, unpubl.):

(1) Zygospores formed within a selfing minus clone yielded no viable progeny:

(2) Zygospores formed between a normal plus and a selfing minus yielded plus and minus progeny (viabilities, 55-75 %) in a 1:1 ratio, with some selfers only in minus progeny:

(3) Zygospores formed between a selfing minus and a normal minus yielded plus and minus progeny (viabilities, 40-50 %) in a 1:1 ratio with no selfers.

Based on these results, we suggest the following hypothesis. The mating type of algal cells in a clonal population can be switched from minus to plus and/or from plus to minus under certain genetic circumstances, as is well known in the yeasts *Saccharomyces cerevisiae* (e. g. Sternberg, 1987) and *Schizosaccharomyces pombe* (e. g. Egel, 1989). In the selfing minus clones of *Closterium*, however, mating-type switching should be unidirectional, from minus to plus (cf. Perkins, 1987). Cells switched from minus to plus can form stable plus clones, which are, however, distinguishable from normal plus clones by their significantly lower viabilities. Therefore, at least one gene that is responsible for yielding viable zygospore progeny should be present on the normal mt^+ chromosome, perhaps closely linked to the mating-type gene locus. But the switched mt^+ chromosome, which was originally an mt^- chromosome and has changed genetic information from mt^- to mt^+ only in the mating-type gene locus, lacks such a viability gene. As regards zygospore viability, some more genes are thought to be located on the mt^+ and mt^- chromosomes (not necessarily linked to the mating-type gene locus) and/or the autosomes, because the selfed zygospores were completely inviable but the zygospores obtained in crosses, even with a normal minus clone, were viable.

Since a homothallic clone forms many zygospores within a clonal population and all their progeny do the same, it should switch mating types in both ways, from minus to plus and *vice versa*. Genetic studies, for example those by Van Winkle-Swift and Hahn (1986), clearly show the presence of mt^+ and mt^- gametes in a homothallic alga, *Chlamydomonas monoica*. Therefore, it is possible that a homothallic clone also may outbreed in some cases. We may gain an insight into breeding patterns in nature for homo- and heterothallic organisms from the following discussion. Selfed zygospores in homothallic clones, such as those of *Closterium moniliferum* (an ancestor of *C. ehrenbergii*), are highly viable, although selfed zygospores in some aberrant heterothallic clones, such as the selfing minus clones of *C. ehrenbergii*, yield no viable progeny. In addition, we have observed occasional parthenospore formation in homothallic clones of *C. moniliferum* but never in heterothallic clones of *C. ehrenbergii*, even when many gametes are released in the vicinity of fused gametes forming a zygospore (Ichimura and Kasai, 1987). We interpret these observations as follows. At many gene loci that control the process from spore wall formation through nuclear fusion and meiosis to formation of germlings, an exclusively outbreeding heterothallic organism might have accumulated deleterious recessive alleles, sheltered by normal alleles in the diploid phase. However, an exclusively inbreeding homothallic organism should have only normal alleles at these zygote-specific genes because all the loci except a few such as the mating-type locus become homozygous in their zygotes. In other words, the haploid complement of a vegetative cell contains all necessary genetic information for the above-mentioned zygotic process; otherwise an exclusive inbreeder would have perished under the harsh conditions. The argument not only explains the potency of parthenospore formation in the homothallic clones of *C. moniliferum* but also suggests a high degree of their inbreeding in nature. Parthenosporism could have evolved directly from homothallism.

VI. Conclusions

Polyploidy, perhaps associated with hybridization between different genotypes of the same and/or closely related species, should have played a major role in evolution of the high chromosome numbers of desmids. Because of their haplontic life cycle and their sex-determining mechanism, genome duplication events could have contributed to the evolution of desmids with much less difficulties than to higher plant evolution. Genomic or chromosomal rearrangements should have occurred during the speciation events associated with their niche differentiation, though we have still no clear insight into the mechanism of speciation. Most desmid species, including those mating groups (biological species) of *C. ehrenbergii*, must be very old polyploids because haploidization of their genome has almost been accomplished already.

Although we know little about the mechanism of meiotic chromosome pairing, it should be controlled by a species-specific intrinsic genetic system because interspecific crosses are known to cause irregular meiotic events accompanied with the lethality of F_1 progeny in the *C. ehrenbergii* species complex. Such genetic systems are known to work well even in triploid zygospores in some species of *Closterium* and *Pleurotaenium* (with haploid chromosome numbers more than n=53), but not in *Cosmarium* spp. (with less than n=26). On the other hand, sexual interactions between mt^+ and mt^- cells, from initiation of sexual agglutination through conjugation to thick-walled zygospore formation, are also controlled by a species-specific genetic system. The subtle but substantial differences in these two genetic systems, as related to pre- and postzygotic isolating mechanisms, enable us to distinguish between even two closely related mating groups with indistinguishable morphological characters. In future studies of speciation in desmids, we should clarify the mechanisms of these two important genetic systems, both of which are supposed to be under the control of the mating-type gene locus (cf. Goodenough and Ferris, 1987). At this moment we must content ourselves with expressing the following views. The mating-type gene locus itself is controlled by some other genetic factors, which can alter the mating systems of desmids, as is well known in the yeasts. The main evolutionary trends in desmids are species of outbreeding populations, whereas species of inbreeding or apomictic populations arise occasionally as branch lines. Sexual reproduction (at least certain component process) and dormant cell (zygospore or parthenospore) formation are fixed together by genetic and developmental constraints in all desmids. Desmid populations incapable of forming any dormant cell would perish sooner or later in an evolutionary time scale. Therefore, no matter how they might differentiate aberrant morphological characters, such populations should not have any evolutionary significance.

Acknowledgements

We appreciate the financial support of grants 58540445 and 61480013 from the Scientific Research Fund of the Ministry of Education, Science and Culture, Japan.

References

Biebel, P. (1964). The sexual cycle of *Netrium digitus*. *Am. J. Bot.* **51**, 697-704.

Blackburn, S. I. and Tyler, P. A. (1980). Conjugation, germination and meiosis in *Micrasterias mahabuleshwarensis* Hobson (Desmidiaceae). *Br. phycol. J.* **15**, 83-93.

Blackburn, S. I. and Tyler, P. A. (1981). Sexual reproduction in desmids with special reference to *Micrasterias thomasiana* var. *notata* (Nordst.) Grönblad. *Br. phycol. J.* **16**, 217-229.

Blackburn, S. I. and Tyler, P. A. (1987). On the nature of eclectic species - a tiered approach to genetic compatibility in the desmid *Micrasterias thomasiana*. *Br. phycol. J.* **22**, 277-298.

Brandham, P. E. (1965). Polyploidy in desmids. *Can. J. Bot.* **43**, 405-417.

Brandham, P. E. (1967). Time-lapse studies of conjugation in *Cosmarium botrytis*. II. Pseudoconjugation and an anisogamous mating behavior involving chemotaxis. *Can. J. Bot.* **45**, 483-493.

Brandham, P. E. and Godward, M. B. C. (1965a). The inheritance of mating type in desmids. *New Phytol.* **64**, 428-435.

Brandham, P. E. and Godward, M. B. C. (1965b). Meiosis in *Cosmarium botrytis*. *Can. J. Bot.* **43**, 1379-1386.

Cavalier-Smith, T. (1985). Cell volume and the evolution of eukaryote genome size. *In* "The Evolution of Genome Size" (T. Cavalier-Smith, ed.), pp. 105-184. John Wiley & Sons, Chichester.

Coesel, P. F. M. (1974). Notes on sexual reproduction in desmids. I. Zygospore formation in nature (with special reference to some unusual records of zygotes). *Acta Bot. Neerl.* **23**, 361-368.

Coesel. P. F. M. (1988). Biosystematic studies in the *Closterium moniliferum/ehrenbergii* complex (Conjugatophyceae, Chlorophyta) in western Europe. II. Sexual compatibility. *Phycologia* **27**, 421-424.

Coesel, P. F. M. (1989). Biosystematic studies in the *Closterium moniliferum/ehrenbergii* complex (Conjugatophyceae, Chlorophyta) in western Europe. IV. Distributional aspects. *Cryptogamie Algol.* **10**, 133-141.

Coesel, P. F. M. and de Jong, W. (1986). Vigorous chemotactic atraction as a sexual response in *Closterium ehrenbergii* Meneghini (Desmidiaceae, Chlorophyta). *Phycologia* **25**, 405-408.
Coleman, A. W. (1977). Sexual and genetic isolation in the cosmopolitan algal species *Pandorina morum*. *Am. J. Bot.* **64**, 361-368.
Egel, R. (1989). Mating-type genes, meiosis, and sporulation. *In* "Molecular Biology of the Fission Yeast" (A. Nasim, P. Young and B. F. Johnson, eds), pp. 31-73. Academic Press, London.
Franke, J. A. and Coesel, P. F. M. (1985). Isozyme variation within and between Dutch populations of *Closterium ehrenbergii* and *C. moniliferum* (Chlorophyta, Conjugatophyceae). *Br. phycol. J.* **20**, 201-209.
Fritsch, F. E. (1930). Über Entwicklungstendenzen bei Desmidiaceen. *Zeit. f. Bot.* **23**, 402-418.
Godward, M. B. E. (1966). The Chlorophyceae. *In* "The Chromosomes of the Algae" (M. B. E. Godward, ed.), pp. 1-77. Edward Arnold Ltd. London.
Goldstein, M. (1964). Speciation and mating behaviour in *Eudorina*. *J. Protozool.* **11**, 317-344.
Goodenough, U. W. (1985). An essay on the origins and evolution of eukaryotic sex. *In* "The Origin and Evolution of Sex" (H. O. Halvorson, ed.), pp. 123-140. Alan R, Liss, Inc. New York.
Goodenough, U. W. and Ferris, P. J. (1987). Genetic regulation of development in *Chlamydomonas*. *In* "Genetic Regulation of Development" (W. Loomis, ed.), pp. 171-189. Alan R. Liss, Inc. New York.
Hutchinson, G. E. (1968). When are species necessary? *In* "Population Biology and Evolution" (R. C. Lewontin, ed.), pp. 177-186. Syracuse University Press, Syracuse.
Ichimura, T. (1971). Sexual cell division and conjugation-papilla formation in sexual reproduction of *Closterium strigosum*. *Inter. Seaweed Symp.* **7**, 208-214.
Ichimura, T. (1981). Mating types and reproductive isolation in *Closterium ehrenbergii* Meneghini. *Bot. Mag. Tokyo* **94**, 325-334.
Ichimura, T. (1982). Isolating mechanisms in speciation of *Closterium*. *Jpn. J. Phycol.* **30**, 332-343. (in Japanese)
Ichimura, T. (1983). Hybrid inviability and predominant survival of mating type minus progeny in laboratory crosses between two closely related mating groups of *Closterium ehrenbergii*. *Evolution* **37**, 252-260.
Ichimura, T. (1985). Geographical distribution and isolating mechanisms in the *Closterium ehrenbergii* species complex (Chlorophyceae, Closteriaceae). *In* "Origin and Evolution of Diversity in Plants and Plant Communities" (H. Hara, ed.), pp. 295-303. Academia Scientific Book Inc. Tokyo.
Ichimura, T. and Kasai, F. (1984a). Post-zygotic isolation between allopatric mating groups of *Closterium ehrenbergii* Meneghini (Conjugatophyceae). *Phycologia* **23**, 77-85.
Ichimura, T. and Kasai, F. (1984b). Time lapse analyses of sexual reproduction in *Closterium ehrenbergii* (Conjugatophyceae). *J. Phycol.* **20**, 258-265.
Ichimura, T. and Kasai, F. (1985). Studies on the life cycle of *Spinoclosterium cuspidatum* (Bailey) Hirano (Conjugatophyceae). *Phycologia* **24**, 205-216.
Ichimura, T. and Kasai, F. (1987). Time-lapse analyses of sexual isolation between two closely related mating groups of the *Closterium ehrenbergii* species complex (Chlorophyta). *J. Phycol.* **23**, 523-534.
Ichimura, T. and Kasai, F. (1989a). Life cycle of homothallic and heterothallic clones of *Triploceras gracile* Bailey (Desmidiaceae, Chlorophyta). *Phycologia* **28**, 212-221.
Ichimura, T. and Kasai, F. (1989b) Genome differentiation in speciation of desmids (Chlorophyta). *Jpn. J. Phycol.* **30**, 305-319. (in Japanese)
Ichimura, T. and Watanabe, M. (1974). The *Closterium calosporum* complex from the Ryukyu Islands. Variation and taxonomical problems. *Mem. Nat. Sci. Mus. Tokyo* **7**, 89-102.
Kasai, F. and Ichimura, T. (1983). Zygospore germination and meiosis in *Closterium ehrenbergii* Meneghini (Conjugatophyceae). *Phycologia* **22**, 267-275.
Kasai, F. and Ichimura, T. (1984). Meiotic anomalies in intergroup crosses between closely related mating groups of *Closterium ehrenbergii* Meneghini (Conjugatophyceae). *Phycologia* **23**, 508-

510.
Kasai, F. and Ichimura, T. (1986). Morphological variabilities of three closely related mating groups of *Closterium ehrenbergii* Meneghini (Chlorophyta). *J. Phycol.* **22**, 158-168.
Kasai, F. and Ichimura, T. (1987). Stable diploids from intragroup zygospores of *Closterium ehrenbergii* Menegh. (Conjugatophyceae). *J. Phycol.* **23**, 344-351.
Kasai, F. and Ichimura, T. (1990) A sex determining mechanism in the *Closterium ehrenbergii* (Chlorolphyta) species complex. *J. Phycol.* **26**, 195-201.
Kasai, F. and Ichimura, T. Temperature optima of three closely related mating groups of the *Closterium ehrenbergii* species complex. *Phycologia* (in press).
Kies, L. (1964). Über die experimentelle auslösung von Fortpflanzungsvorgangen und die Zygotenkeimung bei *Closterium acerosum* (Schrank) Ehrenberg. *Arch. Protistenk.* **107**, 331-350.
Ling, H. U. and Tyler, P. A. (1974). Interspecific hybridity in the desmid genus *Pleurotaenium. J. Phycol.* **10**, 225-230.
Ling, H. U. and Tyler, P. A. (1976). Meiosis, polyploidy and taxonomy of the *Pleurotaenium mamillatum* complex. *Br. phycol. J.* **11**, 315-330.
Lippert, B. E. (1967). Sexual reproduction in *Closterium moniliferum* and *Cl. ehrenbergii. J. Phycol.* **3**, 182-198.
Mayr, E. (1963). "Animal Species and Evolution". The Belknap Press of Harvard University Press, New York, 797 pp.
Miyake, A. (1981). Physiology and biochemistry of conjugation in ciliates. *In* "Biochemistry and Physiology of Protozoa, vol. 4" (M. Levandowsky and S. H. Hutner, eds), pp. 125-198. Academic Press, New York.
Perkins, D. D. (1987). Mating-type switching in filamentous ascomycetes. *Genetics* **11**, 215-216.
Růžička, J. (1977). "Die Desmidiaceen Mitteleuropas. Band 1, 1. Lieferung", E. Schweizerbart'sche Verlagsbuchhandlung, Stuttgart. 291 pp.
Sarma, Y. S. R. K. (1983). Algal karyology and evolutionary trend. *In* "Chromosomes in Evolution of Eukaryotic Groups. Vol. 1" (Sharma, A. K. and Sharma, A., eds), pp. 177-223. CRC Press, Boca Raton, Florida.
Starr, R. C. (1954). Inheritance of mating type and a lethal factor in *Cosmarium botrytis* var. *subtumidum* Wittr. *Proc. Natl. Acad. Sci.* **40**, 1016-1063.
Starr, R. C. (1958). The production and inheritance of the triradiate form in *Cosmarium tupinii. Am. J. Bot.* **45**, 243-248.
Starr, R. C. (1959). Sexual reproduction in certain species of *Cosmarium. Arch. Protistenk.* **104**, 155-164.
Stebbins, G. L. (1971). "Chromosome Evolution in Higher Plants", Addison-Wesley, Reading, Massachusetts. 216 pp.
Sternberg, P. W. (1987). Control of cell type and cell lineage in *Saccharomyces cerevisiae. In* "Genetic Regulation of Development" (W. Loomis, ed.), pp. 83-108. Alan R. Liss, Inc. New York.
Therezien, Y. (1985). Contribution a l'etude des algues d'eau douce de la Bolivie. Les Desmidiales. *Nova Hedwigia* **41**, 505-576.
Thomasson, K. (1986). Algal vegetation in North Australian billabongs. *Nova Hedwigia* **42**, 301-378.
Van Winkle-Swift, K. P. and Hahn, J. -H. (1986). The search for mating-type-limited genes in the homothallic alga *Chlamydomonas monoica. Genetics* **113**, 601-619.
Watanabe, M. (1978). A taxonomic study of the *Closterium calosporum* complex (1). *Bull. Nat. Sci. Mus., Ser. B (Bot.)* **4**, 133-154.
Watanabe, M. M. and Ichimura, T. (1978). Biosystematic studies of the *Closterium peracerosum-strigosum-littorale* complex II. Reproductive isolation and morphological variation among several populations from the northern Kanto area in Japan. *Bot. Mag. Tokyo* **91**, 1-10.
West, G. S. (1899). On variation in the Desmidieae, and its bearings on their classification. *J. Linnean Soc., London, Bot.* **34**, 366-416.

White, M. J. D. (1978). "Mode of Speciation" W. H. Freeman, San Francisco. 453 pp.
Wiese, L. (1984). Mating systems in unicellular algae. *In* "Encyclopedia of Plant Physiology. New Series, Vol. 17, Cellular Interactions" (H.-F. Linskens and J. Heslop-Harrison eds.), pp. 238-260. Springer-Verlag, Berlin.
Wiese, L. and Wiese, W. (1977). On speciation by evolution of gametic incompatibility: A model case in *Chlamydomonas*. *Am. Nat.* **111**, 733-742.

Part IV

POPULATION BIOLOGY AND LIFE HISTORY EVOLUTION

(2) DEMOGRAPHY AND LIFE HISTORY EVOLUTION OF PLANTS

19 Optimal Growth Schedule of Terrestrial Plants

YOH IWASA

Department of Biology, Faculty of Science, Kyushu University, Fukuoka 812, Japan

I. Introduction

Plant growth can be regarded as the adaptive strategy through which the plant allocates the material obtained by photosynthesis each day of a growing season to various organs, such as leaves, stems, roots, flowers, fruits, and storage. The diverse patterns of plant growth and life history are observed in nature. They are presumably the results of the plant's adaptation to each environment. In this paper, I illustrate the use of optimization models in the study of plant growth by three examples: (1) shoot / root balance of a plant, (2) growth over multiple seasons, and (3) the plasticity of growth schedule in a stochastically changing environment.

II. Shoot/Root Balance

Plants show great phenotypic plasticity in growth. The numbers and relative sizes of organs often change with local environmental conditions. If a plant grows in a moist and nutrient-rich habitat, the size of root (below ground part) relative to shoot (above ground part) is small, while in a dry and open environment, the root to shoot ratio of the same species is larger (Chapin, 1980; see references in Iwasa and Roughgarden, 1984).

To understand the observed balance of shoot and root sizes as part of a plant's adaptation, Iwasa and Roughgarden (1984) analyzed a dynamic optimization model of resource allocation.

BIOLOGICAL APPROACHES AND EVOLUTIONARY TRENDS IN PLANTS ISBN 0-12-402960-4

A. Growth of a system with multiple vegetative growth

Let X_1, X_2, and R be the shoot size, the root size, and the accumulated reproduction. The daily net photosynthesis $g(X_1,X_2)$ increases with the shoot size X_1 because photosynthesis occurs in leaves. However, it also increases with the root size X_2 as water intake ability is necessary in maintaining high photosynthetic rate.

A plant starts from the initial sizes $(X_1(0), X_2(0))$ given by the material in a seed, and it grows by allocating the photosynthate to shoot growth, root growth, and reproductive activity:

$$\frac{dX_1}{dt} = u_1 g(X_1,X_2), \tag{1a}$$

$$\frac{dX_2}{dt} = u_2 g(X_1,X_2), \tag{1b}$$

$$\frac{dR}{dt} = u_0 g(X_1,X_2), \tag{1c}$$

where the allocation ratios $\{u_1(t), u_2(t), u_0(t)\}$ are functions of time, and satisfy constraints $u_1+u_2+u_0=1$, and $u_i \geq 0$ (i = 1, 2, and 0).

The material used to construct vegetative organs is discarded at the end of the season, and only the investment in reproduction (the maintenance and construction of flowers and fruits) contributes to the next generation. Now we are searching for the optimal growth schedule which maximizes the total reproductive activity:

$$\phi = R(T) \longrightarrow \text{maximum}. \tag{2}$$

Justification of this assumption lies in natural selection. A phenotype having greater reproductive success than others would increase its frequency in the next generation, and after many generations the population is expected to include only those individuals realizing the maximal reprodutive success under given constraints.

B. Maximum principle: dynamics of marginal values

The optimization problem formulated as above can be solved by using Pontryagin's maximum principle (Pontryagin *et al.*, 1962). The basic concept underlying the technique is *marginal value.* Suppose that the leaf area X_1 happens to increase by unit amount at time t. Then the enhanced photosynthetic rate produces additional photosynthate from t until the end of the season T, and it finally improves total reproduction R(T). The increment of reproductive performance R(T) caused by the unit increase of shoot size is called *the marginal value of shoot size.* It decreases with time t and becomes zero at the end of the season, because the vegetative part can contribute to the reproductive success only by improving the photosynthesis until the end of the season. Given the marginal values of shoot size, root size, and reproduction, the optimal strategy each day in the season is to invest the photosynthate only in the organ with the highest marginal value. The maximum principle is the mathematical technique of systematically calculating the optimal strategy based on these ideas (for details see Iwasa and Roughgarden, 1984).

C. Switch from vegetative to reproductive growth

In the optimal growth schedule, there is a clear switch from vegetative to reproductive growth. Before the switching time no reproduction occurs, and after then the vegetative growth stops and the photosynthetic product should be allocated only to reproduction. A clear switch was predicted by models with a single variable for the size of vegetative organs (Cohen, 1971, 1976; Vincent and Pulliam, 1980; King and Roughgarden, 1982a, 1982b; Schaffer *et al.*, 1982; Chiariello and Roughgarden, 1984; Hara, Kawano, and Nagai, 1988), and the predicted timing was tested by comparative and experimental data (Penning de Vries *et al.* 1974; King and Roughgarden, 1983).

D. Balanced growth

The trajectory of shoot and root sizes during vegetative growth is illustrated in Fig. 1. We can prove that the optimally growing plant should grow along on *balanced growth path* indicated by a solid curve in Fig. 1 (calculated by $\partial g/\partial X_1 = \partial g/\partial X_2$). Along the balanced growth path, a plant has the root/shoot ratio that maximizes the daily net photosynthesis for a given total biomass. Specifically, if the daily net photosynthsis is

$$g(X_1, X_2) = 1/(a_1/LX_1^{b_1} + a_2/WX_2^{b_2}), \qquad (3)$$

where L is the light intensity at canopy and W is the moisture in soil. Then along the balanced growth path, root and shot sizes satisfy the allometric relation:

$$[\text{Root size}] \propto \left(\frac{[\text{light}]}{[\text{moisture}]}\right)^{1/(1+b_2)} [\text{Shoot size}]^{(1+b_1)/(1+b_2)}, \qquad (4)$$

which shows that the root/shoot ratio should increase with the light intensity and decrease with moisture availability. Also the simple ratio of root to shoot may change with plant size.

How should the plant grow when its shape deviates from the balanced growth path due either to pruning or to environmental changes ? For example, after shoot pruning, the plant should immediately stop root growth and should invest all its

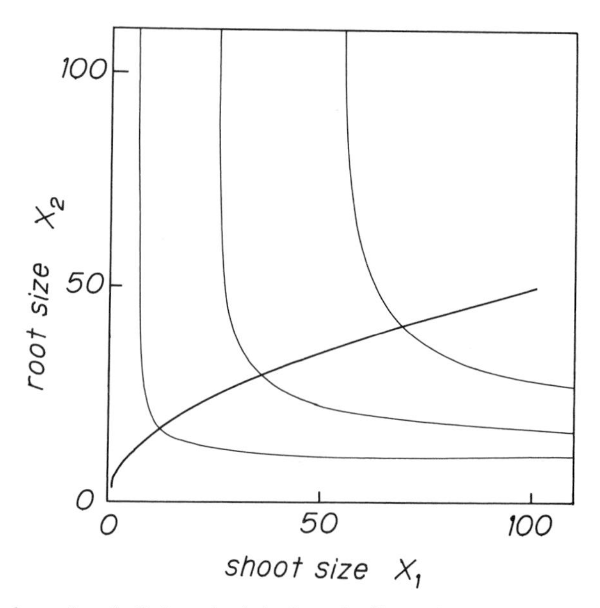

Fig. 1 Balanced growth path. Horizontal axis is shoot size X_1; vertical axis, root size X_2. Fine curves are contours of the daily net photosynthetic rate $g(X_1,X_2)$ given by Eq. (3) with $a_1=2$, $a_2=60$, $b_1=0.5$, $b_2=2$, $L=1$, and $W=1$. The solid curve indicates the balanced growth path. A typical optimal growth schedule of a plant is the quickest convergence to the balanced growth path, followed by simultaneous growth of root and shoot along the path, and then reproductive growth.

photosynthate in shoot recovery, and the simultaneous growth of shoot and root restarts only when their ratio returns to a balanced growth curve. This is also in accord with the observed pattern of shoot/root balance (Russell, 1977; reference in Iwasa and Roughgarden, 1984).

III. Deciduous Perennials

A plant may store material with which it can reconstruct its photosynthetic system at the beginning of the next season. What is the condition under which such a perennial life history is more advantageous than an annual one ? Of the material produced in a year, how much should be allocated to reproduction in the same year and how much should be saved for the following years ? How should these results depend on parameters such as the length of the growing season, the stability of the habitat, and the photosynthetic rate ?

To answer these questions, Iwasa and Cohen (1989) developed an optimal allocation model for a deciduous plant having two parts: a *production part* including vegetative organs working for photosynthesis (leaves, stems, roots, etc.) and a *storage part* including both stored material and accumulated reproductive investment for the year (also see Schaffer 1983; Pugliese, 1988). The model is the combination of continuous-time models for a within-season growth schedule and discrete-time models for between-season allocation (Cole, 1954; Gadgil and Bossert, 1970; Charnov and Schaffer, 1973; Schaffer, 1974; Schaffer and Schaffer, 1977; Taylor *et al.*, 1974; Leon, 1976; Hirose & Kachi, 1982).

A. Growth over multiple seasons

The year is indicated by a subscript n (= 1,2,3..) and time within a season by a continuous parameter, t ($0 \leq t \leq T$). Let $F_n(t)$ be the size of the production part, and $S_n(t)$ be that of the storage part. The daily net photosynthetic rate increases with the production part size F but saturates as F becomes very large because of self-shading in the canopy and nutrient depletion in the soil. Photosynthetic product is allocated between the production part and the storage part.

$$\frac{dF_n}{dt} + \frac{dS_n}{dt} = \frac{fF_n}{(1+hF_n)}, \qquad (n=1,2,..;\ 0 \leq t \leq T). \qquad (5)$$

The production part is lost at the end of each growing season (the plant is deciduous), and needs to be reconstructed at the following season's beginning: $F_n(0) = 0$, (n = 1,2,3,..). The production part F does not include trunks of trees.

The size of the storage part starts with a positive value if a fraction of material produced in the previous year is saved:

$$S_n(0) = \gamma(S_{n-1}(T) - R_{n-1}), \qquad (n = 2,3,..), \qquad (6)$$

where R_n is the total reproductive investment made in the nth season by constructing and maintaining flowers and fruits, and satisfies the constraint: $0 \leq R_n \leq S_n(T)$. The final storage part size $S_{n-1}(T)$ minus reproductive investment R_{n-1} is the material stored for the nth year. Only fraction γ of the stored material can be recovered from storage. The initial size for the first year is given by seed size: $S_1(0) = s_{seed}$. The growth rate of the production part may also be constrained by an upper limit, e.g. $dF_n/dt \leq aF_n+b$.

The optimal growth schedule is the one which maximizes the lifetime investment to reproduction:

$$\phi = \sum_{n=1}^{\infty} \sigma^n R_n \longrightarrow \text{maximum}, \qquad (7)$$

where σ is the annual survivorship ($0 \leq \sigma \leq 1$). The strategy that a plant chooses is its growth schedule within a season and the allocation of resources between reproductive investment and saving for the following season.

B. Growth schedule within a season

The optimal growth schedule within a season (say, year n), when the initial storage size $S_n(0)$ is given, is the one which maximizes the final size of storage part $S_n(T)$, because reproduction both in the same year and in the following years is affected only by $S_n(T)$ (see Yokoi (1976)). The problem is solved by the maximum principle, as in the previous model.

The optimal growth schedule in a season is composed of three phases. In the first phase, the plant constructs its production part at the maximum speed, mainly using the material stored in the previous year. After the storage organ becomes empty, the plant continues to grow vegetatively, using all the daily net photosynthate. At a certain time, vegetative growth stops, and the plant starts accumulating photosynthetic products in the storage part.

Let $\psi(S_o)$ be the size of the storage part at the end of a season as a function of the initial storage size S_o:

$$\psi(S_o) = \max S(T), \tag{8}$$

where "max" indicates that the plant follows the optimal allocation schedule during the season. $\psi(S_o)$ is an increasing function but with diminishing rate of increase.

C. Allocation between reproduction and storage

In each season, the plant must decide how much of the accumulated storage product should be used for reproduction and how much should be saved for the construction of production parts in the following year. The problem can be solved by dynamic programming (Bellman, 1957). Let V(S) be the expected total reproductive investment of a plant from the nth season until its death, provided that the plant has a storage part of size S at the end of the nth season:

$$V(S_n(T)) = \max\, [R_n + \sigma R_{n+1} + \sigma^2 R_{n+2} + \ldots], \tag{9}$$

where the symbol "max" indicates that the allocation schedule is chosen optimally. Equation (9) satisfies the recursive equation:

$$V(S_n(T)) = \max_{0 \leqq R_n \leqq S_n(T)} [R_n + \sigma V(\psi(\gamma(S_n(T) - R_n)))]. \tag{10}$$

The optimal reproductive investment R_n is the one which attains the maximum of Eq. (10). It is determined by considering the tradeoff between reproduction in the current year and that in future years, indicated by the first and the second terms, respectively. The optimal solution is:

$$R_n = \begin{cases} 0, & \text{if } S_n(T) < S^*, \\ S_n(T) - S^*, & \text{if } S_n(T) > S^*, \end{cases} \tag{11}$$

where the critical storage size S^* is the value of S which maximizes $\sigma \psi(\gamma S) - S$.

If the annual survival σ is small, the optimal storage size S^* is zero. Then all the photosynthate in the storage part produced in the first year is used for reproduction, indicating annual life-history. In contrast, if the optimal storage size S^* is positive, the plant takes polycarpic perennial life-history -- it experiences several immature years during which $S_n(T)$ is less than the optimal storage size S^*, and saves all of the resources for the next season. When $S_n(T)$ becomes greater than S^*, the plant invests the excess $S_n(T) - S^*$ in reproductive activities, and repeats the same growth schedule thereafter.

D. *Optimal growth schedule*

The model of an optimal growth schedule over multiple seasons explains many regularities of plant growth and leaf phenology which have been observed in nature.

1. *Annual versus perennial*

Annuality is favored over perenniality if the habitat reliability σ is low, i.e. if the habitat of a plant has a high chance of being destroyed before the end of the next season, due to such factors as herbivores, pathogens, or catastrophic physical disturbances. A low storage efficiency γ has a similar effect. A short growing season T and a small net photosynthetic rate f also favor annual life.

2. *Reproductive effort*

The ratio of reproductive investment to the total annual productivity $R/(S(T)+F(t_2))$ is called the reproductive effort (Harper *et al.*, 1970; Abrahamson and Gadgil, 1973; Pianka, 1976). Reproductive effort is large for annual plants, which invest all their available resources in current reproduction. If the length of a season T increases and exceeds a threshold, the polycarpic perennial life reaches the optimal stage, in which the reproductive effort for a mature plant is the lowest. However, as the productivity in the environment increases further, the reproductive effort again increases and can become even larger than the value for annuals.

3. *Timing of leaf production*

Comparative studies have revealed that the phenology of leaf production and shoot elongation is correlated with the status of trees along a successional axis and with the habitat characteristics (Marks, 1975; Maruyama, 1978; Bicknell, 1982; Kikuzawa, 1983; 1984). For example, among deciduous trees, species living in forest gaps or early successional species tend to continue to expand their leaves and shoots until mid-season, while climax species often produce most of their leaves at the beginning of a season.

Three predictions are derived from the optimal growth model. First, the timing of termination of leaf production within a season becomes earlier as the plant grows (Fig. 2).

Second, the timing of leaf production for a small immature plant and for a mature plant are controlled by different factors: productivity, measured by the maximum relative growth rate f multiplied by the length of a season T, is the major determinant of the timing for small immature plants (causing a plant in a more productive site to stop leaf production later). Hori and Oshima (1986) demonstrated that a perennial herb shifts the stopping of leaf production within a season according to the light availability.

Third, the timing for mature plants is controlled by the stability of the habitat, measured by the annual survival σ multiplied by the storage efficiency γ (shorter period of leaf production under a more stable environment), but is almost independent of the productivity fT. Intuitive justification is that, in a stable environment, the plant saves resources for the next season so that in the beginning of the next season sufficiently many leaves are produced to fill all the space, resulting in little benefit

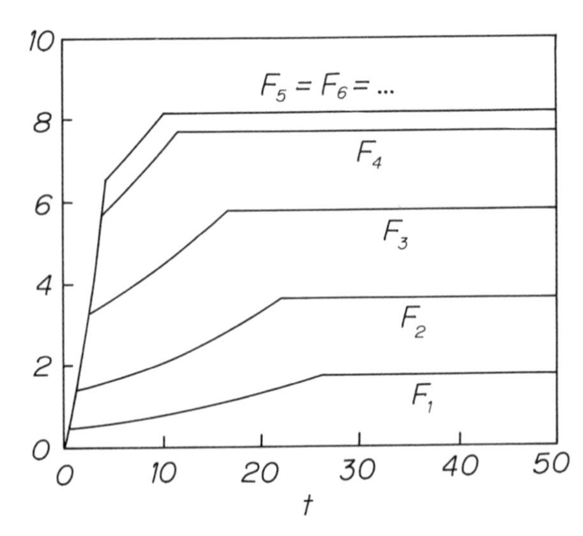

Fig. 2 Leaf production in the immature years. Horizontal axis is the time within a season; vertical axis, the production part size. The numeral indicates the year (n). Parameters are: $S_1(0) = 0.5$, and $a = 0.2$, $b = 1$, $f = 0.05$, $h = 0.05$, $\gamma = 0.7$, and $T = 50$. The plant first reproduces in the fourth year and repeats the steady-state growth thereafter. The timing of the cessation of leaf production becomes earlier as a plant grows.

for making additional leaves. In contrast, since the next year's reproduction is not dependable in an unstable habitat, the plant tends to use more of its resources for the current year's reproduction. At the beginning of the next season, after the initial leaf flush, there is still some space left, in which the plant continues to produce leaves over a considerable fraction of the season.

However, the effects of these two factors are difficult to separate because a larger productivity (large f) caused by a higher resource availability is often accompanied by environmental disturbance, and hence by a low stability (small σ) (Tilman, 1982).

IV. Growth in Stochastic Environments

Plant often grows in a heterogeneous environment, changing drastically with time. The optimally growing plant continues vegetative growth until quite late in the season in a constantly productive environment, while it starts reproduction earlier in an unproductive environment. Hence the plant should change its growth schedule according to the resource availability in the local environment, implying the importance of plasticity.

If the resource level changes day to day stochastically, the optimal decision on allocation between vegetative growth and reproductive activity can be analyzed by stochastic dynamic programming.

A. *Growth under stochastic evironment*

Consider an annual plant having a growth period of length T. Denote the vegetative part size by F. The daily net photosynthesis increases with X, the environmental resource level; e.g. light intensity at canopy or nutrient availability in soil. X may fluctuate with time. Each day in the growing season, the material obtained by photosynthesis is allocated either to vegetative growth or to reproductive activity with the ratio of 1–u:u. The vegetative part size increases as

$$\frac{dF}{dt} = (1-u)\frac{aXF}{1+bF}, \tag{12}$$

starting from $F(0)=F_0$ given by the seed size. The expected total reproductive activity done in the whole season is

$$\phi = E\left[\int_0^T u\frac{aXF}{1+bF}\,dt\right] \longrightarrow \text{maximum.} \tag{13}$$

I then search for the growth schedule which maximizes ϕ.

As the simplest model of stochastic change, I first consider 2-state Markovian processes, in which environmental resource level X jumps between high (x_1) and low (x_2) levels a random number of times. Let p_1 be the rate of transition from x_1 to x_2, and p_2 be the rate of transition from x_2 to x_1. The optimal allocation u depends not only on the time t but also depends on the size F(t) and on the current resource level X(t).

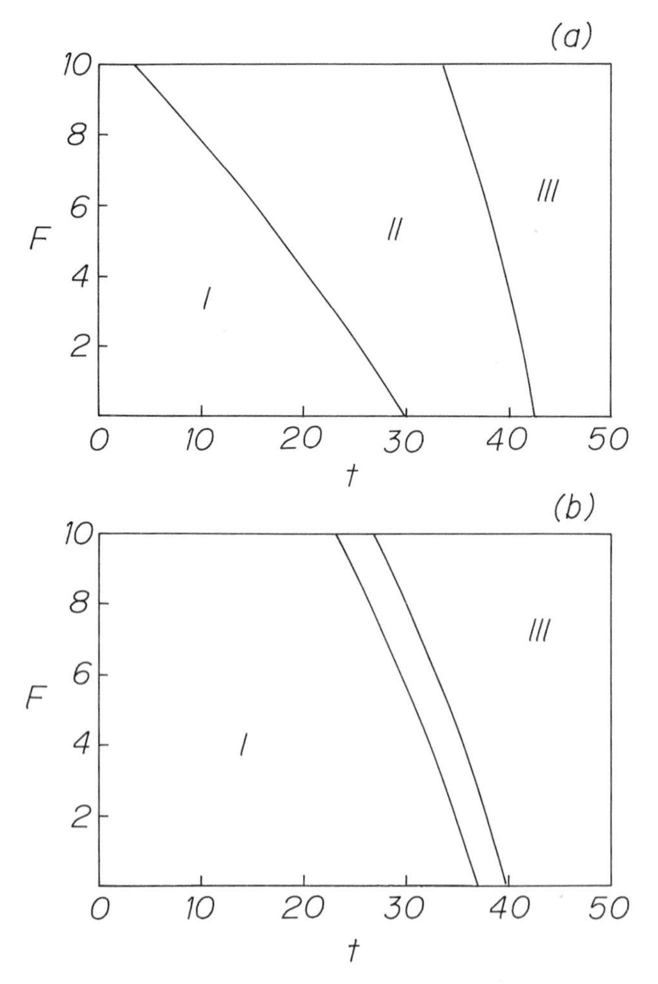

Fig. 3 The optimal growth rule for the plant living in a fluctuating environment. Horizontal axis is the time of the season t, vertical axis is the plant size F. In Region I, the plant should grow vegetatively in both environments; in Region III it should reproduce. In Region II the plant should grow vegetatively if the current resource level X is high, and should reproduce if X is low. Region II is wider if the plant is adapted to the slowly changing environment, as in (a) (p_1=p_2=0.01), than if it is adapted to a quickly changing environment as in (b) (p_1=p_2=0.2). Other parameters are: a = 0.05, b = 0.05, T = 50, x_1 = 3, and x_2 = 0.8.

B. Optimal growth rule

The problem, thus formulated, can be solved using stochastic dynamic programming (Mangel, 1985; for application to animal behavior see Iwasa *et al.*, 1984; Mangel & Clark, 1988). Let $V_i(F, t)$ be the expected reproductive investment to be made before the end of the season, provided that the vegetative size is F and the resource level is x_i at time t and that the plant follows the optimal schedule thereafter. By considering events that possibly can happen during a short time interval of length Δt, we can derive a pair of partial differential equations for $V_i(F, t)$ $(i=1, 2)$, and thereby calculate the optimal allocation (Iwasa, MS).

The optimal growth rule is illustrated on the F-t place (Fig. 3a). At the beginning of a season (Region I), the plant should grow vegetatively at both resource levels. Toward the end of a season (Region III), the plant should reproduce. In Region II, which lies between the other two, the plant should use different strategy depending on the current resource level: it should grow vegetatively $(u=0)$ if the current resource availability is high $(X=x_1)$, and it should reproduce if X is low (x_2). The plant growth in Region II is *tracking* the fluctuation of environmental resource availability.

Figure 3b illustrates the optimal growth rule if the environmental changes are more frequent than in Fig. 3a. Region II becomes narrower, and the optimal plant should follow an almost fixed schedule of phenology independent of the resource level that it experiences. If the environmental resource supply changes quickly, the current level does not give a reliable estimate of the future resource availability.

C. Continuously changing environment

The optimal growth schedule can be calculated also for the case in which the resource availability may change continuously with time, following diffusion processes, which are characterized by the standard level K and the exponential rate r of decay in autocorrelation (Iwasa, MS).

Figure 4 illustrates the growth rate of plants averaged over the whole season if the resource level is kept at a constant level X in the laboratory. The plant, however, uses the decision rule, which is optimal under fluctuating environments. A curve labled O indicates that the average growth rate increases with the resource supply level. This is an *optimistic plant*, which produces large vegetative organs for a high rate of resource supply. In contrast, the other curve, labeled P, is for a *pessimistic plant*, which does not produce large vegetative organs even under favorable conditions. This plant is adapted to the environment in which the resource level fluctuates with time very quickly, and hence the plant *suspects* that the currently high resource supply will be reduced very soon to the low standard level. Both plants have the same photosynthetic ability and the same resource supply, but differ in the scheme of environmental changes to which each plant is adapted.

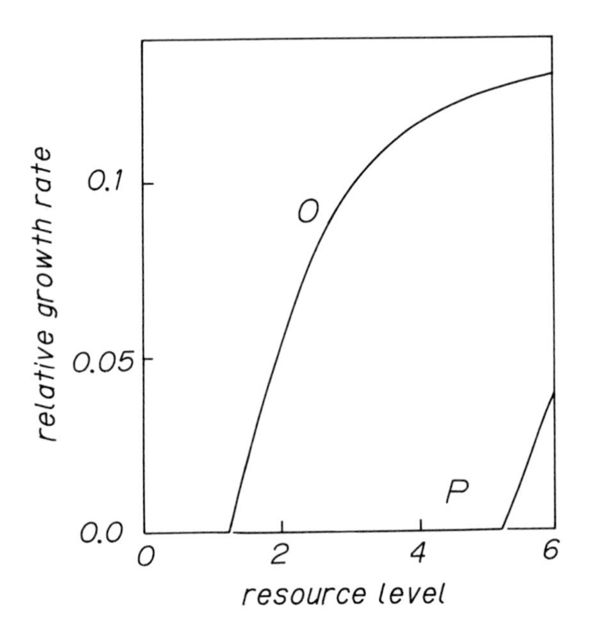

Fig. 4 The average growth rate for the whole growing period (1/T)log{F(T)/F(0)} when the plant is given a constant resource supply X indicated by the horizontal axis. The plant follows the optimal growth rule under a fluctuating environment. An optimistic plant (labeled O) adapted to the slowly changing environment makes large vegetative organs in responding to the high resource level, while a pessimistic plant (labeled as P) adapted to quickly changing environment would not respond to the favorable environmental conditions. Standard resource level is K = 0.2, and the exponential decay rate of autocorrelation is r = 0.1 for curve O, and r = 0.5 for curve P. Other parameters are: a = 0.05, b = 0.01, T = 50, and F_o = 0.1.

D. *Slow growing plants*

Plant species growing under severely restricted nutrient supply tend to grow slowly in experimentally given environments with high nutrient availability (Chapin, 1980; Kachi and Rorison, 1989). This is a charactersitic of *stress-tolerant* plants (Grime, 1979, 1988; Grime *et al.*, 1986; Crick & Grime, 1987; Sibly and Grime, 1986).

The apparent paradox of slow-growing plants can be understood in the light of stochastic dynamic optimization studied here. Production of large vegetative organs is accompanied by large costs in construction and maintenance, which are paid to the plant only when the environmental resource level remains high for a sufficiently long time to pay back the investment. A *pessimistic plant*, adapted to the environment with a constantly low resource level and with occasional short flushes of resource supply, would not respond to the high resource level experimentally given.

V. Discussion

Other problems of plant growth can also be studied by dynamic optimization models. For example, plants with a storage organ can recover the photosynthetic organs after their destruction by herbivory or fire. The optimal size of storage for insurance can be examined by stochastic dynamic programming (Iwasa and Cohen, MS). Still other examples are the production of defense substances by plants and insect herbivore dynamics.

An important aspect not discussed here is the competitive resource capture between plants. Studies of dynamic game models have shown that competition for light makes each plant produce a shoot larger than in the solitary optimal model (Mirmirani and Oster, 1978; Makela 1985; Iwasa *et al.*, 1985).

Although all the models examined here deal with the pattern of vegetative growth, the measure of adaptation to the plant is expected lifetime reproductive success (or fitness), rather than total production, maximum biomass, or net carbon gain. This is the straightforward way of modelling the adaptation of plant growth from the viewpoint of evolutionary population ecology.

Acknowledgement

This work was supported in part by a Grant-in-Aid (Bio Cosmos Program) from the Ministry of Agriculture, Forestry and Fisheries. (BCP90-III-A-2)

References

Abrahamson, W.G. and Gadgil, M.D. (1973). Growth form and reproductive effort in golden rods (*Solidago*, Compositae) *Am. Nat.* **107**, 651-661.

Bellman, R. (1957). "Dynamic programming." Princeton, Princeton Univ. Press.

Bicknell, S.H. (1982). Development of canopy stratification during early succession in northern hardwoods. *Forest Ecol. Manag.* **4**, 41-51.

Charnov, E. and Schaffer, W.M. (1973). Life history consequences of natural selection: Cole's result revisited. *Am. Nat.* **107**, 791-793.

Chapin, F.S. III. (1980). The mineral nutrietion of wild plants. *Annu. Rev Ecol. Syst.* **11**, 233-260.

Chiariello, N. and Roughgarden, J. (1984). Storage allocation in seasonal races of an annual plant: optimal versus actual allocation. *Ecology* **65**, 1290-1301.

Cohen, D. (1971). Maximizing final yield when growth is limited by time or by limiting resources. *J. theor. Biol.* **33**, 299-307.

Cohen, D. (1976). The optimal timing of reproduction. *Am. Nat.* **110**, 201-207.

Cole, L. (1954). The population consequences of life history phenomena. *Q. Rev. Biol.* **29**, 103-137.

Crick, J.C. and Grime, J.P. (1987). Morphorogical plasticity and mineral nutrient capture in two herbaceous species of contrasted ecology. *New Phytol.* **107**, 403-414.

Gadgil, M. and Bossert, W.H. (1970). Life history consequences of natural selection. *Am. Nat.* **104**, 1-24.

Grime, J.P. (1979). "Plant strategies and vegetational processes." Wiley, Chichester.

Grime, J.P. (1988). The C-S-R model of primary plant strategies -- origins, implications and test.

In "Evolutionary plant biology" (L.D. Gottlieb and S. Jain, eds) pp. 371-393. Chapman Hall.

Grime, J.P., Crick, J.C. and Rincon, J.E. (1986). The ecological significance of plasticity. "Plasticity in plants." (D.H. Jennings and A.J. Trewavas, eds), Company of Biologists, Cambridge, pp.5-29.

Hara, T., Kawano, S. and Nagai, Y. (1988). Optimal reproductive strategy of plants, with special reference to the modes of reproductive resource allocation. *Plant Species Biol.* **3**, 43-59.

Harper, J.L., Lovell, P.H. and Moore, K.G. (1970). The shape and sizes of seeds. *Annu. Rev. Ecol. Syst.* **1**, 327-356.

Hirose, T. and Kachi, N. (1982). Critical plant size for flowering in biennials with special reference to their distribution in a sane dune system. *Oecologia* (Berl.) **55**, 281-284.

Hori, Y. and Oshima, Y. (1986). Life history and population dynamics of the Japanese Yam, *Dioscorea japonica* Thunb. I. Effect of initial plant size and light intensity on growth. *Bot. Mag. Tokyo* **99**, 407-418.

Iwasa, Y. (MS). Pessimistic plants: the optimal growth schedule of a plant under fluctuating environment.

Iwasa, Y. and Cohen, D. (1989). Optimal growth schedule of a perennial plant. *Am. Nat.* **133**, 480-505.

Iwasa, Y. and Cohen, D. (MS). The storage for recovery after disturbance.

Iwasa, Y. and Roughgarden, J. (1984). Shoot/root balance of plants: optimal growth of a system with many vegetative organs. *Theor. Pop. Biol.* **25**, 78-105.

Iwasa, Y., Suzuki, Y. and Matsuda, H. (1984). Theory of oviposition strategy of parasitoid. I. Effect of mortality and limited egg number. *Theor. Pop. Biol.* **26**, 205-227.

Iwasa, Y., Cohen, D. and Leon, J.A. (1985). Tree height and crown shape as result of competitive game. *J. theor. Biol.* **112**, 279-297.

Kachi, N. and Rorison, I.H. (1989). Optimal partitioning between roots and shoot in plants with contrasted growth rates in response to nitrogen availability. *Func. Ecol.* **3**, 549-559.

Kikuzawa, K. (1983). Leaf survival of woody plants in deciduous broad-leaved forests. 1. Tall trees. *Can. J. Bot.* **61**, 2133-2139.

Kikuzawa, K. (1984). Leaf survival of woody plants in deciduous broad-leaved forests. 2. Small trees and shrubs. *Can. J. Bot.* **62**, 2551-2556.

King, D. and Roughgarden, J. (1982a). Multiple switches between vegetative and reproductive growth in annual plants. *Theor. Pop. Biol.* **21**, 194-204.

King, D. and Roughgarden, J. (1982b). Graded allocation between vegetative and reproductive growth for annual plants in growing seasons of random length. *Theor. Pop. Biol.* **22**, 1-16.

King, D. and Roughgarden, J. (1983). Energy allocation patterns of the California grassland annuals *Plantago erecta* and *Clarkia rubicunda*. *Ecology* **64**, 16-24.

Leon, J.A. (1976). Life histories as adaptive strategies. *J. theor. Biol.* **60**, 301-335.

Makela, A. (1985). Differential games in evolutionary theory, height growth strategies of trees. *Theor. Pop. Biol.* **27**, 239-267.

Mangel, M. (1985). "Decision and control in uncertain resource systems." Academic Press.

Mangel, M. and Clark, C.W. (1988). "Dynamic modelling of behavior." Princeton Univ. Press, Princeton.

Marks, P. (1975). On the relation between extension growth and successional status of deciduous trees of the northeastern United States. *Bull. Torrey Bot. Club.* **102**, 172-177.

Maruyama, K. (1978). Shoot elongation characteristics and phenological behavior of forest trees in natural beech forest: ecological studies on natural beech forest (32). *Bull. Niigata Univ. Forest.* **11**, 1-30. (in Japanese with English summary).

Mirmirani, M. and Oster, G. (1978). Competition, kin selection, and evolutionary stable strategies. *Theor. Pop. Biol.* **13**, 304-339.

Penning de Vries, F.W.T., Brunsting, A. and van Laar, H. (1974). Products, requirements and efficiency of biosynthesis: a quantitative approach. *J. theor. Biol.* **45**, 339-377.

Pianka, E. (1976). Natural selection of optimal reproductive tactics. *Am. Zool.* **16**, 775-784.

Pontryagin, L.S., Boltyanskii, V.G., Gamkrelidze, R.V. and Mischenko, E.F. (1962). "The mathematical theory of optimal processes." trans. by K.N. Trirogoff. Interscience Pub., John Wiley

& Sons, New York.
Pugliese, A. (1988). Optimal resource allocation in perennial plants: a continuous-time model. *Theor. Pop. Biol.* **34**, 215-247.
Russell, R.S. (1977). "Plant root systems: their function and interaction with soil." London: McGraw Hill
Schaffer, W.M. (1974). Selection for optimal life histories: the effects of age structure. *Ecology* **55**, 291-303.
Schaffer, W.M. (1983). The application of optimal control theory to the general life history problem. *Am. Nat.* **121**, 418-431.
Schaffer, W.M. and Schaffer, M.V. (1977). The adaptive significance of variations in reproductive habit in the Agavaceas. *In* "Evolutionary ecology" (B. Stonehouse and C.M. Perrins, eds), pp. 26-276.
Schaffer, W.M., Inouye, R.S. and Whittam, T.S. (1982). The dynamics of optimal energy allocation for an annual plant in a seasonal environment. *Am. Nat.* **120**, 787-815.
Sibly, R.M. and Grime, J.P. (1986). Strategies of resource capture by plants -- evidence for adversity selection. *Am. Nat.* **118**, 247-250.
Taylor, H.M., Gourley, R.S., Lawrence, C.E. and Kaplan, R.S. (1974). Natural selection of life history attributes: an analytical approach. *Theor. Pop. Biol.* **5**, 104-122.
Tilman, D. (1982). "Resource competition and community structure." Princeton Univ. Press, Princeton.
Vincent, T.L. and Pulliam, H.R. (1980). Evolution of life history strategies for an asexual plant model. *Thoer. Pop. Biol.* **17**, 215-231.
Yokoi, Y. (1976). Growth and reproduction in higher plants. I. Theoretical analysis by mathematical models. *Bot. Mag. Tokyo* **89**, 1-14.

20 Annual Plants: A Life-History and Population Analysis

ANDREW R. WATKINSON

School of Biological Sciences, University of East Anglia, Norwich NR4 7TJ, U.K.

I. Introduction

Annual plants are typically thought of as breeding once and having a simple life-cycle that lasts for less than a year. But this is nothing more than a caricature. As a result of seed dormancy the true life-span of annual plants lasts at least a year and for many it can be measured in tens of years. Indeterminate and determinate growth also produce a range of breeding schedules with some species showing relatively synchronous reproduction whilst others flower and set seed over an extended period either as a result of the production of a succession of monocarpic shoots or the persistent reproductive activity of polycarpic shoots. The timing of the life cycle may also vary. Some plants are truly ephemeral, completing their life cycle whenever the conditions are suitable and the resources available within a year, whilst others are either summer or winter annuals.

With such a range of life-histories annual plants can be expected to exhibit a wide range of dynamical behaviour. Symonides (1988) categorized these into three major groups. The first group includes those ephemeral species which occur primarily as a seed bank, appearing above ground only when conditions are favourable. Many desert ephemerals and weeds belong to this group. Clearly the unpredictability of the environment in time is a major factor influencing the population dynamics of such species. The second includes those species which are very short-lived at a particular site, and which rely on long distance dispersal for the colonization of ephemeral patches in time and space. They are fugitive species. The third group encompasses all those species that are present on a site for many years. All of these

BIOLOGICAL APPROACHES AND EVOLUTIONARY TRENDS IN PLANTS ISBN 0-12-402960-4

plants show fluctuations and oscillations in abundance from year to year, some large and some small. The sites on which they occur are open but relatively predictable in time and space.

In this paper I firstly examine briefly the influence of environmental variation in time and space on the life-history characteristics of annual plants. I then go on to explore the temporal and spatial dynamics of annual plant populations.

II. Life History Characteristics

It is widely recognized that annuals are associated with disturbance. Cole (1954) focused our attention on the question of what factors favour annual over perennial reproduction by demonstrating that an annual would have to produce only one more seed than a perennial to have the same rate of population increase. It was on this basis that he questioned why the world was not dominated by annuals. But Schaffer and Gadgil (1975) demonstrated that if juvenile and adult survivorship differ then annual plants can be expected to predominate only in those environments where the there is low adult survivorship. Typically annual plants will then only predominate in environments where disturbance prevents a full cover of perennials from developing whether this is, for example, a result of ploughing in productive habitats or drought in unproductive habitats.

The timing of the disturbance has a major impact on the phenology of plants. Predictable disturbance in either the summer or the winter can be expected to select for either winter or summer annuals respectively. A consideration of the annuals that occur in coastal habitats (Watkinson and Davy, 1985) shows that summer annuals predominate on the strandline and saltmarsh where disturbance occurs primarily in the winter as a result of wave action. In contrast winter annuals predominate in the dry dune habitats where summer drought is the major disturbance preventing a full cover of perennials from developing.

The predictability of the timing of disturbance in relation to the length of the growing season also has a major impact on phenology. Where the length of the growing season is typically unpredictable, plants can be expected to start growing early and flower over an extended period - they are uniseasonally iteroparous (Kirkendall and Stenseth, 1985). In contrast where the length of the growing season is much more predictable, big bang reproduction predominates. The plant puts all its resources into vegetative growth at first but then switches to reproduction (Cohen, 1971; King and Roughgarden, 1982). But why do some annual species within the same habitat have much longer juvenile periods than others. Delayed flowering poses the significant risk of encountering those conditions which favour the annual habit, such as summer drought, so that successful setting of seed becomes more uncertain (Ritland, 1983). Presumably in such cases those species which delay flowering are more tolerant of these very conditions. Mooney *et al.*(1986), for example, found in a study of co-occurring grassland annuals that the late flowering species did not experience the same degree of water stress as the early flowering ones. By tapping deeper soil water than the shorter lived species, the longer lived annuals were able to extend their growth period and attain considerably greater biomass (Gulmon *et al.*, 1983). The variability in plant size was, however, much greater as result of not all of the plants reaching the deeper water supply.

Delayed flowering can be viewed as a riskier, but on average more productive mode of reproduction in a variable environment. Seed germination similarly holds a risk in an unpredictable environment where there is a chance of reproductive failure, whether this occurs at the time of germination or just prior to flowering. Environmental unpredictability can, therefore, be expected to have been a major selective force in determining the germination and dormancy strategies of annuals. Some annuals produce only a transient seed bank - all the seeds germinate in the year that they are produced. In contrast others have a persistent seed bank. Cohen (1966) demonstrated that the predictability of the environment had a major impact on the fraction of seeds that could be expected to germinate. On the basis of an optimization model he predicted that that where the probability of successful reproduction was high so would be the optimal germination fraction. Conversely, in an environment in which a species has a high probability of reproductive failure he predicted the necessity of having a high yield when successful, seed longevity in the soil and a low yearly germination fraction. Venable and Brown (1988) have extended this analysis by considering the effect of increasing the probability of favourable conditions on germination, dispersal and seed size. They again showed that increasing the probability of favourable conditions increases the germination fraction but they also showed that it tends to increase seed size and to a lesser extent dispersal. Increasing the temporal autocorrelation of the environment also increases germination as there is nothing to be gained by remaining dormant to germinate under similar conditions next year.

Seed dormancy is clearly an evolved response to environmental predictability in time and can be regarded as a bet-hedging tactic. Similarly seed dispersal has evolved in relation to environmental predictability in space, although it should be noted that selection is likely to favour dispersal even in stable environments (Hamilton and May, 1977). All environments are subject to change and those organisms that exploit them must either be able to deal with the fluctuations in time or disperse elsewhere. Dormancy and dispersal can, therefore, to a certain extent be regarded as alternatives: escape in time or space (Venable and Lawlor, 1980; Klinkhamer *et al.*, 1987). But in considering the selective interactions of dispersal, dormancy and seed size Venable and Brown (1988) found that this was not always the case. Dormancy, seed size and dispersal all reduce year to year variance in fitness in slightly different ways such that the balance of costs and benefits in different environments creates selective correlations between them. The effect of increasing spatial autocorrelation in the environment is, however, to decrease dispersal and increase dormancy as seeds are likely to be dispersed to patches that are either similarly favourable or unfavourable for germination.

III. Population Dynamics

A. *Temporal patterns*

Observations on natural populations of annual plants show that when conditions are suitable and resources are freely available (as in the early stages of secondary succession) , all populations have the potential to increase exponentially. With time, however, the rate of population growth slows and eventually a maximum size is reached. For example, in the Botanic Gardens at Liverpool, where a derelict piece of land was laid bare in late December, Law (1981) found that the first colonists of the weedy annual *Poa annua* appeared in the following April (Fig. 1). Most of these colonists survived and produced large numbers of seeds that germinated from August to September, leading to a massive increase in recruitment and a corresponding increase in population size. Following this very rapid increase in population size, the growth rate of the population declined. In this case the growth rate was density-dependent as the seedling recruits to the population at high densities germinated in dense aggregations around the parent plant where they had to compete for a limited supply of resources. As a consequence the plants grew more slowly, produced fewer seeds and were more likely to die at an early age. In a dif-

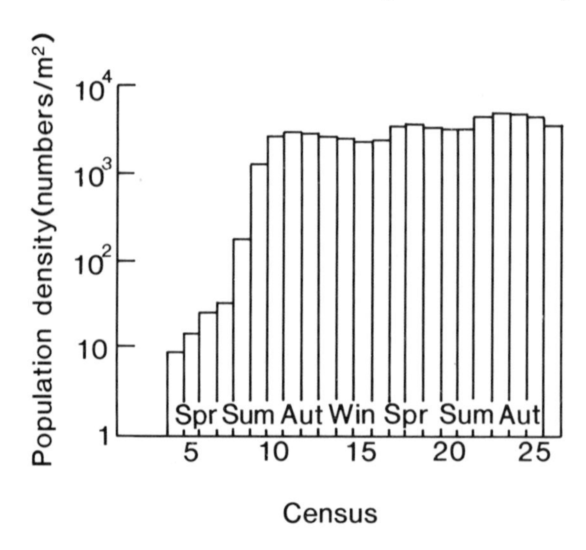

FIG. 1 The population density at each census during colonization of derelict land by *Poa annua*. From Law (1981).

ferent environment, the colonization of bare mud flats by the annual *Suaeda maritima* shows a similar pattern (Joenje, 1978). In this case, however, the time taken for the population to reach its maximum density was longer, because *S. maritima* has discrete generations and produces only one seed crop per year.

The population growth of both of these species appears to be very similar to that which might be expected on the basis of simple population models. Such models are based on two fundamental processes: that populations have the potential to increase exponentially and that there is density-dependent feedback that progressively reduces the actual rate of increase. The major difference between the two species is that *Poa annua* has overlapping generations and the population changes in a continuous manner whereas *S. maritima* has discrete generations and population growth, therefore, occurs in discrete steps.

But even in a closed, homogeneous world where the growth rate of the population is determined entirely by population density there is the potential for a wide range of dynamical behaviour including exponential and oscillatory damping towards a stable equilibrium point, stable limit cycles and chaotic behaviour (May and Oster, 1976). Which type of dynamics a population will exhibit depends upon the magnitude of the intrinsic rate of natural increase and the severity of the density-dependence which determines the extent of the nonlinearity in the relationship between population growth rate and density. Clearly if populations fluctuate in size it is essential to know whether or not the fluctuations are due to such intrinsic properties of the population itself or some form of environmental noise. Watkinson (1980) has argued that annual plants are unlikely to exhibit stable limit cycles or chaotic behaviour as the nature of intraspecific competition which involves contest competition and self-thinning is likely to promote stability. The presence of a seed bank can also be expected to have a stabilizing effect on the dynamical behaviour of populations (MacDonald and Watkinson, 1981; Pacala, 1986a). Crawley and May (1987) reluctantly agree! However, if the allometric relationship between the number of seeds set per plant and plant weight is such that reproductive effort declines as plant size decreases and in particular if there is a threshold size below which plants will not flower then a parabolic relationship between seed production and density can be expected and the potential for complex dynamical behaviour then exists. Similarly time delays in the regulatory effects can be expected to produce a more complicated range of behaviour.

Clearly the nature of density-dependence in populations of *Poa annua* and *Suaeda maritima* is such that the populations tend towards a stable equilibrium point. Analyses of population data where fluctuations occur are, however, generally insufficient to allow any conclusions to be drawn on whether or not nonlinearities in density-dependence contribute towards those fluctuations. An exception (but see also Wilkon-Michalska, 1976) is the analysis of Symonides, Silvertown and Andreasen (1986) on population fluctuations in the annual crucifer *Erophila verna*. A seven year study of *E. verna* on dunes in Poland showed that the populations of seedlings underwent two year cycles in abundance. These fluctuations can be explained in terms of the nature of the density-dependent control of fecundity and the level of seed germination which affects the finite rate of population increase. In plots with a high density of seedlings a large proportion of individuals never flowered but died or remained 'juvenile' rosettes resulting in a humped relationship between the density of seedlings and the number of seeds produced per plot (Fig. 2). The consequences of this relationship for the dynamics of the population are profound. Depending on the proportion of seeds that germinate, which is relatively independent of density, the population may either show monotonic damping towards a stable equilibrium point, stable limit cycles or oscillations of increasing amplitude. On the basis of the model it would appear that 2 year cycles of the kind observed in the field will occur only when germination is between 0.5% and 1%.

It remains unclear how important density-dependent factors are generally in determining fluctuations in abundance. In most cases the dynamical behaviour of populations will be the outcome of some mixture of density-dependent factors (tending to produce stasis or perhaps stable cycles or chaos) and density-independent factors tending to produce unpredictable fluctuations. Analysis of fluctuations in the relative abundance of annual plants from year to year in the annual grasslands of northern California shows that they depend strongly on the weather. Using multiple regression Pitt and Heady (1978) showed, for example, that *Bromus mollis* the most common annual grass in terms of percent botanical composition increased in abundance when the period following germination in the autumn was dry, the winter was relatively warm and the spring wet. This analysis also revealed numerous subtleties in the correlations between species' abundance and various weather variables but unfortunately made no distinction between the size and number of individuals. Nevertheless the study shows that the weather plays a crucial role in determining the botanical composition of the sward from year to year. Correlations between weather and abundance, however, provide no indication of how weather affects abundance through its effects on mortality and fecundity and provide no information on population regulation.

Analysing data (Symonides, 1979) on fluctuations in the number of flowering individuals in populations of *Androsace septentrionalis* using key factor analysis Silvertown (1982) concluded that the key factor in causing population change from one year to the next was seed mortality in the soil. The only apparent regulatory factor was seedling mortality, which tended to undercompensate for changes in population size. The analysis, however, failed to reveal the strong dependence of fecundity on density reported by Symonides (1979) perhaps because the relationship was only evident in years when the period for vegetative growth was long enough to allow the plants to grow to a size when competition would occur between individuals. Alternatively it might be that density-dependence acts spatially within a generation and is thus generally undetectable by key factor analysis (Hassell,

1985). Nevertheless we can conclude that it is the density-independent variation in seed mortality that is responsible for fluctuations in the abundance of *A. septentri-*

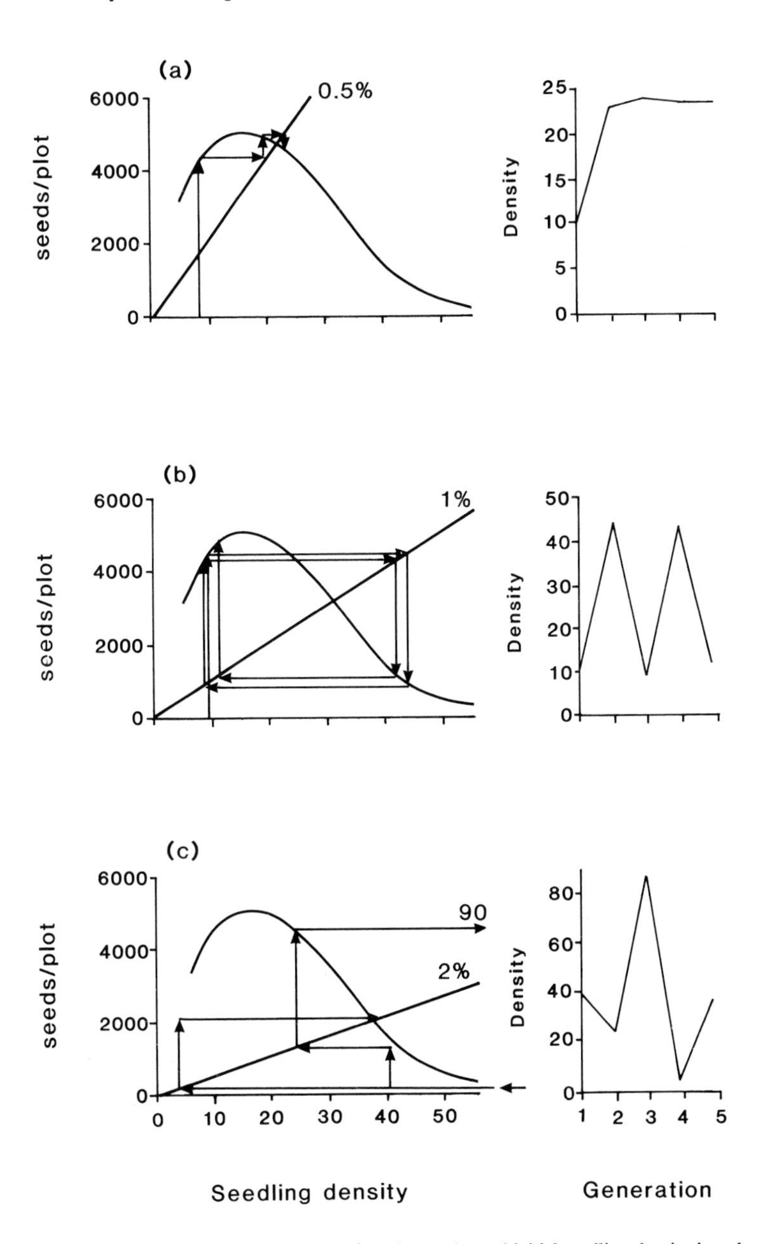

FIG. 2 The relationship between the number of seeds per plot and initial seedling density in a dune population of *Erophila verna* in Poland and the dynamics of the population predicted when the germination success of seeds is a) 0.5%, b) 1% and c) 2%. From Symonides, Silvertown and Andreasen (1986).

onalis.

Rarely is it possible to relate correlations between environmental changes and abundance to demography. Both Klemow and Raynal (1983) and Weiss (1981), however, have shown how the mortality of seedlings in populations of *Erucastrum gallicum* and *Emex australis* is dependent upon the pattern of rainfall. Similarly the detailed analysis of Mack and Pyke (1983, 1984) on the demography of *Bromus tectorum* at a range of sites on the steppe in eastern Washington over three years demonstrated that the dynamics of populations depended on the vagaries of the effective precipitation in autumn and the winter temperature regime. Desiccation during September and October was in particular a major source of mortality amongst the late summer and early autumn germinating plants. Other significant causes of mortality were frost heaving, grazing by voles and infection by the smut, *Ustilago bullata.*

Whilst the above studies show the considerable variation in mortality and fecundity that may occur from year to year they do not allow a picture to be built up of how the various density-dependent and density-independent processes operating in populations interact to determine the dynamical behaviour of the populations. In a study of *Sorghum intrans* in the wet-dry tropics Watkinson, Lonsdale and Andrew (1989) showed that the primary regulating process was the negatively density-dependent relationship between fecundity and plant density, and that abundance was determined in the main by the interaction between the density-dependent control of fecundity and various sources of density-independent mortality (Fig. 3). Some of the sources of mortality could be attributed to a specific cause such as fire or seed predation, but in other cases it was only known that a specific level of mortality occurred over a given period such as over the wet season. However, form a limited range of data on the variability of mortality and fecundity it was possible to show how environmental variation in time could be expected to affect the dynamics of the population. The picture that emerges from a stochastic population model is one of a strongly regulated population in which fluctuations occur around an equilibrium (Fig. 3). Undoubtedly the picture that emerges from the model is limited by our knowledge of the variation in the demographic parameters and how these are related to environmental variables. Nevertheless it is clear that in this case the density-dependent control of fecundity tends to promote stasis whilst density-independent factors tend to produce unpredictable fluctuations. There is no support for the notion of density-vague regulation outlined by Strong (1986) in which density effects are apparent only at very high and low densities.

Populations of annual plants such as *Sorghum intrans* may occur at a site for many years but others often occupy a site for only a relatively short period of time before perennials encroach and replace them (Bazzaz, 1968; Hogeweg *et al,* 1985; Tilman, 1988). Sometimes this replacement may occur over a very short period (Whigham 1984) but in other cases it may be much longer. At Cedar Creek in

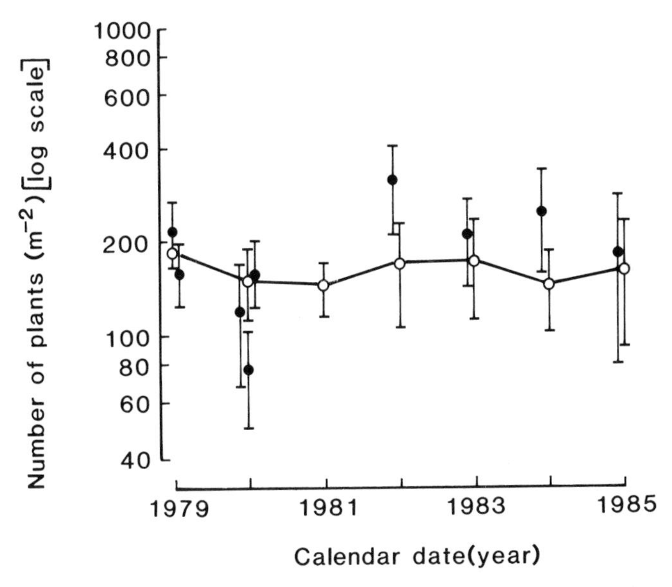

FIG. 3 The mean observed (•) population densities of *Sorghum intrans* in the wet-dry tropics of Northern Australia over a seven year period together with the predicted population densities (o) derived from a stochastic model in which fecundity was negatively density-dependent and mortality density-independent. The 95% confidence limits for population size are also given. From Watkinson, Lonsdale and Andrew (1989).

Minnesota Tilman (1988) found that annuals would often cover 10-40% of the ground in newly abandoned fields and that they remained a major component of the vegetation for over 40 years before being replaced by perennials. A critical feature of the succession in this environment was the slow rate of nitrogen accumulation. On the basis of a series of observations and experiments Tilman suggested that a major determinant of the successional dynamics was the abilities of the species to compete for nitrogen and light. Indeed the annuals were dominant at low levels of nitrogen supply and declined at greater rates of nitrogen supply. Similarly Whigham (1984) concluded that *Ipomoea hederacea* was rapidly eliminated from early successional sites because it cannot compete for nutrients. But whilst it is tempting to ascribe the decline of annuals to resource competition when the abundance of perennials increases this need not always be the case.

In comparing a series of stands of the shrub *Baccharis pilularis* in California Hobbs and Mooney (1986) found that the abundance of all herbaceous species including numerous annuals declined greatly after *Baccharis* formed a complete canopy after 2-3 years. This suggests that competition for light may have had a dramatic effect on the annuals, but an experiment with herbivore exclosures showed that much of the decline in the abundance of the herbaceous plants could be attributed to the activity of small mammals. It would appear in this grassland system at least that the effects of herbivores on the abundance of annuals is only significant where there is a closed shrub canopy and sufficient shelter from predators.

Competition from perennials would at first sight also appear to be implicated in the decline of *Vulpia fasciculata* (Fig. 4) on the dunes at Aberffraw and Newborough Warren in North Wales (Watkinson, 1990). Pattern analysis shows that there is a negative correlation between *V. fasciculata* and the dominant perennial species *Festuca rubra* (Pemadasa, Greig-Smith and Lovell, 1974) and competition experiments show that vigorously growing dense populations of perennials considerably reduce both the survival and performance of *V. fasciculata* (Pemadasa and Lovell, 1974). However, despite an inverse correlation between the abundance of *V. fasciculata* and the cover of perennials Watkinson (1990) could find no evidence for any specific competitive interactions. Neither was there any evidence for an increase in diffuse competition with the cover of perennials as the density-dependent relationship between fecundity and density in *V. fasciculata* remained the same over time. Rather it would appear that the decline of *V. fasciculata* was associated with an increase in seedling mortality.

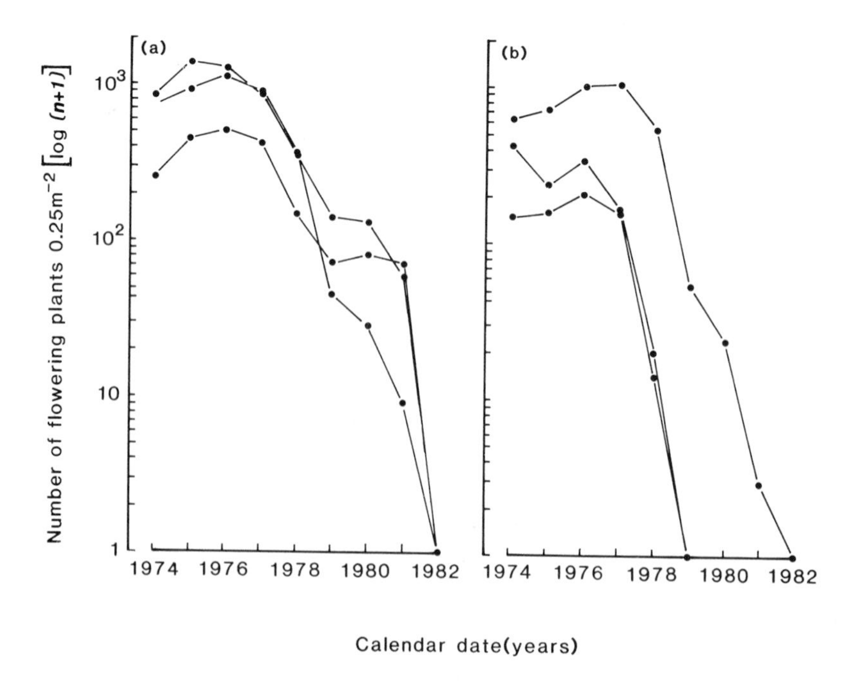

FIG. 4 The number of flowering plants of *Vulpia fasciculata* on a) three plots at Aberffraw and b) three plots at Newborough Warren in North Wales between June 1974 and June 1982. From Watkinson (1990).

Wind drag on seedlings from seeds which germinate on the sand surface is one of the principal causes of mortality during the life cycle of the plant, especially among the later germinating individuals, whilst early germinating individuals on the sand surface are liable to desiccation. The likelihood of both of these causes of mortality is diminished where the vegetation cover is low and shifting sand buries most of the seeds (Watkinson, 1978, 1990). But where the cover of perennials is higher, the local movement of surface blown sand is reduced and a crust often forms on the sand surface. Under such circumstances most seeds can be expected to germinate on the sand surface and these individuals are much more likely to die than those which are buried. Thus it would seem that the local disappearance of *Vulpia fasciculata* on some of the fixed dunes at Aberffraw and Newborough Warren can best be explained by the indirect effect of an increase in the abundance of perennials on the germination and establishment of *V. fasciculata* rather than resource competition from perennials.

The only regulatory factor so far detected in populations of *Vulpia fasciculata* is the density-dependent control of fecundity. The dynamical behaviour of *Vulpia fasciculata* would thus appear to be determined by an interaction of the density-dependent control of fecundity which tends to promote stasis and density-independent mortality. In the mid 1970's the level of mortality was such that the populations appeared to be close to some form of equilibrium (Fig. 4) but with the encroachment of perennials onto the plots the level of mortality steadily increased to bring about a continuous decline in abundance. It would be interesting to know whether the dynamics of any of the other annuals in these dune communities were determined in a similar way. It is intriguing to contemplate a community of annual plants in which interspecific competition for resources plays no role and where the influence of perennials, through their impact on the the stability of the sand, is an indirect one on germination and establishment. This can only be viewed as competition and highly asymmetric competition at that in the population sense (Law & Watkinson, 1989).

B. *Spatial and temporal dynamics*

But whatever the reason for the disappearance of annuals from a patch of ground, in a successional environment all species are doomed and their continued survival depends on dispersal and establishment elsewhere (Harper, 1977). A large number of studies have examined the local dispersal of plants and shown that most seeds are dispersed not far from the parent plant. Typically the dispersal curve for wind- or animal-dispersed seeds tends to be leptokurtic (Howe and Westley, 1986).

It is in the colonization of areas that the role of dispersal in determining population size is most clear (Joenje, 1978; Law, 1981) with the pattern of dispersal affecting the spatial and temporal dynamics of populations (Harper, 1977). But for most populations the role of dispersal in determining population size is unclear. An exception is provided by the study of *Cakile edentula* which Keddy (1981) found to vary in abundance along a sand dune gradient stretching across a dune ridge from the seaward end inland. Along this gradient the levels of fecundity and mortality were such that the populations could only be maintained at the seaward end of the gradient. On the middle and landward sites the level of mortality exceeded reproductive output, so that the populations would not have been able to persist in isolation. A computer simulation of the populations (Watkinson, 1985) shows that the populations only exist at the landward and middle ends of the gradient because of the immigration of seeds from the seaward end, and indeed there is a high dispersal of seeds (> 50%) landward from the highly fecund plants on the beach (Keddy, 1982).

Spatially explicit population models that include dispersal have been developed by Pacala and Silander (1985) and Pacala (1986a, b). In the computer simulation models it is typically assumed that the distance between a seedling and its mother is an exponential random variable. Plants are included as points on a modeled plot and following dispersal and germination it is assumed that the plants interfere with each other at a local neighbourhood level. Density-dependence thus operates at the local level through crowding. But if the mean dispersal distance of seeds is sufficiently large relative to the neighbourhood size of plants then the spatial location of seedlings can be expected to be an approximately Poisson process (Pacala and Silander, 1985). Analytically tractable versions of the neighbourhood population dynamics models are then possible. The implication is that simple nonspatial models (e.g. Watkinson, 1980) are sufficient to describe the population dynamics of plants when the dispersal distance is relatively large in comparison with the neighbourhood size (Pacala, 1989).

Short dispersal distances will, however, lead to aggregation which can be expected to affect the mean survivorship and fecundity of species (Pacala, 1986b) and thus dynamics. It might be expected that aggregation would lead to reduced population size but in fact there is a complex relation between population density and dispersal distance. This is because clumping produces two opposing tendencies: on the one hand there is a reduction in seed production in the clumps which tends to reduce population size and on the other an increase in the fecundity of isolated plants between the clumps which tends to increase population size.

Extensions of the neighbourhood models to more than one species (Pacala, 1986b, 1987) indicate that the spatial distribution of plants can markedly affect the outcome of competition between plant species. Pacala (1986b) for example demonstrates that short dispersal may facilitate coexistence by increasing the degree of interspecific spatial segregation. In this case limited dispersal essentially creates "biotic" spatial heterogeneity (Pacala, 1989). Crawley and May (1987) have also shown that an annual will be able to persist in a community dominated by perennials even if competition is completely one-sided in favour of the perennials, as long as the product of the annual's fecundity and the proportion of empty sites that can be potentially colonized by annuals exceeds unity. The proportion of empty sites depends in their model on the birth and death rates of perennial ramets.

On a larger scale Carter and Prince (1981) have shown how epidemiological theory can be used to quantify the population dynamics of an annual plant in a patchy environment using four variables: the number of unoccupied colonizable sites, the number of colonizable sites that are occupied, the rate at which unoccupied sites become occupied as a result of dissemination from occupied sites, and the rate at which occupied sites become unoccupied. The usefulness of such models depends critically on understanding the dispersal process and the dynamics of gaps. Applying the model to species at their distribution limits leads to the conclusion that small changes in plant performance may lead to abrupt distribution limits. This prediction is consistent with the observations made by Prince, Carter and Dancy (1985) on the abrupt distribution limits of *Lactuca serriola* in southern Britain.

IV. Concluding Remarks

In comparison with just over a decade ago when most population theory was developed specifically in relation to animals, there is now a consistent body of theory emerging that is relevant to annual plants. This theory takes into account autotrophy, sedentary habit, spatially local interactions and plastic growth (Pacala, 1989). Much of the data that has been collected to date on the demography of annual plants has, however, been concerned primarily with temporal dynamics and biased heavily towards annuals with determinate growth. Despite the pioneering work of Watt (1947), population biologists are only just beginning to collect significant data on both the temporal and spatial dynamics of populations. The study of Symonides (1983) is exceptional in providing detailed data on the spatial and temporal flux of *Erophila verna*, The picture that emerged from that study was one of a variable number of seedlings oscillating in abundance on a two year cycle (see earlier) in small gaps in a compact grassland. But because the cycling of populations in the gaps was out of phase averaging the high and low densities produced a picture of apparent stability in the population as a whole. The scale at which a population is viewed clearly has a major impact on the way we perceive its dynamics.

References

Bazzaz, F.A. (1968). Succession on abandoned fields in the Shawnee Hills, Southern Illinois. *Ecology* **49**, 924-936.

Carter, R.N. and Prince, S.D. (1981). Epidemic models used to explain biogeographical distribution limits. *Nature* **293**, 644-645.

Cohen, D. (1966). Optimizing reproduction in a randomly varying environment. *J. theor. Biol.* **33**, 299-307.

Cohen, D. (1971). Maximizing final yield when growth is limited by time or by limiting resources. *J. theor. Biol.* **12**, 119-129.

Cole, L.C. (1954). The population consequences of life-history phenomena. *Q. Rev. Biol.* **29**, 103-137.

Crawley, M.J. and May, R.M. (1987). Population dynamics and plant community structure: competition between annuals and perennials. *J. theor. Biol.* **125**, 475-489.

Gulmon, S.L., Chiariello,N.R., Mooney, H.A. and Chu, C.C. (1983). Phenology and resource use in three co-occurring grassland annuals. *Oecologia* (Berl.) **58**, 33-42.

Hamilton, W.D. and May, R.M. (1977). Dispersal in stable habitats. *Nature* **269**, 578-581.

Harper, J.L. (1977). "Population Biology of Plants". Academic Press, London.

Hassell, M.P. (1985). Insect natural enemies as regulating factors. *J. Anim. Ecol.* **54**, 323-334.

Hobbs, R.J. and Mooney, H.A. (1986). Community changes following shrub invasion of grassland. *Oecologia*(Berl.) **70**, 508-513.

Hogeweg, P., Hesper, B., van Schaik, C.P. and Beeftink, W.G. (1985). Patterns in vegetation succession, an ecomorphological study. *In* "The Population Structure of Vegetation" (J. White, ed.), pp. 638-666. Dr W. Junk Publishers, Dordrecht.

Howe, H.F. and Westley, L.C. (1986). Ecology of pollination and seed dispersal. *In* "Plant Ecology" (M.J. Crawley, ed.), pp. 185-215. Blackwell Scientific Publications, Oxford.

Joenje, W. (1978). "Plant colonization and succession on embanked sand flats: a case study in the Lauwerszeepolder". PhD thesis, University of Groningen.

Keddy, P.A. (1981). Experimental demography of the sand-dune annual *Cakile edentula*, growing along an environmental gradient in Nova Scotia. *J. Ecol.* **69**, 615-630.

Keddy, P.A. (1982). Population ecology on an environmental gradient: *Cakile edentula* on a sand dune. *Oecologia* (Berl.) **52**, 348-355.

King, D. and Roughgarden, J. (1982). Multiple switches between vegetative and reproductive growth in annual plants. *Theor. Popul. Biol.* **21**, 194-204.

Kirkendall, L.R. and Stenseth, N.C. (1985). On defining 'breeding once'. *Am. Nat.* **125**, 189-204.

Klemow, K.M. and Raynal, D.J. (1983). Population biology of an annual plant in a temporally variable habitat. *J. Ecol.* **71**, 691-703.

Klinkhamer, P.G.L., de Jong, T.J., Metz, J.A.J. and Val, J. (1987). Life history tactics of annual organisms: the joint effects of dispersal and delayed germination. *Theor. Popul. Biol.* **32**, 127-156.

Law, R. (1981). The dynamics of a colonizing population of *Poa annua*. *Ecology* **62**, 1267-1277.

Law, R. and Watkinson, A.R. (1989). Competition *In* "Ecological Concepts" (M. Cherrett, ed.), pp. 243-284. Blackwell Scientific Publications, Oxford.

MacDonald, N. and Watkinson, A.R. (1981). Models of an annual plant with a seed bank. *J. theor. Biol.* **93**, 643-653.

Mack, R.N. and Pyke, D.A. (1983). The demography of *Bromus tectorum*: variation in time and space. *J. Ecol.* **71**, 69-93.

Mack, R.N. and Pyke, D.A. (1984). The demography of *Bromus tectorum*: the role of microclimate, grazing and disease. *J. Ecol.* **72**, 731-748.

May, R.M. and Oster, G.F. (1976). Bifurcations and dynamic complexity in simple ecological models. *Am. Nat.* **110**, 573-599.

Mooney, H.A., Hobbs, R.J., Gorham, J. and Williams, K. (1986). Biomass accumulation and resource utilization in co-occurring grassland annuals. *Oecologia* (Berl.) **70**, 555-558.

Pacala, S.W. (1986a). Neighborhood models of plant population dynamics. 4. Single-species and multi-species models of annuals with dormant seeds. *Am. Nat.* **110**, 573-599.

Pacala, S.W. (1986b). Neighborhood models of plant population dynamics. 2. Multi-species models of annuals. *Theor. Popul. Biol.* **29**, 262-292.

Pacala, S.W. (1987). Neighborhood models of plant population dynamics. 3. Models with spatial heterogeneity in the physical environment. *Theor. Popul. Biol.* **31**, 359-392.

Pacala, S.W. (1989). Plant population dynamic theory. *In* "Perspectives in Theoretical Ecology" (R. May, S. Levin and J. Roughgarden, eds), pp. 54-67. Princeton University Press, Princeton.

Pacala, S.W. and Silander, J.A.(1985). Neighborhood models of plant population dynamics. 1. Single-species models of annuals. *Am. Nat.* **125**, 385-411.

Pemadasa, M.A., Greig-Smith, P. and Lovell, P.H. (1974). A quantitative description of the distribution of annuals in the dune system at Aberffraw, Anglesey. *J. Ecol.* **62**, 379-402.

Pemadasa, M.A. and Lovell, P.H. (1974). Interference in populations of some dune annuals. *J. Ecol.* **62**, 855-868.

Pitt, M.D. and Heady, H.F. (1978). Responses of annual vegetation to temperature and rainfall patterns in Northern California. *Ecology* **59**, 336-350.

Prince, S.D., Carter, R.N. and Dancy, K.J. (1985). The geographical distribution of the prickly lettuce (*Lactuca serriola*) II. Characteristics of populations near its distribution limit in Britain. *J. Ecol.* **73**, 39-48.

Ritland, K. (1983). The joint evolution of seed dormancy and flowering time in annual plants living in variable environments. *Theor. Popul. Biol.* **24**, 213-243.

Schaffer, W.M. and Gadgil, M.D. (1975). Selection for optimal life-histories in plants. *In* "Ecology and Evolution of Communities" (M.L. Cody and J.M. Diamond, eds), pp. 142-157. Harvard University Press, Cambridge, Massachusetts.

Silvertown, J.W. (1982). "Introduction to Plant Population Ecology". Longman, London.

Strong, D.R. (1986). Density vagueness: abiding the variance in the demography of real populations. *In* "Community Ecology" (J. Diamond and T.J. Case, eds), pp. 257-268. Harper and Row, New York.

Symonides, E. (1979). The structure and population dynamics of psammophytes on inland dunes. II. Loose-sod populations. *Ekol. Pol.* **27**, 191-234.

Symonides, E. (1983). Population size regulation as a result of intra-population interactions. I. Effect of density on the survival and development of individuals of *Erophila verna* (L.) C.A.M. *Ekol. Pol.* **31**, 839-881.

Symonides, E. (1988). Population dynamics of annual plants. *In* "Plant Population Ecology" (A.J. Davy, M.J. Hutchings and A.R. Watkinson, eds), pp. 221-248. Blackwell Scientific Publications, Oxford.

Symonides, E., Silvertown, J.W. and Andreasen, V. (1986). Population cycles caused by overcompensating density-dependence in an annual plant. *Oecologia* (Berl.) **71**, 156-158.

Tilman, D. (1988). "Plant Strategies and the Dynamics and Structure of Plant Communities". Princeton University Press, Princeton.

Venable, D. and Brown, D.L. (1988). The selective interactions of dispersal, dormancy, and seed size as adaptations for reducing risk in variable environments. *Am. Nat.* **131**, 360-384.

Venable, D.L. and Lawlor, L. (1980). Delayed germination and dispersal in desert annuals: escape in space and time. *Oecologia* (Berl.) **46**, 272-282.

Watkinson, A.R. (1978). The demography of a sand dune annual: *Vulpia fasciculata*. III. The dispersal of seeds. *J. Ecol.* **66**, 483-498.

Watkinson, A.R. (1980). Density-dependence in single-species populations of plants. *J. theor. Biol.* **83**, 345-357.

Watkinson, A.R. (1985). On the abundance of plants along an environmental gradient. *J. Ecol.* **73**, 569-578.

Watkinson, A.R. (1990). The population dynamics of *Vulpia fasciculata*: a nine year study. *J. Ecol.* (in press).

Watkinson, A.R. and Davy, A.J. (1985). Population biology of salt marsh and sand dune annuals. *Vegetatio* **62**, 487-497.

Watkinson, A.R., Lonsdale, W.M. and Andrew, M.H. (1989). Modelling the population dynamics of an annual plant: *Sorghum intrans* in the wet-dry tropics. *J. Ecol.* **77**, 162-181.

Weiss, P.W. (1981). Spatial distribution and dynamics of the introduced annual *Emex australis* in south-eastern Australia. *J. Appl. Ecol.* **18**, 849-864.

Whigham, D.F. (1984). The effect of competition and nutrient availability on the growth and reproduction of *Ipomoea hederacea* in an abandoned field. *J. Ecol.* **72**, 721-730.

Wilkon-Michalska, J. (1976). Struktura i dynamika populacji *Salicornia patula* Duval-Jouve. Rozprawy UMK, Torun.

21 Evolution of Size-dependent Reproduction in Biennial Plants: A Demographic Approach

NAOKI KACHI

Division of Environmental Biology, National Institute for Environmental Studies, Tsukuba, Ibaraki 305, Japan

I. Introduction

Based on the length of generations, herbaceous plant species are often classified as annuals, biennials and perennials. Biennials are monocarpic species which germinate and grow vegetatively in the first growing season and flower and die in the next. However, demographic studies on natural populations of species classified traditionally as biennials have revealed that only a minority of them are strict biennials (Harper, 1977; Silvertown, 1984). In the field, a majority of them behave as short-lived monocarpic perennials, although under productive conditions, they are still capable of flowering in the second or even the first year (Reinartz, 1984; Kachi and Hirose, 1983; Lee and Hamrick, 1983; Klinkhamer, de Jong and Meelis, 1987b). These 'biennials' have been called facultative biennials or delayed biennials, in contrast to strict biennials which complete the life-cycle within two years irrespective of the growing conditions (Kelly, 1985a).

In 1975, Werner demonstrated the priority of plant size over chronological age in predicting population dynamics of a facultative biennial species (*Dipsacus sylvestris*). It has since been established that in most biennial species, the timing of flowering is determined primarily by size rather than age (Gross, 1981; van der Meijden and van der Waals-Kooi, 1979). In these species, vegetative rosettes must grow above a certain critical size before reproduction occurs. The size-dependency in flowering causes delay of reproduction when rosette growth is restricted by competition or low productivity of the habitat or both, because it may take three years or more for seedlings to attain the critical size for flowering (Werner, 1977; Gross and Werner, 1983). For example, the year of flowering of *Daucus carota* in

BIOLOGICAL APPROACHES AND EVOLUTIONARY TRENDS IN PLANTS ISBN 0-12-402960-4

old-fields in North America is increasingly delayed as succession proceeds (Holt, 1972; Gross and Werner, 1982; Lacey, 1988). This delay is explained by an environmentally-induced reduction in the growth rate of vegetative rosettes.

Size-dependent flowering has been recognised in all facultative biennials hitherto investigated. Life-histories of facultative biennials differ fundamentally from those of strict biennials, because the timing of reproduction of the former is changed plastically, depending on the growing conditions, while the timing of reproduction of the latter is fixed genetically, irrespective of rosette growth. Several questions arise about these life-histories. (i) What are the physiological mechanisms underlying the size-dependency of reproduction? (ii) Is the critical size genetically controlled? (iii) Do plants show a critical size for flowering which maximizes the population growth rate or individual fitness? (iv) Why is the timing of reproduction in most biennials determined by size, and not by age? (v) Why is the strict biennial life-history so rare? The aim of this review is to answer these questions, emphasising the importance of demographic studies for an understanding of the life-history variations exhibited among biennial species.

II. Physiological Basis of Size-dependent Reproduction

Most biennials require vernalization (chilling) or a long-day photoperiod or both for developing a bolting shoot and/or flower primodia (Lang, 1965). If only rosettes larger than a critical size can be vernalized or sense the photoperiod, flowering must be size-dependent. Kachi and Hirose (1983) reported that in *Oenothera glazioviana*, which colonises coastal sand-dunes in Japan, bolting is restricted to size classes with rosette diameters greater than 9 cm, regardless of the chronological age (Fig. 1). They also found that even in rosettes sufficiently larger than the critical size, when only a central part of the rosettes (5 cm in diameter) was exposed to a long-day photoperiod, bolting was not induced. This result suggests that in *Oenothera glazioviana*, more than a certain amount of a leaf-area (corresponding to the rosette diameter of 9 cm) is necessary to respond to a photoperiodic stimulus for bolting induction. This species also requires vernalization for bolting, although this process is not size-dependent (Kachi and Hirose, 1983). Klinkhamer, de Jong and Meelis (1987b) reported similar physiological requirements for flowering of *Cirsium vulgare*, a biennial growing on Dutch coastal dunes.

In productive conditions, these species can be winter annuals, because vernalization is completed even in small seedlings during winter and the seedlings can grow fast to reach the critical size during the first spring before they respond to a photoperiod. Accordingly, in biennials which show size-dependency in photo-induction but not in vernalization, plant growth or rosette size in spring (not plant size at the end of the previous growing season or in winter) directly affects the probability of flowering in that year. *Carduus nutans*, a common member of an early stage of

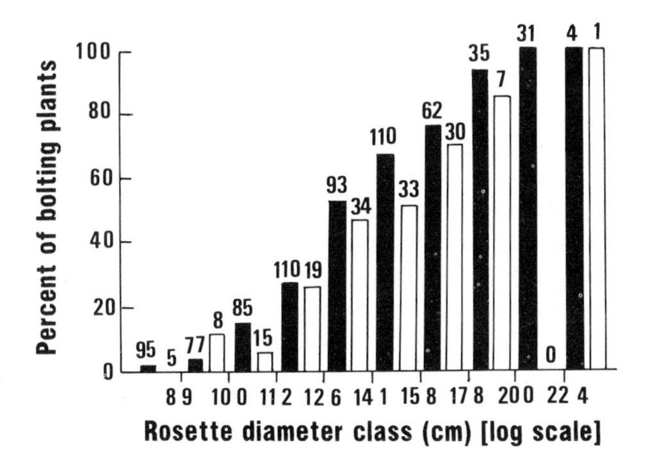

FIG. 1 Percentage of bolting of *Oenothera glazioviana* in relation to rosette-diameter classes in May in a natural population (■) and in an population treated with a compound chemical fertilizer (□). The natural population includes individuals of different ages ranging 2-6 years, while all individuals of the fertilized population were germinated in October in the previous year. Figures on each column indicate the number of plants in each size class (from Kachi and Hirose, 1983).

old-field succession in North America, may be another example, because it is a winter annual on productive soils, while on unproductive habitats it is a monocarpic perennial (Lee and Hamrick, 1983). If plants show size-dependency only in photo-induction, but no requirement of vernalization, they may even behave as summer annuals. *Lactuca virosa* might be an example, because spring seedlings raised in a glasshouse flowered and set seed by the end of the same year (Boorman and Fuller, 1984).

Some other species show size-dependency in vernalization. de Jong, Klinkhamer and Prins (1986) reported that flowering of the monocarpic perennial, *Cynoglossum officinale,* is size-dependent, because development of flower primordia during cool periods is restricted to individuals larger than a critical size. Baskin and Baskin (1979a, b) studied flowering habits of *Pastinaca sativa,* a tall umbelliferous weed in North America, and *Grindelia lanceolata,* a Compositae found in cedar (limestone) glades in Tennessee in the United States. They suggested that in both species, plants which have reached a critical size by the end of a growing season, will be vernalized during winter and flower in spring. In these species, a long-day photoperiod had only secondary effects on flowering (e.g. delay of bolting dates).

If size-dependency is related to vernalization, rosette size at the end of a growing season or during winter will be a reasonable measure for predicting flowering the next season. However, if size-dependency is related to photo-induction, the size or rosette growth in spring should be a more appropriate predictor, because the final decision concerning whether or not a rosette flower is made in spring rather than in the previous autumn/winter. If size-dependency is related to vernalization, plants will be able to flower earlier in spring than plants in which size-dependency is related to photo-induction, because they start reproduction during winter. The early start of reproduction will be advantageous, if a period of time in spring favourable for plant growth is limited. Therefore, physiological information on the size-dependency of flowering is essential, not only for using plant size as a predictor of population dynamics, but also for examining life-historical consequences of size-dependent reproduction in biennial plants (Kachi and Hirose, 1985; Klinkhamer, de Jong and Meelis, 1987a; Lacey, 1986a).

It has been supposed by several authors that size-dependent flowering observed in biennial species is related to the amount of resources (e.g. carbohydrates and nutrients) available for reproduction (Harper and Ogden, 1970; Mooney and Chiariello, 1984). However, in biennials which show a size-requirement for photo-induction, the amount of resources in plants (especially stored in a tap root) has no direct physiological effects on the induction of flowering, because a long-day photoperiod is sensed by leaves (Lang, 1965). For example, in *Oenothera glazioviana,* bolting induction is related directly to the leaf-area rather than the total dry weight (Kachi and Hirose, 1983). On the other hand, Glier and Caruso (1973, 1974) reported that in *Verbascum thapsus*, chilling causes the induction of enzymes which convert stored starch to soluble sugars. Calder and Cooper (1961) observed a positive effect of nitrogen nutrition in promoting vernalization in *Dactylis glomerata.* These findings may suggest that concentrations of carbohydrates and/or nitrogen could affect the process of vernalization. However, de Jong, Klinkhamer and Prins (1986) found no significant effects of nitrogen concentration of plants and light intensity on flowering of *Cynoglossum officinale*, a biennial showing a size-requirement for vernalization.

III. Genetic Basis of Size-dependent Reproduction

There is some evidence for genetic basis of a flowering year in several monocarpic species (Clausen and Hiesey, 1958; Quinn, 1969; Lacey, 1986b). Different populations of a single species from different regions vary in the year of flowering even when they are grown in a common garden. Usually, the length of the prereproductive period tends to be short in plants from low latitudes and long from high latitudes. Reinartz (1984) observed that in *Verbascum thapsus* distributed in North America, plants originating from the northern part of the range have a greater tendency to delay reproduction than did plants of the same size from the central and southern parts of the range. Lacey (1986b) found the same tendency for populations of *Daucus carota.*

Genetic variations in flowering may exist within local populations (Law, Bradshaw and Putwain, 1977; Lacey, 1988). Natural populations of *Daucus carota* ssp. *carota* in Michigan in the United States include annual, biennial and triennial plants. Lacey (1988) demonstrated that individuals originated from annual plants tended to flower earlier at smaller sizes than those originated from biennial or triennial mothers, when they were grown under controlled environments simulating seasonal changes in temperatures and day-lengths.

There is also evidence for genetic control over the physiological requirements of vernalization and/or photo-induction for flowering (Lang, 1965). An annual race of *Verbascum thapsus* lacks a vernalization requirement for flowering (Reinartz, 1984). *Arabidopsis thaliana* has strains which differ in their vernalization requirements (Chouard, 1960). In *Hyoscyamus niger*, vernalization characteristics are determined at a single locus (Smith, 1927), while several loci might be involved in controlling the flowering year of *Melilotus alba* (Clausen and Hiesey, 1958).

IV. Demographic Approach to Life-history Strategy

Individual fitness, which is defined as the contribution of a given genotype to the subsequent generation relative to that of other genotypes, can be measured by the expected number of offspring or expected population growth rate of individuals with a particular phenotype which corresponds to a particular genotype (Cole, 1954). The number of offspring in the next generation is determined by the schedules of survival and reproduction in the current generation. Demography is the study of populations, which deals with statistics of these life-history schedules (Hutchings, 1986). The schedules are usually presented by a life-table or survivorship (l_x) curve, which summarizes age-specific survival, and a fecundity-table or fecundity (m_x) curve, which summarizes age-specific reproduction. Fig. 2 is an example for a dune-population of *Oenothera glazioviana* (Kachi and Hirose, 1985). From l_x and m_x, we can determine the intrinsic rate of natural increase of a population (r) using the Euler equation:

$$1 = \sum_{x=0}^{\infty} e^{-rx}\, l_x m_x$$

where e is the base of the natural logarithm and x is the age of individuals. When a population grows with a constant schedules of survival and fecundity, it eventually reaches a stage of stable age-distribution and increases (or decreases) exponentially at a rate of r specified by the above equation (Lotka, 1922). In other words, r is a measure of population growth averaged over a long time without density-dependent regulation (Charlesworth, 1980). For the *Oenothera glazioviana* population shown in Fig. 2, the r is 0.04 year^{-1}, which means the population at steady-state exponential stage will increase at a rate of 1.04 (= $e^{0.04}$) a year.

Another useful approach to the analysis of life-history schedules is the use of matrix projection models based either on age-specific classification or on size(or stage)-specific classification, or both (Leslie, 1945, 1948; Lefkovitch, 1965; Law, 1983; Caswell, 1986). The matrix models are powerful tools with which we can determine marginal values of each element of the matrix (a transient probability from one category to another) in affecting the intrinsic rate of natural increase (Caswell and Werner, 1978; de Kroon *et al.*, 1986; van Groenendael, de Kroon and Caswell, 1988; Lacey and Pace, 1983).

In a given environment, individuals with different life-histories may exhibit different schedules of survival and fecundity and hence different values of *r*. Relative advantages of different life-histories are examined by comparing the expected *r* (or its antilog, the finite rate of population growth) for individuals with contrasted life-histories (Cole, 1954; Gadgil and Bossert, 1970; Schaffer and Gadgil, 1975; Schaffer, 1983). In fact, in most of the previous discussion on life-history evolution in annuals, biennials and perennials, it has been assumed that the current generation is followed immediately by the next, so that the generation times of annuals, biennials and perennials are one, two and more than two years. However, this is not always true in the real world, because the length of the generation time is also influenced whether or not the population has a persistent seed-bank or only a transient seed-bank. This problem will be discussed in the next section.

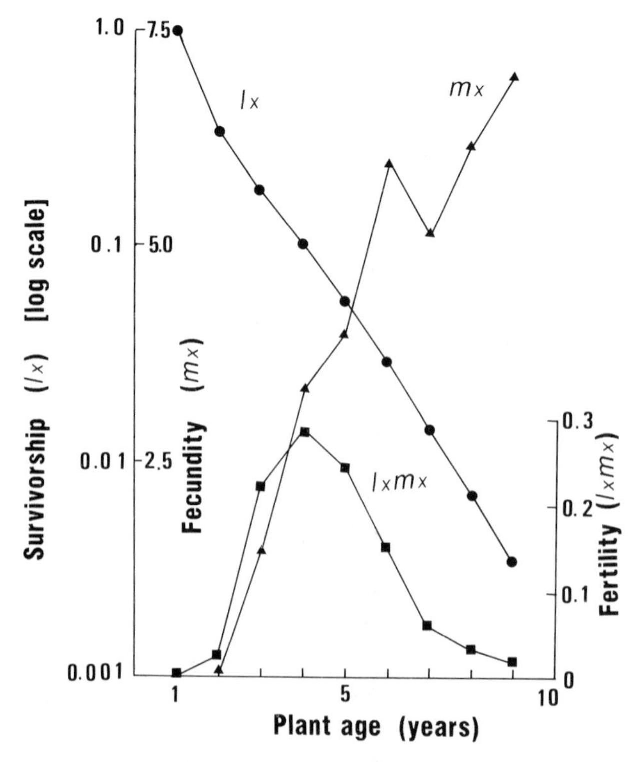

FIG. 2 Survivorship (l_x), and fecundity (m_x) and l_xm_x in rela tion to age for a dune-population of *Oenothera glazioviana* calculated by a simulation model (from Kachi and Hirose, 1985).

V. R_0 - An Alternative Criterion of Life-history Strategy

As mentioned in the previous section, intrinsic rates of natural increase *(r)* have commonly been used as a criterion of fitness or a measure of adaptiveness of different life-histories. For a population at a steady-state exponential growth stage, *r* is determined by the equation,

$$r = (\ln R_0)/T,$$

where R_0 is the net reproductive rate (expected total number of offsprings per newborn individual), and *T* is the generation time. The equation indicates that R_0 affects *r* through its natural logarithm, while the generation time (*T*) affects *r* through its reciprocal. This is why the generation time has a profound effect on *r* compared with R_0 (Lewontin, 1965) and hence, an annual life-history has an advantage over biennial and delayed biennial life-histories. However, Reinartz (1984) pointed out that if the seed-bank stage is included, differences in the lengths of the generation time between annuals, biennials and delayed biennials could be eliminated. This is particularly true, when the regeneration of seedlings from the seed-bank depends on intermittent disturbances, and the time-interval between consecutive disturbances is longer than the time from germination to seed-set. In this case, the generation time including a seed-bank stage is identical for annuals, biennials and delayed biennials and *r* depends only on R_0 (or the expected total seed production for a seed). Kelly (1985b) proposed a similar conclusion in a situation where germination occurs in 'gaps' which are intermittently created in vegetation by small-scale disturbance. However, Silvertown (1986) noted that if the probabilities of failing to exploit gaps are taken into account, annuals could still attain higher *r* than biennials and delayed biennials.

When a population is under density-dependent regulation, a life-history which maximizes *r* is not necessarily identical to the evolutionarily stable strategy (Boyce, 1984). de Jong, Klinkhamer and Metz (1987) showed that if recruitment is limited by the availability of seeds and also by the probability at which 'safe sites' (or 'gaps') for establishment are created, the evolutionarily stable strategy is to maximize expected total seed production (which is exactly identical to R_0), irrespective of the length of the generation time. In this case, delay of flowering is favoured if the relative gain in seed production through prolonged vegetative growth compensates for the increased mortality. In other words, when a population is under density-dependent regulation, through gap-creation by death of adult plants for example, a larger critical size for flowering could be selected than that expected from a *r*-maximization model. The importance of gaps created by small-scale disturbance in affecting seedling establishment has been demonstrated for dune-populations of *Cirsium vulgare* and *Cynoglossum officinale* in semi-open habitats (Klinkhamer and de Jong, 1988).

In conclusion, if we emphasise density-dependent regulation, as assumed in safe-site models, R_0 will be an appropriate fitness criterion. If we emphasise density-independent processes such as large-scale disturbance of vegetation or climatic hazards, r could still be a suitable measure of life-history evolution, even if the average r is close to zero or negative. In the field, population levels should be determined through both density-dependent and density-independent processes, and the relative importance of both processes must depend on the habitat type and the biology of individual species. In this context, it is noteworthy that sparse vegetation does not necessarily mean that the population is not under density-dependent regulation (Klinkhamer and de Jong, 1989).

VI. The Optimal Size for Reproduction

If size-requirement for flowering in biennial plants has a genetic basis and if genetic variations in the critical size exist, then evolution of size-dependent reproduction will occur through selection on the critical size of reproduction (Hirose, 1983; Kachi and Hirose, 1985; de Jong *et al.*, 1989). Individuals with a larger critical size reproduce in later years than those with a smaller critical size (Lacey, 1986a). Delayed reproduction itself has, if other life-history parameters are not changed, no positive effects on population growth. However, it may be associated with an increase in the reproductive capacity or seed production, because larger plants can produce a larger number of seeds (Salisbury, 1942). Thus, delayed reproduction, which is caused by an increase in the critical size, is advantageous when the increased seed production overcompensates for the increased prereproductive mortality and prolonged generation time (Harper, 1967; Schaffer and Gadgil, 1975; Hart, 1977; Silvertown, 1983).

In order to predict the optimal size for reproduction, Kachi and Hirose (1985) developed a demographic model for a dune-population of *Oenothera glazioviana*. In the model, yearly survival and growth of vegetative rosettes and seed production of an individual were determined as a function of rosette diameter (Fig. 3). Size-hierarchy in a cohort was generated by varying stochastically the relative growth rates of rosette diameters around the mean values which was determined by a negative regression against the rosette diameter [Fig. 3 (b)]. The authors showed that there is an optimal size for reproduction which maximizes the intrinsic rate of natural increase, and that the mean size at reproduction observed in the field (14 cm) was close to the expected optimum (16.6 cm).

The optimal reproductive size is determined through the trade-off relationships between survival and fecundity affected by the critical size of reproduction. Reproducing at a smaller size results in a higher prereproductive survival and shorter generation time, but also leads to lower fecundity. Conversely, reproducing at a larger size allows greater fecundity, but leads to higher mortality during the

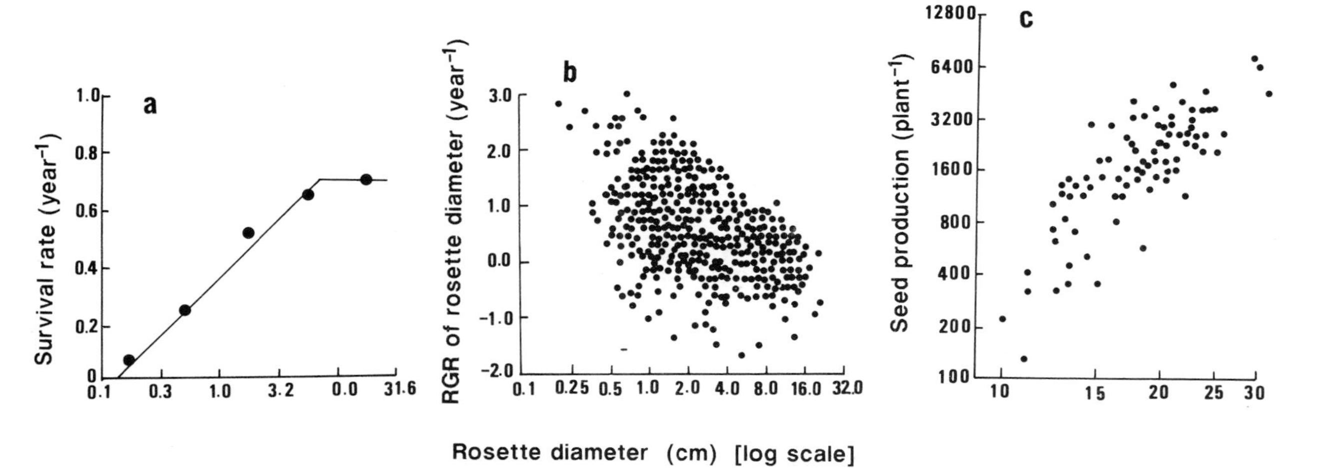

FIG. 3 Probability of yearly survival (a), relative growth rate (*RGR*) of rosette diameters for the period from May to the next (b) and seed-production (c) in a dune-population of *Oenothera glazioviana* in relation to rosette diameters in May (data from Kachi, 1983 and Kachi and Hirose, 1985).

TABLE 1

Summary of the effects of varying critical size for reproduction on the life-history schedules and the intrinsic rate of natural increase (r) for simulated populations of *Oenothera glazioviana* in a sand-dune system. Size is critical rosette diameter for flowering. The optimal size is indicated by *. Prereproductive survivorship is based on the probability of established seedlings surviving to reproduction. In the calculation of r, no perennial seed-bank and 0.48 seedling survival were assumed (from Kachi and Hirose, 1985)

Diameter (cm)	Prereproductive survivorship	Seed production	Generation time (years)	r ($year^{-1}$)
1.0	0.586	17	1.46	−1.370
3.2	0.240	154	2.41	−0.400
5.0	0.163	357	2.92	−0.186
10.0	0.087	1240	4.01	0.014
16.6*	0.044	3140	5.09	0.058
32.0	0.010	11000	6.59	0.004

prolonged vegetative period (Table 1). The same conclusion was obtained by de Jong *et al.* (1989) with a simulation model for two dune-biennials, *Cirsium vulgare* and *Cynoglossum officinale*. However, the optimal sizes predicted from their model were larger than the critical sizes observed in real populations.

VII. Factors Favouring a Larger Critical Size for Reproduction

Since an increase in a critical size causes delay of reproduction, factors which favour a larger critical size are those which favour delayed reproduction. These factors are 1) higher rosette survival relative to seedling survival (Schaffer and Gadgil, 1975; Hart, 1977; Hirose, 1983; de Jong et al., 1989) or relative to seed survival in the seed-bank (Caswell and Werner, 1978; Baskin and Baskin, 1979b, Kachi and Hirose, 1985; Klemow and Raynal, 1985), 2) a large increase in potential fecundity associated with later flowering through increasing an adult size (Hirose and Kachi, 1986; Werner and Caswell, 1977) and/or through increasingly successful pollination in larger individuals (Schaffer and Schaffer, 1979) and 3) decreased relative costs for producing reproductive organs, such as flowering stalks, with increasing size (Schaffer and Gadgil, 1975). However, there have been very few studies which demonstrate the quantitative relationships between these factors and the intrinsic rate of natural increase based on demographic data from natural populations (Hirose and Kachi, 1986; Kachi and Hirose, 1985: de Jong *et al.*, 1989).

Figure 4 shows an example, in which the intrinsic rates of natural increase are plotted as a function of critical rosette sizes for flowering at various seedling emer-

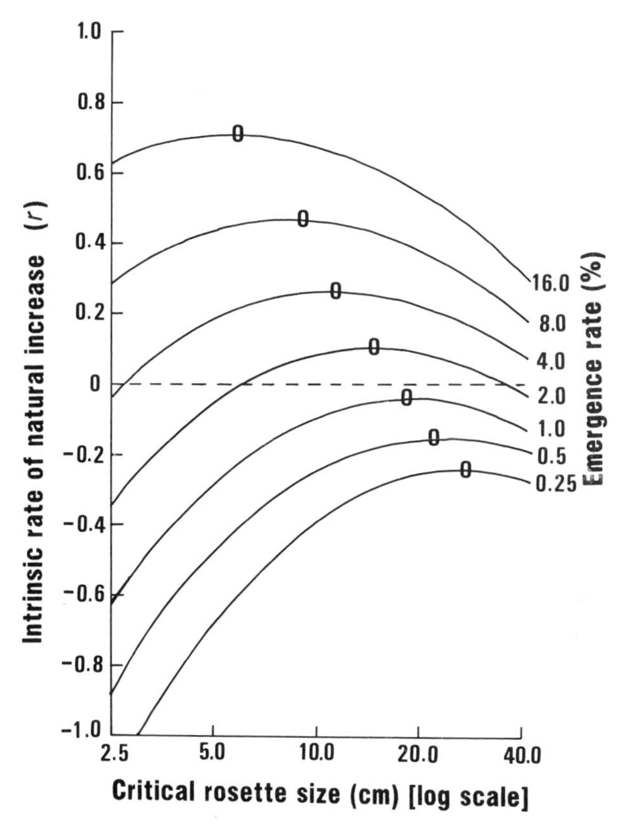

FIG. 4 Intrinsic rates of natural increase (r) calculated for hypothetical populations of *Oenothera glazioviana* as a function of rosette sizes at reproduction assuming various seedling emergence rates. Optimal values are shown as symbols on each curve (from Kachi and Hirose, 1985).

gence rates for hypothetical populations of *Oenothera glazioviana* (Kachi and Hirose, 1985). Since the seeds have no accessories for dispersal and there is no persistent seed-bank, it is likely that most of the dispersed seeds remain in the same population and those which fail to germinate are dead within a year after dispersal. In this case, the emergence rate of seedlings, which is based on the number of dispersed seeds, is equivalent to the seed-survival during the seed-bank stage. It is obvious that the optimal critical size increases with decreasing emergence rates (or decreasing seed-survival during the seed-bank stage).

VIII. Size- vs. Age-dependent Reproduction

If plant size is related deterministically to the chronological age, the problem of de-

TABLE 2
Comparison of size- and age-specific model in relation to the effects of varying seedling emergence rates on the maximal intrinsic rate of natural increase (r_{max}) for hypothetical populations of *Oenothera glazioviana* (from Kachi and Hirose, 1985)

	Size-specific model		Age-specific model	
Emergence rate (%)	Optimal size (cm)	r_{max} ($year^{-1}$)	Optimal age (year)	r_{max} ($year^{-1}$)
0.5	24.5	–0.179	6	–0.287
1.0	20.4	–0.074	6	–0.171
2.0	16.6	0.054	5	–0.035
4.0	14.5	0.209	4	0.125
8.0	11.0	0.384	3	0.306
16.0	8.9	0.593	3	0.537

termining the optimal critical size for reproduction is exactly identical to that of determining the optimal age of reproduction. However, when there are size-variations within a cohort, as is the case in real populations, size- and age-dependent flowering results in different schedules of survival and reproduction and hence, different *r*. A simulation model was used to examine the relative advantages of size- versus age-dependent reproduction (Kachi and Hirose, 1985).

Table 2 shows comparisons of maximal intrinsic rates of natural increase (r_{max}) between size- and age-specific models for hypothetical populations of *Oenothera glazioviana*. The optimal age and optimal size of reproduction is increased with decreasing rates of seedling emergence, because in the model a lower seedling emergence rate corresponds to lower seed-survival relative to rosette-survival (cf. Fig. 4). It is obvious that size-dependent reproduction always results in a higher r_{max} than age-dependent reproduction. This is because survival and seed production are directly related to size rather than age. In the age-specific model, all plants of the same age must reproduce irrespective of the size. This means that for smaller plants, the reproductive timing is too early, because remaining vegetative can considerably increase seed production in future, and for larger plants, the opposite case holds.

IX. Evolution of Size-dependent Reproduction in Fluctuating Environments

If flowering is determined by age, all the plants in a cohort reproduce in the same year. However, if flowering is determined by size, differences in rosette growth cause variations in years of flowering. In a fluctuating environment, spreading the time of reproduction results in higher fitness than synchronised reproduction. This argument is analogous to that proposed as evolutionary advantages of dispersal and delayed germination in fluctuating environments, because both dispersal and delayed life-cycle (either in reproduction or in germination) bring an arithmetic component in averaging population growth rates over space and time under varying survival and reproduction (Lacey *et al.*, 1983; Levins, 1968; Venable and Lawlor, 1980; Klinkhamer and de Jong, 1983). This is in contrast to a synchronous life-cycle, in which the averaged population growth rate should be a geometrical mean of growth rates. Therefore, the role of a rosette-bank, which has been proposed by several authors (e.g. Caswell and Werner, 1978; Baskin and Baskin, 1979b; Klemow and Raynal, 1985; de Jong *et al.* 1989), is similar to that of seed-bank (Cohen, 1966; Bulmer, 1984).

However, the maintenance of a rosette-bank has the bonus of allowing increasing reproductive capacity of individuals, whereas the maintenance of a seed-bank, in itself, has no positive contribution to make for population growth, if there were no environmental variations (de Jong and Klinkhamer, 1988). In fact, relative advantages of a rosette-bank to a seed-bank are also influenced by the relative rates of mortality during rosette- and seed-bank stage and this will be discussed in the next section.

In an previous argument on delayed reproduction in monocarpic species under fluctuating environments, the optimal fraction of reproduction (rather than the year of flowering) was investigated (Klinkhamer and de Jong, 1983). In this case, we must assume genetic similarities between individuals, or parental control on flowering (or otherwise, group selection), because each individual actually either flowers or remains vegetative, although the fraction of flowering can be set at any value between 0 to 1. On the other hand, in the case of size-dependent reproduction, above conditions like genetic similarities are not necessary, because selection operates through the expected number of offsprings affected by the reproductive size of individuals, and not through the expected population growth rate affected by the fraction of flowering plants.

X. Seed-bank vs. Rosette-bank Oriented Life-histories

Some biennial species maintain large persistent seed-banks, while others have transient seed-banks (Ødum, 1978). Longevity of buried seeds of species with a transient seed-bank is expected to be much shorter than that of species with a persistent large seed-bank. Interestingly, a shorter life-span of buried seeds or lower seed-survival relative to rosette-survival is one of the factors which favour delayed reproduction (or larger critical sizes for reproduction)(Kelly, 1985b). This leads us to suspect that species with a transient seed-bank show a tendency to have a relatively large critical size for reproduction compared with those with persistent seed-banks. A larger critical size causes extension of the vegetative stage, which results in a higher proportion of individuals in a rosette-bank. In other words, it is expected that there are typically two extremes of life-histories in biennial plants; rosette-bank oriented and seed-bank oriented life-histories.

In environments where intermittent disturbance and interspecific competition cause greater probabilities of rosette mortality than seed-bank mortality, seed-bank oriented life-histories will be advantageous compared with rosette-bank oriented life-histories. An early stage of secondary succession, which is initiated by disturbance of vegetation on relatively productive soils, could provide such environments. *Oenothera biennis*, a common component of old-field succession, may be one example, because its critical size is relatively small (Hirose and Kachi, 1982) and the longevity of its buried seeds is more than 10 years (Goss, 1924; Kachi, personal observation).

On the other hand, rosette-bank oriented life-histories will be favoured in chronically stressful, but relatively stable environments, such as unproductive sand-dunes or semi-open grasslands. *Oenothera glazioviana*, which typically colonises sparsely vegetated coastal sand-dunes (Kachi and Hirose, 1979), may be included in this category. In a dune-population of *Oenothera glazioviana*, there is practically no persistent seed-bank (Kachi and Hirose, 1985) and plants show a relatively large critical size for flowering compared with *Oenothera biennis* (Hirose and Kachi, 1982). However, these facts provide only circumstantial evidence. To test the hypothesis, further demographic studies on populations including seed-banks at contrasted habitats will be required.

XI. Strict- vs. Facultative-biennials

Although most biennials are facultative biennials, there are a few strict biennials which flower consistently in their second year; e.g. *Linum catharticum* (Bradshaw and Doody, 1978); *Pedicularis palustris* and *Pedicularis sylvatica* (ter Borg, 1979), *Melilotus alba* (Klemow and Raynal, 1981), *Gentianella germanica* (Verkaar and Schenkeveld, 1984) and *Gentianella amarella* (Kelly, 1985a). Theoretically, species which require vernalization for seed germination as well as for flowering, but which show no size-dependency in flowering, will exhibit a strict biennial life-cycle, because seeds which remain dormant and vernalized during winter can germinate in spring, and the rosettes, which are vernalized during the second winter, will flower the following spring. If the vernalization requirement for seed germination disappears, the plants may behave as winter annuals, because seeds can germinate in autumn after seed dispersal and the seedlings are vernalized during the first winter. In this sense, the life-history of strict biennials is like that of extended annuals rather than precocious short-lived monocarpic species (Kelly, 1985a).

Then what environment favours strict biennials rather than winter or summer annuals and facultative biennials? If the potential mortality of autumn seedlings during winter is significantly higher than that of dormant seeds, spring germination would be selected. Furthermore, if delay of reproduction to the second year allows an increased seed production which compensates for increased mortality and prolonged generation time, but delay to the third year does not (or in other words, survival and fecundity are related to age rather than size), the strict biennial life-cycle would be favoured. In the field, these conditions seems to be rarely satisfied, but could be realised in unproductive habitats where small-scale, light disturbance occurs so frequently as to reduce rosette survival and hence diminishes the merit of remaining vegetative for three years or more. It is noteworthy that all strict biennials reported hitherto occur in unproductive, permanently semi-open habitats such as infertile grasslands, sand-dunes or limestone quarries. However, more demographic studies on natural populations are required to quantify the advantages (or disadvantages) of the strict biennial life-cycle relative to those of annuals and facultative biennials [cf. Lee and Hamrick (1983) cited by Kelly (1985b)].

XII. Perspectives

Although biennial species contribute minority of plant kingdom (Hart, 1977), they have been good subjects in the study of life-history evolution in monocarpic species, because 'biennials' show actually a variety of life-cycles of annuals, biennials and monocarpic perennials. As shown in this review, optimization models are useful and attractive tools for investigating life-history strategy.

However, a fundamental problem still remains to be answered. It is "what is an appropriate measure of the fitness?". In other words, we have not yet had any general criterion for maximization models. This is particularly true, when we consider about the real world, in which every environmental condition is variable in space and over time and furthermore, survival and fecundity of individuals are influenced by density-dependent interactions. In a fluctuating environment or in that where density-dependent processes affect the population dynamics, the intrinsic rate of natural increase (r) is not necessarily an suitable measure of fitness, because r is defined for a population with stable age-distribution, which is never realised by a fluctuating population. Evolutionarily stable strategy (ESS) could be an alternative measure under density-dependent situations, but it also holds at a 'stable' state, which is often not the case in the real world. The real world is more or less transient. What population ecologists have to do is to identify what factors contribute the dynamic aspect of population dynamics and to demonstrate how transient populations could be treated in the study of life-history evolution.

Acknowledgments

I thank Mike Hutchings, Tom de Jong, Peter Klinkhamer, Elizabeth Lacey, Richard Law, Akio Takenaka, Jonathan Silvertown, Andrew Watkinson and Pat Werner for discussion and critical comments on an earlier draft. I also thank the members of

Unit of Comparative Plant Ecology (Natural Environment Research Council) at the University of Sheffield for their hospitality during the preparaion of the manuscript.

References

Baskin, J. M. and Baskin, C. M. (1979a). Studies on the autecology and population biology of the weedy monocarpic perennial, *Pastinaca sativa*. *J. Ecol.* **67**, 601-610.

Baskin, J. M. and Baskin, C. M. (1979b). Studies on the autecology and population biology of the weedy monocarpic perennial, *Grinde lia lanceolata*. *Am. Midl. Nat.* **102**, 290-299.

Boorman, L. A. and Fuller, R. M. (1984). The comparative ecology of two sand dune biennials: *Lactuca virosa* L. and *Cynoglossum officinale* L. *New Phytol.* **69**, 609-629.

Borg, S. J. ter (1979). Some topics in plant population biology. *In* "The Study of Vegetation" (M. J. A. Werger, ed.), pp. 11-56. W. Junk, The Hague.

Boyce, M. S. (1984). Restitution of r- and K-selection as a model of density-dependent natural selection. *Annu. Rev. Ecol. Syst.* **15**, 427-447.

Bradshaw, M. E. and Doody, J. P. (1978). Plant population studies and their relevance to nature conservation. *Biol. Conserv.* **14**, 223-242.

Bulmer, M. G. (1984). Delayed germination of seeds: Cohen's model revisited. *Theor. Popul. Biol.* **26**, 367-377.

Calder, D. M. and Cooper, J. P. (1961). Effect of spacing and nitrogen level on floral initiation in cocksfoot (*Dactylis glomerata* L.). *Nature* **191**, 195-196.

Caswell, H. (1986). Life cycle models for plants. *In* "Some Mathematical Questions in Biology", Lectures on Mathematics in the Life Sciences 18 (L. J. Gross and R. M. Miura, eds), pp. 171-233. The American Mathematical Society, Rhode Island.

Caswell, H. and Werner, P. A. (1978). Transient behavior and life history analysis of teasel (*Dipsacus sylvestris* Huds.). *Ecology* **59**, 53-66.

Charlesworth, B. (1980). "Evolution in Age Structured Populations". Cambridge University Press, Cambridge.

Chouard, P. (1960). Vernalization and its relations to dormancy. *Annu. Rev. Plant Physiol.* **11**, 191-238.

Clausen, J. and Hiesey, W. M. (1958). Experimental studies on the nature of species. IV. Genetic structure of ecological races. Carnegie Institution of Washington, Publication number 615.

Cohen, D. (1966). Optimizing reproduction in a randomly varying environment. *J. theor. Biol.* **12**, 119-129.

Cole, L. C. (1954). The population consequences of life history phenomena. *Quart. Rev. Biol.* **29**, 103-137.

Gadgil, M. and Bossert, W. H. (1970). Life historical consequences of natural selection. *Am. Nat.* **104**, 1-24.

Glier, J. H. and Caruso, J. L. (1973). Low-temperature induction of starch degradation in roots of a biennial weed. *Cryobiology* **10**, 328-330.

Glier, J. H. and Caruso, J. L. (1974). The influence of low temperatures on activities of starch degradative enzymes in a cold-requiring plant. *Biochem. Biophys. Res. Commun.* **58**, 573-578.

Goss, W. L. (1924). The vitality of buries seeds. *Journal of Agricultural Research* **29**, 349-362.

Groenendael, J. van, Kroon, H. de and Caswell, H. (1988). Projection matrices in population biology. *Trends Ecol. & Evol.* **3**, 264-269.

Gross, K. L. (1981). Predictions of fate from rosette size in four "biennial" plant species: *Verbascum thapsus*, *Oenothera biennis*, *Daucus carota*, and *Tragopogon dubius*. *Oecologia*(Berl.) **48**, 209- 213.

Gross, K. L. and Werner, P. A. (1982). Colonizing abilities of 'biennial' plant species in relation to ground cover: implications for their distribution in a successional sere. *Ecology* **63**, 921-931.

Gross, R. S. and Werner, P. A. (1983). Probability of survival and reproduction relative to rosette size in the common burdock (*Arctium minus*: Compositae). *Am. Midl. Nat.* **109**, 184-193.
Harper, J. L. (1967). A Darwinian approach to plant ecology. *J. Ecol.* **55**, 247-270.
Harper, J. L. (1977). "Population Biology of Plants". Academic Press, New York.
Harper, J. L. and Ogden, J. (1970). The reproductive strategy of higher plants. I. The concept of strategy with special reference to *Senecio vulgaris* L. *J. Ecol.* **58**, 681-698.
Hart, R. (1977). Why are biennials so few? *Am. Nat.* **111**, 792-799.
Hirose, T. (1983). A graphical analysis of life history evolution in biennial plants. *Bot. Mag.* Tokyo **96**, 37-47.
Hirose, T. and Kachi, N. (1982). Critical plant size for flowering in biennials with special reference to their distribution in a sand dune system. *Oecologia*(Berl.) **55**, 281-284.
Hirose, T. and Kachi, N. (1986). Graphical analysis of the life history evolution of *Oenothera glazioviana* Micheli. *Oecologia*(Berl.) **68**, 490-495.
Holt, B. R. (1972). Effects of arrival time on recruitment, mortality and reproduction in successional plant populations. *Ecology* **53**, 668-673.
Jong, T. J. de and Klinkhamer, P. G. L. (1988). Population ecology of the biennials *Cirsium vulgare* and *Cynoglossum officinale* in a sand-dune area. *J. Ecol.* **76**, 366-382.
Jong, T. J., Klinkhamer, P. G. L. and Metz, J. A. J. (1987). Selection for biennial life histories in plants. *Vegetatio* **70**, 149-156.
Jong, T. J. de, Klinkhamer, P. G. L. and Prins H. A. (1986). Flower ing behaviour of the monocarpic perennial *Cynoglossum officinale* L. *New Phytol.* **103**, 219-229.
Jong, T. J. de, Klinkhamer, P. G. L., Geritz, S. A. H. and Meijden, E. van der (1989). Why biennials delay flowering: an optimization model and field data on *Cirsium vulgare* and *Cynoglossum officinale*. *Acta Bot. Neerl.* **38**, 41-55.
Kachi, N. (1983). Population dynamics and life-history strategy of *Oenothera erythrosepala* in a sand-dune system. Doctoral thesis, University of Tokyo.
Kachi, N. and Hirose, T. (1979). Multivariate approaches to the plant communities related with edaphic factors in the dune system at Azigaura, Ibaraki Pref. I. Association-analysis. *Jpn. J. Ecol.* **29**, 17-27.
Kachi, N. and Hirose, T. (1983). Bolting induction in *Oenothera erythrosepala* Borbás in relation to rosette size, vernalization and photoperiod. *Oecologia*(Berl.) **60**, 6-9.
Kachi, N. and Hirose, T. (1985). Population dynamics of *Oenothera glazioviana* in a sand-dune system with special reference to the adaptive significance of size-dependent reproduction. *J. Ecol.* **73**, 887-901.
Kelly, D. (1985a). On strict and facultative biennials. *Oecologia* (Berl.) **67**, 292-294.
Kelly, D. (1985b). Why are biennials so maligned? *Am. Nat.* **125**, 473-479.
Klemow, K. M. and Raynal, D. J. (1981). Population ecology of *Melilotus alba* in a limestone quarry. *J. Ecol.* **69**, 33- 44.
Klemow, K. M. and Raynal, D. J. (1985). Demography of two facultative biennial plant species in an unproductive habitat. *J. Ecol.* **73**, 147-167.
Klinkhamer, P. G. L. and Jong, T. J. de (1983). Is it profitable for biennials to live longer than two years? *Ecol. Modell.* **20**, 223-232.
Klinkhamer, P. G. L. and Jong, T. J. de (1988). The importance of small-scale disturbance for seedling establishment in *Cirsium vulgare* and *Cynoglossum officinale*. *J. Ecol.* **76**, 383- 392.
Klinkhamer, P. G. L. and Jong, T. J. de (1989). A deterministic model to study the importance of density-dependence for regulation and the outcome of intra-specific competition in populations of sparse plants. *Acta Bot. Neerl.* **38**, 57-65.
Klinkhamer, P. G. L, Jong, T. J. de and Meelis, E. (1987a). Life-history variation and the control of flowering in short-lived monocarps. *Oikos* **49**, 309-314.
Klinkhamer, P. G. L., Jong, T. J. de and Meelis, E. (1987b). Delay of flowering in the 'biennial' *Cirsium vulgare*: size effects and devernalization. *Oikos* **49**, 303-308.
Kroon, H. de, Plaisier, A, Groenendael, J van and Caswell, H. (1986). Elasticity: the relative contribution of Demographic parameters to population growth rate. *Ecology* **67**, 1427-1431.

Lacey, E. P. (1986a). Onset of reproduction in plants: size- versus age-dependency. *Trends Ecol. & Evol.* **1**, 72-75.
Lacey, E. P. (1986b). The genetic and environmental control of reproductive timing in a short-lived monocarp *Daucus carota* (Umbelliferae). *J. Ecol.* **74**, 73-86.
Lacey, E. P. (1988). Latitudinal variation in reproductive timing of a short-lived monocarp, *Daucus carota* (Apiaceae). *Ecology* **69**, 220-232.
Lacey, E. P. and Pace, R. (1983). Effect of parental flowering and dispersal times on offspring fate in *Daucus carota* L. (Apiaceae). *Oecologia*(Berl.) **6**, 274-278.
Lacey, E. P., Real, L., Antonovics, J. and Heckel, D. G. (1983). Variance models in the study of life histories. *Am. Nat.* **122**, 114-131.
Lang, A. (1965). Physiology of flower induction. *In* "Encyclopedia of Plant Physiology" 15 (W. Ruhland, ed.), pp. 1380-1536. Spring er, Berlin.
Law, R. (1983). A model for the dynamics of a plant population containing individuals classified by age and size. *Ecology* **64**, 224-230.
Law, R., Bradshaw, A. D. and Putwain, P. D. (1977). Life history variation in *Poa annua*. *Evolution* **31**, 233-246.
Lee, J. M. and Hamrick, J. L. (1983). Demography of two natural populations of musk thistle (*Carduus nutans*). *J. Ecol.* **71**, 923-936.
Lefkovitch, L. P. (1965). The study of population growth in organisms grouped by stages. *Biometrics* **21**, 1-18.
Leslie, P. H. (1945). On the use of matrices in certain population mathematics. *Biometrika* **33**, 183-212.
Leslie, P. H. (1948). Some further notes on the use of matrices in population dynamics. *Biometrika* **35**, 213-245.
Levins, R. (1968). "Evolution in Changing Environments: Some Theoretical Explorations". Princeton University Press, Princeton.
Lewontin, R. C. (1965). Selection for colonizing ability. The Genetics of Colonizing Species. (H. G. Baker and G. H. Stebbins, eds), pp. 77-94. Academic Press, New York.
Lotka, A. J. (1922). The stability of the normal age distribution. *Proc. Natl. Acad. of Sci.* **8**, 339-345.
Meijden, E. van der and Waals-Kooi, R. E. van der (1979). The population ecology of *Senecio jacobaea* in a sand dune system. I. Reproductive strategy and the biennial habit. *J. Ecol.* **67**, 131-153.
Mooney, H. A. and Chiariello (1984). The study of plant function: the plant as a balanced system. *In* "Perspectives on Plant Popula tion Ecology" (R. Dirzo and J. Sarukhán, eds), pp. 305-323. Sinauer, Sunderland, MA.
Ødum, S. (1978). Dormant Seeds in Danish Ruderal Soils, an experimental study of Relations between Seed Bank and Pioneer Flora. The Royal Vet. and Agricultural University, Horsholm Arboretum, Denmark.
Quinn, J. F. (1969). Variability among high plains populations of *Panicum virgatum*. *Bull. Torrey Bot. Cl.* **96**, 20- 41.
Reinartz, J. A. (1984). Life history variation in the common mullein (*Verbascum thapsus*). 1. Latitudinal differences in population dynamics and timing of reproduction. *J. Ecol.* **72**, 897-912.
Salisbury, E. J. (1942). "The Reproductive Capacity of Plants". Bell, London.
Schaffer, W. M. (1983). The application of optimal control theory to the general life history problem. *Am. Nat.* **121**, 418-431.
Schaffer, W. M. and Gadgil, M. D. (1975). Selection for optimal life histories in plants. *In* "Ecology and Evolution of Communities" (M. L. Cody and J. M. Diamond, eds), pp. 142-157. Harvard University Press, Cambridge, Massachusetts.
Schaffer, W. M. and Schaffer, M. V. (1979). The adaptive significance of variations in reproductive habit in the Agavaceae. II: pollinator foraging behaviour and selection for increased repro ductive expenditure. *Ecology* **60**, 1051-1069.
Silvertown, J. W. (1983). Why are biennials sometimes not so few? *Am. Nat.* **121**, 448-453.

Silvertown, J. W. (1984). Death of the elusive biennial. *Nature* **310**, 271.
Silvertown, J. W. (1986). "Biennials": a reply to Kelly. *Am. Nat.* **127**, 721-724.
Smith, H. B. (1927). annual versus biennial growth habit and its inheritance in Melilotus alba. *Am. J. Bot.* **14**, 129-146.
Venable, D. L. and Lawlor, L. (1980). Delayed germination and dispersal in desert annuals: escape in space and time. *Oecologia* (Berl.) **46**, 272-282.
Verkaar, H. J. and Schenkeveld, A. J. (1984). On the ecology of short-lived forbs in chalk grasslands: life-history characteristics. *New Phytol.* **98**, 659-672.
Werner, P. A. (1975). Predictions of fate from rosette size in teasel (*Dipsacus fullonum* L.). *Oecologia*(Berl.) **20**, 197-201.
Werner, P. A. (1977). Colonization succession of a "biennial" plant species: Experimental field studies of species cohabitation and replacement. *Ecology* **58**, 840-849.
Werner, P. A. and Caswell, H. (1977). Population growth rates and age versus stage-distribution models for teasel (*Dipsacus sylvestris* Huds.). *Ecology* **58**, 1101-1111.

22 The Use of Functional as Opposed to Phylogenetic Systematics: A First Step in Predictive Community Ecology

PAUL A. KEDDY

Department of Biology, University of Ottawa, Ottawa, Canada

I. Introduction

> "That the plant-species may be classified not only according to their taxonomical relationship into genera, families etc., but also according to their types, has been clear to botanists ever since the childhood of botanical science".
>
> du Rietz 1931

Plant classification could have two objectives: constructing groups with similar evolutionary histories in order to construct phylogenies, or constructing groups with similar ecological traits for predictive ecology. The former approach has had a major impact upon the historical development of ecology: many of the most high profile research questions in ecology can be traced back to the phylogenetic basis of species taxonomy. The objective of this paper is to argue for an increased emphasis upon the latter approach and to illustrate both some procedures and possible consequences for biosystematics and ecology. A re-direction of emphasis towards ecological classification could have a major impact upon the future development of ecology. I will offer a general research strategy for the latter approach, illustrate its application in the study of wetland plant communities, and then briefly discuss its broader implications for predictive ecology and the management of the biosphere.

BIOLOGICAL APPROACHES AND EVOLUTIONARY TRENDS IN PLANTS ISBN 0-12-402960-4

A. Two views of reality

To return to the dichotomy presented in the first paragraph, let us consider the two contrasting approaches to classifying plants. In one case we could emphasize traits that are conservative to allow us to postdict -- that is, to reconstruct the history of life on this planet. For whatever reasons, this has been the major emphasis of recent systematics. The alternative approach is to emphasize traits that have to do with function: nutrient uptake, competitive ability, stress tolerance, dispersal ability, etc, in order to classify species according to their function in ecological communities. There are two reasons why Kyoto is a particularly auspicious city in which to initiate discussion of this dichotomy and its consequences.

First, Kyoto is a city closely associated with the world's Zen Buddhist heritage. One of the axioms of buddhist psychology is that human beings become caught in habitual ways of interpreting reality, and begin to confuse these habitual views with reality itself. This bears directly upon the above point about biosystematics: there are two (at least) models for organizing biological classifications, but a majority of work has concentrated upon one view. While neither view is 'right', in that both are merely human thought systems imposed upon nature, each view may have intrinsic value.

Second, a recent international symposium in Kyoto (Kawano *et al.*, 1987) illustrated the logical endpoint of the phylogenetic, postdictive approach to biosystematics (Figure 1). The logic appears to go in the following manner. We begin with species classifications. One of the most dramatic results of enumerating the species composition of the biosphere is the discovery of the planet's biological diversity. This leads to the first major question: how did so many species arise? Darwin provided an answer, and stimulated a century of research into the mechanisms and consequences of evolution through natural selection. This leads to the second major question: how do all these species co-exist (May, 1986)? This of course, raises the question of coexistence, which has been a central theme of ecology at least since the Hutchinson's 1959 paper entitled 'Homage to Santa Rosalia'. It is not at all obvious that this is the most important or the most tractable question for ecologists to address. Its popularity may arise in part from the logical sequence described above, and equally from the human fascination with collections, be they lists of species or shelves of Elvis Presley memorabilia. Connell (1987) in Kawano *et al.* (1987) reviews the mechanisms of co-existence as they are currently understood and marks the end point of this logical progression of investigation. Another paper in Kawano *et al.* (1987), Raven (1987), marks the other logical end point -- the applied problem of protecting this biological diversity from the ravages of human population growth and overconsumption of resources.

Where does this path lead next? Actually, it seems that both of these endpoints can be dealt with best by returning to the starting point. That is, the problems they

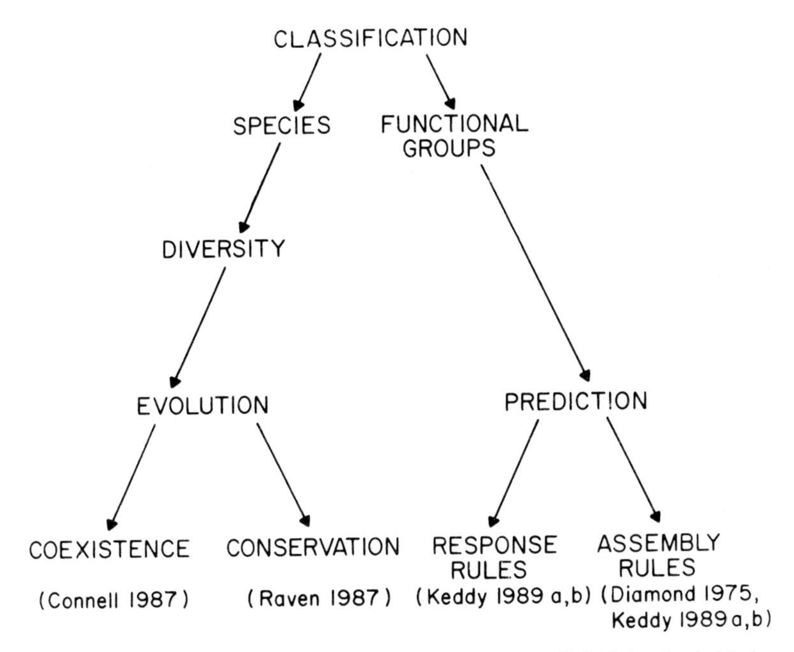

Fig. 1 Two research paths growing out of classification. The well-travelled left hand path (phylogenetic classification) leads to questions of diversity and coexistence. The less-travelled right hand path (functional classification) leads to predictive community ecology.

raise for ecology are inherent in the view that is chosen, and cannot be addressed by technical tinkering. Rather, it is necessary to fundamentally re-structure our research strategy. Let me illustrate why this is the case.

B. Classification and prediction

Suppose we take a simple plant community with S species, and make two simplifying assumptions (1) that this community can be reconstructed from knowledge of pairwise interactions and (2) that these interactions do not change as the environment changes. The number of interactions we need to study is then simply $\binom{S}{2}$ where S is the number of species. In wetlands in the Ottawa Valley, a very small part of the world, we can easily find 160 species, which yields 12,720 pairwise interactions. Assume we run a large experiment each year to look at 100 pairwise interactions. At this rate, it would take more than a century to develop simple predictions for this single vegetation type. If we assume following Colinvaux (1986) that there are more than 250,000 species of vascular plants on the planet, models for sustainable development will be very far off indeed (unless ecologists wait until a sufficient number are extinct, not a pleasing prospect). Rigler (1982) and Wimsatt (1982) provide other examples of the limitations of models built upon species nomenclature. Species oriented models therefore cannot succeed in providing general prediction. The question, therefore, is how do we devise models that are both accurate and simple enough to be tractable? This directs attention to the second path in Figure 1.

This second path of inquiry (Figure 1, right) has a different logic. While on one hand there is the overwhelming species diversity of the biosphere, there is also dramatic and obvious repetition of certain themes. That is, there are recurring melodies as well as different notes. The convergence in plant growth forms representing very different families in deserts is but one example: 'succulents' occur in groups including the Cactaceae, Liliaceae, Euphorbiaceae, Crassulaceae, and Asclepidaceae. At the opposite extreme, in wetland habitats, 'mud flat annuals' are found in families including the Juncaceae, Cyperaceae, Asteraceae and Polygonaceae. If we develop this view, it is the convergence we emphasize. It is of course of interest to know what constraints phylogeny places upon convergence (Hodgson and Mackey, 1986; Givnish, 1987). But from the point of view of predictive ecology, the important questions start with what are the major convergent groups and how many are there? How many do we need recognize for a specific level of precision in our models? Growing out of this are other questions. What are the traits which they share? How do we use a knowledge of these traits to predict how a particular plant community will change after an external perturbation (response rules, Keddy, 1989a,b)? How can we use a knowledge of these traits to predict the group of species that will be present in a specified environment (assembly rules, sensu Diamond 1975, but modified according to Box (1981), Haefner (1981) and Keddy, 1989a,b)? These two endpoints represent two important questions facing ecologists today. On the theoretical side the question is whether we know enough to be able to develop rules to put communities together from their constituent pieces. On the applied side the question is whether we can predict state changes in ecological systems.

From the perspective of biosystematics, it would appear that biosystematists could stimulate this entire research path. This might generate a development for ecology as significant as species classifications did for evolution. The remainder of this paper therefore explores three aspects of this process: (1) a review of some recent ideas about functional classifications of plants, (2) a discussion of the sorts of traits necessary for recognizing functional groups, and (3) results from applying this approach to wetland plants.

II. Functional Groups

The idea that organisms fall into functional groups is not a new one. The idea is particularly well-developed in the zoological literature, where the term 'guilds' (Root, 1967) is used to describe groups of functionally similar species in a community (Pianka, 1983). This terminology is particularly well-established in the study of bird and mammal communities (e.g. Severinghaus, 1981; Pianka, 1983; Diamond and Case, 1986; Terborgh and Robinson, 1986). Similarly, the concept of 'functional feeding groups' has been successfully applied to aquatic invertebrates (Cummins, 1973; Cummins and Klug, 1979). This approach is less well-established in the study of plant communities (but see Platt and Weiss, 1977; Beattie and Culver, 1981; Cody, 1986; Fitter, 1987; Givnish, 1987; Day *et al.*, 1988). The use of guilds in plant communities has often been primarily to split off one group of species from the rest of the community, such as the ruderal species which colonize animal mounds in herbaceous prairie vegetation (e.g., Platt and Weiss, 1977) or species which are dispersed by ants in deciduous forest (Beattie and Culver, 1981). The limited application to plants probably is the result of one major obstacle: functional groups of animals are recognized largely by the different foods they consume. In stream invertebrates, for example, functional groups include 'shredders', 'collectors', 'scrapers' and 'piercers'. This approach cannot be extrapolated to plants since all plants use the same few basic resources, light, water and mineral nutrients. This has been a principal reason why concepts of niche-differentiation have been so difficult to apply to plant communities (Harper, 1977; Grubb, 1977). There have been some attempts to apply concepts of niche-differentiation to resources such as the regeneration-niche hypothesis (Grubb, 1977) and resource ratio-hypothesis (Tilman, 1982, 1986), but these do not lead easily to functional classifications either. Are there some general procedures for constructing functional groups in plants? Before answering this, let us look at attempts that have been used to date.

The idea that plants can be grouped naturally into functional groups is not a new one. Du Rietz (1931) traced the idea back to Threophrastos (ca. 300 B.C.). Du Rietz also provided an extensive review of early classification schemes, noting that some emphasized only 'morphological' traits, whereas others emphasized 'biological' ones. More recently, there has been a proliferation of ecological classification schemes (e.g. Grime, 1974, 1977, 1988; van der Valk, 1981; Grubb, 1985; Cody, 1986; Givnish, 1987; Day *et al.*, 1988). Some of the best examples of guild classification come from studies on the photosynthetic phenology of temperate woodland plants (Kawano, 1985; Givnish, 1987) where the seasonal patterns of leaf development can be used to build a classification of photosynthetic guilds.

There are three principal trends which these recent studies illustrate. The first is to place greater emphasis upon the ecological role of species in communities (e.g., Noble and Slatyer, 1980; Givnish, 1987; Day *et al.*, 1988). The second is to emphasize functional traits which can only be determined by screening (e.g, Grime and Hunt, 1975; Grime *et al.*, 1981). The last is a shift from using these schemes to describe vegetation to using them to predict future states of vegetation (e.g. van der Valk, 1981). Most of these recent trends seem to have been developed ad hoc for specific systems and the objective of the next section is to uncover the basic principles. Figure 2 summarizes the process and these principles. It also presents the structure of the remainder of this paper.

III. Trait Selection

Du Rietz reviewed the many traits considered in early classifications, including life form, life span, method of vegetative propagation and position of overwintering shoots. These are all traits which can be determined upon inspection of the plant form. However, if we consider the processes which occur in vegetation, such as nutrient uptake, competition, and interaction with agents of disturbance or stress,

TABLE 1

Some recent classifications which emphasize function of species in communities rather than their morphology

functional group (number)	types	reference
strategies	ruderal	Grime, 1977
(3)	competitor	
	stress tolerator	
life history type	annual	van der Valk, 1981
(12)	perennial[1]	
---	matrix	
(2)	interstitial	Grubb, 1985
guild	winter annual	
(7)	spring ephemeral	
	early summer	
	later summer	
	wintergreen	
	evergreen	
	dimorphic	
functional group	clonal dominants	Day *et al*., 1988
(6)	gap colonizers	
	ruderals	
	stress tolerators	
	reeds	
	ruderals	

[1] van der Valk's system recognizes twelve different life history types by also considering whether or not the perennials vegetatively propagate, establishment requirements (flooded, unflooded) and propagule longevity.

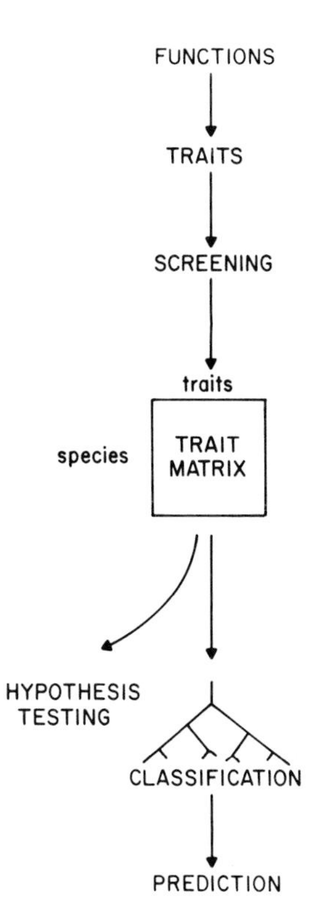

Fig. 2 A general research strategy for constructing functional groups (guilds). Key steps are selecting the traits which are related to different plant functions in communities, screening large numbers of species for these traits, constructing a trait matrix, and building a classification from the trait matrix. The classfication of functional groups is then used to make simple predictive ecological models.

there are traits which may not be obvious upon inspection, but may be none-theless closely related to the role the plant plays in a community. We could begin by looking for traits that measure performance in these three areas: (1) ability to forage for essential elements in the absence of neighbours, (2) traits associated with interaction with other plants and (3) traits associated with ability to withstand agents such fire, flooding, grazing, etc. Table 2 illustrates some traits which have been measured to date.

Since many of these traits are not obvious upon inspection, we need to consider the process of screening (e.g., Grime and Hunt, 1975; Grime *et al.*, 1981). The objective of screening is to develop a simple bioassay for a particular attribute, and then apply it systematically to the entire flora of interest. There is still a great deal of opportunity to develop new and innovative bioassays for plant traits ranging from foraging ability to competitive ability to stress tolerance (Table 2).

TABLE 2
Some recent examples of traits which reflect plant function and for which screening has been conducted

trait	examples
germination	Grime *et al.* (1981) Shipley *et al.* (1990)
relative growth rate	Grime and Hunt (1975) Sheldon (1987) Shipley *et al.* (1990) Shipley and Peters (in press)
nutrient foraging	Crick and Grime (1987)[1]
shoot extension rates	Boutin and Keddy (ms)
evergreenness	Givnish (1987) Wisheu and Keddy (1989)
competitive ability	Gaudet and Keddy (1988)
stress tolerance	Shipley and Keddy (1989)
biomass partitioning	Shipley and Peters (in press)
drought tolerance	Raynal *et al.* (1985) Boot *et al.* (1986[1])
palatability	Sheldon (1987) Southwood *et al.* (1986) McCanny *et al.* (ms.)
canopy form	Givnish (1987) Shipley *et al.* (1990)
genomic form	Bennett (1987)
sensitivity to acid rain	Percy (1986)

[1] Only two species were `screened', but these are included because it is an idea that merits further work

Screening for traits is a research strategy almost completely orthogonal to the traditional population approach (Harper, 1977, 1982) which emphasizes accumulating information on many aspects of one population (rows) rather than one aspect of many populations (columns). The recent literature would suggest that the latter is much less common than the former, but it may be argued that general predictive relationships are more likely to emerge from systematic comparative studies (Peters, 1980; Rorison *et al.*, 1987; Keddy, 1989b).

In conclusion, the three steps in producing an ecological trait matrix are (1) to decide the functions which the plant must perform to survive in a community, (2) to measure the traits which measure these functions, and (3) to devise the screening technique(s) for those traits.

IV. Traits in Wetland Plants

The following studies illustrate the progress to date in applying the scheme in Figure 2 to the study of wetland plant communities. While the central objective remains predicting future states of these communities, there are important evolutionary and ecological sub-questions which can be asked along the way. I will therefore begin by illustrating the side branch labelled hypothesis testing. This is partly because this where we have completed most work to date, and partly because there are many important evolutionary questions which can be answered using such data.

A. Correlation of adult and juvenile traits

One of the early questions which can be posed in deciding upon traits to measure is whether juvenile and adult traits are strongly correlated. If they are, fewer traits need be measured. If they are not, then there can be different adult and juvenile strategies. There are models which predict that adult and juvenile traits are correlated (r-K) and those which predict they are uncorrelated (C-S-R). To test between these, 7 juvenile traits and 13 adult traits were measured on 25 species of wetland plants (Shipley, 1987;Shipley *et al.*, 1990). Using the test developed by Lefkovitch (1984), Shipley was able to show that the adult and juvenile traits are uncorrelated. Functional classifications must consider both juvenile and adult traits (Figure 3). In the case of wetland plants, the principal adult trait axis appears to be capacity of adult plant to occupy space and intercept light; this is consistent with other studies on plant competition (e.g., Givnish, 1982, 1987; Tilman, 1988; Keddy, 1989a,b; Keddy and Shipley, 1989). The major juvenile axis is traits associated with regeneration. This yields four functional types of species (Fig. 3), which can be differentiated on relatively few traits.

B. Traits in wetland plants

Figure 4 shows four different pairs of traits we have measured in wetland plant communities, and relationships among them.

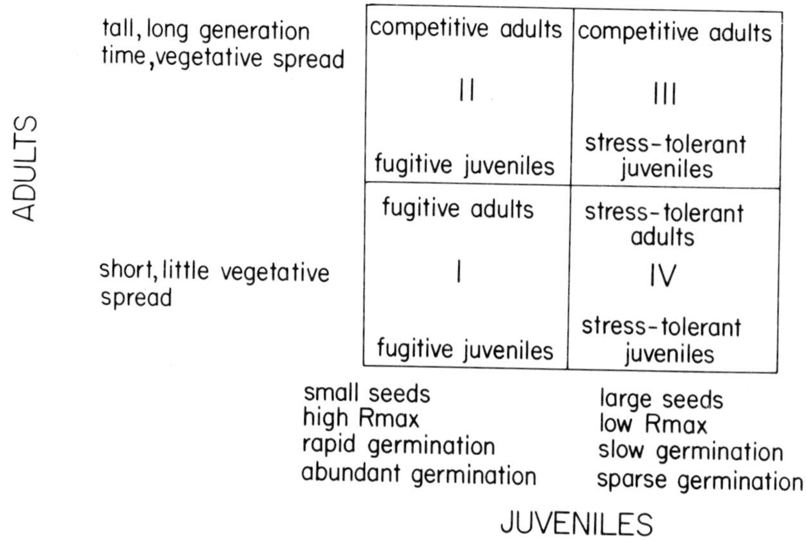

ADULTS	small seeds, high Rmax, rapid germination, abundant germination	large seeds, low Rmax, slow germination, sparse germination
tall, long generation time, vegetative spread	competitive adults II fugitive juveniles	competitive adults III stress-tolerant juveniles
short, little vegetative spread	fugitive adults I fugitive juveniles	stress-tolerant adults IV stress-tolerant juveniles

JUVENILES

Fig. 3 Juvenile traits are uncorrelated with adult traits, so both must be screened for. In the case of wetland plants, the combination of adult traits associated with holding space (vertical axis) and juvenile traits associated with regeneration (horizontal axis) yield four basic functional groups of wetland plants (after Shipley, 1987; Shipley *et al.*, 1990).

Size and competitive ability

There is general agreement that competition is one of the major forces involved in the assembly of plant communities (Grime, 1979; Tilman, 1982, 1988; Keddy, 1989a). But there is consensus neither on what traits confer competitive ability, nor on how competition assembles communities. In Gaudet and Keddy (1988) we developed a bioassay for competitive ability and showed across 44 wetland plant species that it was predictable from above ground biomass. We attributed this to the importance of competition for light. Whether this relationship occurs in other plants and in other sets of environmental conditions can only be tested by using the bioassay design with other species in other environments.

Investment in plant defenses

Grazing is another factor known to be important in plant communities (Harper, 1977; Grime, 1979; Sheldon, 1987). There are good evolutionary reasons for expecting the chemical defenses in plants to be negatively correlated with relative growth rate (Coley *et al.*, 1985; Southwood *et al.*, 1986). In McCanny *et al.* (manuscript) we developed a simple bioassay for palatability of wetland plants to generalist herbivores. In this case, we used extracts from field-collected plant tissue

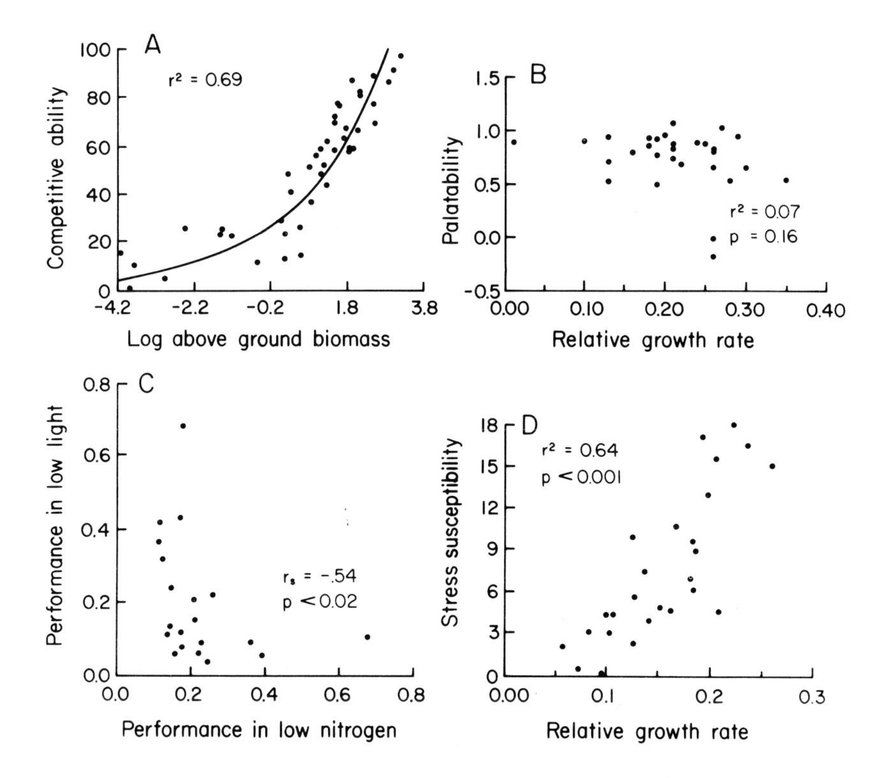

Fig. 4 Once ecological traits have been measured in large numbers of species, it is possible to look for relationships among them. These examples all come from wetland plants.

A. Competitive ability plotted against plant biomass (Gaudet and Keddy, 1988)

B. Palatability plotted against relative growth rate (McCanny *et al.*, manuscript)

C. Performance in low light plotted against performance in low nitrogen (Fricker, unpublished data)

D. Stress susceptibility plotted against relative growth rate (Shipley and Keddy, 1988).

in diet cubes fed to an herbivorous insect. We found no significant relationship between palatability and relative growth rate. There are other traits which may be associated with resistance to herbivory such as rate of resprouting, position of meristems, etc, so there are many opportunities for extending this work.

Trade-offs in competitive ability

In studies of plant competitive ability, Tilman (1982) has argued that there should be a negative relationship between above and below ground competitive ability. Fricker (unpublished data) has tested for this relationship in 20 wetland plant species by screening for their ability to grow under low nutrient and light regimes. We found a significant negative correlation in these traits. Apparently traits which confer the ability to tolerate low light levels are negatively correlated with those which confer ability to tolerate low nutrient levels.

Stress tolerance and relative growth rate

Shipley and Keddy (1989) screened 28 wetland plant species for ability to tolerate low nutrient conditions. They found that those species which grew fastest under high nutrient levels (e.g. *Lythrum salicaria*, *Leersia oryzoides*, *Scirpus acutus*) were those species most sensitive to reductions in nutrient supply. Species with slower growth rates (e.g. *Triadenum fraseri*, *Scirpus americanus*, *Juncus filiformis*, *Iris versicolor*) were more stress-tolerant; that is, their growth rates were proportionally less-depressed by low nutrient levels.

These four examples illustrate the possibility for testing general hypotheses about the evolution of plant traits using trait matrices constructed from screening. While the ultimate goal may be to predict how species possessing these traits will respond to different environments (e.g. increased grazing, decreased fertility, etc.), in the short term they provide valuable information on the co-evolution of plant traits. In the long run, only a subset of such traits will probably need to be measured in order to make ecological predictions.

V. Functional Groups in Wetland Vegetation

To construct functional groups we need systmetically collected data on the traits of many species in a specific vegetation type.

Boutin and Keddy (manuscript) have recently finished a functional groups classification based upon a matrix of 43 species by 27 traits. The species were selected to represent wetland habitats and functional groups from across eastern North America. Species included rare or endangered taxa from infertile lakeshores (*Coreopsis rosea*, *Panicum longifolium*), annuals typical of mud flats (*Bidens cernua*, *Cyperus aristatus*), large perennials (*Phalaris arundinacea*, *Typha glauca*), reeds from river banks (*Scirpus acutus*, *Eleocharis calva*), and a wide array of other species which represented other life forms and habitats. They were grown in a large outdoor experiment. Additional traits were measured in the field and in more controlled conditions.

Classification revealed three major groups (Figure 5). The first separation was between perennials and ruderals. Ruderals (sensu Grime, 1977) were species with the capacity to behave as annuals, which included facultative annuals (perennials with nearly 100 percent flowering the first year). They also had proportionally less biomass allocated to below ground structures. The perennials then split into two groups based largely upon traits associated with ability to hold space: a group of large space-holding species with clonal spread, and a group of species without vig-

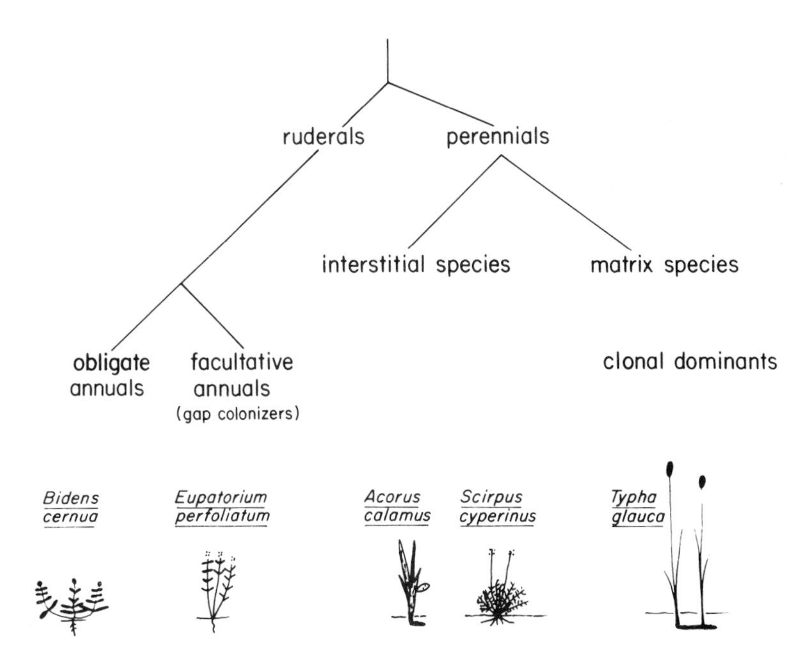

Fig. 5 A preliminary classification of guilds in wetland plants based upon clustering of a matrix of 27 traits measured upon 43 wetland plant species from across eastern North America. They are arranged along an axis of increasing ability to hold space (after Boutin and Keddy ms).

orous clonal spread and with shallow rhizomes. This group apparently occupies gaps in the large clonal species. Grubb (1986) has termed these two groups 'matrix' and 'interstitial species'. Let us explore these four groups without at the moment concerning ourselves with the other subgroups of interstitial or matrix species. It is noteworthy that these functional groups all appear to fall along a continuum of life histories adjusted to different light regimes. This is consistent with the growing body of literature showing that standing crop gradients can predict species richness and life form in plant communities (Grime, 1973, 1979; Al-Mufti *et al.,* 1977; Moore *et al.,* 1989; Wisheu and Keddy, 1989).

A. Obligate annuals

Obligate annuals (e.g. *Bidens cernua*, *Cyperus aristatus*, *Eleocharis obtusa*, *Echinochloa weigandii*) lived for only one growing season, and for this reason shared many other traits such as low rates of shoot extension, high rates of tiller production, and high rates of biomass production. These species emerge en masse from buried seeds after the adult vegetation has been killed (van der Valk and Davis, 1978; van der Valk, 1981; Keddy and Reznicek, 1982, 1986).

B. Facultative annuals

Facultative annuals (e.g., *Lythrum salicaria, Verbena hastata, Mimulus ringens, Eupatorium perfoliatum*) all flowered at the end of the first year and therefore could, if necessary, function as annuals. They normally overwinter and send up several slender stalks from the base of the previous year's shoot. The seeds are borne at the very top of a tall stalk, presumably assisting with locating new gaps. Other traits included the highest rates of extension of the main shoot, and lowest rates of tiller production. Although these species can be found on mud flats, they also regenerate in smaller gaps and along animal trails. This group of species could also be called 'gap colonizers' (Day *et al.,* 1989) and similar functional groups have been described in prairie vegetation (Platt and Weiss, 1977, 1985).

C. Interstitial species

The interstitial species (e.g. *Carex crinita, Scirpus cyperinus, Asclepias incarnata, Iris versicolor, Alisma plantago-aquatica*) were the most heterogeneous group of species. However, almost all lacked the ability for clonal spread, and had shallow root systems compared with the following group of matrix species. They also tended to have compact arrangements of foliage, suggesting that they hold small areas of space against invasion from neighbours.

D. Matrix species

Matrix species (e.g. *Phalaris arundinacea, Typha glauca, Scirpus americanus*) were the tallest group with the most deeply buried rhizomes and extensive clonal spread. They dominate fertile, undisturbed environments (Day *et al.,* 1989).

VI. Guilds and Prediction

Two possible approaches to prediction are 'assembly rules' and 'response rules' (Keddy, 1989a,b). Let us consider them in turn.

A. Assembly rules

In the case of assembly rules, we can ask which subset of the species pool for a region will be found in a specified environment (Diamond, 1975; Keddy, 1989a). If we know the species pool for any particular system, we can picture a series of sieves which progressively eliminate species or guilds until only a specific subset are left (Figure 6). If an entire trait matrix is available for a flora, then such sieve

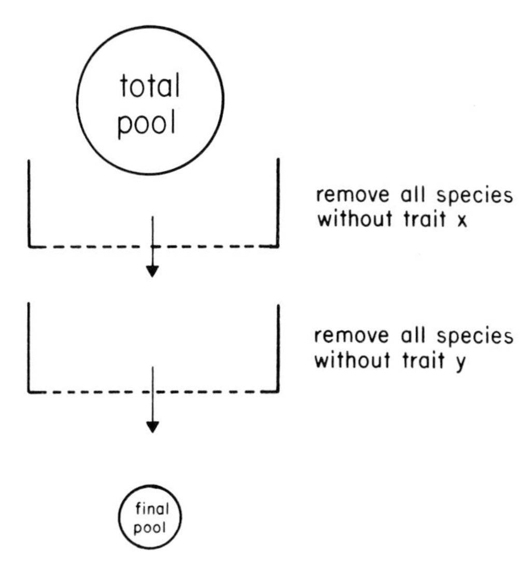

Fig. 6 By starting with the total species pool for an area and then subtracting species lacking the traits to tolerate the local environment, one is left with the pool of species which will comprise the local vegetation.

models could be based upon actual measured traits for each species. But a far more efficient approach is to develop guild classifications based upon the minimum number of traits necessary and then consider which guilds are filtered out by the environment. The remaining functional groups will form the vegetation. Box (1981) has developed such a model to predict world vegetation types. The model starts with the total pool of plant life forms, and then progressively subtracts those which cannot tolerate the local climate. A second step is to examine dominance relations (competitive interactions) among life forms, with position in dominance hierarchy determined largely by size. This is an example of assembly rules for plant communities at the largest scale; a similar approach should be possible at local scales.

B. Response rules (transformation rules)

In the case of response rules we wish to predict how a specified perturbation will convert one vegetation type (that is, one vector of species abundances) to another (Lewontin, 1974; Keddy, 1989a). Again, this would be a complex process if attempted on a species by species basis, but it is greatly simplfied by using guilds. Figure 7 illustrates one approach to response rules. Here the 43 species we studied (which represent the range of variation found in hundreds of other wetland plant species), can be reduced to a column vector of only 4 guilds. We can then explore how changes in different sets of environmental factors will transform the abun-

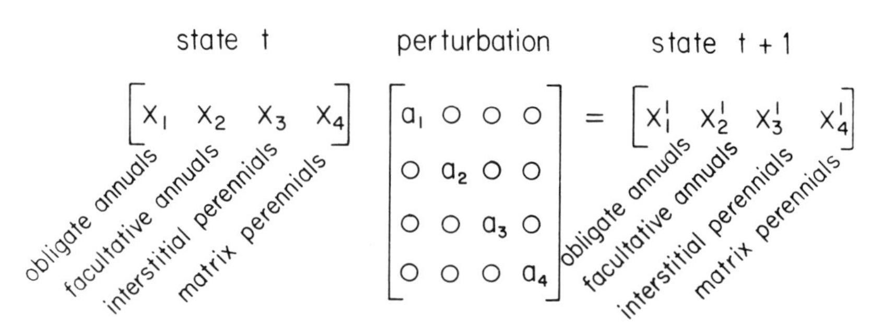

Fig. 7 One method of using guilds to make ecological predictions. In this case, the 43 wetland plant species can be represented by a vector of only four guilds, greatly simplifying the process of predicting future states of the system.

dances of these guilds through time. For example, any factor which increases productivity or decreases disturbance will reduce light availability and reduce regeneration opportunities; we would then predict a gradual shift away from annuals and facultative annuals to clonal dominants. That is, the matrix A would have negative coefficients for all but element a_4. Conversely, if the wetland were subjected to fluctuating water levels, grazing, or recreational development, we would expect a shift towards the ruderal and gap colonizer end of the continuum, in which case only a_1 and a_2 would have positive coefficients. If the disturbances created only small gaps -- say due to increased population sizes of a small mammal, then only the interstitial species and facultative annuals would be predicted to increase. The process of estimating these coefficients and predicting future states has been briefly addressed in Keddy *et al.* (1989) but much remains to be done. Van der Valk (1981) has illustrated one approach to response rules for prairie pothole vegetation; in this case he recognized 12 life history types and showed how knowledge of their traits (e.g. germination requirements) and changes in the environment (e.g. flooding) could predict future vegetation states. Once guilds are recognized, and key environmental factors determined, this approach could presumably be used in most other vegetation types.

The potential for using functional classifications and trait matrices in predictive ecology has barely been explored. They may offer both a new avenue of work for biosystematists, and a research path that will maximize our rates of progress in achieving predictive community ecology.

C. Questions remaining

Since we are in the early phase of classifying plants into functional groups, and in developing predictive models based upon them, there are many opportunities for research. Unanswered questions would include:

(1) What are the best traits for measuring the functional roles of plants in vegetation?

(2) What are the minimum number of traits we need to measure to produce accurate and useful classifications of functional groups?

(3) What are the most efficient methods for screening for the above traits?

(4) How many functional groups are necessary for particular levels of accuracy of prediction?

(5) Across how many vegetation types can one extrapolate with a particular model? Is it possible to develop one or two simple models that could be distributed and used around the world?

Acknowledgements

It is important to begin by acknowledging the pioneering work of J.P. Grime in developing some of the fundamentals of comparative plant ecology (e.g. Grime, 1979; Rorison *et al.*, 1987; Grime *et al.*, 1988); I have borrowed freely from his ideas and techniques here. I thank also C. Boutin, P. Fricker, C. Gaudet, S. McCanny, D. Moore, B. Shipley, and I. Wisheu for stimulating ideas and hard work. Thanks in particular to C. Boutin, P. Fricker, and S. McCanny for allowing the use of unpublished data. I also thank Anita Payne for preparing the manuscript and Jacques Helie for preparing the figures. This research was supported by a Strategic Grant from the Natural Sciences and Engineering Research Council of Canada.

References

Al-Mufti, M. M., Sydes, C. L., Furness, S. B., Grime, J. P. and Band, S. R. (1977). A quantitative analysis of shoot phenology and dominance in herbaceous vegetation. *J. Ecol.* **65**, 759-791.

Beattie, A. J. and Culver, D. C. (1981). The guild of myrmecochores in the herbaceous flora of West Virginia forests. *Ecology* **62**, 107-115.

Bennett, M. D. (1987). Variation in genomic form in plants and its ecological implications. *New Phytol.* **106**, 177-200.

Boot, R., Raynal, D. J. and Grime, J. P. (1986). A comparative study of the influence of drought stress on flowering in *Urtica Dioica* and *U. urens*. *J. Ecol.* **74**, 485-495.

Box, E. O. (1981). "Macroclimate and plant forms: an introduction to predictive modeling in phytogeography". Junk, The Hague.

Boutin, C. and Keddy, P. A. (manuscript). Guilds and predictive plant community ecology.

Cody, M. L. (1986). Structural niches in plant communities. *In* "Community Ecology" (J. Diamond and T. J. Case, eds), pp. 381-405. Harper and Row, New York.

Coley, P. D., Bryant, J. P., Chapin, F. S. III (1985). Resource availability and plant antiherbivore defense. *Science* ***230***, 895-899.

Colinvaux, P. (1986). "Ecology". Wiley, New York.

Connell, J. H. (1987). Maintenance of species diversity in biotic communities. *In* "Evolution and Coadaptation in Biotic Communities" (S. Kawano, J. H. Connell and T. Hidaka, eds), pp. 201-218. University of Tokyo Press.

Crick, J. C. and Grime, J. P. (1987). Morphological plasticity and mineral nutrient capture in two herbaceous species of contrasted ecology. *New Phytol.* **107**, 403-414.

Cummins, K. W. (1973). Trophic relationships of aquatic insects. *Annu. Rev. Entomol.* **18**, 183-206.

Cummins, K. W. and Klug, M. J. (1979). Feeding ecology of stream invertebrates. *Annu. Rev. Ecol. Syst.* **10**, 147-172.

Day, R.T., Keddy, P.A., McNeill, J. and Carleton, T.(1988). Fertility and disturbance gradients: a summary model for riverine marsh vegetation. *Ecology* **69**, 1044-1054.

Diamond, J. M. (1975). Assembly of species communities. *In* "Ecology and Evolution of Communities" (M. L. Cody and J. M. Diamond, eds), Belknap Press, Harvard University Press, Cambridge, U.S.A.

Diamond, J. M. and Case, T. J. (1986). "Community Ecology". Harper and Row, New York.

du Rietz, G. E. (1931). Life-forms of terrestrial flowering plants. *Acta Phytogeogr. Suec. III.* Almqvist and Wiksells, Uppsala.

Fitter, A.H. (1987). An architectural approach to the comparative ecology of plant root systems. *New Phytol.* **106**, 61-77.

Gaudet, C. L. and Keddy, P. A. (1988). Predicting competitive ability from plant traits: a comparative approach. *Nature* **334**, 242-243.

Givnish, T. J. (1982). On the adaptive significance of leaf height in forest herbs. *Am. Nat.* **120**, 353-381.

Givnish, T. J. (1987). Comparative studies of leaf form: assessing the relative roles of selective pressures and phylogenetic constraints. *New Phytol.* **106**, 131-160.

Grime, J. P. (1974). Vegetation classification by reference to strategies. *Nature* **250**, 26-31.

Grime, J. P. (1977). Evidence for the existence of three primary strategies in plants and its relevance to ecological and evolutionary theory. *Am. Nat.* **111**, 1169-1194.

Grime, J. P. (1979). "Plant Strategies and Vegetation Processes". Wiley, Chichester.

Grime, J. P (1988). The C-S-R model of primary plant strategies - origins, implications and tests. *In* "Plant Evolutionary Biology" (L. D. Gottlieb and S. K. Jain, eds), pp. 371-393. Chapman and Hall, London.

Grime, J. P., Hodgson, J. G. and Hunt, R. (1988). "Comparative Plant Ecology". Unwin Hyman, London.

Grime, J. P. and Hunt, R. (1975). Relative growth rate: its range and adaptive significance in a local flora. *J. Ecol.* **63**, 393-422.

Grime, J.P., Mason, G., Curtis, A.V., Rodman, J., Band, S.R., Mowforth, M.A.G., Neal, A.M. and Shaw, S.(1981). A comparative study of germination characteristics in a local flora. *J. Ecol.* **69**, 1017-1059.

Grubb, P. J. (1977). The maintenance of species richness in plant communities: the importance of the regeneration niche. *Biol. Rev.* **52**, 107-145.

Grubb, P. J. (1977). The maintenance of species richness in plant communities: the importance of the regeneration niche. *Biol. Rev.* **52**, 107-145.

Grubb, P. J. (1985). Plant populations and vegetation in relation to habitat, disturbance and competition: problems of generalization. *In* "The Population Structure of Vegetation" (J. White, ed.), pp.595-621. Junk, Dordrecht.

Haefner, J. W. (1981). Avian community assembly rules: the foliage-gleaning guild. *Oecologia* **50**, 131-142.

Harper, J. L. (1977). "Population Biology of Plants". Academic Press, London.

Harper, J. L. (1982). After description. *In* "The Plant Community as a Working Mechanism" (E. I. Newman, ed.), pp.11-25. Blackwell, Oxford.

Hodgson, J. G. and Mackey, J. M. L. (1986). The ecological specialization of dicotyledonous families within a local flora: some factors constraining optimization of seed size and their possible evolutionary significance. *New Phytol.* **104**, 497-515.

Hutchinson, G. E. (1959). Homage to Santa Rosalia or why are there so many kinds of animals? *Am. Nat.* **93**, 145-149.

Kawano, S. (1985). Life history characteristics of temperate woodland plants in Japan. *In* "The Population Structure of Vegetation" (J.White, ed.), pp.515-549. Junk, Dordrecht.

Kawano, S., Connell, J. H. and Hidaka, T. (1987). "Evolution and Coadaptation in Biotic Communities". University of Tokyo Press.

Keddy, P. A. (1989a). "Competition", Chapman and Hall, London.

Keddy, P. A. (1989b). Competitive hierarchies and centrifugal organization in plant communities. *In* "Plant Competition" (J. Grace and D. Tilman, eds). Academic Press, New York.

Keddy, P.A. and Reznicek, A.A. (1982). The role of seed banks in the persistence of Ontario's coastal plain flora. *Am. J. Bot.* **69**, 13-22.

Keddy, P. A. and Reznicek, A. A. (1986). Great Lakes vegetation dynamics: the role of fluctuating water levels and buried seeds. *Journal of Great Lakes Research* **12**, 25-36.

Keddy, P. A. and Shipley, B. (1989). Competitive hierarchies in herbaceous plant communities. *Oikos* **49**, 234-241.

Keddy, P. A., Wisheu, I. C., Shipley, B. and Gaudet, C. (1989). Seed banks and vegetation management for conservation: towards predictive community ecology. *In* "Ecology of Soil Seed Banks" (M.A. Leck, V.T. Parker and R. L. Simpson, eds). Academic Press, San Diego.

Lefkovitch, L. P. (1984). A nonparametric method for comparing dissimilarity matrices, a general measure of biogeographical distance, and their application. *Am. Nat.* **123**, 484-489.

Lewontin, R. C. (1974) "The Genetic Basis of Evolutionary Change". Columbia University Press, New York.

May, R. M. (1986). The search for patterns in the balance of nature: advances and retreats. *Ecology* **67**, 1115-1126.

McCanny, S.J., Keddy, P.A., Arnason, T.J. and Shipley, B. (manuscript). Fertility and the food quality of wetland plants: a test of the resource availability hypothesis.

Moore, D.R.J., Keddy, P.A., Gaudet, C.L. and Wisheu, I.C. (1989). Conservation of wetlands: do infertile wetlands deserve a higher priority? *Biol. Conserv.* **47**, 203-217.

Noble, I.R. and Slatyer, R.O. (1980). The use of vital attributes to predict successional changes in plant communities subject to recurrent disturbances. *Vegetation* **43**, 5-21.

Percy, K. (1986). The effects of simulated acid rain on germinative capacity, growth and morphology of forest tree seedlings. *New Phytol.* **104**, 473-484.

Peters, R. H. (1980). From natural history to ecology. *Perspect. Biol. Med.* **23**, 191-203.

Pianka, E. R. (1983). "Evolutionary Ecology". 3rd. ed. Harper & Row, N.Y.

Platt, W. J. and Weiss, I. M. (1977). Resource partitioning and competition within a guild of fugitive prairie plants. *Am. Nat.* **111**, 479-513.

Platt, W. J. and Weiss, I. M. (1985). An experimental study of competition among fugitive prairie plants. *Ecology* **66**, 708-720.

Raven, P. H. (1987). Biological resources and global stability. *In* "Evolution and Coadaptation in Biotic Communities" (S. Kawano, J.H. Connell and T. Hidaka, eds) University of Tokyo Press.

Raynal, D. J., Grime, J. P. and Boot, R. (1985). A new method for the experimental droughting of plants. *Ann. Bot.* **55**, 893-897.

Rigler, F. H. (1982). Recognition of the possible: an advantage of empiricism in ecology. *Can. J. Fish. Aquat. Sci.* **39**, 1323-1331.

Root, R. (1967). The niche exploitation pattern of the blue-grey gnatcatcher. *Ecol. Monogr.* **37**, 317-350.

Rorison, I.H., Grime, J.P., Hunt, R., Hendry, G.A.F. and Lewis, D.H. (1987). Frontiers of Comparative Plant Ecology. *New Phytol.* **106** (supplement), 1-317.

Severinghaus, W. D. (1981). Guild theory development as a mechanism for assessing environmental impact. *Environ. Manage.* **5**, 187-190.

Sheldon, S. P. (1987). The effects of herbivorous snails on submerged macrophyte communities in Minnesota lakes. *Ecology* **68**, 1920-1931.

Shipley, B. (1987). Pattern and mechanism in the emergent macrophyte communities along the Ottawa River (Canada). Ph.D. Thesis, University of Ottawa, Ottawa, Canada.

Shipley, B. and Keddy, P. A. (1988). The relationship between relative growth rate and sensitivity to nutrient stress in twenty-eight species of emergent macrophytes. *J. Ecol.* **76**, 1101-1110.

Shipley, B., Keddy, P.A., Moore, D.R.J. and Lemky, K. (1990). Regeneration and establishment strategies of emergent macrophytes. *J. Ecol.* **77**, 1093-1110.

Shipley, B. and Peters, R. H. A test of the Tilman model of plant strategies: relative growth rates and biomass partitioning. *Am. Nat.* (in press).

Southwood, T.R.E., Brown, V.K., Reader, P.M. (1986). Leaf palatability, life expectancy and herbivore damage. *Oecologia* (Berl.) **70**, 544-548.

Terborgh, J. and Robinson, S. (1986). Guilds and their utility in ecology. *In* "Community Ecology: Pattern and Process" (J. Kikkawa and D. J. Anderson, eds), pp.65-90. Blackwell, Melbourne.

Tilman, D. (1982). "Resource Competition and Community Structure". Princeton University Press, Princeton, N.J.

Tilman, D. (1986). Evolution and differentiation in terrestrial plant communities: the importance of the soil resource: light gradient. *In* "Community Ecology". (J. Diamond and T. J. Case, eds), pp. 359-380.

Tilman, D. (1988). "Plant Strategies and the Structure and Dynamics of Plant Communities". Princeton University Press, Princeton, New Jersey.

van der Valk, A. G. (1981). Succession in wetlands: a Gleasonian approach. *Ecology* **62**, 688-696.

van der Valk, A. G. and Davis, C. B. (1978). The role of seed banks in the vegetation dynamics of prairie glacial marshes. *Ecology* **59**, 322-335.

Wimsatt, W. C. (1982). Reductionistic research strategies and their biases in the units of selection controversy. *In* "Conceptual Issues in Ecology" (E. Saarinen, ed.), pp. 155-201. D. Reidel, Dordrecht.

Wisheu, I. C. and Keddy, P. A. (1989). Species richness-standing crop relationships along four lakeshore gradients: constraints on the general model. *Can. J. Bot.* **67**, 1609-1617.

Index

abscisic acid 280
Abutilon theophrastii 9, 10
achenia 44
Achillea millefolium 7
adaptation 335
adaptive characteristics 81
adaptive responses 82
adaptive responses to flooding 89
adaptive strategy 335
adult traits 395
adventitious embryony 278
adventitious roots 86
Aegilops 112
aerenchyma 87
agamic complexes 293
agamospermy 277
agmatoploidy 312, 322
Agrostemma githago 68
Alisma plantago-aquatica 400
Alismatales 147
Allium 58
Allium altaicum 61
Allium oschaninii 61
Allium pskemense 61
Allium cepa 61
Allium fistulosum 61
Allium galanthum 61
Allium roylei 61
Allium sativum 51
Allium vavilovii 61
allometric relationship 337, 355
allopatric 319
allopolyploid 97, 322
allopolyploidy 323
allozyme 8, 28, 231
alpine flora 207
Amaranthus retroflexus 4, 41
amphibious 88
Amphicarpum purshii 288
anaerobic soils 86
Andropogonae 146
Androsace septentrionalis 356
anemochory 288
anemophilous 207
Angiospermae 180
angiosperms 147, 273, 277
Anguria subgroup 59
animal-pollinated species 235
annual plants 351
annuals 205, 367
Antennaria 295
Antennaria chilensis 209, 212
Antennaria parlinii 298
Antennaria parvifolia 295
Apera spica-venti 11, 12, 23
Aphanothese sacrum 163
Apiaceae 177
apomictic 212
apomictic plants 276
apomictic populations 329
apomixis 42, 55, 277, 293
apospory 278
aquatic invertebrates 391
Arabidopsis thaliana 371
Arales 147
Araliaceae 177
arm length asymmetry 185
arm ratio 185
Arrhenatheretum 83
Asclepiadaceae 173
Asclepias 255
Asclepias amplexicaulis 256, 261
Asclepias curassavica 260
Asclepias exaltata 258, 259
Asclepias incarnata 262, 400
Asclepias purpurascens 258
Asclepias quadrifolia 259

Asclepias syriaca 258
Asclepias tuberosa 258
Asclepias verticillata 259, 260, 262
Asclepidaceae 390
ascorbate peroxidase 46
asexual reproduction 273
Asplenium 112
assembly rules 400
Asteraceae 125, 172, 177, 207, 295, 390
Asteridae 119
atrazine 33
Australian plate 180
autopolyploidy 323
autogamy 231
automatic selection 232
autonomous agamospermy 278
autopolyploid 97
Avena 67

Baccharis pilularis 359
backcrossing 52
balanced growth path 337
barley 51, 68, 141
Barnedesiinae 125
bee-pollination 207
Bellis perennis 7
bentazon 34
Benthamiella 216
Bidens cernua 398, 399
biennials 367
big bang reproduction 352
biological individual 276
biparental inbreeding 230, 233
bipyridyl 33
bird and mammal communities 391
Boehmeria 298
bootstrap method 125
boreal countries 316
bottlenecks 232
Brachiaria 76
bradysporous plants 289
Brassica 67, 98, 143
Brassica napus 98
breeding systems 21, 205, 229, 231
broad bean 141
Bromus mollis 23, 356
Bromus tectorum 358
Bryophyta 164
Bryophytes 247
Bryopsis maxima 163
butterfly pollination 207
Buxaceae 177

^{14}C 262
C bands 196
Cabomba 148
Cactaceae 390
Caesalpiniaceae 174
Cakile edentula 362
Calyceraceae 129
Camelina 79
Camelina linicola 67
Camelina sativa 67
Campanulaceae 177
Caprifoliaceae 175, 177
Capsella bursa-pastoris 4, 19
Capsella grandiflora 20
Capsella rubella 20
Carduus nutans 368
Carex crinita 400
Caryophyllaceae 213, 214
Caryophyllales 130
caryopses 68
cell cycle 184
Centrospermae 177
Cephalaria syriaca 67
Ceratonia 175
Ceratophyllum 148
chaotic behaviour 355
character divergence 315
character polarity 122
Charophyceae 149
chemotactic attraction 312
Chenopodietum glauco-rubri 84
Chenopodium 81
Chenopodium album 4
Chenopodium rubrum 82
Chlamydomonas 150, 313
Chlamydomonas humicola 150
Chlamydomonas moewusii 150
Chlamydomonas monoica 327
Chlamydomonas reinhardtii 150
Chlorophyceae 149
Chlorophyta 164
chloroplast DNA 97, 119
chloroplasts 46
chromosomal rearrangements 55, 324
chromosome 192
chromosome number 42
chronological age 367
Chrysobalanaceae 177
Cirsium vulgare 368, 373, 376
Citrus 279
cladistic 249
cladistic analysis 122
Claytonia 143

Claytonia virginica 98
cleistogamous flowers 230, 288
Clematis fremontii 142
clonal plants 275
clonal spread 398
clone fragmentation 279
Closteriaceae 310
Closterium 312, 313
Closterium ehrenbergii 310, 313
cluster analysis 57
co-evolution 398
coastal habitats 352
Codiaceae 163
Coffea 57
Coffea arabica 57, 112
Coffea canephora 57
colchicin 57
collectors 391
colonization ability 299
competition 294
competition for light 396
competitive ability 388, 393
Compositae 33, 101, 213
computer simulation 362
concerted evolution 136
Coniferales 147
conifers 237
consistency index 122
convars 53
Convolvulaceae 175, 177
Conyza bonariensis 42
Conyza sumatrensis 37
Coreopsis rosea 398
Coriariaceae 177
corpusculum 256
Cosmarium botrytis var. *subtumidum* 324, 325
cost of sex 293, 295
costs and benefits 353
cot reassociation experiments 194
Cotula 245
Crassulaceae 390
Crepis 197, 293
critical size before reproduction 367
crop-weed complex 67
crop-weed interaction 5
crops 67
crossing experiments 52
Crossosomataceae 177
Cruciferae 106, 207, 213
cryptobiotic phase 274
Cucumis 58
Cucumis aculeatus 59
Cucumis africanus 59
Cucumis anguria subsp. *anguria* 59
Cucumis anguria subsp. *longipes* 59
Cucumis diniae 59
Cucumis dipsaceus 59
Cucumis ficifolius 59
Cucumis figarei 59
Cucumis heptadactylus 59
Cucumis myriocarpus subsp. *leptodermis* 59
Cucumis myriocarpus subsp. myriocarpus 59
Cucumis prophetarum 59
Cucumis sativus 51
Cucumis zeyheri 59
cucurbitacins 59
cultivar-groups 53
cultivated plants 51
cyanobacterium 160
cycad 145
Cycadales 147
Cynoglossum officinale 369, 370, 373, 376
Cyperaceae 207, 212, 390
Cyperus aristatus 398 399
cytogenetics 52

Dactylis glomerata 370
Datura stramonium 9, 10
Daucus carota 367, 370
deciduous forest 391
deciduous perennials 338
deciduous trees 342
Decodon verticillatus 241, 242
defense substances 347
delayed biennials 367
delayed self-fertilization 233
deletions 120, 121, 129, 163, 194
demographic 296
demographic studies 368
dendrogram 59
density gradient analysis 184, 196
density-dependent 354
Desmidiaceae 310
desmids 310
dichogamy 241
dicot 256
dicotyledons 171, 180
differential gene probing 141
dinitroaniline 33
dioecious 221, 295
dioecious agamospermous species 278
dioecism 229, 243
dioecy 231
diploid zygotic 310

Dipsacus sylvestris 367
Dipterans 207
diquat 34
dispersal ability 388
dispersal patterns 288
displospory 278
distance matrix 127
DNA amounts 189
DNA analysis 52
DNA content 183
DNA-DNA hybridization 140
DNA sequencing 127, 143
Dollo parsimony 122
domestication 5, 53
dominant Y type 326
dormancy 321
dormancy strategies 353
double digestions 141
Draba fladnizensis 104
Draba lactea 104
dry dune habitats 352
Dryopteridaceae 103
Dutch river ecosystems 81
dynamic programming 340

Ecballium elaterium 245
Echinochloa crus-galli 70, 72
Echinochloa frumentacea 72
Echinochloa utilis 70
Echinochloa weigandii 399
Echinodorus 148
ecocline 27
ecological niches 310, 315
ecotypes 27
Eichhornia paniculata 8, 241
Elaeagnaceae 177
electrophoresis 52
electrophoretic analyses 268
electrophoretic techniques 229
Eleocharis calva 398
Eleocharis obtusa 399
Eleusine corocana 112
embryo sac development 52
Emex australis 358
environmental homogeneity 4
environmental noise 355
Epacridaceae 177
Equisetum arvense 164
Equisetum telmateia 164
Ericaceae 177
Erigeron annuus 42
Erigeron philadelphicus 33
Erigeron strigosus 42
Erigeron sumatrensis 37
Erophila verna 363
Eruca sativa 67
Erucastrum gallicum 358
Eschericia coli 145
ethylene 89
Euastrum 312
Eucalyptus 235
euchromatin 197
Euler equation 371
Eupatorium perfoliatum 400
Euphorbiaceae 177, 390
evolutionarily stable strategy (ESS) 24
  373, 381
evolutionary unit 275

F_1 viability 322
facultative annuals 398, 400
facultative biennials 367, 380
Fagaceae 172, 179
Fagopyrum tataricum 67
fecundity 297, 356, 374
female success 266
fern 162
ferredoxin 160
Festuca rubra 360
fitness 40, 353, 368
fixed heterozygosity 4
fixed heterozygotes 10, 20
fixed heterozygous 104
flagella 312
flavonoids 59
fleshy fruits 211
flooding 81
floral morphology 230
flower color polymorphism 241
flower primordia 369
flower size 241
fluctuating environments 378
foraging ability 393
founder effects 4, 23
French vetch 188
fugitive species 351
function 388
functional feeding groups 391
functional groups 391
fungicides 33
fusion 185

G+C content 193
gamete fusion (postzygotic) 55
gamete fusion (prezygotic) 55
gametic incompatibility 310
gametophyte 278
gametophytic apomixis 277

gap colonizers 400
gaps 373
geitonogamous pollination 246
gender polymorphisms 243
gene amplification 136
gene conversion 136
gene duplication 164
gene flow 21, 28, 44, 263
general-purpose genotypes 5, 299
generation time 184
genet 275
genetic bottlenecks 4
genetic differentiation 82, 90
genetic drift 55
genetic foundation 273
genetic individual 275, 276
genetic make-up 274
genetic relatedness 229
genetic structure 265
genetic variation 5
Gentianaceae 213
Gentianella amarella 380
Gentianella germanica 380
geographical parthenogenesis 294, 304
Geraniaceae 131
Geranium 131
germ plasm 53
germination behavior 27
germlings 326
Geum reptans 281, 288
glutathion reductase 46
glyceraldehyde-3 phosphate dehydrogenase 146
Glycine 143
Glycine tabacina 112
glyphosate 34
Gnetales 147
Gonatozygaceae 310
Gondwanaland 179
Goodeniaceae 129, 177
Gossypium 112
Gramineae 207, 212
grasses 141, 147, 237
grassland 83
grazing 396
green algae 149, 162
Grindelia lanceolata 369
group selection 294
guilds 390
gymnosperms 171, 173
gynodioecious 245
gynodioecism 241, 243
gynodioecy 231
gynostegium 256

habitats 398
Haematococcus 150
Haloragaceae 177
Hamadryas 209
Hamadryas kingii 213, 221
Hamamelidaceae 175, 177
haploid gametic 310
haplontic unicellular algae 309
Hemionitis palmata 112
herbicide-resistant 33
herbicides 33
herbivores 341
heritability 245, 248
herkogamy 241
hermaphrodite 221
heterochromatin 192, 197
heterosis 56
heterostyly 229, 241
heterothallism 309
heterozygotes 189
Heuchera grossulariifolia 107
Heuchera micrantha 106
hexaploid 42
Hieracium villosum 277, 288
highly repetitive DNA 194
homoplasy 238
homozygotes 189
Hordeum glaucum 41
hormonal regulations 89
homothallism 309
homozygous cultivar lines 56
horsetails 162
Hoya 175
Humulus 173, 175, 177
hybrid inviability 310
hybrid melting point analyses 140
hybridization 113, 140, 299
Hydrophyllaceae 177
hymenopterans 207
Hyoscyamus niger 371
hypervariable minisatellite sequences 248

inbreeding 229, 328, 329
inbreeding depression 231, 232, 234
incompatibility 56
incongruity 56
individual fitness 371
inflorescence size 265
insecticides 33
insertions 120, 121, 129, 194
in situ hybridization 184
inter-arm transposition 185

interstitial species 399, 400
intrinsic rates of natural increase 373
introgression 113
introgressive hybridization 52, 146
intron 129
inversions 120
inverted repeat 119
Ipomoea hederacea 359
Ipomoea purpurea 241
Iridaceae 213
Iris hexagona 147
Iris fulva 147
Iris versicolor 398
isolation barriers 52
isozyme 28
iteroparous 352

Juncaceae 390
Juncus filiformis 398
juvenile traits 395

K-selection 321

Lactuca sativa 51
Lactuca virosa 369
Larix laricina 235
Lathyrus 194
Lathyrus sativus 67
Lecythidaceae 177
Leersia oryzoides 398
Leguminosae 131, 219
Lemnaceae 280
lemon 145
length mutations 121
lepidopterans 207
Leucaena 175
life forms 207, 220, 392, 398
life span 392
life cycle 351
life histories 22, 213, 235, 351, 368
life history characteristics 90
life table 371
Liliaceae 390
line-intersect method 86
Linum catharticum 380
Linum usitatissimum 67
Lisianthius 98
liverwort 164
Loasaceae 177
Loganiaceae 173
Lolio-Cynosuretum 83
Lolium temulentum 68
long-day photoperiod 368
longevity 205, 220, 234
lottery models 294
Lupinus 246
Lupinus nanus 237, 246
Lycopersicon 57, 122, 123
Lycopersicon cheesmanii 123
Lycopersicon chilense 123
Lycopersicon esculentum 57, 123
Lycopersicon peruvianum 57, 123
Lycopersicon pimpinellifolium 57, 123, 241
Lycopodium clavatum 164
Lythraceae 241
Lythrum salicaria 398, 400

Magnolia 148
magnoliids 148
maize 138, 141, 145
male mating 230
male success 266
Malva moschata 246
Marchantia polymorpha 127, 164
marginal value 336
mating systems 4, 229, 309
matrix 399
matrix projection models 371
matrix species 400
maximum likelihood methods 127
MCPA 34
MDA reductase 46
meiosis 58, 184
meiotic chromosome pairing 328
Melilotus alba 371, 380
Mendelian laws 51
Mesotaeniaceae 310
methylation 141
Micrasterias 312
Micromonadophyceae 149
middle repetitive DNA 194
milkweeds 255
Mimosaceae 174
Mimulus 241
Mimulus ringens 400
minisatellite sequences 248
minority advantage 294
mitochondrial genome 120
mixed mating systems 230, 235
modular organisms 275
monocarpic perennial 369
monocarpic shoots 351
monochorous 288
monocotyledons 147, 171, 256
monozygotic twins 276
Moraceae 173

morphological and physiological responses to flooding 82
morphological divergence 319
morphological traits 391
morphology 52
mud flat annuals 390
multilocus estimates 231
multivariate selection models 248
mutation rates 184, 234
mutations 53, 194
Mutisieae 125
Myriocarpus subgroup 59
Myrtaceae 238

Nanocyperion alliance 83
nectar production 259
Nelumbo 148
net photosynthesis 36
net reproductive rate 373
niche differentiation 328, 391
Nicotiana 123
Nicotiana acuminata 131
Nicotiana attenuata 131
Nicotiana excelsior 124
Nicotiana exigua 124
Nicotiana gossei 124
Nicotiana megalosiphon 124
Nicotiana rotundifolium 124
Nicotiana sylvestris 124
Nicotiana tabacum 124, 127
Nicotiana velutina 124
Nigritella nigra 279
nonlinearity 355
Nothoscordum fragrans 279
nuclear DNA 120, 183
nuclear gene 44
nucleolar organizing region 135
nucleotide substitutions 120, 121
nutrient uptake 388
Nymphaea 148
Nymphaeales 147

oat 68
obligate annuals 399
Oenothera biennis 380
Oenothera glazioviana 368, 370, 371
Onagraceae 172
open reading frames 130
optimal allocation 345
optimal growth rule 345
optimal size for reproduction 374
optimistic plant 345
optimization models 335
Oreopolus glacialis 216
Oryza rufipogon 79
Oryza sativa 127
oscillatory damping 355
outbreeding 55, 209, 329
outcrossing rates 21
outgroup 122
ovules 282
Oxalidaceae 216
Oxalis loricata 216
Oxalis patagonica 216

palatability 396
Panicum longifolium 398
Panicum miliaceum 9, 10
Panicum miliare 72
panmixia 263
Papilionaceae 172, 174, 207, 213
parapatric differentiation 44
paraquat 34
parasites 294
parthenospore 329
parthenosporism 309, 328
Paspalum scrobiculatum 72
Pastinaca sativa 369
paternity 230
path analysis 248
pathogens 294, 341
Pedicularis palustris 380
Pedicularis sylvatica 380
Pedinomonas 150
Pelargonium 131
Peniaceae 310
perennial ramets 363
perennials 90, 367, 398
pericentric inversion 185, 196
persistent seed bank 353, 372
pessimistic plant 345
petioles 85
Petunia 123
Phalaris arundinacea 398, 400
phenology 352
phenotype 336
phenotypic 245
phenotypic plasticity 5, 23, 90, 274
Phlox 142
photo-induction 369
photoacoustic CO_2-wave guide laser 89
photosynthesis 335
photosynthetic guilds 391
phragmoplast 150
phylogenetic reconstruction 122
phylogenies 387, 390
physiological plasticity 82
Phytolacca americana 164

Phytolacca esculenta 164
phytotrons 82
piercers 391
pin-board method 86
Pinaceae 131, 238
Piperales 147
Pisum arvense 67
Pisum sativum 67
Plagiomnium medium 112
plant defenses 396
plant geography 52
plant size 220, 367
Plantago 81
Plantago lanceolata 7
Plantago major 7, 14
Plantago major ssp. *pleiosperma* 82
plasticity of growth schedule 335
plantae linicolae 67
Pleurastrophyceae 149
Pleurotaenium 312, 323
Pleurotaenium coronatum 323
Pleurotaenium ehrenbergii 323
Pleurotaenium mamillatum 323
ploidy level 295
Plumbaginaceae 177
Poa alopecurus 209, 221
Poa alpina 288, 289
Poa annua 7, 354
Poaceae 146, 147
point mutations 120, 121
Polemoniaceae 175, 177
pollen discounting 232
pollen donation hypothesis 265
pollen grains 282
pollen limitation 297
pollination 205, 207
pollination mode 235
pollinator-limitation 205
pollinium 256
polycarpic perennial life-history 341
polycarpic shoots 351
polychory 288
Polygonaceae 177, 390
Polygonum lapathifolium 11, 12
Polygonum viviparum 288, 289
polyploid 295
polyploidization 55
polyploidy 4, 321
Polystichum californicum 103
Polystichum dudleyi 103
Polystichum imbricans 103
Polystichum munitum 103
Polytoma 150
Pontederiaceae 241
Pontryagin's maximum principle 336
population biology 229
population dynamics 273
population genetic structure 229
population growth rate 371
Potamogeton 148
prairie vegetation 391, 400
predators 294
predictive ecology 387
principal component analysis 57
principal coordinate analysis 57
production part 338
propagules 274
Proteaceae 172, 177, 179
protein sequence 159, 171
protoplast 46
provars 53
Prunella vulgaris 7
Prunus domestica 51
pseudogamous 278
Pteridophyta 164
Pteridophytes 247
Pteris cretica 112
Pyrus 175

quantum jumps 190

r- and *K*-selection 321
ramets 44, 275
Ranunculaceae 207, 209, 213, 214
Ranunculo-Alopecuretum geniculatis
Raphanus sativus 166
rapid speciation 197
*rbc*L 127
reciprocal transplantation experiment 9
red algae 162
regeneration 395
regeneration-niche hypothesis 391
repeated sequences 184
repetitive sequences 194
replicative transposition 136
reproduction mode 42
reproductive assurance 233
reproductive behaviour 274
reproductive biology 273
reproductive efficiency 281
reproductive effort 281, 341, 355
reproductive isolation 55
reproductive offer 281
reproductive success 260, 282, 336, 3
resource allocation 335
resource ratio-hypothesis 391
resource limitation 297
resource ratio-hypothesis 391

response rules 400
restriction endonuclease 121, 140
restriction fragment analysis 97
restriction fragment length polymorphisms 140, 247
restriction maps 122
restriction site analyses 121, 140
Rhamnaceae 177
Rhizirideum 61
rhizomes 399
Rhizophoraceae 177
ribosomal gene families 136
ribosomal RNA genes (rDNA) 97, 135
ribulose bisphosphate carboxylase 127
rice 138, 145
RNA synthesis 192
Robertsonian fusions 185, 191, 196
root architecture 86
root cortex 87
root respiration 87
Rosaceae 174, 213
rosette-bank 379
rosettes 84, 367
Rosidae 128
Rubiaceae 216
Rubisco 20, 127, 171
ruderal species 391, 398
Rumex 81, 82
Rumex acetosa 83
Rumex crispus 83
Rumex maritimus 83
Rumex palustris 83
Rumex thyrsiflorus 83
Rumicetum maritimi 83
rye 141

s-triazine 33
Sabatia angularis 246
Saccharomyces cerevisiae 327
Saccharum 146
Sagittaria 148
Sagittaria latifolia 245
satellite DNA 196
Saxifragaceae 108, 213
Schisandraceae 173
Schizosaccharomyces pombe 327
Scirpus acutus 398
Scirpus americanus 398, 400
Scirpus cyperinus 400
scrapers 391
Scrophulariaceae 213
Secale 67
seed bank 21, 351, 379
seed dispersal 21
seed dormancy 289
seed germination 242, 353
seed plants 147, 229
seed weight 184
seedling emergence 27
seedling establishment 242
selective force 353
self-compatibility 55, 205
self-compatible 90, 212, 262
self-fertilization rate 90
self-incompatibility 205
self-incompatible 212, 268
self-pollination 21, 42, 262
selfing 9
selfing minus clone 327
selfing rate 230
Senecio vulgaris 33, 41, 241
sequencing studies 143
Setaria 5
Setaria glauca 74
Setaria italica 70
Setaria viridis 70, 72
sex allocation 245
sexual dimorphism 209, 217
sexual reproduction 273, 293
sexual selection 230, 265
sexuality 293
shredders 391
shrub life-form 218
sibling competition 294
simazine 33
Simmondsia 177
Sinapis 67
Sinapis alba 67
size-dependency 367
size-hierarchy 374
Solanaceae 121, 125, 172, 175, 177, 179, 216
Solanum pennellii 122
Sorghum 143, 146
Sorghum halepense 9
Sorghum intrans 358
soybean 138, 145
spatial autocorrelation 354
speciation 229, 245, 313, 328
Spirodela polyrrhizza 288, 289
Spirulina platensis 160
split-up clone 275
sporophyte 277
stable equilibrium point 355
stable limit cycles 355
Stenactis 42
stochastic dynamic programming 343, 345

stochastic environments 343
stochastically changing environment. 335
storage part 338
stress tolerance 388, 393
stress-tolerant plants 346
strict biennials 380
structural individual 276
structural rearrangements 120, 129
stylar canal 256
Suaeda maritima 355
subdioecious 245
successional axis 341
summer annuals 351
superoxide detoxification 46
superoxide dismutase 46
survival 374
survival rate 90
survivorship 371
switching time 337
sympatric populations 319
synapomorphies 143
Synechococcus 162

tachysporous plants 289
Taraxacum alpinum 288
Tellima grandiflora 98
temperate latitudes 220
temperate woodland 391
teosintes 146
Tetraselmis 149
tetrasomic inheritance 108
tetrasomic segregation 106
tetrazolium test 289
thermal denaturation analysis 184
Thymelaeaceae 177
Tolmiea menziesii 98, 108
tomato 168
topocline 27
trade-off relationships 374
Tragopogon dubius 99
Tragopogon mirus 99
Tragopogon miscellus 99
Tragopogon porrifolius 51, 99
Tragopogon pratensis 51, 99
transient seed bank 353, 372
transitions 127, 143
translocation 185
transversions 127, 143
tree life form 218
Trentepohliales 150
Triadenum fraseri 398
triennial plants 371
Trillium flexipes 166
Trillium grandiflorum 166
Trillium kamtschaticum 165
Trillium ovatum 166
Trillium smallii 165
Triploceras 312
Tripsacum 146
tristyly 241
Triticale 184
Triticum 112, 140
Triticum monococcum 51
tropical latitudes 217, 220
turions 280
Turnera 241
Turnera ulmifolia 58, 241
Typha glauca 398, 400

Ulvophyceae 149
Umbelliferae 207, 213
unequal crossing over 136
unequal segmental interchange 196
uniparental (maternal) inheritance 44
urbanization 81
Ustilago bullata 358

Valerianaceae 175, 177
vegetative organs 338
vegetative propagation 55, 392
Verbascum thapsus 370, 371
Verbena hastata 400
vernalization 368
Vicia 175, 184
Vicia amoena 187
Vicia bithynica 188
Vicia eriocarpa 194
Vicia faba 187, 188, 189, 196
Vicia grandiflora 189
Vicia hybrida 196
Vicia hyrcanica 193
Vicia johannis 194
Vicia lutea 193
Vicia melanops 194, 196
Vicia monantha 187, 189, 197
Vicia narbonensis 186, 188, 196, 197
Vicia sativa 188, 197
Vicia tenuifolia 187
Vitaceae 177
viviparous plants 274
Vulpia fasciculata 360

Wagner parsimony 122
Wallace's line 180
water lilies 148
weed control 33
weeds 3, 19, 67
wetland habitats 390

wetland plant communities 387
wheat 51, 68, 141
wild barley 14
wind pollination 207, 211
wind-pollinated species 235
winter annuals 351, 368
Winteraceae 179
woody species 218
Wurmbea dioica 245

Xenopus 145

yeasts 329
Youngia japonica 37

Zea 143, 146
Zea diploperennis 112
Zea perennis 112
Zygnemataceae 312
zygospore 329